D1328194

# NUTRITIONAL AND TOXICOLOGICAL ASPECTS OF FOOD SAFETY

# ADVANCES IN EXPERIMENTAL MEDICINE AND BIOLOGY

Recent Volumes in this Series

# NUTRITIONAL AND TOXICOLOGICAL ASPECTS OF FOOD SAFETY

Edited by

## Mendel Friedman

U.S. Department of Agriculture
Agricultural Research Service
Western Regional Research Center
Berkeley, California

PLENUM PRESS • NEW YORK AND LONDON

Library of Congress Cataloging in Publication Data

Main entry under title:

Nutritional and toxicological aspects of food safety.

(Advances in experimental medicine and biology; v. 177)
"Based on the Symposium of Food Safety: Metabolism and Nutrition, sponsored by
the Pacific Conference on Chemistry and Spectroscopy, held October 27–29, 1982, in
San Francisco, California"—T.p. verso.
Includes bibliographical references and index.
1. Nutritionally induced diseases—Congresses. 2. Food contamination—Congresses.
3. Food poisoning—Congresses. I. Friedman, Mendel. II. Symposium on Food Safety:
Metabolism and Nutrition (1982: San Francisco, Calif.) III. Pacific Conference on
Chemistry and Spectroscopy (1982: San Francisco, Calif.) IV. Series. [DNLM: 1.
Diet—congresses. 2. Food Contamination—congresses. W1 AD559 v. 177/WA 701
N975 1982]
RC622.N893   1984                           363.1'92                           84-8418
ISBN 0-306-41708-1

Based on the symposium on Food Safety: Metabolism and Nutrition, sponsored by
the Pacific Conference on Chemistry and Spectroscopy, held October 27–29, 1982,
in San Francisco, California, with additional invited contributions

©1984 Plenum Press, New York
A Division of Plenum Publishing Corporation
233 Spring Street, New York, N.Y. 10013

Printed in the United States of America

PREFACE

Naturally occurring antinutrients and food toxicants, and those formed during food processing, adversely affect the nutritional quality and safety of foods. Because of the need to improve food quality and safety by plant breeding, fortification with appropriate nutrients, and processing methods, and because of the growing concern about possible direct relationships between diet and diseases, research is needed to:

(1) evaluate the nutritive quality and safety of crops and fortified, supplemented, and processed foods;

(2) define conditions that favor or minimize the formation of nutritionally antagonistic and toxic compounds in foods; and

(3) define the toxicology, metabolism, and mechanisms of the action of food ingredients and their metabolites.

As scientists interested in improving the safety of the food supply, we are challenged to respond to the general need for exploring:

(1) possible adverse consequences of antinutrients and food toxicants; and

(2) factors which contribute to the formation and inactivation of undesirable compounds in foods.

Medical research offers an excellent analogy. Studies on causes and mechanisms of disease processes are nearly always accompanied by parallel studies on preventive measures and cures. Such an approach offers the greatest possible benefits to the public.

Although much work is needed to define the basic mechanisms of toxic action of food ingredients, enough is known to permit rational research approaches to develop new ways to minimize

effects of antinutrients and toxicants in foods.  Such approaches
include:

(1)  inactivating deleterious compounds with site-specific
     reagents to prevent them from interacting with living
     cells;

(2)  eliminating the responsible food ingredients from our
     diet;

(3)  breeding new plant varieties which are both nutritious
     and safe to consume;  and

(4)  identifying dietary constituents that protect against
     the adverse action of antinutrients and food toxicants.

For example, fiber in the diet appears to reduce the incidence
of cancer.  This effect could arise because fiber strongly binds
certain carcinogens and food toxicants, thus minimizing or preventing their absorption by the intestine into the blood stream.
Since other dietary components affect absorption and excretion
of deleterious compounds in foods, dietary protein, carbohydrate,
fat, mineral, vitamin, or polyphenol content, as well as fiber,
probably may influence the biological action of nutrients, antinutrients, and toxicants in foods.  The paper by H. F. Stich and
M. P. Rosen on antimutagenic and anticarcinogenic effects of naturally occurring phenolic compounds and that by T. K. Smith and
M. S. Carson on the effect of diet on toxicosis of trichothecenes
deserve special mention for pointing to promising future research
directions on beneficial effects of diet on food safety.

     The most important function of a symposium, I believe, is
dissemination of insights and cross-fertilization of ideas that
will catalyze progress by permitting synergistic interaction among
related disciplines.  Therefore, in organizing the symposium
FOOD SAFETY:  METABOLISM AND NUTRITION, sponsored by the Pacific
Conference on Chemistry and Spectroscopy, San Francisco, California,
October 27-29, 1982, I invited both reviews and reports of recent
work.  In addition, a number of scientists who did not take part
in the symposium accepted my invitation to contribute papers to
this volume.  This book is, therefore, a hybrid between symposium
proceedings and a collection of invited contributions.

     I am particularly grateful to Glen A. Bailey of the Pacific
Conference for inviting me to organize the symposium, and to all
contributors for a well-realized meeting of minds and for excellent
collaboration.

     The papers are being published under the title NUTRITIONAL
AND TOXICOLOGICAL ASPECTS OF FOOD SAFETY as a volume in the series

Advances in Experimental Medicine and Biology.  This book is intended to complement the following published volumes in the same series: *Protein-Metal Interactions (1974); Protein Crosslinking: Biochemical and Molecular Aspects (1977); Protein Crosslinking: Nutritional and Medical Consequences (1977); and Nutritional Improvement of Food and Feed Proteins (1978).*

I very much hope that these volumes will be a valuable record and resource for further progress in animal and human nutrition, food safety and toxicology, medicine, and protein chemistry.

Mendel Friedman
Moraga, California
January, 1984

*The eyes of all look to you expectantly,*

*and you give them their food when it is due.*

*You give it openhandedly, feeding every*

*creature to its heart's content.*

*Psalm 145: 15-16*

*Eat nothing that will prevent you from eating.*

*Ibn Tibbon, c. 1190*

# CONTENTS

NATURALLY OCCURRING PHENOLICS AS ANTIMUTAGENIC

AND ANTICARCINOGENIC AGENTS

Hans F. Stich and Miriam P. Rosin

Environmental Carcinogenesis Unit
British Columbia Cancer Research Centre
Vancouver, B.C., Canada

INTRODUCTION

Epidemiological evidence points to an inverse relationship between the consumption of vegetables and the incidence of cancer at various sites (Hirayama, 1979, 1981; Graham et al., 1978; Mettlin et al., 1981). The search for the protective components in these vegetables has focused on β-carotene and vitamin A (e.g., Bjelke, 1975; Shekelle et al., 1981; Cambien et al., 1980; Peto et al., 1981; Doll and Peto, 1981; Marshall et al., 1982) and ascorbic acid (e.g., Haenszel and Correa, 1975; Kolonel et al., 1981). However, the inverse relationship observed between the ingestion of green/yellow vegetables and the incidence of human cancers could conceivably be due to many other plant components. At present, the percentage contribution of vitamins to the cancer-protective activity of vegetables or fruits is unknown. In this paper, we present results suggesting an involvement of naturally occurring phenolics in the prevention of genotoxicity and carcinogenicity. Since the number of phenolics in various plants is staggering and the discussion of their beneficial or toxic effects is beyond the scope of any short review, we have focused on non-flavonoid simple phenolics (C6), phenolic acids (C6-C1), cinnamic acid and related compounds (C6-C3).

PHENOLICS AS INHIBITORS OF DIRECT-ACTING AND S9-REQUIRING MUTAGENS

Recently, many *in vitro* test systems have been successfully applied to estimate the mutagenic, clastogenic and recombinogenic properties of manmade and naturally occurring chemicals. These rapid and economic bioassays can also be used to uncover the anti-genotoxic and by implication anticarcinogenic activity of compounds

(e.g., Rosin, 1982; Rosin and Stich, 1978a, 1979). The chemical
to be examined can be added concurrently, prior to, or after
exposure of the test organism to the carcinogen. Results of such
studies may help to reveal the mechanism involved in antimutagenic
and anticarcinogenic activities of a compound and the better design
of treatment protocols. Three examples involving phenolics may
exemplify this approach.

## Inhibition of Aflatoxin $B_1$ ($AFB_1$)-Induced Mutagenesis by Suppression of its Metabolism

AFB$_1$ and many of its metabolites require activation by the
microsomal mixed function oxidase system into a mutagenic (Garner
et al., 1972; Wong and Hsieh, 1976) or clastogenic (Stich and
Laishes, 1975) species. The ultimate reactive molecule is believed
to be $AFB_1$-2,3-oxide (Lin et al., 1978; Neal and Colley, 1978).
The addition of caffeic acid, chlorogenic acid or gallic acid to
the S9 activation mixture strongly reduced the mutagenicity of AFB$_1$
(Fig. 1). Phenolics could exert an inhibitory effect either by
suppressing the metabolism of AFB$_1$ or by trapping the $AFB_1$-2,3-
oxide. The experimental design to find an answer to this question
consisted of tracing the formation of AFB$_1$ metabolites by high
pressure liquid chromatography (HPLC) following the addition of
phenolics to a mixture containing S9 and AFB$_1$. The results of this
study suggest that the above-mentioned phenolics suppressed the
mutagenicity of AFB$_1$ by interfering with its metabolic activation
(Chan, 1982). However, in addition to this action, phenolics could
trap the reactive $AFB_1$-2,3-oxide if it should indeed be formed in
their presence. The present study does not exclude this possibility.

## Inhibition of Benzo(a)Pyrene Diol Epoxide-Induced Mutagenesis Through Direct Interaction with Phenolics (Wood et al., 1982)

The mutagenicity of the direct-acting benzo(a)pyrene-7,8-diol-
9,10-epoxide-2 in *Salmonella typhimurium* and cultured Chinese
hamster V79 cells was strongly reduced following the addition of
ferulic, caffeic, chlorogenic and ellagic acids. This inhibition
of the mutagenicity of the reactive species of a polycyclic aromatic
hydrocarbon is assumed to be due to its direct interaction with the
phenolics resulting in the formation of complexes.

## Inhibition of the Direct-Acting Carcinogen and Mutagen, N-Methyl-N'-Nitro-N-Nitrosoguanidine (MNNG)

Mutagenicity of MNNG was only reduced when phenolics were
administered to *S. typhimurium* concurrently with the mutagen
(Fig. 2). The tested phenolics (gallic acid, caffeic acid, chlorogenic acid and a commercial preparation of tannins) had no detectable inhibitory effects when added to Salmonella cultures prior to
or after their exposure to MNNG (Table 1). It appears likely that

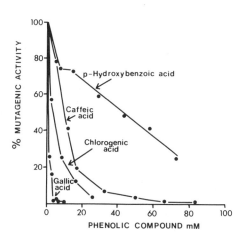

Fig. 1.  Effect of several simple phenolics on the mutagenic
         activity of $AFB_1$ ($3 \times 10^{-5}$M) in logarithmically growing
         *S. typhimurium* TA98 cultures.  Mutation frequency of $AFB_1$
         alone was 64 his[+] revertants per $10^7$ survivors.

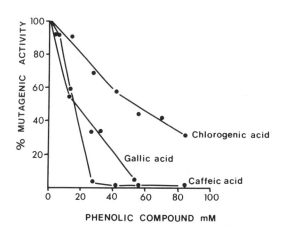

Fig. 2.  Effect of chlorogenic acid, gallic acid and caffeic acid
         on the mutagenic activity of *S. typhimurium* TA1535
         cultures exposed to MNNG ($3 \times 10^{-5}$M).  Mutation frequency
         of MNNG alone was 138 his[+] revertants per $10^7$ survivors.
         Values plotted are percentage of this mutagenic activity
         which remains when treatments are carried out in the
         presence of phenolic at concentration indicated.  All
         treatments were non-toxic.

Table 1.  Effect of Timing on Inhibition by Chlorogenic Acid of
          MNNG-Induced Mutagenesis in Salmonella Cultures[a]
          (Chan, 1982)

| | Mutagenic Activity (His[+] Revertants/$10^7$ Survivors) When Chlorogenic Acid (20 mg/ml) Applied | | |
|---|---|---|---|
| No Chlorogenic Acid | Prior to MNNG ($3 \times 10^{-5}$M) | Concurrent to MNNG ($3 \times 10^{-5}$M) | After MNNG ($3 \times 10^{-5}$M) |
| 183 | 175 | 69 | 188 |

[a]*S. typhimurium* TA1535 logarithmically growing cultures;  control
frequency, 1 revertant/$10^7$ survivors.  All treatments performed in
suspension.


the inhibitory effect is due to a scavenging action by the phenolics
of one of the electrophilic decomposition products of MNNG.

PHENOLICS AS INHIBITORS OF MUTAGENICITY RESULTING FROM NITROSATION
REACTIONS

     Nitrosation of methylurea leads to the formation of direct-
acting mutagens which can be readily detected by the *S. typhimurium*
mutagenicity assay.  This well-defined system was used as a model
to examine the effect of several common plant phenolics on nitrosation reactions (Stich et al., 1982a).  Gallic acid, caffeic acid,
chlorogenic acid, catechin and tannic acid reduced the formation of
mutagenic compounds (Table 2).  The inhibitory effect of these
plant phenolics was similar to or even greater than that of ascorbic
acid, which has been widely used to prevent nitrosamine formation
*in vitro* and *in vivo* (reviewed by Newmark and Mergens, 1981).  The
simplest explanation of the inhibitory action of the phenolics is
to assume that the yield of mutagenic nitrosation products of
methylurea is reduced due to the trapping of the nitrosating species.
It is also conceivable that phenolics could interact with nitrosation products and in this way inhibit their mutagenicity (Rosin
and Stich, 1978b).  However, at the examined doses of phenolics,
this mechanism for the inhibitory action appears unlikely, since
the phenolics did not significantly reduce mutagenicity when they
were added after completion of the nitrosation of methylurea.  The
mutagenicity results are in general agreement with chemical studies
showing an inhibition of N-nitroso compound formation by tannins
(Bogovski et al., 1972;  Gray and Dugan, 1975), gallic acid
(Pignatelli et al., 1976) and several other phenolics (Gray and
Dugan, 1975;  Groenen, 1977;  Kawabata et al., 1979).

Table 2.   Inhibition by Phenolics of Bacterial[a] Mutagenicity Due to
            Nitrosation Products of Methylurea (Stich et al., 1982a)

| Phenolics | Concentration | |
| --- | --- | --- |
| | Active Range[b] | 50% Inhibition |
| Catechol | 0.3–1.0 mM | 1.0 mM |
| Gallic acid | 0.7–6.9 mM | 2.1 mM |
| Chlorogenic acid | 0.6–6.4 mM | 2.1 mM |
| Tannic acid | 0.1–0.9 mg/ml | 0.3 mg/ml |
| Catechin | 0.2–3.5 mM | 1.0 mM |
| Resorcinol | 0.3–6.4 mM | 2.1 mM |
| p-Hydroxybenzoic acid | No effect | – |

[a]$S.$ $typhimurium$ TA1535 logarithmically growing cultures were exposed
to nitrosation products for 20 min before plating for mutagenicity
and toxicity.  Phenolics were present throughout the nitrosation
reaction.  Mutagenicity of nitrosation products formed in the
absence of phenolics was 205 his[+] revertants/$10^7$ survivors.
[b]The range given is the concentration of phenolics in the nitrosation
reaction mixture which results in 10-100% inhibition of mutagenicity.


PHENOLIC-CONTAINING FOOD ITEMS AS INHIBITORS OF MUTAGENICITY IN
$IN$ $VITRO$ TEST SYSTEMS

     Human populations are exposed daily to thousands of chemicals
in the form of complex mixtures.  Their potential or actual health
hazard is unknown due to the difficulties of extrapolating $in$ $vitro$
results to human populations.  Moreover, reliable mutagenicity and
carcinogenicity data are usually only available on single pure
compounds.  Thus the behaviour of a mixture, which is not a simple
product of the activation of mutagenic (carcinogenic) and antimutagenic (anticarcinogenic) activities, cannot be predicted from
data on single chemicals.  Complex interactions between chemicals
can lead to "additive", "synergistic", "potentiating" and "antagonistic" effects.  The number of combinations and permutations among
the hundreds of chemicals which are present in even the simplest
food products is so large that an analysis of all the possible
interactions just cannot be accomplished.  A way out of this dilemma
is to simulate $in$ $vitro$ the conditions prevailing in man or to
examine the genotoxic effects directly in human beings.

     The following attempts were made to gain at least some insight
into the possible role of phenolic-containing food items as anti-
genotoxic agents.

     1.   Three North-American coffee samples were tested for their

capacity to modify the development of mutagenic activity resulting
from the nitrosation of methylurea.  All three examined instant
coffees (one was a decaffeinated brand) suppressed the induction of
mutations at and even below concentrations which are customarily
used in the preparation of these coffees (Stich et al., 1982a).  A
similar inhibitory effect was observed when three different teas,
including a Japanese, Chinese and Indian brand, were added to the
nitrosation mixture consisting of nitrite and methylurea (Table 3).
The examined teas and coffees did not reduce the mutagenicity when
added after the completion of the nitrosation reaction at concentrations which exerted a strong inhibitory effect when present during
nitrosation.  Such observations seem to be of particular importance
when protocols for epidemiological studies are designed.  It is
becoming apparent that the sequence of intake of various food items
strongly influences the mutagenic and carcinogenic hazard.  A
simply protective effect of trapping and scavenging food ingredients
can only manifest itself when they can intermingle with the reactive
forms of mutagens and carcinogens.  Only rarely can one find data
on this issue, the importance of which seems to be underestimated
in epidemiological surveys.

Table 3.    Inhibition by Phenolic-Containing Products of Bacterial[a]
            Mutagenicity Due to Nitrosation Products of Methylurea

| | | Concentration | |
| --- | --- | --- | --- |
| Phenolic-Containing Product | | Active Range[b] | 50% Inhibition |
| Coffee: | Instant | 0.1-18 mg/ml | 1.8 mg/ml |
| | Instant, decaffeinated | 0.1-59 mg/ml | 2.6 mg/ml |
| | Roasted, ground | 0.1-60 mg/ml | 2.7 mg/ml |
| Tea: | Japanese | 3.2-33 mg/ml | 10 mg/ml |
| | Indian | 3.6-102 mg/ml | 21 mg/ml |
| | Chinese | 0.8-37 mg/ml | 5.3 mg/ml |
| | Mate | 0.6-36 mg/ml | 4.7 mg/ml |
| | Chinese green | 0.7-33 mg/ml | 4.9 mg/ml |
| Wine: | White (5 types) | 7-510 μl/ml | 60 μl/ml |
| | Red (5 types) | 9-350 μl/ml | 55 μl/ml |
| Beer: | 4 brands | 6-220 μl/ml | 34 μl/ml |
| Betel | Tannin fraction | 0.1-1 mg/ml | 0.3 mg/ml |
| nut | Catechin fraction | 0.1-2 mg/ml | 0.3 mg/ml |
| extracts: | Flavonoid fraction | 0.1-1 mg/ml | 0.2 mg/ml |

[a,b]See Table 2 for description.

2.  Our second attempt to simulate *in vitro* conditions which actually occur during the consumption of a meal dealt with Chinese salt-preserved fish.  This particular fish preparation was chosen because of its large consumption in virtually all Asiatic countries, its mutagenic activity following nitrosation (Marquardt et al., 1977), its carcinogenic effect in rodents (Weisburger et al., 1980) and its possible involvement in human cancer (Ho et al., 1978; Huang et al., 1978).  The test system consisted of nitrosating (pH 2.0, 1 hr, 37°C) an aqueous fraction of a salt-preserved Chinese fish (Pak Wik) and estimating the frequency of his[+] revertants per survivor of *S. typhimurium* (strain TA1535).  Several dietary phenolics and three teas were added to the nitrosation mixture (Fig. 3).  Catechin, chlorogenic acid, gallic acid and pyrogallol suppressed the formation of mutagenic nitrosation products.  The efficiency of inhibition was comparable to that of ascorbic acid. A Japanese, Chinese and Ceylon tea also prevented the formation of mutagenic nitrosated fish products at doses which are usually consumed by man.  Such studies on food products which are actually ingested may, despite the difficulties in handling "messy" mixtures, bridge the gap between epidemiological evidence pointing to a link between consumption of this fish and an elevated risk for gastric cancer and biochemical studies on nitrosation reactions.

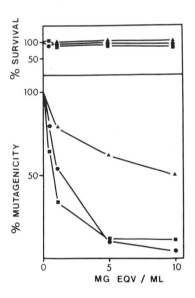

Fig. 3.   Inhibitory effect of catechin (■), chlorogenic acid (●) and Chinese tea (▲) on mutagenicity of the nitrite-treated salted fish extract.

ANTICARCINOGENIC ACTIVITY OF PHENOLICS

Several derivatives of cinnamic acid have been reported to
inhibit chemical carcinogenesis when present in the diet of treated
animals. p-Hydroxycinnamic acid, o-hydroxycinammic acid, caffeic
acid and ferulic acid each caused an inhibition of benzo(a)pyrene-
induced neoplasia of the forestomach in female ICR/Ha mice
(Wattenberg, 1979; Wattenberg et al., 1980). These phenolics were
present from 8 days prior to exposure of the mice to benzo(a)pyrene
by oral intubation and continued to be added to the food until 3
days after the last dose of the carcinogen. A reduction was observed
in both the percentage of mice with tumours of the forestomach and
the number of such tumours per mouse among the cinnamic acid-fed
mice. p-Hydroxycinnamic acid has also been reported to suppress the
development of tumours in the forestomach of ICR/Ha mice receiving
the direct-acting carcinogen β-propiolactone (Wattenberg, 1979).

There is a great need for extending these studies in order to
obtain a broader understanding of the overall modulating action of
naturally occurring phenolics. The specificity of these phenolics
for particular carcinogens is unknown. Two structurally related
synthetic phenolics, butylated hydroxyanisole (BHA) and butylated
hydroxytoluene (BHT), have been shown to be very versatile inhibitors.
Their presence leads to a suppression of the activity of a range of
carcinogens, including polycyclic aromatic hydrocarbons, diethylnitrosamine, urethane, uracil mustard, 4-nitroquinoline-1-oxide,
N-2-fluorenylacetamide, p-dimethylaminoazobenzene, methylazoxymethanol acetate, and trans-4-amino-3-[2-(5-nitro-2-furyl)-vinyl]-
1,2,4-oxadiazole (Wattenberg, 1976, 1979). However, the inhibitory
capacity of phenolic compounds is very dependent upon chemical
structure, with substituted groups on the phenolic ring affecting
the potency of the compound (Wattenberg et al., 1980). Another
unresolved issue is the ability of naturally occurring phenolics to
affect carcinogen activity at sites removed from the digestive
tract. Once again, previous studies have shown that BHT and BHA
can suppress dimethylbenz(a)anthracene-induced neoplasms of the
lung in mice and of the breast in rats (Wattenberg, 1972, 1973).
The effect of the exposure sequence of phenolics and carcinogens to
the overall frequency of carcinogenesis in exposed animals is yet
another area requiring consideration. Careful studies are now
underway for BHA and BHT (King et al., 1983), and should be extended
to naturally occurring phenolics.

PHENOLICS AS NITRITE TRAPPERS

Phenolics are known to react with nitrite to form C-nitroso
phenolic compounds (Mirvish, 1981; Walker et al., 1982) (Fig. 4).
The strength of this reaction depends on the number of hydroxy
groups and their position. A fast way to estimate the nitrite

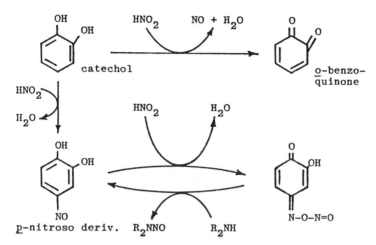

Fig. 4.  Mechanism whereby phenols can either inhibit nitrosation
(by using up the nitrite) or enhance nitrosation (by
forming an active nitrosating species).  The reactions
shown have not all been demonstrated specifically for
catechol (Mirvish, 1981).

trapping capacity is to measure the depletion of nitrite from a
solution following the addition of various compounds.  An example
of the trapping capacity of mono-, di- and trihydroxy phenolics is
shown in Fig. 5.  The efficiency of some common naturally occurring
phenolics is equal to or exceeds that of ascorbic acid, which is
one of the dietary products generally believed to be the main
inhibitor of endogenous or exogenous nitrosation reactions (Fig. 6).
The nitrite depletion caused by caffeic acid, gallic acid and
ferulic acid may exemplify the reaction capacity of many other
phenolics found in vegetables and fruits.  A depletion of nitrite
can also be obtained by phenolic-containing beverages, including
Chinese, Japanese or Indian teas and instant coffees.  However,
whether the trapping of nitrite by these complex mixtures is due
mainly to phenolics (C6, C6-C1, C6-C3), flavonoids or other
products is unknown.

When phenolics are added to nitrosation mixtures, C-nitroso
derivatives could be formed.  The question must be raised as to
their genotoxic or carcinogenic properties.  It would be an act of
folly to use phenolics as inhibitors of mutagenic nitrosamines,
nitrosamides or nitrosoalkylureas and, in the course of this process, create equally or even more active C-nitroso phenolics.  Only
scanty information is available on this issue.  The mutagenicity
of *p*-nitrosophenol is higher than that of phenol in the *S. typhi-*

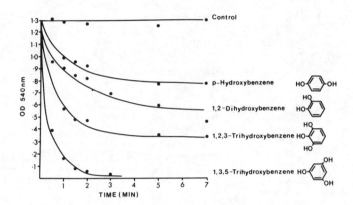

Fig. 5.   The nitrite depleting ability of mono-, di- and trihydroxy-
          benzenes.  Nitrite depletion by phenolic compounds was
          determined by the procedure according to Sen and Donaldson
          (1978) using N-N-naphthylethylenediamine dihydrochloride
          and sulfanilic acid in 20% acetic acid;  absorbance read
          at 540 nm.   All reactions in nitrite depletion assays were
          carried out at pH 2.5 at room temperature ($NaNO_2$, 4 mM;
          hydroxybenzenes, 4 mM).

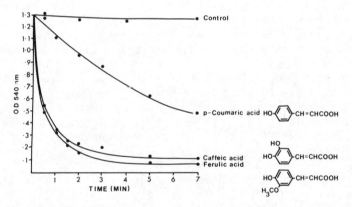

Fig. 6.   The nitrite depleting ability of mono- and dihydroxy
          cinnamic acids as determined by the Griess colour reaction
          for nitrite.  For methods, see legend to Fig. 5 ($NaNO_2$,
          4 mM;  cinnamic acids, 4 mM).

*murium* test system (Gilbert et al., 1980), and the clastogenicity
of nitrosated derivatives of phloroglucinol exceeds that of the
parental compounds (Table 4). Undoubtedly, the genotoxic activity
of the examined C-nitroso phenolics must be lower than that of N-
nitrosomethylurea, or otherwise we would not have obtained an
inhibitory effect during the nitrosation of methylurea (Table 2).
However, how much lower the mutagenicity of C-nitroso phenolics will
be compared to an equimolar concentration of N-nitrosamines or N-
nitrosoureas remains to be seen, assuming that they will indeed
prove to be genotoxic.

## PHENOLICS AND PHENOLIC-CONTAINING BEVERAGES AS INHIBITORS OF NITROSAMINE FORMATION WITHIN MAN

Considering the relatively large daily ingestion of nitrosable
as well as nitrosating agents and a favourable acid pH in the human
gastric juice, a considerable amount of N-nitroso compounds should
be formed endogenously. Contrary to expectation, the amounts of
nitrosamines in various body fluids are low, frequently below
detection levels (Eisenbrand et al., 1981). Eiether nitrosamines
are formed and thereafter rapidly metabolized, or their endogenous
formation is inhibited by food compounds consumed jointly with
nitrosating and nitrosable molecules. Dietary phenolics could be
involved in the latter process. To yield information on this topic,
we applied a modified version of the newly designed procedure of
administering nitrate and proline to individuals and estimating the
urinary excretion of N-nitrosoproline over a 24-hr period (Ohshima
and Bartsch, 1981; Ohshima et al., 1982). This method appeared to
be ideally suited to answer the question as to whether the intake
of phenolics and beverages which contain relatively high concentrations of phenolics will enhance or suppress endogenous nitrosation.

The administration patterns of nitrate, proline and inhibitor
are shown in Fig. 7. Following ingestion of nitrate and proline,
the levels of nitrosoproline in the urine increased (Fig. 8A). An
inhibitory effect on nitrosoproline formation was observed when
caffeic acid (Fig. 8B), an instant decaffeinated coffee (Fig. 8C)
and a Chinese tea (Fig. 8D) were consumed with and after the intake
of proline. The results point to dietary phenolics as being good
inhibitors of endogenous nitrosation within man. In agreement with
this conclusion are results showing that betel nut extracts rich in
(+)catechin, (-)epicatechin, polymeric pelargonidin and cyanidin
also suppressed the endogenous formation of nitrosoproline (Stich
et al., 1983a).

Table 4.   Clastogenic Activity of Phloroglucinol and Nitrosated
           Phloroglucinol

| | mg/ml 6 | 3 | 1.5 | 0.75 | 0.37 |
|---|---|---|---|---|---|
| | \multicolumn{5}{c}{Percent Metaphase Plates with Chromosome Aberrations[a]} | | | | |
| Phloroglucinol | 60.5 | 30.0 | 11.7 | 0.4 | 0.2 |
| Nitrosated phloroglucinol | MI[b] | MI | 46.5 | 8.9 | 0.4 |

[a]CHO cells, 3 hr exposure;  samples taken 20 hr post-treatment.
 Chromosome aberrations include chromatid and chromosome breaks
 and chromatid exchanges.
[b]MI, mitotic inhibition.

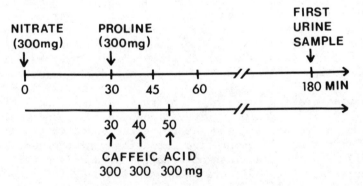

Fig. 7.   Sequence in which compounds were ingested following
          onset of experiment (i.e., intakes of nitrate) and prior
          to taking the first urine sample.  For each dose of
          nitrate, proline and caffeic acid, 300 mg of the compound
          was dissolved in 20 ml water.

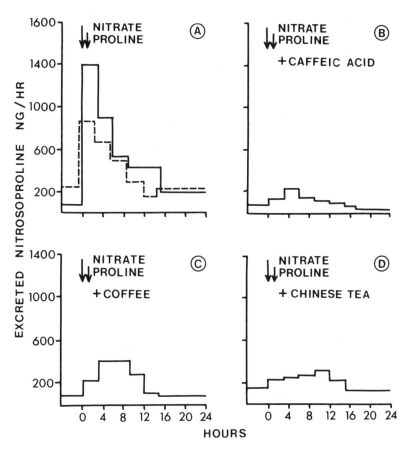

Fig. 8.  Inhibitory effects of three different foods on nitrosoproline formation *in vivo*.  Arrows indicate intake of
nitrate followed by proline.  Fig. 8A represents a control
(2 subjects), where no inhibitory substances were
ingested;  8B, 8C and 8D show inhibition (see text for
details).

GENOTOXICITY, CARCINOGENICITY AND PROMOTING ACTIVITY OF PHENOLICS

We would be remiss if we did not include in this short overview the capacity of various plant phenolics to induce an array of
genotoxic effects (Table 5) (reviewed by Stich and Powrie, 1982).
In this connection, it is of particular interest to note the mutagenic and DNA-breaking activities of endogenous catecholamines,
which are an integral part of a mammalian metabolism (Table 6).  As
in the case of exogenous phenolics (Korycka-Dahl and Richardson,
1978), the genotoxic effect appears to be due to the oxidation of
the catechol group which generates hydroxyl and superoxide radicals
and hydrogen peroxide (Cohen and Heikkila, 1974;  Graham et al.,
1978).  It is likely that $H_2O_2$, $O_2^-$ and $OH^-$ rather than the phenolic
molecules are responsible for the various genotoxic activities
(Table 7) (Yamaguchi, 1981;  Hanham et al., 1982).  The cytotoxic
and genotoxic activities of $H_2O_2$ are well established (e.g., Freese,
1971;  Stich et al., 1979;  MacRae and Stich, 1979;  Speit et al.,
1982).  Thus it should not be surprising that many naturally
occurring compounds that can generate superoxide radicals and $H_2O_2$
also have the capacity to induce a wide spectrum of DNA, gene and
chromosome aberrations.  The above-mentioned catecholamines, dietary
phenolics (Stich and Powrie, 1982;  Morimoto and Wolff, 1980),
ascorbic acid (Stich et al., 1979) or ferritin (Whiting et al.,
1981) belong in this group of compounds.  The crucial question is
not whether they can or cannot produce a genotoxic effect (this
capacity is beyond any doubt), but whether the conditions prevailing
in human tissues and body fluids permit the generation of free
radicals or $H_2O_2$.

At present, statements dealing with the carcinogenicity of
phenolics for rodents or man can only be speculative in nature.
The polyphenolic fractions of betel nuts (*Areca betle*) have been
implicated in the development of oral, pharyngeal and esophageal
cancers among betel quid or betel nut chewers (reviewed by Stich et
al., 1983b).  High incidences of esophageal cancers among populations
of the Caribbean islands have been linked to the excessive use of
herbal teas containing relatively high concentrations of phenolics
(Kapadia et al., 1976, 1983).  There are other epidemiological
pointers for the involvement of phenolic-rich beverages in human
carcinogenesis.  Coffee was implemented as an etiological factor in
the carcinogenesis of the urinary bladder (Cole and MacMahon, 1971;
Simon et al., 1975;  Howe et al., 1980), pancreas (MacMahon et al.,
1981) and ovary (Trichopoulos et al., 1981).  The consumption of
"chagayu", a rice gruel boiled with tea leaves, has been associated
with an elevated risk for esophageal cancer (Segi, 1975;  Hirayama,
1979).  On the other hand, a negative association was observed
between tea drinking and stomach carcinomas (Stocks, 1970) and
coffee consumption and, for example, kidney cancers (Jacobsen and
Bjelke, 1982).

Table 5.  Genotoxicity of Non-Flavonoid Phenolics or Catecholamines in Various *In Vitro* Test Systems

| Endpoint | Organism | Phenolics (Samples Only) | Reference |
| --- | --- | --- | --- |
| DNA breaks | HeLa, rat fetal; lung | Epinephrine | Yamada et al. (1979) |
| Point mutations | *S. typhimurium* | Gallic acid; pyrogallol; epinephrine | Yamaguchi (1981) |
| | | Pyrogallol | Ben-Gurion (1979) |
| | *E. coli* | Pyrogallol; hydroquinone | Bilimoria (1975) |
| | *S. cerevisiae* | Catechol | Kunz et al. (1980) |
| Mitotic crossingover | *S. cerevisiae* | Catechol | Kunz et al. (1980) |
| Gene conversion | *S. cerevisiae* | Caffeic acid | Stich and Powrie (1982) |
| Chromosome aberrations | CHO cells | Simple phenols; phenolic acids; cinnamic acids | Stich et al. (1981a) |
| | | Chlorogenic acid | Stich et al. (1981b) |
| | Onion root tips | Several phenolics | Levan and Tjio (1948) |
| Sister chromatid exchanges | Human lymphocytes | Catechol | Morimoto and Wolff (1980) |
| DNA repair synthesis (UDS) | CHO cells | Dopa; dopamine; epinephrine | Wei et al. (1983) |
| DNA inhibition test | Mice | Resorcinol | Seiler (1977) |
| Micronucleus test *in vivo* | Mice | Pyrogallol; resorcinol | Mitra and Manna (1977) Gocke et al. (1981) |

Table 6.  The Capacity of *l*-Dopa to Induce a Clastogenic Effect and
an Unscheduled DNA Synthesis (UDS) in CHO Cells Following
a 3 hr Exposure

$$CH-NH_2 \quad\text{COOH}$$

*l*-Tyrosine                    *l*-Dopa

| Modifying agents: | Chromosome Aberrations | UDS | Chromosome Aberrations | UDS |
|---|---|---|---|---|
| None | − | − | + | + |
| + Mn$^{++}$ (10$^{-5}$M) | − | − | +++ | ++ |
| + Mn$^{++}$ and catalase | − | − | ± | ± |

Table 7.  Clastogenic Activity of Gallic Acid in the Presence of
Catalase, Superoxide Dismutase and Catalase Plus
Superoxide Dismutase (Hanham, 1983)

| | Percent Metaphase Plates with Chromosome Aberrations[a] | | | |
|---|---|---|---|---|
| Gallic acid (mg/ml) | Without Enzyme | Catalase | Superoxide Dismutase | Catalase + Superoxide Dismutase |
| 0.032 | 51.2 | 21.2 | T/MI[b] | 7.5 |
| 0.016 | 31.6 | 5.0 | MI | 0.0 |
| 0.008 | 0.0 | 0.0 | 10.2 | − |

[a]CHO cells were exposed to freshly prepared solutions of gallic
acid, pH 7.0, in the presence and absence of enzyme (0.5 µg/ml);
they were sampled 20 hr later for chromosome analysis.
[b]T, toxic.
MI, mitotic inhibition.

Such contradictory results also plague the experimental studies
on rodents.   Polyphenolic betel nut fractions induce oral and gastric
lesions in rodents (Bhide et al., 1979;  Ranadive et al., 1979;
Shivapurkar et al., 1980).   Various phenolics have cocarcinogenic
(Mori and Hirono, 1977;  Van Duuren, 1981) or tumour-promoting
activities (e.g., Boutwell, 1967;  Van Duuren et al., 1973;  Hecht
et al., 1981;  Van Duuren, 1982).  However, non-flavonoid phenolics
were also found to possess strong anticarcinogenic properties (see
preceding section) and lack the capacity of tumour induction (e.g.,
Hirono et al., 1980).

The oxidation of phenolics and thus the formation of superoxide, $H_2O_2$ and free radicals is catalyzed by several transition
metals and alkaline pH levels.   The influence of alkalinity on the
extent of genotoxicity is readily seen when using yeast cells which
can survive in a wide pH range (Fig. 9).   In this connection, it is
of particular interest to note the widely distributed use of lime
among betel nut or betel quid chewers of India, Malaysia, Indonesia,
The Philippines and Papua New Guinea, as well as the mixing of lime
with tobacco among "nass" chewers in Iran and the "Khaini" chewers
of Bihar.   The possibility must be considered that the lime-induced
alkalinity creates a condition within the oral cavity which favours
the oxidation of phenolics and thus the formation of $H_2O_2$ and free
radicals, which in turn could cause a genetic damage in the buccal
mucosa and initiate carcinogenesis.   A cytogenetic damage can
actually be revealed by an elevated frequency of micronucleated
buccal mucosa cells among betel nut and betel quid chewers (Stich
et al., 1982b).   The difficulty with these explanations lies in the
fact that the carcinogenicity of $H_2O_2$ is still a debatable issue
(Ito et al., 1982).

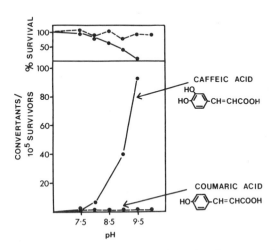

Fig. 9.   The induction of gene conversion in *S. cerevisiae* strain D7
         by exposure to caffeic acid and coumaric acid at the pH
         values indicated.

DISCUSSION

The lower incidence of cancers among high consumers of vegetables as well as an elevated cancer incidence among groups with a diet deficient in vegetables and fruits have been repeatedly used as an argument for the protective role of vitamin C and vitamin A.  These conclusions were supported by chemopreventive studies showing a regression of preneoplastic lesions following local or systemic administration of ascorbic acid (DeCosse et al., 1975;  Bussey et al., 1982) or retinoids (Bollag, 1979;  Ryssel et al., 1971;  Brocq et al., 1956;  Edvard and Bollag, 1972;  Koch, 1978;  Gouveia et al., 1982).  Moreover, the two vitamins and synthetic retinoids were efficient inhibitors of carcinogenesis in several different animal models (reviewed by the Committee on Diet, Nutrition and Cancer, 1982).  However, the observed inverse relationship between the intake of vegetables and cancer risk must not necessarily be due to vitamin C and vitamin A, but could result from the intake of other protective factors (Peto et al., 1981).  In this paper, we have tried to gather evidence for a possible involvement of naturally occurring phenolics in preventing the development of cancer.  The results show that several non-flavonoid phenolics may meet the criteria demanded from a chemopreventive agent (Table 8).  They can inhibit the mutagenic activities of both direct-acting carcinogens and precarcinogens which require metabolic activation by mixed function oxidases.  They can trap nitrite, thus reducing nitrosating species and the formation of mutagenic and carcinogenic N-nitroso compounds.  In a rodent test system, they can reduce the carcino-genic activity of benzo(a)pyrene and $\beta$-propiolactone (Wattenberg et al., 1980).  In man, phenolics can inhibit the endogenous formation of nitrosamines, as judged by the reduction of urinary nitroso-proline following administration of nitrate, proline and caffeic acid, ferulic acid or phenolic-containing beverages (teas, instant coffees).  This paper focuses on reviewing the antimutagenic and anticarcinogenic potential of simple phenolics (C6), phenolic acids (C6-C1), cinnamic acid and related compounds (C6-C3).  An inhibition of carcinogenesis and mutagenesis is by no means restricted to the above-mentioned groups of phenolics.  Several naturally occurring flavonoids show comparable activities (e.g., Wattenberg, 1979; Buening et al., 1981;  Ohta et al., 1983).  Population groups on a Western diet consume daily milligram to gram quantities of numerous phenolic compounds with antimutagenic and anticarcinogenic capacities (Table 9).  This group of chemicals can therefore not be ignored in a search for those factors which are responsible for the cancer-protective effect of vegetables.

Whether naturally occurring phenolics or synthetic ones such as, for example, BHT, BHA and propyl gallate, will be suitable in a chemopreventive regime is difficult to predict at this stage.  Several crucial questions remain unanswered.  Many phenolics of vegetable and fruit origin will be ingested as glycosides.  Little

Table 8.   Inhibitory Effects of Naturally Occurring Phenolics on Mutagenicity and Carcinogenicity

1.   Inhibitory effect on mutagenesis induced by direct-acting or S9-requiring mutagens

*S. typhimurium* mutagenicity test
Tested phenolics:   chlorogenic acid, caffeic acid, gallic acid, salicylic acid, coumaric acid,
                    *p*-hydroxybenzoic acid, *p*-cinnamic acid, tannic acid, catechol

2.   Inhibitory effect on the formation of mutagenic nitrosation products

Nitrosation of methylurea
Tested phenolics:   catechol, gallic acid, chlorogenic acid, tannic acid, catechin, resorcinol
Tested phenolic-containing products:   roasted ground coffee, instant coffee, decaffeinated
                    coffee, Japanese tea, Indian tea, Chinese tea, Maté tea, white wine, red wine,
                    beer, betel nut extracts

3.   Inhibitory effect on carcinogenesis

Benzo(a)pyrene-induced stomach carcinomas
Tested phenolics:   *o*-hydroxycinnamic acid, *p*-hydroxycinnamic acid, caffeic acid, ferulic acid

4.   Trapping of nitrite (nitrite depletion assay)

Tested phenolics:   caffeic acid, catechin, gallic acid, chlorogenic acid, ferulic acid,
                    β-resorcyclic acid, *p*-coumaric acid, *o*-pyrocatechuic acid, hydroquinone,
                    phloroglucinol, gallacetophenone, pyrogallol
Tested phenolic-containing products:   betel nut extracts, roasted ground coffee

5.   Inhibitory effect of endogenously formed nitrosoproline in man

Tested phenolics:   caffeic acid, ferulic acid
Tested phenolic-containing products:   Chinese tea, Indian tea, Japanese tea, instant coffee,
                    decaffeinated coffee, betel nut extracts

Table 9.  Amounts of Non-Flavonoid Phenolics in Several Food Products

| Food Product | Phenolics | Amount | Reference |
|---|---|---|---|
| **Vegetables:** | | | |
| Brussel sprouts | Caffeic acid | 3.4 mg/100 g | Schmidtlein and Herrmann (1975a) |
| | Ferulic acid | 1.0 mg/100 g | |
| | Coumaric acid | 1.2 mg/100 g | |
| | Sinapic acid | 10.7 mg/100 g | |
| Potatoes | Caffeic acid | 28.0 mg/100 g | Schmidtlein and Herrmann (1975b) |
| | Ferulic acid | 2.8 mg/100 g | |
| | Coumaric acid | 0.4 mg/100 g | |
| | Sinapic acid | 0.3 mg/100 g | |
| Lettuce | Caffeic acid | 16-90 mg/100 g | Herrmann (1978) |
| | Hydroxybenzoic acid | 20-40 mg/100 g | |
| **Fruits:** | | | |
| Apples | Caffeic acid | 8.5-127 mg/100 g | Mosel and Herrmann (1974) |
| | Ferulic acid | 0.4-9.5 mg/100 g | |
| | Coumaric acid | 1.5-46 mg/100 g | |
| Strawberries | Caffeic acid | 1.5-3.9 mg/100 g | Stöhr and Herrmann (1975) |
| | Coumaric acid | 6.9-17.5 mg/100 g | |
| | Gallic acid | 8.0-12.1 mg/100 g | |
| | Hydroxybenzoic acid | 0.0-10.8 mg/100 g | |
| **Beverages:** | | | |
| Wine | Caffeoyl tartaric acid | 8-116 mg/1 | Okamura and Watanabe (1981) |
| | Coumaroyl tartaric acid | 2-33 mg/1 | |
| | Non-flavonoid phenols | 180-320 mg/1 | |
| Coffee | Chlorogenic acid | 260 mg/cup | Challis and Bartlett (1975) |
| | Catechol | ±0.7 mg/cup | Tressl et al. (1978) |
| | Pyrogallol | ±0.4 mg/cup | |

is known about whether glycoside and glucuronate derivatives of phenolics retain their capacity to trap nitrite, form complexes with carcinogens and inhibit the formation of ultimate carcinogens. The conditions which favour the generation of genotoxic $H_2O_2$ and free radicals during oxidation of phenolics must also be clearly delineated. However, this is an issue which is not restricted to the group of phenolics. Many antioxidants can, depending on the redox potential, either adsorb or donate electrons. These dual, apparently contradictory functions of several antioxidants make it difficult to simply categorize them as either antigenotoxic agents or genotoxic compounds.

Before electrophiles or free radicals are formed and induce genetic changes in the target tissue, there is a long chain of complex chemical interactions outside and inside of an organism. Phenolics could conceivably exert a protective effect at various stages. Catechin, protocatechuic acid, chlorogenic acid and the flavonoids, quercetin and rutin, at a dose of 0.2 g per 1 gram of albumin, reduced the mutagenic level of pyrolyzed albumin (Fukuhara et al., 1981). Similarly, chlorogenic acid inhibited the formation of mutagenic products when added prior to the frying of beef patties (Wang et al., 1982). These two examples point to a preventive role of phenolics which has not been thoroughly explored. Within an organism, the non-flavonoid phenolics could act as trapping agents in various body fluids (e.g., gastric juice, urine, blood) or within cells. At present, the main unresolved issues deal with the metabolism of phenolics within the human body and the design of a harmless, simple and acceptable way to increase the concentration of phenolics in various tissues to levels which will have anticarcinogenic activity. In spite of the many unanswered questions, the possible usefulness of non-flavonoid phenolics in intervention programmes seems to be worth a more detailed consideration.

## REFERENCES

Ben-Gurion, R., 1979, Mutagenic and colicine-inducing activity of two antioxidants: pyrogallol and purpurogallin, Mutation Res., 68:201.

Bhide, S.V., Shivapurkar, N.M., Gothoskar, S.V., and Ranadive, K.J., 1979, Carcinogenicity of betel quid ingredients: feeding mice with aqueous extracts and the polyphenol fraction of betel nut, Br. J. Cancer, 40:922.

Bilimoria, M.H., 1975, The detection of mutagenic activity of chemicals and tobacco smoke in a bacterial system, Mutation Res., 31:328.

Bjelke, E., 1975, Dietary vitamin A and human lung cancer, Int. J. Cancer, 15:561.

Bogovski, P., Castegnaro, M., Pignatelli, B., and Walker, E.A.,
    1972, The inhibiting effect of tannins on the formation of
    nitrosamines, in: "N-Nitroso Compounds, Analysis and
    Formation", P. Bogovski, R. Preussmann and E.A. Walker, eds,
    IARC Scientific Publications No. 3, International Agency for
    Research on Cancer, Lyon.

Bollag, W., 1979, Retinoids and cancer, Cancer Chemother. Pharmacol.,
    3:207.

Boutwell, R.K., 1967, Phenolic compounds as tumor-promoting agents,
    in: "Phenolic Compounds and Metabolic Regulation", B.J.
    Finkle and V.C. Runeckles, eds, Appleton-Century Crofts,
    New York.

Brocq, P., Stora, C., and Bernheim, L., 1956, De l'emploi de la
    vitamine A dans le traitement des mastoses, Ann. Endocrinol.,
    17:193.

Buening, M.K., Chang, R.L., Huang, M-T., Fortner, J.G., Wood, A.W.,
    and Conney, A.H., 1981, Activation and inhibition of benzo-
    (a)pyrene and aflatoxin $B_1$ metabolism in human liver micro-
    somes by naturally occurring flavonoids, Cancer Res., 41:67.

Bussey, H.J.R., DeCosse, J.J., Deschner, E.E., Eyers, A.A., Lesser,
    M.L., Morson, B.C., Ritchie, S.M., Thomson, J.P.S., and
    Wadsworth, J., 1982, A randomized trial of ascorbic acid in
    polyposis coli, Cancer, 50:1434.

Cambien, F., Ducimetiere, P., and Richard, J., 1980, Total serum
    cholesterol and cancer mortality in a middle-aged population,
    Am. J. Epidemiol., 112:388.

Challis, B.C., and Bartlett, C.D., 1975, Possible cocarcinogenic
    effects of coffee constituents, Nature (Lond.), 254:532.

Chan, R.I.M., 1982, Inhibitory effect of browning reaction products
    and phenolic compounds on carcinogen-induced mutagenesis,
    M.Sc. thesis, University of British Columbia, Vancouver, B.C.

Cohen, G., and Heikkila, R.E., 1974, The generation of hydrogen
    peroxide, superoxide radical, and hydroxyl radical by 6-
    hydroxydopamine, dialuric acid, and related cytotoxic
    agents, J. Biol. Chem., 249:2447.

Cole, P., and MacMahon, B., 1971, Attributable risk percent in
    case-control studies, Br. J. Prev. Soc. Med., 25:242.

Committee on Diet, Nutrition and Cancer, 1982, "Diet, Nutrition,
    and Cancer", Assembly of Life Sciences, National Research
    Council, National Academy Press, Washington D.C.

DeCosse, J.J., Adams, M.B., Kuzma, J.F., LoGerfo, P., and Condon,
    R.E., 1975, Effect of ascorbic acid on rectal polyps of
    patients with familial polyposis, Surgery, 78:608.

Doll, R., and Peto, R., 1981, The causes of cancer: quantitative
    estimates of avoidable risks of cancer in the United States
    today, J. Natl. Cancer Inst., 66:1191.

Edvard, J.P., and Bollag, W., 1972, Konservative Behandlung der
    rezidivierenden Harnblasenpapillomatose mit Vitamin-A-Säure,
    Schweiz. Med. Wschr., 102:1880.

Eisenbrand, G., Spiegelhalder, B., and Preussmann, R., 1981, Analysis of human biological specimens for nitrosamine contents, in: "Gastrointestinal Cancer: Endogenous Factors", W.R. Bruce, P. Correa, M. Lipkin, S.R. Tannenbaum and T.D. Wilkins, eds, Banbury Report 7, Cold Spring Harbor Laboratory, Cold Spring Harbor, N.Y.

Freese, E., 1971, Molecular mechanisms of mutation, in: "Chemical Mutagens: Principles and Methods for Their Detection", A. Hollaender, ed., Plenum Publ. Corp., New York.

Fukuhara, Y., Yoshida, D., and Goto, F., 1981, Reduction of mutagenic products in the presence of polyphenols during pyrolysis of protein, Agric. Biol. Chem., 45:1061.

Garner, R.C., Miller, E.C., and Miller, J.A., 1972, Liver microsomal metabolism of aflatoxin $B_1$ to a reactive derivative toxic to *Salmonella typhimurium* TA1530, Cancer Res., 32:2058.

Gilbert, P., Rondelet, J., Poncelet, F., and Mercier, M., 1980, Mutagenicity of *p*-nitrosophenol, Food Cosmet. Toxicol., 18:523.

Gocke, E., King, M.-T., Eckhardt, K., and Wild, D., 1981, Mutagenicity of cosmetics ingredients licensed by the European Communities, Mutation Res., 90:91.

Gouveia, J., Hercend, T., Lemaigre, G., Mathe, G., Gros, F., Santelli, G., Homasson, J.P., Angebault, M., Lededente, A., Parrot, R., Gaillard, J.P., Bonniot, J.P., Marsac, J., and Pretet, S., 1982, Degree of bronchial metaplasia in heavy smokers and its regression after treatment with a retinoid, Lancet, 1:710.

Graham, D.G., Tiffany, S.M., Bell, W.R., Jr, and Gutknecht, W.F., 1978, Autoxidation versus covalent binding of quinones as the mechanism of toxicity of dopamine, 6-hydroxydopamine, and related compounds towards C1300 neuroblastoma cells *in vitro*, Mol. Pharmacol., 14:644.

Gray, J.I., and Dugan, L.R., 1975, Inhibition of N-nitrosamine formation in model food systems, J. Food Sci., 40:981.

Groenen, P.J., 1977, A new type of N-nitrosation inhibitor, in: "Proc. 2nd Int. Symp. Nitrite in Meat Products", B.J. Tinbergen and B. Krol, eds, Pudoc, Wageningen, The Netherlands.

Haenszel, W., and Correa, P., 1975, Developments in the epidemiology of stomach cancer over the past decade, Cancer Res., 35:3452.

Hanham, A.F., 1983, The biological activity of phenolics, Ph.D. thesis, University of British Columbia, Vancouver, B.C., Canada (to be conferred).

Hanham, A.F., Dunn, B.P., and Stich, H.F., 1982, Clastogenic activity of caffeic acid and its relationship to hydrogen peroxide generated during autooxidation, Mutation Res., in press.

Hecht, S.S., Carmella, S., Mori, H., and Hoffmann, D., 1981, A study of tobacco carcinogenesis. XX. Role of catechol as a major cocarcinogen in the weakly acidic fraction of smoke condensate, J. Natl. Cancer Inst., 66:163.

Herrmann, K., 1978, Übersicht über nichtessentielle Inhaltsstoffe der Gemüsearten. III. Möhren, Sellerie, Pastinaken, rote Rüben, Spinat, Salat, Endiven, Treibzichorie, Rhabarber und Artischocken, Z. Lebensm. Unters.-Forsch., 167:262.

Hirayama, T., 1979, Epidemiological evaluation of the role of
        naturally occurring carcinogens and modulators of carcino-
        genesis, in: "Naturally Occurring Carcinogens-Mutagens and
        Modulators of Carcinogenesis", E.C. Miller, J.A. Miller,
        T. Sugimura, S. Takayama and I. Hirono, eds, Jap. Sci. Soc.
        Press, Tokyo/Univ. Park Press, Baltimore.
Hirayama, T., 1981, A large-scale cohort study on the relationship
        between diet and selected cancers of digestive organs, in:
        "Gastrointestinal Cancer: Endogenous Factors", W.R. Bruce,
        P. Correa, M. Lipkin, S.R. Tannenbaum and T.D. Wilkins, eds,
        Banbury Report 7, Cold Spring Harbor Laboratory, Cold Spring
        Harbor, N.Y.
Hirono, I., Hosaka, S., Uchida, E., Takanashi, H., Haga, M., Sakata,
        M., Mori, H., Tanaka, T., and Hikino, H., 1980, Safety
        examination of some edible or medicinal plants and plant
        constituents, part 3, J. Food Safety, 4:205.
Ho, J.H.C., Huang, D.P., and Fong, Y.Y., 1978, Salted fish and
        nasopharyngeal carcinoma in southern China, Lancet, 2:626.
Howe, G.R., Burch, J.D., Miller, A.B., Cook, G.M., Esteve, J.,
        Morrison, B., Gordon, P., Chambers, L.W., Fodor, G., and
        Winsor, G.M., 1980, Tobacco use, occupation, coffee, various
        nutrients, and bladder cancer, J. Natl. Cancer Inst., 64:701.
Huang, D.P., Saw, D., Teoh, T.B., and Ho, J.H.C., 1978, Carcinoma
        of the nasal and paranasal regions in rats fed Cantonese
        salted marine fish, in: "Nasopharyngeal Carcinoma: Etiology
        and Control", G. de-The and Y. Ito, ed., IARC Scientific
        Publications No. 20, International Agency for Research on
        Cancer, Lyon.
Ito, A., Naito, M., Naito, Y., and Watanabe, H., 1982, Induction and
        characterization of gastro-duodenal lesions in mice given
        continuous oral administration of hydrogen peroxide, Gann,
        73:315.
Jacobsen, B.K., and Bjelke, E., 1982, Coffee consumption and cancer:
        a prospective study, in: "Proc. 13th Int. Cancer Congress",
        Seattle, WA (Abstract).
Kapadia, G.J., Paul, B.D., Chung, E.B., Ghosh, B., and Pradhan, S.N.,
        1976, Carcinogenicity of Camellia sinensis (tea) and some
        tannin-containing folk medicinal herbs administered subcu-
        taneously in rats, J. Natl. Cancer Inst., 57:207.
Kapadia, G.J., Rao, G.S., and Morton, J.F., 1983, Herbal tea con-
        sumption and esophageal cancer, in: "Carcinogens and Mutagens
        in the Environment", Vol. III, "Naturally Occurring Compounds:
        Epidemiology and Distribution", H.F. Stich, ed., CRC Press,
        Boca Raton, Florida, in press.
Kawabata, T., Ohshima, H., Vibu, J., Nakamura, M., Matsui, M., and
        Hamano, M., 1979, Occurrence, formation and precursors of
        N-nitroso compounds in Japanese diet, in: "Naturally
        Occurring Carcinogens-Mutagens and Modulators of Carcinogene-
        sis", E.C. Miller, J.A. Miller, T. Sugimura, S. Takayama and
        I. Hirono, eds, Jap. Sci. Soc. Press, Tokyo/Univ. Park Press,
        Baltimore.

King, M.M., McCoy, P.B., and Russo, I., 1983, Dietary fat influences on the effectiveness of dietary antioxidants as tumor inhibitors, in: "Diet and Cancer: From Basic Research to Policy Implications", Alan R. Liss, New York, in press.

Koch, H.F., 1978, Biochemical treatment of precancerous oral lesions: the effectiveness of various analogues of retinoic acid, J. max.-fac. Surg., 6:59.

Kolonel, L.N., Nomura, A.M.Y., Hirohata, T., Hankin, J.H., and Hinds, M.W., 1981, Association of diet and place of birth with stomach cancer incidence in Hawaii Japanese and Caucasians, Am. J. Clin. Nutr., 34:2478.

Korycka-Dahl, M.B., and Richardson, T., 1978, Activated oxygen species and oxidation of food constituents, Crit. Rev. Food Sci. Technol., 10:209.

Kunz, B.A., Hannan, M.A., and Haynes, R.H., 1980, Effect of tumor promoters on ultraviolet light-induced mutation and mitotic recombination in Saccharomyces cerevisiae, Cancer Res., 40:2323.

Levan, A., and Tjio, J.H., 1948, Induction of chromosome fragmentation by phenols, Hereditas, 34:453.

Lin, J.-K., Kennan, K.A., Miller, E.C., and Miller, J.A., 1978, Reduced nicotinamide adenine dinucleotide phosphate-dependent formation of 2,3-dihydro-2,3-dihydroxy-aflatoxin $B_1$ from aflatoxin $B_1$ by hepatic microsomes, Cancer Res., 38:2424.

MacMahon, B., Yen, S., Trichopoulos, D., Warren, K., and Nardi, G., 1981, Coffee and cancer of the pancreas, N. Engl. J. Med., 304:630.

MacRae, W.D., and Stich, H.F., 1979, Induction of sister chromatid exchanges in Chinese hamster cells by the reducing agents bisulfite and ascorbic acid, Toxicology, 13:167.

Marquardt, H., Rufino, F., and Weisburger, J.H., 1977, On the aetiology of gastric cancer: mutagenicity of food extracts after incubation with nitrite, Food Cosmet. Toxicol., 15:97.

Marshall, J., Graham, S., Mettlin, C., Shedd, D., and Swanson, M., 1982, Diet in the epidemiology of oral cancer, Nutr. Cancer, 3:145.

Mettlin, C., Graham, S., Priore, R., Marshall, J., and Swanson, M., 1981, Diet and cancer of the esophagus, Nutr. Cancer, 2:143.

Mirvish, S.S., 1981, Inhibition of the formation of carcinogenic N-nitroso compounds by ascorbic acid and other compounds, in: "Cancer 1980: Achievements, Challenges, and Prospects for the 1980's", J.H. Burchenal and H.F. Oettgen, eds, Vol. 1, Grune and Stratton, New York.

Mitra, A.B., and Manna, G.K., 1977, Effect of some phenolic compounds on chromosomes of bone marrow cells of mice, Ind. J. Med. Res., 59:1442.

Mori, H., and Hirono, I., 1977, Effect of coffee on carcinogenicity of cycasin, Br. J. Cancer, 35:369.

Morimoto, K., and Wolff, S., 1980, Increase of sister-chromatid exchanges and perturbation of cell division kinetics in human lymphocytes by benzene metabolites, Cancer Res., 40:1189.

Mosel, H.-D., and Herrmann, K., 1974, The phenolics of fruits.
    III. The contents of catechins and hydroxycinnamic acids in
    pome and stone fruits, Z. Lebensm. Unters.-Forsch., 154:6.
Neal, G.E., and Colley, P.E., 1978, Some high-performance liquid-
    chromatographic studies of the metabolism of aflatoxins by
    rat liver microsomal preparations, Biochem. J., 174:839.
Newmark, H.L., and Mergens, W.J., 1981, Blocking nitrosamine forma-
    tion using ascorbic acid and α-tocopherol, in: "Gastro-
    intestinal Cancer: Endogenous Factors", W.R. Bruce, P.
    Correa, M. Lipkin, S.R. Tannenbaum and T.D. Wilkins, eds,
    Banbury Report 7, Cold Spring Harbor Laboratory, Cold Spring
    Harbor, N.Y.
Ohshima, H., and Bartsch, H., 1981, Quantitative estimation of
    endogenous nitrosation in humans by monitoring N-nitroso-
    proline excreted in the urine, Cancer Res., 41:3658.
Ohshima, H., Bereziat, J-C., and Bartsch, H., 1982, Measurement of
    endogenous N-nitrosation in rats and humans by monitoring
    urinary and faecal excretion of N-nitrosamino acids, in:
    "N-Nitroso Compounds: Occurrence and Biological Effects",
    H. Bartsch et al., eds, IARC Scientific Publications No. 41,
    International Agency for Research on Cancer, Lyon.
Ohta, T., Watanabe, K., Moriya, M., Shirasu, Y., and Kada, T.,
    1982, Antimutagenic effects of cinnamaldehyde and its
    structural analogues on mutagenesis in E. coli., Mutation
    Res., in press.
Okamura, S., and Watanabe, M., 1981, Determination of phenolic
    cinnamates in white wine and their effect on wine quality,
    Agric. Biol. Chem., 45:2063.
Peto, R., Doll, R., Buckley, J.D., and Sporn, M.B., 1981, Can
    dietary beta-carotene materially reduce human cancer rates?
    Nature (Lond.), 290:201.
Pignatelli, B., Castegnaro, M., and Walker, E.A., 1976, Effects of
    gallic acid and of ethanol on formation of nitrosodiethyl-
    amine, in: "Environmental N-Nitroso Compounds, Analysis and
    Formation", E.A. Walker, P. Bogovski and L. Griciute, eds,
    IARC Scientific Publications No. 14, International Agency
    for Research on Cancer, Lyon.
Ranadive, K.J., Ranadive, S.N., Shivapurkar, N.M., and Gothoskar,
    S.V., 1979, Betel quid chewing and oral cancer: experimental
    studies on hamsters, Int. J. Cancer, 24:835.
Rosin, M.P., 1982, Inhibition of genotoxic activities of complex
    mixtures by naturally occurring agents, in: "Carcinogens
    and Mutagens in the Environment", Vol. I, "Food Products",
    H.F. Stich, ed., CRC Press, Boca Raton, Florida, in press.
Rosin, M.P., and Stich, H.F., 1978a, The inhibitory effect of
    reducing agents on N-acetoxy- and N-hydroxy-2-acetylamino-
    fluorene-induced mutagenesis, Cancer Res., 38:1307.
Rosin, M.P., and Stich, H.F., 1978b, The inhibitory effect of
    cysteine on the mutagenic activities of several carcinogens,
    Mutation Res., 54:73.

Rosin, M.P., and Stich, H.F., 1979, Assessment of the use of the
    Salmonella mutagenesis assay to determine the influence of
    antioxidants on carcinogen-induced mutagenesis, Int. J.
    Cancer, 23:722.
Ryssel, H.J., Brunner, K.W., and Bollag, W., 1971, Die perorale
    Anwerdung von Vitamin A Säure bei Leukoplakien.  Hyperkeratosen
    und Plattenepithelkarzinomen:  Ergebnisse und Verträglichkeit,
    Schweiz. Med. Wschr., 101:1027.
Schmidtlein, H., and Herrmann, K., 1975a, Uber die Phenolsäuren des
    Gemüses.  I.  Hydroxyzimtsäuren und Hydroxybenzoesäuren der
    Kohlarten und anderer Cruciferen-Blätter, Z. Lebensm. Unters.-
    Forsch., 159:139.
Schmidtlein, H., and Herrmann, K., 1975b, Uber die Phenolsäuren des
    Gemüses.  IV.  Hydroxyzimtsäuren und Hydroxybenzoesäuren
    weiterer Gemüsearten und der Kartoffeln, Z. Lebensm. Unters.-
    Forsch., 159:255.
Segi, M., 1975, Tea-gruel as a possible factor for cancer of the
    esophagus, Gann, 66:199.
Seiler, J.P., 1977, Inhibition of testicular DNA synthesis by
    chemical mutagens and carcinogens, Mutation Res., 46:305.
Sen, N.P., and Donaldson, B., 1978, Improved colourimetric method
    for determining nitrate and nitrite in foods, J. Assoc.
    Off. Anal. Chem., 61:1389.
Shekelle, R.B., Liu, S., Raynor, W.J., Jr, Lepper, M., Maliza, C.,
    and Rossof, A.H., 1981, Dietary vitamin A and risk of cancer
    in the Western Electric Study, Lancet, 2:1185.
Shivapurkar, N.M., Ranadive, S.N., Gothoskar, S.V., Bhide, S.V.,
    and Ranadive, K.J., 1980, Tumorigenic effect of aqueous and
    polyphenolic fractions of betel nut in Swiss strain mice,
    Ind. J. Exp. Biol., 18:1159.
Simon, D., Yen, S., and Cole, P., 1975, Coffee drinking and cancer
    of the lower urinary tract, J. Natl. Cancer Inst., 54:587.
Speit, G., Vogel, W., and Wolf, M., 1982, Characterization of sister
    chromatid exchange induction by hydrogen peroxide, Environ.
    Mutagen., 4:135.
Stich, H.F., and Laishes, B.A., 1975, The response of Xeroderma
    pigmentosum cells and controls to the activated mycotoxins,
    aflatoxins and sterigmatocystin, Int. J. Cancer, 16:266.
Stich, H.F., and Powrie, W.D., 1982, Plant phenolics as genotoxic
    agents and as modulators for the mutagenicity of other food
    components, in:  "Carcinogens and Mutagens in the Environ-
    ment", Vol. I, "Food Products", H.F. Stich, ed., CRC Press,
    Boca Raton, Florida, in press.
Stich, H.F., Wei, L., and Whiting, R.F., 1979, Enhancement of the
    chromosome-damaging action of ascorbate by transition metals,
    Cancer Res., 39:4145.
Stich, H.F., Rosin, M.P., Wu, C.H., and Powrie, W.D., 1981a, The
    action of transition metals on the genotoxicity of simple
    phenols, phenolic acids and cinnamic acids, Cancer Lett.,
    14:251.

Stich, H.F., Rosin, M.P., Wu, C.H., and Powrie, W.D., 1981b, A
    comparative genotoxicity study of chlorogenic acid (3-0-
    caffeoylquinic acid), Mutation Res., 90:201.

Stich, H.F., Rosin, M.P., and Bryson, L., 1982a, Inhibition of
    mutagenicity of a model nitrosation reaction by naturally
    occurring phenolics, coffee and tea, Mutation Res., 95:119.

Stich, H.F., Stich, W., and Parida, B.B., 1982b, Elevated frequency
    of micronucleated cells in the buccal mucosa of individuals
    at high risk for oral cancer: betel quid chewers, Cancer
    Lett., in press.

Stich, H.F., Ohshima, H., Pignatelli, B., Michelon, J., and Bartsch,
    H., 1983a, Inhibitory effect of betel nut extracts on endo-
    genous nitrosation in man, J. Natl. Cancer Inst., in press.

Stich, H.F., Bohm, B., Chatterjee, K., and Sailo, J., 1983b, The
    role of saliva-borne mutagens and carcinogens in the etiology
    of oral and esophageal carcinomas of betel nut and tobacco
    chewers, in: "Carcinogens and Mutagens in the Environment",
    Vol. III, "Naturally Occurring Compounds: Epidemiology and
    Distribution", H.F. Stich, ed., CRC Press, Boca Raton,
    Florida, in press.

Stocks, P., 1970, Cancer mortality in relation to national consump-
    tion of cigarettes, solid fuel, tea and coffee, Br. J.
    Cancer, 24:215.

Stöhr, H., and Herrmann, K., 1975, Die phenolischen Inhaltsstoffe
    des Obstes. VI. Die phenolischen Inhaltsstoffe der Johannis-
    beeren, Stachelbeeren und Kulturheidelbeeren. Veränderungen
    der Phenolsäuren und Catechine während Wachstum und Reife
    von schwarzen Johannisbeeren, Z. Lebensm. Unters.-Forsch.,
    159:31.

Tressl, R., Bahri, D., Köppler, H., and Jensen, A., 1978, Diphenole
    und Caramelkomponenten in Röstkaffees verschiedener Sorten.
    II. Z. Lebensm. Unters.-Forsch., 167:111.

Trichopoulos, D., Papapostolou, M., and Polychronopoulou, A., 1981,
    Coffee and ovarian cancer, Int. J. Cancer, 28:691.

Van Duuren, B.L., 1981, Cocarcinogens and tumor promoters and their
    environmental importance, J. Environ. Pathol. Toxicol., 5:
    959.

Van Duuren, B.L., 1982, Cocarcinogens and tumor promoters and their
    environmental importance, J. Am. Coll. Toxicol., 1:17.

Van Duuren, B.L., Katz, C., and Goldschmidt, B.M., 1973, Cocarcino-
    genic agents in tobacco carcinogenesis, J. Natl. Cancer Inst.,
    51:703.

Walker, E.A., Pignatelli, B., and Friesen, M., 1982, The role of
    phenols in catalysis of nitrosamine formation, J. Sci. Food
    Agric., 33:81.

Wang, Y.Y., Vuolo, L.L., Spingarn, N.E., and Weisburger, J.H.,
    1982, Formation of mutagens in cooked foods. V. The mutagen
    reducing effect of soy protein concentrates and antioxidants
    during frying of beef, Cancer Lett., 16:179.

Wattenberg, L.W., 1972, Inhibition of carcinogenic and toxic effects of polycyclic hydrocarbons by phenolic antioxidants and ethoxyquin, J. Natl. Cancer Inst., 48:1425.

Wattenberg, L.W., 1973, Inhibition of chemical carcinogen-induced pulmonary neoplasia by butylated hydroxyanisole, J. Natl. Cancer Inst., 50:1541.

Wattenberg, L.W., 1976, Inhibition of chemical carcinogenesis by antioxidants and some additional compounds, in: "Fundamentals in Cancer Prevention", P.N. Magee, ed., University Park Press, Baltimore.

Wattenberg, L.W., 1979, Naturally occurring inhibitors of chemical carcinogenesis, in: "Naturally Occurring Carcinogens-Mutagens and Modulators of Carcinogenesis", E.C. Miller, J.A. Miller, I. Hirono, T. Sugimura and S. Takayama, eds, Jap. Sci. Soc. Press, Tokyo/Univ. Park Press, Baltimore.

Wattenberg, L.W., Coccia, J.B., and Lam, L.K.T., 1980, Inhibitory effects of phenolic compounds on benzo(a)pyrene-induced neoplasia, Cancer Res., 40:2820.

Wei, L., Whiting, R.F., and Stich, H.F., 1983, The capacity of catecholamines to induce chromosome aberrations in DNA repair synthesis in mammalian cells (unpublished).

Weisburger, J.H., Marquardt, H., Hirota, N., Mori, H., and Williams, G.M., 1980, Induction of cancer of the glandular stomach in rats by an extract of nitrite-treated fish, J. Natl. Cancer Inst., 64:163.

Whiting, R.F., Wei, L., and Stich, H.F., 1981, Chromosome-damaging activity of ferritin and its relation to chelation and reduction of iron, Cancer Res., 41:1628.

Wong, J.J., and Hsieh, D.P.H., 1976, Mutagenicity of aflatoxins related to their metabolism and carcinogenic potential, Proc. Natl. Acad. Sci. USA, 73:2241.

Wood, A.W., Huang, M-T., Chang, R.L., Newmark, H.L., Lehr, R.E., Yagi, H., Sayer, J.M., Jerina, D.M., and Conney, A.H., 1982, Inhibition of the mutagenicity of bay-region diol epoxides of polycyclic aromatic hydrocarbons by naturally occurring plant phenols: exceptional activity of ellagic acid, Proc. Natl. Acad. Sci. USA, 79:5513.

Yamada, K., Murakami, H., Nishiguchi, H., Shirahata, S., Shinohara, K., and Omura, H., 1979, Varying responses of cultured mammalian cell lines to the cellular DNA breaking activity of epinephrine, Agric. Biol. Chem., 43:901.

Yamaguchi, T., 1981, Mutagenicity of low molecular substances in various superoxide generating systems, Agric. Biol. Chem., 45:327.

# 2

# SULFHYDRYL GROUPS AND FOOD SAFETY

Mendel Friedman

Western Regional Research Center, ARS-USDA
Berkeley, California  94710

Sulfhydryl (thiol) compounds such as cysteine, N-acetylcysteine, and reduced glutathione interact with disulfide bonds of plant protease inhibitors _via_ sulfhydryl-disulfide interchange and oxidation reactions.  Such interactions with inhibitors from soy and lima beans facilitate their heat inactivation of inhibitors, resulting in beneficial nutritional effects.  Since thiols are potent nucleophiles, they have a strong avidity for unsaturated electrophilic centers of several dietary toxicants, including cyclopropene fatty acids, aflatoxins, sesquiterpene lactones, and trichothecenes.  Such interactions may be used _in vitro_ to lower the toxic potential of our food supply, and _in vivo_ for prophylactic and therapeutic effects against antinutrients and food toxicants.  A number of examples are cited to illustrate the concept of applying site-specific nucleophilic reagents such as thiols to reduce antinutritional and toxic manifestations of food ingredients by modifying pharmacophoric and toxicophoric electrophilic sites.  Several related interactions between sulfhydryl groups and toxic compounds are also briefly mentioned to illustrate the generality of this approach.

## INTRODUCTION

Naturally occurring antinutrients and food toxicants and those formed during food processing adversely affect the nutritional quality and safety of foods.  Because of a growing concern about direct relationships between diet and diseases (Ames, 1983; Doll and Peto, 1981) and because of a growing need to improve the quality and safety of our food supply, research is needed to define conditions that favor or minimize the formation of nutritionally antagonistic and toxic compounds in foods.

31

Most naturally occurring food toxicants and many antinutrients possess specific sites that are responsible for their deleterious effects. Therefore, by modifying such sites with site-specific reagents, such as thiols or other compounds, in a manner that will alter structural integrity and thus prevent them from interacting with receptor sites in vivo, it should be possible to lessen their toxic potential.

Sulfhydryl (thiol) groups in amino acids, peptides, and proteins participate in anionic, cationic, and free radical reactions both in vitro and in vivo (Friedman, 1973). The chemical reactivities of negatively charged sulfur anions ($RS^-$) are much greater than would be expected from their basicities. This great reactivity presumably results from (1) polarizabilities of outer shell sulfur electrons and (2) the availability of d-orbitals in the electornic structure of sulfur permitting d-orbital overlap during the formation of transition states. Besides acting as a precursor for disulfide bonds that stabilize proteins, sulfhydryl groups participate directly in many and varied chemical and biochemical processes.

In this review, I will describe some of the possible approaches to reducing deleterious effects of representative antinutrients and food toxicants, based on this reactivity of the sulfhydryl group with electrophilic centers.

## PROTEASE INHIBITORS

Plant protease inhibitors of trypsin, chymotrypsin, carboxypeptidase, α-amylase, etc. occur in many agricultural products including legumes and cereals (Bozzini and Silano, 1978). Such inhibitors adversely affect the nutritional quality and safety of foods (Liener, 1975; Anderson, et al., 1979; Liener and Kakade, 1980; Rackis and Gumbmann, 1981; Gallaher and Schneeman, 1984). For example, soybeans and other legumes containing active trypsin inhibitors depress growth in rats compared to analogous feeding of inhibitor-free soybeans. Growth inhibition is accompanied by pancreatic hyperplasia and hypertrophy (Anderson et al, 1979).

A need therefore exists to find ways to inactivate the inhibitors before such foods are consumed. Although heat is often used to inactivate inhibitors, such inactivation is often incomplete (Anderson et al., 1979). The use of high temperatures to destroy inhibitors may also destroy lysine and sulfur amino acids, and thus damage protein nutritional quality (Rackis and Gumbmann, 1981).

Most inhibitors of trypsin and other proteolytic enzymes contain disulfide bonds, some or all of which are needed to maintain active conformations. Amino acid sequence studies show that the reactive peptide link responsible for the enzyme inhibiting activity of several protease inhibitors appears as a cyclic substrate in a loop closed by a disulfide bridge. This is the so-called "disulfide loop". Partial loss of disulfide bridges may or may not reduce or eliminate inhibitory activities of enzyme inhibitors. In the case of inhibitors present in lima beans, studies of reductive S-alkylation of disulfide bonds revealed that incomplete reduction may essentially abolish inhibitory activity against trypsin and chymotrypsin when about half of the disulfide bonds have been modified (Friedman et al., 1980b; Zahnley and Friedman, 1982).

The loss of inhibitory activity following reduction and S-alkylation has been postulated to arise from the change in the native conformation of the protein inhibitors so that they are unable to combine with the active site(s) of the enzyme(s) (Zahnley, 1984).

Since reductive S-alkylation of disulfide bonds introduces unnatural amino acid side chains into the protein and, therefore, cannot serve as a useful model for nutritional and toxicological studies, we initiated systematic studies of effects of thiols on the inhibitory process. Such thiols are expected to interact with inhibitor disulfide bonds via sulfhydryl-disulfide interchange and oxidation reactions (Friedman, 1973).

Figures 1 and 2 illustrate the facilitating influence of sulfhydryl compounds in inactivating inhibitors in soybean and lima bean flours.

The described cooperative action of thiols probably involves sulfhydryl-disulfide interchange or oxidation-reduction reactions between added thiols, native inhibitor disulfide bonds and, in the case of flours, disulfide and sulfhydryl groups of the other proteins present, as illustrated:

1. $R-SH$ + $In-S-S-In$ $\longrightarrow$ $R-S-S-In$ + $HS-In$
2. $R-SH$ + $Pr-S-S-Pr$ $\longrightarrow$ $R-S-S-Pr$ + $HS-Pr$
3. $In-SH$ + $Pr-S-S-Pr$ $\longrightarrow$ $In-S-S-Pr$ + $HS-Pr$
4. $In-SH$ + $HS-Pr$ + $\frac{1}{2} O_2$ $\longrightarrow$ $In-S-S-Pr$ + $H_2O$
5. $In-SH$ + $HS-In$ + $\frac{1}{2} O_2$ $\longrightarrow$ $In-S-S-In$ + $H_2O$

$R-SH$ = added thiols; $In-S-S-In$ = inhibitor (In) disulfide bonds; $Pr-S-S-Pr$ = protein (Pr) disulfide bonds.

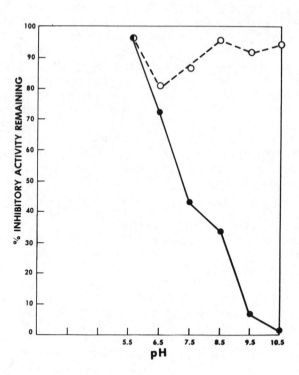

Figure 1.
Effect of pH, heat, and
N-acetylcysteine (NAC)
on trypsin inhibitory
activity of of lima
bean inhibitor at 45°C.
Upper curve, heat only;
lower curve, heat plus
NAC. (Friedman et al.,
1982 b).

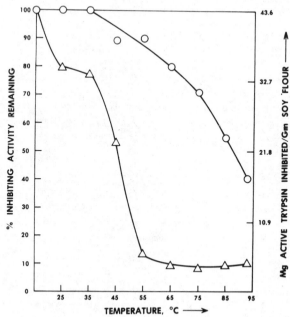

Figure 2.
Effect of heat and N-
acetylcysteine (NAC) on
trypsin inhibitory ac-
tivity of soybean trypsin
inhibitor in soybean flour.
Upper curve, heat only.
Lower curve, heat plus
NAC. (Friedman et al.,
1982 a).

If the inactivation as a thiol-disulfide interchange, then excess thiolate anion is expected to force the equilibrium to the right. Although the inhibitor may tend to regenerate its original conformation, this process may be hindered or suppressed by excess thiol, or the presence of mixed disulfides. Because of altered geometry, the modified inhibitors do not readily combine with the active site of trypsin or chymotrypsin trypsin. Alternatively, if combination does occur, dissociation is rapid because of the change in binding forces needed to stabilize the trypsin complexes.

From both chemical and nutritional viewpoints, inactivation of a protease inhibitor by an added thiol _via_ a sulfhydryl-disulfide interchange differs from the related reductive cleavage of disulfide bonds followed by alkylation of the generated SH groups. The former treatment produces new disulfide bonds, whereas the latter eliminates the original ones from the protein. In the case of soy flour, possibilities for sulfhydryl-disulfide interchange include interaction between inhibitor SH and S-S bonds with soybean protein SH and S-S groups, respectively, to form a complex network that may make it difficult for the original disulfide bonds to be reformed. The relative concentration of SH and S-S bonds in the thiol and protein may dictate whether the thiol is considered a catalyst or stoichiometric reagent in the inactivation. For these reasons, optimum conditions for the inactivation of disulfide-containing protease inhibitors by thiols have to be established in each case.

Recent studies (Friedman et al., 1982 a,b,c; 1984) showed that the presence of cysteine or N-acetylcysteine during treatment of at 45, 65, and 75$^0$C increased the protein efficiency ratio (PER) of soy flour. The treated flours had a lower SH content and a higher disulfide (cystine) content than the starting material. These results confirm the described mechanistic hypotheses.

In summary, nutritional improvement may result from inactivation of trypsin inhibitors and introduction of new disulfide bonds _via_ sulfhydryl-disulfide interchange and oxidation reaction among added thiols, inhibitors, and structural proteins. Such an approach should be useful for inactivating related disulfide-containing toxic compounds such as lectins (hemagglutinins) and ricin, widely distributed in legumes and castor beans, respectively.

Finally, protease inhibitors also appear to have the ability to protect against radiation-induced cell damage (Yavelow et al., 1983).

The discussion of protease inhibitors illustrates one general mode of inactivation of antinutrients or food toxicants by sulfhydryl compounds. The sections that follow illustrate a second general approach based on combination of nucleophilic sulfhydryl groups with electrophilic double bonds.

## CYCLOPROPENE FATTY ACIDS

Cyclopropene fatty acids are present in cotton (Jones, 1981) and other seeds and leaves (Berry, 1980). The incidence of liver tumors in rainbow trout fed diets containing aflatoxin is increased by addition of cyclopropenoid fatty acids (Hendricks, et al. 1980; Cf. also, Eisele et al., 1982). This co-carcinogenic effect may perhaps be due to reaction of enzyme SH groups with the double bond of the cyclopropene rings, as illustrated in Figure 3. Although the acids themselves are apparently not carcinogenic, they do interact with SH groups of thiols, aminothiols, and enzymes (Friedman, 1973).

A key question is: "Will chemical modification of the double bond of cyclopropene fatty acids by thiols or other nucleophiles prevent cancer promotion?"

## AFLATOXINS AND RELATED COMPOUNDS

Aflatoxin $B_1$ ($AFB_1$) is a precarcinogen that is transformed in vivo to an active epoxide (Patterson, 1977; Swenson et al., 1977). Prior treatment with a site-specific reagent should modify $AFB_1$ in a manner that will prevent formation of the epoxide and suppress its mutagenic and carcinogenic activity. Because thiols are potent nucleophiles both in vitro and in vivo (Friedman et al., 1965; Friedman, 1973), they may be expected to react with electrophilic sites of $AFB_1$ and thus competitively inhibit the interaction of these sites with DNA or other biomolecules (Figure 4). The resulting product(s) are, therefore, expected to be inactive in the Ames Salmonella/typhimurium mutagenicity test.

Several relevant studies will be briefly described to test this hypothesis.

Dickens and Cook (1965) determined second-order rate constants ($k_2$) for reaction of the SH group in cysteine with several carcinogenic lactones and related compounds in bicarbonate buffer (initial pH, 7.3). Compounds for which $k_2$ was 0.171 mole$^{-1}$s$^{-1}$ or more proved carcinogenic, except N-ethylmaleimide, which is, however, highly toxic to mammals. Less reactive compounds generally were only slightly or not all carcino-

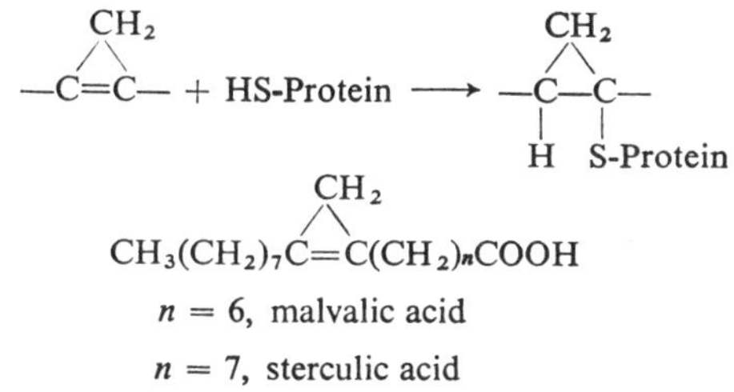

Figure 3. Possible mode of reaction of cyclpropenoid fatty acids with protein thiol groups (Friedman, 1973).

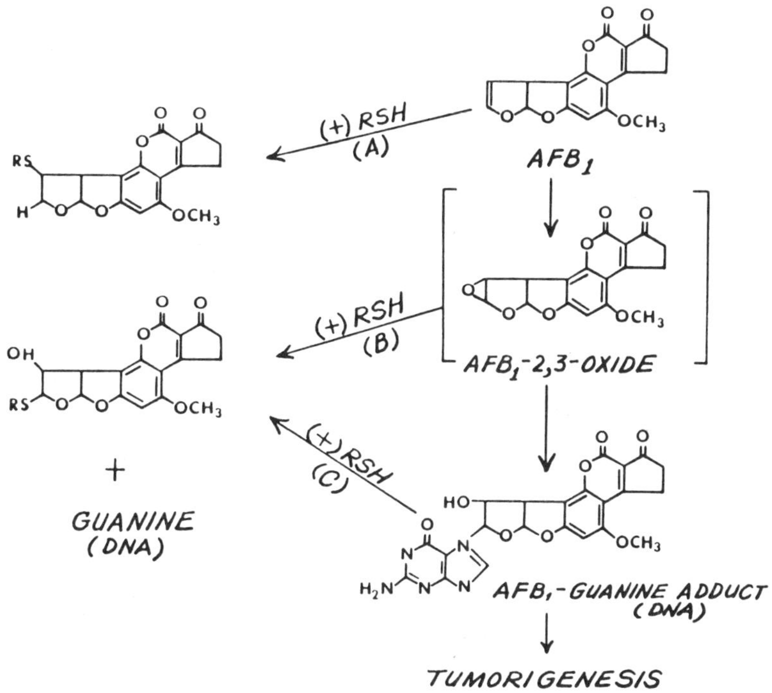

Figure 4. Some possible aflatoxin-thiol interactions:

Pathway A: Postulated interaction of the 2,3-double bond of aflatoxin $B_1$ to form an inactive thiol adduct.

Pathway B: Postulated interaction of a thiol with $AFB_1$-2-3-epoxide which may prevent the epoxide from interacting with DNA.

Pathway C: Postulated displacement of an aflatoxin-DNA (guanine) adduct blocking tumorigenesis.

genic. The striking exception in this series is $AFB_1$, which reacted slowly with cysteine ($k_2$ = 0.008 1 mole$^{-1}$s$^{-1}$) but is highly carcinogenic. On the basis of studies of the reaction of γ-lactones and L-cysteine, Jones and Young (1968) concluded that reaction rate with cysteine is not a reliable criterion for predicting the carcinogenicity of lactones and related compounds. On the other hand, the kinetic and synthetic studies permitted the conclusion that all carcinogenic lactones participated in alkylation reactions with cysteine, while the inactive lactones all gave acylated products (Cf. Kupchan, 1974; Pickman et al., 1979 and Figures 5).

Ganner and Wright (1973) reported that adding of cysteine, glutathione, sodium sulfide, sodium thiosulphate, methionine, or the epoxide hydrase inhibitor, cyclohexene oxide, to the assay medium did not alter the toxic effect of $AFB_1$ to bacteria. However, 2,3-dimercaptopropanol appeared to have a partial protective effect. In contrast, Rosin and Stich (1978a) demonstrated that suspending bacterial cultures of Salmonella typhimurium for 20 minutes at 37°C in various concentrations of cysteine and several carcinogens including $AFB_1$ resulted in a concentration-dependent partial to complete loss of mutagenic activity. Those authors suggested that inhibition of activity might be due to reaction of cysteine with the carcinogen or its metabolite. The same authors (1978b) also noted that concurrent application of cysteine, cysteamine, and reduced glutathione with the carcinogens N-hydroxy- and N-acetoxy-2-acetylaminofluorene to Salmonella typhimurium significantly inhibited formation of revertants. In contrast, mutagenic activity persisted, when cysteine was added before or after exposure of the bacteria to the carcinogens. The authors suggested that the simplest explanation of their results is that cysteine traps the mutagen or a reactive product, thus preventing it from interacting with DNA.

The main object of our study (Friedman et al., 1982d) was to investigate factors expected to influence inhibition of mutagenic activity of $AFB_1$ in the Ames Salmonella/typhimurium test by the model thiol, N-acetyl-L-cysteine (NAC), as a function of thiol concentration, pH and time of treatment. This compound has certain advantages over other thiols, such as cysteine. For instance, the SH group of NAC has a higher pK than the corresponding group in cysteine (9.5 vs. 8.3). The ionized SH group of NAC is therefore a better nucleophile than the corresponding group in cysteine; it is also more stable to oxidation (Snow et al., 1975; Friedman, 1977). In addition, the acetylated $NH_2$ in NAC, in contrast to the free $NH_2$ of cysteine, cannot itself react with $AFB_1$. Finally, NAC is used therapeutically as a spray to

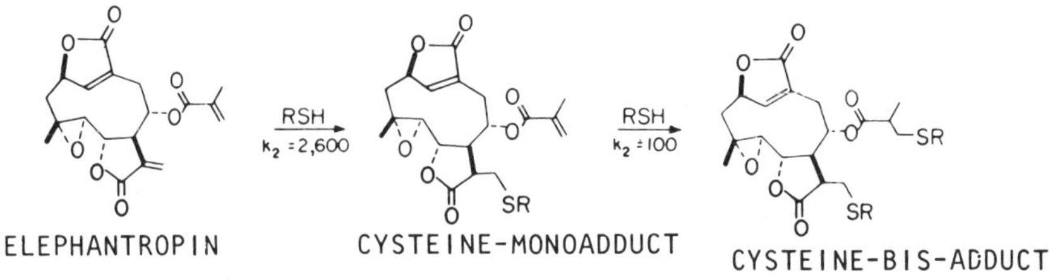

Figure 5 A.  Cysteine adducts of the sesquiterpene lactone, ele-
phantopin (Kupchan, 1974).

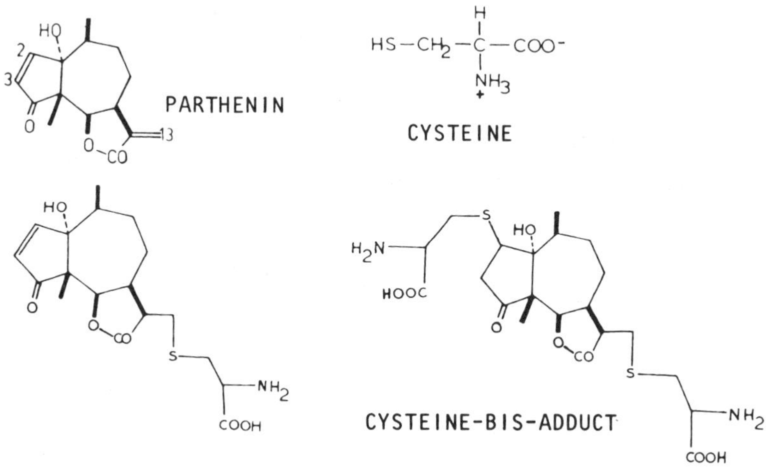

Figure 5 B.  Mono- and bis-cysteine adducts of the sesquiterpene
lactone, parthenin (Pickman et al., 1979).

liquify mucous secretions in pulmonary disorders; it is therefore presumably a safe compound to ingest in small amounts (Anon, 1978).

Table 1 shows results of our comparison of inactivation by eleven different thiols. All of these except penicillamine effectively suppressed the mutagenicity acitivity of $AFB_1$. (Note that L-cysteine was not a very potent inactivator under out conditions). This comparison was then used as the basis for a kinetic study with the three most effective thiols: NAC, reduced glutathione (GSH), and N- mercaptopropionylglycine (MPG). The results in Figure 6 show the trends inaflatoxin inactivation by these three thiols. This figure shows time and thiol-concentration conditions that inactivate $AFB_1$. Inactivation can be accomplished with a wide range of thiols concentrations. These three thiols appear to be equally effective at the higher concentrations, but GSH appears most effective at the lower levels.

In principle, a thiol can react with at least four electrophilic sites in the aflatoxin molecule (Figure 7). In pathway 1, the thiolate anion adds to the double bond of the protonated furan ring of aflatoxin. In pathway 2, a thiol cleaves the lactone ring to $AFB_1$ to produce a thiol ester. Pathway 3 entails nucleophilic addition to the double bond of the lactone ring, which is activated by protonation of the methoxy group. Pathway 4 represents addition of the thiol to the other end of the same double bond where it is activated by a protonated carbonyl group. The last two processes are less likely to occur than the first two because addition to the double bond of the lactone ring (or the central benzene ring) would disrupt conjugation and would therefore not be favored energetically. Cleavage of the lactone ring (pathway 2) would lead to a major change in the ultraviolet spectrum of $AFB_1$. Since the UV spectra of $AFB_1$ and a corresponding solution of $AFB_1$ inactivated by NAC are essentially superimposable (unpublished results), the most likely effect of NAC (and of the other thiols) is interaction with the double bond of the furane ring (pathway 1). Such an interaction with a site-specific reagent would prevent the double bond from forming a presumably mutagenic epoxide. This would explain the effectiveness of thiols in suppressing the mutagenic activity of $AFB_1$. Final proof of this hypothesis awaits isolation and characterization of the postulated product.

High performance liquid chromatography (HPLC) studies showed that the disappearance of $AFB_1$ was accompanined by the appearance of a single new peak on the chromatogram (Figure 8). Since, as mentioned, the 360 nm extinction coefficient of this product is not markedly changed from that of $AFB_1$, the integrated UV

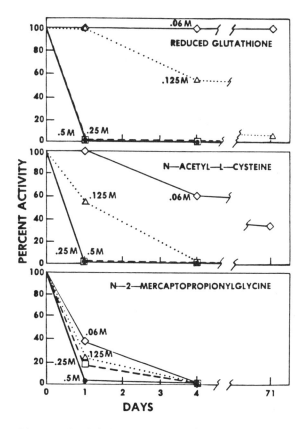

Figure 6. Effect of thiol concentration on time of inactivation of mutagenic activity of aflatoxin $B_1$ by reduced glutathione, N–acetylcysteine, and N-2-mercaptopropionylglycine. (Friedman et al., 1982 d).

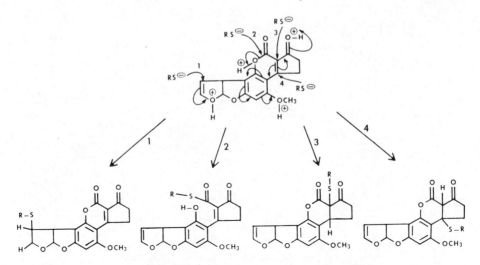

Figure 7.  Theoretical possibilities for interaction between a thiolate anion and four electrophilic sites of aflatoxin $B_1$.

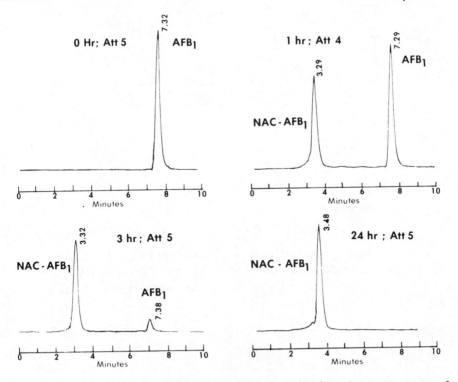

Figure 8.  Reverse phase high pressure liquid chromatogram of aflatoxin $B_1$ (AFB$_1$) and its N-acetyl-cysteine derivative (NAC-AFB$_1$).  Actual elution times in minutes are shown near each peak.  Att = attenuation (Friedman et al., 1982d).

absorbance of this peak indicated that $AFB_1$ was converted nearly quantitatively to this single derivative.

The following arguments may rationalize the fact that inactivation is greatest near pH 4 (Friedman et al., 1982 d). Generally, nucleophilic addition of a thiol to a double bond, as in pathway 1, Figure 7, is proportional to the ionized thiolate ion $(RS^-)$. The concentration of this ion in solution is governed by pH. As a result, the rate of inactivation would be expected to be directly proportional to pH, as previously found for addition of thiols to the double band of vinyl compounds such as acrylonitrile (Friedman et al., 1965).

Since our limited kinetic studies show that this does not appear to be the case with $AFB_1$, another factor must modify this pH effect.


TABLE I.  Inactivation of Aflatoxin $B_1$ by various thiols

| Compound | pK | % original activity | pH |
|---|---|---|---|
| N-Acetyl-L-cysteine (NAC) | 9.5 | 0 | 4.0 |
| Mercaptopropionic acid | 10.4 | <0.2 | 3.4 |
| Mercaptoethanol | 9.6 | <0.3 | 4.5 |
| Reduced glutathione (GSH | 8.8 | <0.6 | 3.9 |
| N-2-Mercaptopropionylglycine (MPG) | 10.4 | <2.6 | 3.8 |
| Mercaptoacetic acid | 10.7 | 7 | 3.2 |
| Mercaptosuccinic acid | 10.9 | 70 | 3.6 |
| L-Cysteine | 8.3 | 79 | 4.2 |
| Acetyl-D,L-homocysteine thiolactone | -- | 85 | 4.2 |
| Cysteine methyl ester | 6.6 | 85 | 4.0 |
| D-Pinicillamine | 7.9 | 94 | 4.3 |
| β-Mercaptoethylamine (cysteamine) | 8.3 | 100 | 4.0 |

(Friedman et al., 1982d).


Another factor that is expected to catalyze nucleophilic addition to the double bond of the furan ring of aflatoxin is protonation of the furan oxygen atom, as shown in Figure 7. Such protonation creates an electron sink that polarized the pi-electrons of the double bond and forms a positive carbonium ion on carbon atom 3. Carbonium ions have strong avidity for nucleophilic species such as thiolate anions (Friedman, 1973). These considerations imply that pH has opposite effects on the two factors expected to facilitate interaction between a thiol and the double of the furan ring of aflatoxin $B_1$. That is, an acid pH would catalyze protonation of the furan oxygen and an alkaline pH the ionization of a thiol to a thiolate anion. The optimum pH for inactivation due to modification of the 2,3- double of $AFB_1$ is therefore a compromise value that would maximize the products of both the carbonium and thiolate ion concentrations. Protonation of the furan oxygen is probably the most important factor, since only a very small fraction of the thiol (pK 8 to 10) is ionized in the pH region 4-4.6. The reaction does proceed at higher and lower pH values, but more slowly.

Although no attempt was made to establish the order of reaction, our results show that the inactivation of mutagenic activity, and presumably the interaction of the thiols with aflatoxin, are concentration dependent. Since addition of thiols to double bonds follows second-order kinetics (Friedman et al., 1965), this is probably also true for addition of a thiol to the 2,3 double bond of aflatoxin.

Finally, our observations may be useful in treating aflatoxin toxicity. Aflatoxins are carcinogenic polycyclic furanoid lactone (coumarone) derivatives produced by various mold species (Aspergillus flavus, Aspergillus parasiticus). Therefore, any agent that would compete sucessfully in vivo at DNA sites for aflatoxin may be useful in treating aflatoxin poisoning. Thus, treatment of poisoned animals and humans with acetyl-L-cysteine may minimize or prevent the toxic (hepatocarcinogenic) effects of aflatoxin $B_1$. Thiols, such as NAC, may also be useful prophylactic agents.

In summary, our observations that many structurally different thiols can suppress the mutagenic activity of aflatoxin $B_1$ suggest that thiols may be useful for inactivating aflatoxins in contaminated foods, as an antidote to treat cases of aflatoxin toxicity, or for prophylaxis to prevent aflatoxin poisoning.

The reader is referred to the following publications for related discussions of the possible beneficial consequences resulting from interactions of sulfur and related compounds with aflatoxin and other dietary constituent: (Schwimmer and Friedman, 1972; Raj et al., 1975; Hayatsu, 1978; Degen and Neumann, 1978; Doyle and Marth, 1978 a, b; Bullerman, 1979; Emerole et al., 1979; Hatch et al., 1979; Campbell, 1980; Busk and Ahlborg, 1980; Loltikar et al., 1980; Shelef and Chinn, 1980; Novi, 1981; Groopman et al., 1981; Meister, 1982; Walther, 1982; Coronogliu et al., 1982; Hagler et al., 1982; Ames, 1983; Wattenberg, 1983; Cabral and Neal, 1983; Moss et al., 1983; Masri, 1984).

TRICHOTHECENES

   Biochemical, pharmacological, physiological, and toxicological effects of Fusarium toxins (trichothecene mycotoxins) have
been -extensively studied (Chung et al., 1974; Bamburg, 1976;
Otsubo and Saito, 1977; Ueno, 1977; 1980; McLaughlin et al.,
1977; Newbern, 1980; Rukmini et al., 1980; Bridges, 1981; Chi
et al., 1981; Doerr et al., 1981; Gentry and Cooper, 1981; Hoerr
et al., 1981; Lutsky and Mor, 1981; Mirocha et al., 1981; Weaver
et al., 1981; Yoshizawa et al., 1981; Hayes and Schieffer, 1982;
Jagadessan et al., 1982; Jarvis and Mazzola, 1982; Thompson and
Wannemacher, 1982; Smith and Carson, 1984).  The cited references
show that these effects include (a) inhibition of protein synthesis; (b) breakdown of polyribosomes to monoribosomes; (c)
causing mycotoxicosis in domestic and farm animals; (d) causing
alimentary aleukia in man; (e) induction of leukopenia, hemorrhage, and respiratory tract infections in monkeys; and (e) suppression of the immune system in monkeys.

   Trichothecenes are responsible for major agricultural losses
(Bamburg, 1976; Kamimura et al., 1979; Koroda et al., 1979;
Gruber and Paoli, 1980; Mirocha et al., 1981; Schmidt et al.,
1981 a,b; Scott et al., 1981; Watson and Lindsay, 1982).  They
may have been used as biological warfare agents (Marshall, 1982;
Kucewich, 1982).  These observations suggest that basic and
applied studies are need to find ways to inactivate the toxic
properties of trichothecenes, both in the pure state and in foods
and feeds.  Such inactivation should lead to better and safer
foods.  It also should contribute to the development of methods
for protecting animals and humans against biological warfare
agents.

   Trichothecenes contain electrophilic sites, including double
bonds and epoxide rings (Figure 9), which make them susceptible
to attack by nucleophilic reagents.  Modification by site-specific reagents may lead to suppression or inactivation of toxic
properties (Bamburg, 1976; Doyle and Bradner, 1980; Tamm, 1974;
1977; Ueno, 1980).  Since sulfhydryl (SH) compounds are among
the most reactive functional groups in chemical and biochemical
systems (Friedman, 1973), effort should be directed towards
establishing the effectiveness of various thiols and other nucleophiles in altering chemical and toxic properties of selected
trichothecenes.  We expect that modified trichothecenes will not
interact in vivo with receptor sites to which native forms bind.
The modified trichothecenes should, therefore, be less toxic.

   Biological effects of trichothecenes depend on their structural integrity. Previous studies (Bamburg, 1976) have shown that
opening the epoxide ring(s) present in all trichothecene molecules essentially abolishes their toxic properties.  Modification

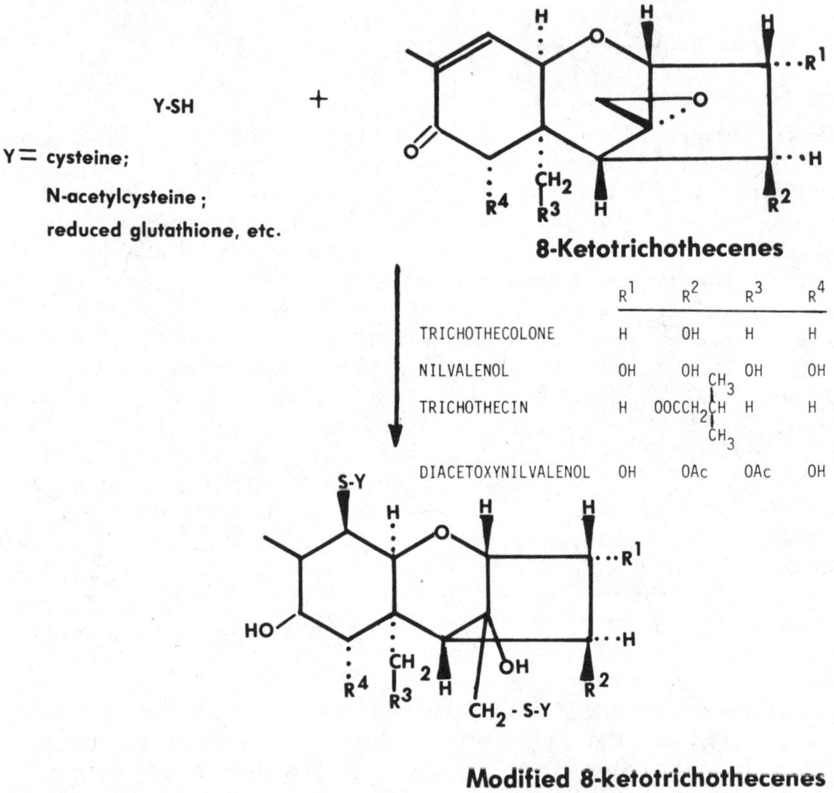

Y-SH    +

Y = cysteine;

N-acetylcysteine ;

reduced glutathione, etc.

**8-Ketotrichothecenes**

|  | $R^1$ | $R^2$ | $R^3$ | $R^4$ |
|---|---|---|---|---|
| TRICHOTHECOLONE | H | OH | H | H |
| NILVALENOL | OH | OH | OH | OH |
| TRICHOTHECIN | H | OOCCH$_2$CH$\genfrac{}{}{0pt}{}{CH_3}{CH_3}$ | H | H |
| DIACETOXYNILVALENOL | OH | OAc | OAc | OH |

**Modified 8-ketotrichothecenes**

Figure 9. Possible modes of interaction of the conjugated double bond and epoxide ring of 8-ketotrichothecenes with sulfhydryl compounds.

of the double bond also appears to alter biological properties. A major objective should be to assess the ability of a series of nucleophilic reagents differing in steric and charge properties to combine with the isolated double bond of T-2 toxin and the conjugated double bond of nilvalenol, and with the epoxide rings of both.

Several studies on reported reactivities of double bonds and epoxide rings with thiols relevant to the theme of this discussion include:

(a)  Ethacrynic acid is a potent diuretic used clinically in the management of edema (Duggan and Noll, 1965). The compound is an α,β-unsaturated ketone derivative (as is nilvalenol) that reacts with SH compounds both in vitro and in vivo. Komorn and Carfuny (1965) report that ethacrynic acid decreases the concentration of protein-bound SH groups in renal cells when renal diuresis is maximal.

(b)  Showdomycin, an antibiotic effective against Gram-negative bacteria, has a conjugated maleimide residue attached to a furanose ring. Roy-Burman et al. (1968) report that showdomycin inhibits UDP-glucose dehydrogenase. Since the inhibitory activity is abolished when showdomycin is allowed to react with cysteine before the enzyme is added, showdomycin may exert pharmacological and toxicological effects by alkylating enzyme SH groups.

(c)  Studies on the interaction of sulfhydryl compounds, with α,β-unsaturated ketone groups of clavicin and penicillic acid (Geiger and Conn, 1945) indicate that the antibacterial properties of these compounds depend on their ability to combine with SH groups of certain enzyme systems.

(d)  Previous studies (Cavins and Friedman, 1968; Friedman and Wall, 1966; Friedman et al., 1965; Friedman, 1967; 1968) investigated factors that govern nucleophilic reactivities of functional groups in aminothiols, thiols, amino acids, peptides, and proteins with conjugated compounds such as acrylonitrile and methyl vinyl ketone. Quantitative estimates of the influence of steric and polar parameters on reaction rates were obtained from linear free-energy relationships. Factors that influenced rates included pH of the reaction medium, pK values of the noncleophilic functional groups, temperature, structure of reactants, and nonaqueous solvents such as dimethyl sulfoxide. Methyl vinyl ketone, whose structural features are present in all 8-ketotrichothecenes such as nilvalenol, was one of the most reactive compounds (Friedman and Wall, 1966).

A variety of biologically active compounds that contain structural features present in trichothecenes readily combine with thiols both in vitro and in vivo. Therefore, trichothecenes probably will behave similarly. My hypothesis is that the double bond of nilvalenol and other 8-ketotrichothecenes will combine readily with thiols and related compounds under appropriate conditions of pH and temperature leading to major changes in toxicological properties of trichothecenes (Figure 9).

Thiols such as cysteine, N-acetyl-cysteine, mercatopropionylglycine, and reduced glutathione inactivate the mutagenic properties of aflatoxin $B_1$ by forming adducts at the electrophilic double bond of the furan ring of the mycotoxin (see above). The same thiols prevents the formation of the unnatural, toxic amino acid lysinoalanine by reacting with the double bond of its dehydroalanine precursor (Friedman, 1982; Friedman and Masri, 1982; Friedman et al., this volume).

These findings support the hypothesis that thiols have a strong avidity for double bonds of a variety of biologically active molecules, and indicate that similar approaches should be effective for trichothecenes.

Transformations of epoxides may also be useful in inactivating trichothecene epoxides, as evidenced by the following relevant observations:

(a) Alkylation of sulfhydryl and other protein functional groups by structurally different epoxides proceeds quite readily under mild conditions (Friedman, 1973; Friedman and Broderick, 1977).

(b) Although toxic sterol oxides formed in eggs and other foods appear to interact with DNA and other nucleophiles in vivo (Finocchiaro and Richardson, 1983), their interaction with thiols or other nucleophiles in vitro to ameliorate their toxicities has apparently not yet been explored.

Since epoxide groups are powerful alkylating agents in vitro, trichothecene epoxides may behave as in vivo alkylating agents. This conclusion is reinforced by the observation of Foster et al. (1975). He devised an enzymatic assay for submicrogram levels of trichothecene mycotoxins in animal feedstuffs, based on the formation of a conjugate or derivative from reduced glutathione and the epoxide ring of a T-2 toxin (diacetoxyscirpenol) in the presence of the enzyme glutathione-S-epoxide transferase. Excess glutathione, determined with 5,5'-dithiobis (2-nitrobenzoic acid) Ellman's reagent), measures the epoxide content of the sample (Figure 10). Although the mechanism of trichothecene activity in vivo (Ueno and Matsumoto, 1975) has not been fully defined, Patterson (1977) suggests that these results could be used to infer the molecular basis for biological and toxicological activity.

Singlet (activated) oxygen is another chemical species that rapidly combines with unsaturated sites of molecules (Korycka-Dahl and Richardson, 1978). Activated oxygen probably also oxidizes trichothecenes, possibly leading to inactivation. Singlet oxygen can be generated under mild conditions from hydrogen peroxide and hypohalite, such as sodium hypochlorite; or by a photosensitized decomposition of hydrogen peroxide catalyzed by riboflavin, chlorophyll, and other photosensitizers (Korycka-Dahl and Richardson, 1978; Rosen and Klebanoff, 1977). Singlet oxygen has a strong affinity for double bonds and is known to induce oxidative changes in a variety of substrates, including unsaturated fatty acids. It also inactivates aflatoxin $M_1$ in milk (Applebaum and Marth, 1982). A need exists to assess the ability of singlet oxygen, ozone, and other oxidizing agents to alter toxic properties of trichothecenes. If effective, such treatments could be applied economically under mild conditions.

In summary, in depth-exploration of the concept of modifying active sites of trichothecenes both in vitro and in vivo to prevent them from interacting with living cells will not only contribute to our understanding of basic mechanisms of trichothecene toxicity at the molecular level, but also should lead to the discovery of new ways to treat contaminated foods and people and to the development of new prophylactic and therapeutic compounds.

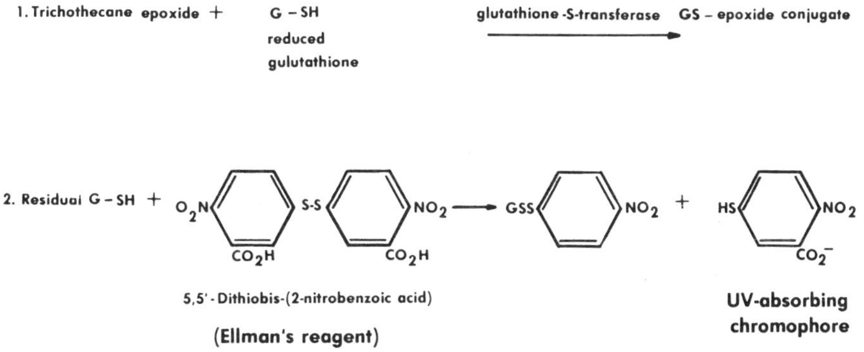

1. Trichothecane epoxide + G – SH reduced gulutathione   ——glutathione-S-transferase——→  GS – epoxide conjugate

2. Residual G – SH + $O_2N$⟨⟩$S-S$⟨⟩$NO_2$ —→ $GSS$⟨⟩$NO_2$ + $HS$⟨⟩$NO_2$
$CO_2H$          $CO_2H$                              $CO_2^-$

5,5'- Dithiobis-(2-nitrobenzoic acid)                              UV-absorbing
(Ellman's reagent)                                              chromophore

Figure 10. Assay of the extent of interaction of the epoxide group in a trichothecane with the sulfhydryl group of reduced glutathione (Adapted from Foster et al., 1975).

OTHER INTERACTIONS

   Sulfhydryl groups in amino acids, peptides, and proteins participate in many other interactions of importance to medicine,
food science, nutrition, and toxicology.  These include interactions with:

   acetaminophen (Lauterburg et al., 1983);

   aldehydes (Sprince et al., 1979);

   bread and other food products (Bruemmel et al., 1980;
   Diamalt, 1974);

   cadmium compounds (Murakami et al., 1981);

   carbon tetrachloride (DeFerreyra et al., 1974);

   carotene (Tajiri et al., 1976);

   cyanide (Westley et al., 1983);

   dehydroascrobic acid (Pribela and Pikulikova, 1969);

   doxorubicin (Unverferth et al., 1983);

   egg albumin (Kato et al., 1974);

   iron compounds (Martinez-Torres et al., 1981); Snedeker,
   and Greger, 1983);

   lipids (Chow, 1979; Gardner and Kleiman, 1981; Tien et al.,
   1982; Yee and Shipe, 1982);

   mercury compounds (MacGregor and Clarkson, 1974;
   Friedman and Masri, 1974; Masri et al., 1974; Aaseth
   and Friedheim, 1978);

   methyl chloride (Dodd et al., 1982);

   milk (Arnold, 1969; Badings et al., 1978);

   muscle proteins (Shih et al., 1977);

   mutagenic dyes (Cisela and Filutowich, 1979; Friedman,
   1980; Friedman et al., 1980a; MacGregor et al., 1980);

   nitrites (Schoental and Rive, 1965; Walker, et al.
   1982; Williams and Aldred, 1982; Byler et al., 1983);

toxic compounds (Williamson et al., 1982);

tyrosinase (Lindbladh et al., 1983); and

zinc salts (Hsu and Smith, 1983).

In conclusion, SH groups seem conclusively implicated in ameliorating the action of deleterious food ingredients both _in vitro_ and _in vivo_. To account in detail for their specific effects, future research should be directed towards understanding the mechanisms of interaction of food ingredients with SH compounds and the metabolism of the resulting products. With such understanding, effective means to prevent the deleterious action of many antinutrients and food toxicants can be reasonably hoped for.

oxidants generated by human neutrophils (Weiss et al., 1983);

patulin (Ciegler et al., 1976; Lindroth and Wright, 1978; Wright and Lindroth, 1978; Lindroth, 1980; Cooray et al., 1982);

penicillamine (Friedman, 1977);

pesticides (Moriya et al., 1978);

pyrrolizidine alkaloids (Miranda et al., 1982);

quinazolines (Pitman et al., 1979);

selenium compounds (Vernie et al., 1975; 1979; 1981);

tin ions (Gruenwedel and Patnaik, 1973);

REFERENCES

Aaseth, J. and Friedheim, E.A.H. (1978). Treatment of methyl-
    mercury poisoning in mice with 2,3-dimercaptosuccinic acid
    and other complexing thiols. Acta Pharmacol. Toxicol., 42,
    248-252.
Ames, B.N. (1983). Dietary carcinogens and anticarcinogens.
    Science, 221, 1256-1264.
Anderson, R.L., Rackis, J.J. and Tallent, W.H. (1979). Biologi-
    cally active substances in soy proteins. "Soy Protein and
    Human Nutrition" H.L. Wilcke, D.T. Hopkins and D.H. Waggle,
    Eds., Academic Press, New York, p. 209.
Anon., (1979), Mucomyst, R., Physicians Desk Reference, Medical
    Economic Co., Oradell, New Jersey, p. 1115.
Applebaum, R.S. and Marth, E.H. (1982). Inactivation of afla-
    toxin $M_1$ in milk using hydrogen peroxide and hydrogen pero-
    xide plus riboflavin or lactoperoxidase. J. Food Protection,
    45, 557-560.
Arnold, R.G. (1969). Inhibition of heat-induced browning of milk
    by L-cysteine. J. Dairy Sci., 52, 1857-1859.
Badings, H.T. and Neeter, R. (1978). Reduction of cooked flavor
    in heated milk products by cystine. Lebensmmittelwissenshaft
    und Technologie, 11, 237-240.
Bamburg, J.R. (1976). Chemical and biochemical studies of the
    trichothecene mycotoxins. In "Mycotoxins and Other Fungal
    Related Food Problems", J.V. Rodericks, Ed., Advances in
    Chemistry Series 149, 144-162. American Chemical Society,
    Washington, D.C.
Berry, S.K. (1980). Cyclopropene fatty acids in Gnetum. genom.
    seeds and leaves. J. Sci. Food Agric., 31, 657-662.
Bozzini, A, Silano, V. (1978). Control through breeding methods
    of factots affecting nutritional quality of cereals and
    legumes. In "Nutritional Improvement of Food and Feed
    Proteins, M. Friedman, Ed., Plenum, New York.
Bridges, J.W. (1981). The use of hepatocytes in toxicological
    investigations. In "Testing for Toxicity", J.W. Gorrod, Ed.,
    Taylor & Francis Ltd. London, England, pp. 125-143.
Bruemmer, J.M., Seibel, W. and Stephan, H. (1980). Effect of L-
    cysteine and L-cystine on the quality of bread and rolls.
    Getreide, Mehl und Brot., 42(7), 173-178.
Bullerman, L.B. (1979). Significance of mycotoxins to food
    safety and human health. J. Food Protection, 42, 65-86.
Busk, L. and Ahlborg, U.G. (1980). Retinol (vitamin A) as an
    inhibitor of the mutagenicity of aflatoxin $B_1$. Toxicol
    Letters, 6, 243-249.
Byler, D.M., Gosser, D.K. and Susi, H. (1983). Spectroscopic
    estimation of the extent of S-nitrosothiol formation by
    nitrite action of sulfhydryl groups. J. Agric. Food Chem.,
    31, 523-527.

Cabral, J.R. and Neal, G.E. (1983). The inhibitory effect of ethoxyquin on the carcinogen action of aflatoxin $B_1$ in rats. Cancer Letters, 19, 125-132.

Campbell, T.C. (1980). Chemical carcinogens and human risk assessment. Fed. Proc., 39, 2467-2484.

Cavins. J.F. and Friedman, M. (1968). Specific modification of protein sulfhydryl groups with $\alpha$, $\beta$-unsaturated compounds. J. Biol. Chem. 243, 3357-3360.

Chi, M.S., El-Halwani, M.E., Waibel, P.E. and Mirocha, C.J. (1981). Effect of T-2 toxin on brain catecholamines and selected components in growing chickens. Poultry Science. 60, 137-141.

Chow, C.K. (1979). Nutritional influence on cellular antioxidant defense systems. Am. J. Clin. Nutr., 32, 1066-1081.

Chung, C.W., Truckess, M.W., Giles, A.L., Jr. and Friedman, L. (1974). Rabbit skin test for estimation of T-2 toxin and other skin-irritating toxins in contaminated corn. J. Ass. Off. Anal. Chem., 57, 1121-1127.

Ciegler, A., Beckwith, A.C. and Jackson, L.K. (1976). Teratogenicity of patulin and patulin adducts formed with cysteine. Applied Environ. Microbiol., 31, 664-667.

Cisela, Z. and Filutowich, M. (1979). Azide mutagenesis in gram-negative bacteria. Reversion of the mutagenic effect by L-cysteine. Mutation Res., 66, 301-305.

Corongiu, F.P. and Millia, A. (1982). Rise of hepatic glutathione concentration induced in rats by chronic lead nitrate treatment. Its role in aflatoxin $B_1$ intoxication. Res. Commun. Chem. Pathol. Pharmacol., 38, 97-112.

Cooray, R., Kiessling, K.H. and Lindahl-Kiessling, K. (1982). The effects of patulin and patulin-cysteine mixtures on DNA synthesis and the frequency of sister-chromatid exchanges in human lymphocytes. Food Chem. Toxicol., 20. 893-898.

DeFerreyra, E.C. et al. (1974). Prevention and treatment of carbon tetrachloride hepatotoxicity by cysteine: studies about its mechanism. Toxicol. Applied Pharmacol. 271, 558-568.

Degen, G. and Neumann, H.G. (1978). The major metabolite of aflatoxin $B_1$ in the rat is a glutathione conjugate, Chem.-Biol. Interactions. 22, 239-255.

Diamalt, A.G. (1974). Effect of L-cysteine and L-cystine on the manufacture and improvement of foods. Lebensm. Wiss. Technol. 7, 64-69.

Dickens, F., and Cooke, J. (1965). Rates of hydrolysis and interaction with cysteine of some carcinogenic lactones and related substances, Br. J. Cancer, 19, 404-410.

Dodd, D.E., Bus, J.S. and Barrow, C.S. (1982). Nonprotein sulf-hydryl alteration in F-344 rats following acute methyl chloride inhalation. Toxicol. Applied Pharmacol. 62, 228-236.

Doerr, J.A., Hamilton, P.B. and Burmeister, H.R. (1981). T-toxicosis and blood coagulation in young chickens. Toxicol. Applied Pharmacol. 60, 157-162.

Doll, R. and Peto, R. (1981). The causes of cancer: quantitative estimation of avoidable risks of cancer in the United States. Natl. Cancer Inst., 66, 1191.

Doroshow, J.H., Locker, G.Y., Ifrim, I. and Myers, C.E. (1980). Prevention of doxorubicin cardiac toxicity in the mouse by N-acetyl-cysteine. J. Clin. Invest. 68, 1053-1064.

Doyle, T.W. and Bradner, W.T. (1980). Trichothecanes. In "Anticancer Agents Based On Natural Products Models," J.M. Cassaday and J.D. Douros, Eds., pp. 43-72, Academic Press New York.

Doyle, M.P. and Marth E.H. (1978a). Bisulfite degrades aflatoxin: effect of temperature and concentration of bisulfite, J. Food Protection 41, 774-780.

Doyle, M.P. and Marth E.H. (1978b). Bisulfite degrades aflatoxin: effect of citric acid and methanol and possible mechanism of degradation, J. Food Protection, 41, 891-896.

Dugan, D.E. and Noll, R.M. (1965). Effects of ethacrynic acid and cardiac glycosides upon a membrane adenosinetriphosphatase of renal complex. Archiv. Biochem. Biophys. 109, 388-396.

Eisele, T.A., Loveland, R.M., Kruk, D.L., Myers, T.R., Sinnhuber, R.O. and Nixon, J.E. (1982). Effect of cyclopropenoid fatty acids on the hepatic microsomal mixed function-oxidase system and aflatoxin metabolism in rabbits. Food Chem. Toxicol., 20, 407-412.

Emerole, G.O., Neskovic, N., and Dixon, R.L. (1979). The detoxication of aflatoxin $B_1$ with glutathione in the rat. Xenobiotica, 9, 737-743.

FDA (1974). Food additives. L-Cysteine: affirmation of GRAS status as direct human food ingredients. Federal Register, 39, p. 185 (Sept. 23).

Finnocchiaro, E.T. and Richardson, T. (1983). Sterol oxides in foodstuffs: a review. J. Food Protection, 46, 917-925.

Foster, P.M.D., Slater, T.F. and Patterson, D.S.P. (1975). A possible enzymic assay for trichothecene mycotoxins in animal feedstuffs, Biochemical Society Transactions, 3, 875-878.

Friedman, M. (1967). Solvent effects in reactions of amino groups in amino acids, peptides, and proteins with α, β-unsaturated compounds. J. Amer. Chem. Soc. 89, 4709- 4713.

Friedman, M. (1968). Solvent effects in reactions of protein functional groups. Quarterly Reports on Sulfur Chemistry, 3, 125-144.

Friedman, M. (1973). "The Chemistry and Biochemistry of the Sulfhydryl Group in Amino Acids, Peptides, and Proteins," Pergamon Press, Oxford, England and Elmsford, New York, Chapter 16.

Friedman, M. (1977).  Chemical basis for pharmacological and therapeutic actions of penicillamine.  Proc. Roy. Soc. Med. (Supplement 3), 70, 50-60.

Friedman, M. (1980).  Chemical basis for biological effects of wool finishing treatments.  Proc. Sixth International Wool Textile Research Conference, Pretoria, South Africa, Vol. V, 323-348.

Friedman, M. (1982).  Lysinoalanine formation in soybean proteins:  kinetics and mechanisms.  In "Mechanism of Food Protein Deterioration", J.P. Cherry Ed., American Chemical Society Symposium Series, Washington, D.C., 206, 231-272.

Friedman, M. and Broderick, G.A. (1977).  Protected proteins in ruminant nutrition.  In "Protein Crosslinking:  Nutritional and Medical Consequences,"  M. Friedman, Ed. pp. 545-558, Plenum Press, New York.

Friedman, M., and Masri, M.S. (1974).  Interaction of mercury compounds with wool and related biopolymers.  Adv. Exp. Med. Biology 48, 505-550.

Friedman, M. and Wall, J.S. (1966).  Additive linear free-energy relationships in reaction kinetics of amino groups with $\alpha$, $\beta$-unsaturated compounds.  J. Org. Chem. 31, 2888-2894.

Friedman, M., Cavins, J.F. and Wall, J.S. (1965).  Relative nucleophilic reactivities of amino groups and mercaptide ions in addition reaction with $\alpha$, $\beta$-unsaturated compounds.  J. Amer.Chem. Soc. 87, 3572-3582.

Friedman, M., Diamond, M.J. and MacGregor, J.T. (1980a).  Mutagenicity of textile dyes.  Environmental Science and Technology, 14, 1145-1146.

Friedman, M., Zahnley, J.C. and Wagner, J.R. (1980b).  Estimation of the disulfide content of trypsin inhibitors as S-$\beta$-(2-pyridylethyl)-L-cysteine.  Analytical Biochemistry 106, 27-34.

Friedman, M., Grosjean, O.K. and Zahnley, J.C. (1982a).  Inactivation soya bean trypsin inhibitors by thiols.  J. Sci. Food Agric., 165-172.

Friedman, M., Grosjean, O.K. and Zahnley, J.C. (1982b).  Cooperative effects of heat and thiols in inactivating trypsin inhibitors in solution and in the solid state.  Nutr. Reports International, 25, 743-751.

Friedman, M., Grosjean, O.K. and Zahnley, J.C. (1982c).  Effect of disulfide bond modification on the structure and activities of enzyme inhibitors.  "Mechanisms of Food Protein Deterioration", J.P. Cherry, Ed., ACS Symp. Series, Washington, D.C., 206, 359-405.

Friedman, M., Wehr, C.M., Schade, J.E. and MacGregor, J.T. (1982d).  Inactivation of aflatoxin $B_1$ mutagenicity.  Food Chemical Toxicol., 20, 887-892.

Friedman, M., Gumbmann, M.R. and Grosjean, O.K. (1984).  Nutritional improvement of soy flour.  Submitted.

Gallaher, D. and Schneeman, B.O. (1984). Nutritional and metabolic response to plant inhibitors of digestive enzymes.
This volume.

Gardner, H.W. and Kleiman, R. (1981). Degradation of linoleic
acid hydroperoxide by a cysteine $FeCl_3$ catalyst as a model
for similar biochemical reactions. II. Specific formation
of fatty acid epoxides. Biochim. Biophys. Acta, 665, 113-
124.

Garner, R.C. and Wright, C.M. (1973). Induction of mutations in
DNA-repair deficient bacteria by a liver microsomal metabolite of aflatoxin $B_1$, Br. J. Cancer Res., 28, 544-551.

Geiger, W.B. and Conn, J.E. (1945). The mechanisms of antibiotic
action of clavicin and penicillic acid. J. Amer. Chem. Soc.
67, 112-116.

Gentry, P.A. and Cooper, M.L. (1981). Effect of Fusarium T-2
toxin on hematological and biochemical parameters in the
rabbit. Canad. J. Comp. Med. 45, 400-405.

Groopman, J.D., Croy, R.G. and Wogan, G.N. (1981). In vitro
reaction of aflatoxin $B_1$ - adducted DNA. Proc. Natl. Acad.
Sci. USA, 78, 5445-5449.

Gruber, S. and Paolis, V.D. (1980). A method for TLC detection
of HT-2 toxin in maize. Landwitschaftliche Forschung, 33,
436-439.

Gruenwedel, D.W. and Patnaik, R.K. (1973). Model studies regarding the internal corrosion of tin-plated food cans. IV.
Polarographic investigation of the interaction of Tin (II) -
ions with L-cysteine. Chem. Mikrobiol. Technol. Lebensm.,
2, 97-101.

Hagler, W.M., Jr., Hutchins, J.E. and Hamilton, P.B. (1982). Destruction of aflatoxin $B_1$ with sodium bisulfite: isolation
of the major product of aflatoxin $B_1$S. J. Food Protection,
46, 295-300.

Hatch, R.C., Clark, J.D., Jain, A.V. and Mahaffey, E.D. (1979).
Experimentally induced acute aflatoxicosis in goats treated
with ethyl maleate, glutathione precursors, or thiosulfate.
Am. J. Vet. Med., 40, 505-511.

Hayatsu, H. (1978). Cooperative mutagenic action of bisulfite
and hydroxylamines. Mut. Res., 54, 211.

Hayes, M.A. and Schiefer, H.B. (1982). Comparative toxicity of
dietary T-2 toxin in rats and mice. J. Applied Toxicol., 20,
207-212.

Hendricks, J.D., Sinnhuber, R.O., Nixon, J.E., Walsh, J.H.,
Masri, J.S., Hsieh, D.P.H. (1980). Carcinogenic response of
rainbow trout to aflatoxin $Q_1$ and synergistic effect of
cyclopropenoid fatty acids. J. National Cancer Inst., 64,
523-527.

Hoerr, F.J., Carlton, W.W. and Yagen, B. (1981). Mycotoxicosis caused by a single dose of T-2 toxin or diacetoxyscirpenol in broiler chickens. Veterinary Pathology, 18, 652-644.

Hsu, J.M. and Smith, J.C., Jr., (1983). Cysteine feeding affects urinary zinc excretion in normal and ethanol-treated rats. J. Nutr., 113, 2171-2177.

Itoh, Y., Yoshinaka, R. and Ikeda, S. (1979). Effect of cysteine and cystine on gel formation of fish meat by heating. Bull. Japan. Soc. Scientific Fisheries, 45, 341-345.

Jagadessan, V., Rukmini, C., Vijayaraghavan, M., and Tulpule, P. (1982). Immune studies with T-2 toxin: effect of feeding and withdrawal in monkeys. Food Chem. Toxicol. 20, 83-87.

Jarvis, B.B. and Mazzola, E.P. (1982). Macrocyclic and other novel trichothecenes: their structure, synthesis, and biological significance. Accounts Chemical Research, 15, 388-395.

Jones, L.A. (1981). Natural antinutrients of cottonseed protein products. In: "Antinutrients and Natural Toxicants in Foods", R.L. Ory, Ed., Food & Nutrition Press, Inc., Westport, Connecticut, pp. 77-98.

Jones, J.B., and Young, J.M. (1968). Carcinogenicity of lactones. III. The reactions of unsaturated γ-lactones with L-cysteine, J. Med. Chem., 11, 1176-1182.

Kamimura, K. et al. (1979). The decomposition of trichothecene mycotoxins during food processing. J. Food Hygienic Soc. Japan, 20, 352-357.

Kato, S., Yano, N., Suzuki, I. Ishii, T., Kurata, T. and Fujimaki M. (1974). Effect of L-cysteine on browning of egg albumin. Agric. Biol. Chem., 38, 2425-2430.

Kolbeck, W. (1976). L-Cysteine: a natural additive for food products. Alimentaria, 69, 59-63.

Komorn, R. and Carfuny, E.J. (1965). Effects of ethacrynic acid on renal protein-bound sulfhydryl groups. J. Pharmacol., 148, 367-372.

Koroda, H., Mori, T., Nishioka, C., Okassaki, H., and Takagi, M. (1979). Gas chromatographic determination of trichothecene mycotoxins in food. J. of the Hygieneic Society of Japan, 20, 137-142.

Korycka-Dahl, M.B. and Richardson, T.H. (1978). Activated oxygen species and oxidation of food constituents. CRC Crit. Rev. Food Sci. Nutr. 10, 209-241.

Kucewicz, W. (1982). U.S. Army studies yellow rain defenses. Wall Street Journal, April 29.

Kupchan, S.M. (1974). Selective alkylation: a mechanisms of tumor inhibition. Intra Science Chem. Repts., 8 (4), 57-66.

Lauterburg, B.H., Corcoran, G.B. and Mitchell, J.R. (1983). Mechanism of action of N-acetylcysteine in the protection against the hepatotoxicity of acetaminophen in rats in vivo. J. Clin. Invest., 71 (4), 980-991.

Lee, S. (1981).  Radioimmunoassay of T-2 toxin in biological fluids.  J. Assoc. Official Analytical Chemistrs, 64, 684-688.

Lindroth, S. (1980).  Occurrence, formation, and detoxificagion of patulin mycotoxins.  Dissertation, University of Helsinki, Finland.

Lindroth, S. and Wright, A. (1978).  Comparison of the toxicities of patulin and patulin adducts formed with cysteine.  Applied Environ. Microbiol., 35, 1003-1007.

Liener, I.E. (1975).  Effect of anti-nutritional and toxic factors on the quality and utilization of legume proteins.  In "Protein Nutritional Quality of Foods and Feeds", M. Friedman Ed., Marcel Dekker, New York, Part 2, pp. 523-550.

Liener, I.E. and Kakade, M. (1980).  Protease inhibitors.  In "Toxic Constituents of Plant Foodstuffs" I.E. Liener, Ed., 2nd edition, Academic Press, New York, p. 7.

Lindbladh, E. et al. (1983).  The effect of catalase on the inactivation of tyrosinase by ascorbic acid and by cysteine or glutathione.  Acta Derm. Venereol. (Stockholm) 63, 209-214.

Loltlikar, P.D. Insetta, S.M., Lyons, P.R. and Jhee, E.C. (1980).  Inhibition of microsome-mediated binding of aflatoxin $B_1$ to DNA by glutathione S-transferase.  Cancer Letters, 9, 143-149.

Lutsky, I.I. and Mor, N. (1981).  Alimentary toxic aleukia (septic angina, endemic panmyelotoxicosis, alimentary hemorrhagic aleukia).  T-toxin induced intoxication of cats.  Amer. J. Pathol., 104, 189-191.

Marshall, E. (1982).  A cloudburst of yellow rain reports.  Science, 218, 1202-1203.

Martinez-Torres, C., Romano, E. and Layrisse, M. (1981).  Effect of cysteine on iron absorption in man.  Am. J. Clin. Nutr., 34, 322-327.

Masri, M.S. (1984).  Defenses against aflatoxin carcinogenesis in humans.  This volume.

Masri, M.S. and Friedman, M. (1982).  Transformation of dehydroalanine residues in casein to S-β-(2-pyridylethyl)-L-cysteine side chains.  Biochem. Biophys. Res. Commun. 104, 321-325.

Masri, M.S., Reuter, F.W., and Friedman, M. (1974).  Binding of metal cations by natural substances.  J. Applied Polym. Sci., 18, 675-681.

McCann, J., Choi, E., Yamasaki, E., and Ames, B.N. (1975).  Detection of carcinogens as mutagens in the Salmonella/ microsome test:  Assay of 300 chemicals, Proc. Natl. Acad. Sci., USA, 72, 5135-5139.

MacGregor, J.T. and Clarkson, T.W. (1974).  Distribution, tissue binding and toxicity of mercurials.  In "Protein-Metal Interactions", M. Friedman, Ed., Plenum Press, New York, New York, Adv. Exp. Med. Biol. 48, 463-503.

MacGregor, J.T., Diamond, M. J. Mazzeno, L.W., and Friedman, M.
    (1980). Mutagenicity test for fabric finishing agents in
    Salmonella/microsome Typhimurium: fiber reactive dyes and
    cotton flame retardants. Environmental Mutagenesis, 2, 405-
    408.
McLaughlin, C.S. et al. (1977). Inhibition of protein synthesis
    by trichothecenes. In "Mycotoxins in Human and Animal
    Health", J.V. Rodricks, C.W. Hesseltine, and M.A. Mehlman,
    Eds., p. 363-373, Pathotox Publishers, Park Forest, Illinois.
Meister, A. (1983). Selective modification of glutathione
    metabolism. Science, 220, 471-476.
Miranda, C.L. et al. (1982). Modification of chronic hepato-
    toxicity of pyrrolizidine alkaloids by butylated hydroxyani-
    sole and cysteine. Toxicol. Lett. 10, 177-182.
Mirocha, C.J., Pathre, S.V., and Christensen, C.M. (1981). Myco-
    toxins. In "Adv. Cereal Science and Technol." Y. Pomeranz,
    Ed., pp. 159-225.
Moriya, M., Kato, K. and Shirasu, Y. (1978). Effects of cysteine
    and a liver metabolic activation system on the activities of
    mutagenic pesticides. Mut. Res., 57, 259-263.
Moss, J.A., Judah, D.J., Przybylski, M. and Neal, G.E. (1983).
    Some mass spectral and NMR analytical studies of a gluta-
    thione conjugate of aflatoxin $B_1$. Biochem. J., 210, 227-
    234.
Murakami. M. et al. (1981). A morphological and biochemical
    study of the effects of L-cysteine on the renal uptake and
    nephrotoxicity of cadmium. Br. J. Exp. Pathol., 62, 115-130.
Newberne, P.M. Naturally occurring food-borne toxicants. In
    "Modern Nutrition in Health and Disease", R.S. Goodhard and
    M.E. Shils, Eds., pp. 463-493. Lea & Febiger, New York.
Novi, A.M. (1981). Regression of aflatoxin $B_1$-induced hepato-
    cellular carcinomas by reduced glutathione. Science, 212,
    541-542.
Ohtsubo, K. and Saiot, M. (1977). Chronic effects of trichothe-
    cene toxins. In "Mycotoxins in Human and Animal Health,"
    J.V. Rodricks, C.W. Hesseltine, and M.A. Mehlman, Eds., pp.
    255-262. Pathotox Publishers, Park Forest, Illinois.
Patterson, D.S.P. (1977). Metabolic activation and detoxifica-
    cation of mycotoxins. Ann. Nutr. Alim., 31, 901-910.
Pickman, A.K., Rodriguez, E. and Towers, G.H.N. (1979). Forma-
    tion of adducts of parthenin and related sesquiterpene
    lactones with cysteine and glutathione. Chem. - Biol. Inter-
    actions, 28, 83-89.
Pitman, I.H. and Morris, J.J. (1979). Covalent addition of glu-
    tathione, cysteine, homocysteine, cysteamine and thiogly-
    colic acid to quinazoline cation. Aust. J. Chem., 32,
    1567-1573.

Pribela, A. and Pikulikova, C. (1969). Reduction of dehydroas-
    scrobic acid with L-cysteine. Z. Lebensmunters. Forsch.,
    139, 219-229.
Rackis, J.J. and Gumbmann, M.R. (1981). Protease inhbitors:
    physiological properties and nutritional significance. In
    "Antinutrients and Natural Toxicants in Foods", R.L. Ory,
    Ed., Food & Nutrition Press, Westport, Connecticut, p. 203-
    237.
Raj, H.G., Santhanam, K., Gupta, R.P. and Venkitasubrtamanian,
    T.A (1975). Oxidative metabolism of aflatoxin $B_1$: obser-
    vations on the formation of epoxide-glutathione conjugate.
    Chem.-Biol. Interactions, 11, 301-305.
Rosen, H. and Klebanoff, S.J. (1977). Formation of singlet oxy-
    gen by the myeloperoxidase-mediated and antimicrobial system.
    J. Biol. Chem. 252, 4803-4810.
Rosin, M.P. and Stich, H.F. (1978a). The inhibitory effect of
    cysteine on the mutagenic activities of several carcingens.
    Mutation Res., 54, 73-81.
Rosin, M.P. and Stich, H.F. (1978b). Inhibitory effect of reduc-
    ing agents on N-acetoxy- and N-hydroxy-2-acetylaminofluorene-
    induced mutagenesis. Cancer Res., 38, 1307-1310.
Roy-Burman, S., Roy-Burman, P. and Visser, D.W. (1968). Showdo-
    mycin, a new nucleoside antibiotic. Cancer Res. 28, 1605-
    1610.
Rukmini, C., Prasad, J.S., and Rao, K. (1980). Effect of feeding
    T-2 toxin to rats and monkeys. Food Cosmet. Toxicol. 18,
    267-269.
Schmidt, R., Ziegenhagen, E., and Dose, K. (1981a). High per-
    formance liquid chromoatorgraphy of trichthecenes. I. Detec-
    tion of T-2 Toxin and HT-2 toxin. J. Chromatography, 212,
    370-373.
Schmidt, R., Bieger, A., Ziegenhagen, E., and Dose, K. (1981b).
    Determination of T-2 toxin in vegetable foods. I. T-2 toxin
    in mouldy rice and maize. Z. Analyt. Chemie, 308, 133-136.
Schoental, R. and Rive, D.J. (1965). Interaction of N-alkyl-
    nitrosourethanes with thiols. Biochem. J., 97, 466-474.
Schwimmer, S. and Friedman, M. (1972). Genesis of volatile sul-
    phur-containing food flavours. Flavour Industry, pp. 137-
    145.
Scott, P.M., Lau, P.Y. and Kanhere, S.R. (1981). Gas chromato-
    graphy with electron capture and mass spectrometric detection
    of deoxynilvalenol in wheat and other grains. J. Ass. Off..
    Analyt. Chem., 64, 1364-1371.
Shelef, L.A. and Chin, B. (1980). Effect of phenolic antioxi-
    dants on the mutagenicity of aflatoxin $B_1$. Applied and.
    Environ. Microbiol., 40, 1039-1043.

Shih, J.C.H., Jonas, R.H. and Scott, L.M. (1977). Oxidative deterioration of muscle proteins during nutritional muscular dystrophy in chicks. J. Nutrition, 107, 1786-1791.

Smith, T.K. and Carson, M.S. (1984). Effect of diet on T-2 toxicosis in rats. This volume.

Snedeker, S.M. and Greger, J.L. (1983). Metabolism of zinc, copper, and iron as affected by dietary protein, cysteine and histidine. J. Nutr., 113, 644-652.

Snow, J.T., Finley, J.W. and Friedman, M. (1975). Oxidation of sulfhydryl groups to disulfides by sulfoxides. Biochem. Biophys. Res. Commun., 64, 441-447.

Sprince, H. et al. (1979). Comparison of protection by L-ascorbic acid, L-cysteine, and adrenergic blocking agents acetaldehyde, acreloin, and formaldehyde toxicity: implication in smoking. Agents Actions, 9, 407-414.

Swenson, D.H., Lin, J.K., Miller, E.C., and Miller, J.A. (1977). Aflatoxin $B_1$-2,3-oxide as a probable intermediate in the covalent binding of aflatoxins $B_1$ and $B_2$ to rat liver DNA and robosomal RNA in vivo. Cancer Res., 37, 172-181.

Tajiri, T., Matsumoto, K. and Hara, K. (1976). Storage stability of freeze-dried carrots. III. The content of carotene as affected by the addition of cysteine hydrochloride at various temperatures. J. Japan. Soc. Hort. Sci., 45, 81-88.

Tamm, C. (1974). In "Progress in the Chemistry of Organic Natural Products", W. Herz. et al. Eds., 31, 63-114, Springler Verlag, Vienna and New York.

Tamm, C. (1977). Chemistry and biosynthesis of trichothecenes. In "Mycotoxins in Human and Animal Health", J.V. Rodricks, C.W. Hesseltine, and M.A. Mehlman, Eds., pp. 209-228, Pathotox Publishers, Park Forest, Illinois.

Tien, M., Bucher, J.R. and Aust, S.D. (1982). Thiol-depended lipid peroxidation. Biochem. Biophys. Res. Commun., 107, 279-285.

Thompson, W.L. and Wannemacher, R.W. (1982). Studies of T-2 toxin inhibition of protein synthesis in tissue culture cells. Fed. Proc. 41, 1390.

Ueno, Y. (1977). Trichothecenes: an overview address. In "Mycotoxins in Human and Animal Health," J.V. Rodricks, C.W. Hesseltine, and M.A. Mehlman, Eds., pp. 189-207. Pathotox Publishers, Park Forest, Illinois.

Ueno, Y. (1980). Trichothecene mycotoxins: mycology, chemistry, and toxicology. In "Advances in Nutritional Research", H.H. Draper, Ed., Vol. 3, pp. 301-353, Plenum Press, New York.

Ueno, Y. and Matsumoto, H. (1975). Inactivation of some thiol-enzymes by trichothecene mycotoxins for Fusarium species. Chem. Pharm. Bulletin, 23, 2439-2442.

Unverfeth, B.J., Magorien, R.D., Balcerzak, S.P., Leier, C.V. and
    Unverfeth, D.V. (1983). Early changes in human myocardial
    nuclei after doxorubicin. Cancer, 52, 215-221.
Vernie, L.N., Bont, W.S., Ginjaar, H.B. and Emmelot, P. (1975).
    Elongation factor 2 as the target of the reaction product be-
    tween sodium selenite and glutathione in the inhibiting of
    amino acid incorporation in vitro. Biochim. Biophys. Acta,
    414, 283-292.
Vernie, L.N., Collard, J.G., Aker, A.P.M., DeWildt, A. and
    Wilders, I.T. (1979). Studies on the inhibition of protein
    synthesis by selenodiglutathione. Biochem. J., 180, 213-218.
Vernie, L.N., Homburg, C.J. and Bont, W.S. (1981). Inhibtion of
    the growth of malignant mouse lymphoid cells by selenoadiglu-
    tathione and selenodicysteine. Cancer Letters, 14, 303-308.
Walker, E.A., Pignatelli, B. and Friesen, M. (1982). The role of
    phenols in catalysis of nitrosamine formation. J. Sci. Food
    Agric., 33, 81-88.
Walther, H.J. (1982). Fast atom bomardment mass spectrometry of
    aflatoxins and reaction products of sodium bisulfite with
    aflatoxin. J. Agr. Food Chem., 31, 168-171.
Watson, D.H. and Lindsay, D.G. (1982). A critical review of bio-
    logical methods for the detection of fungal toxins in foods
    and foodstuffs. J. Sci. Food Agric. 33, 59-67.
Wattenberg, L.W. (1983). Inhibition of neoplasia by minor
    dietary constituents. Cancer Res., 43, 2448-2453s.
Weaver, G.A. Kurtz, H.J., Bates, F.Y., Mirocha, C.J., Behrens,
    J.C. and Hagler, W.M. (1981). Fusarium mycotoxin. Diace-
    toxyscirpenol toxicity in pigs. Res. in Vet. Sci., 31, 131-
    135.
Weiss, S.J., Lampert, M.B. and Test, S.T. (1983). Long-lived
    oxidants generated by human neutrophils: characterization
    and bioactivity. Science, 222, 625-627.
Westley, J., Adler, H., Westley, L. and Nishida, C. (1983). The
    sulfurtransferases. Fundamental and Applied Toxicol., 3,
    377 382.
Williams, D.L.H. and Aldred, S.E. (1982). Inhibition of nitrosa-
    tion of amines by thiols, alcohols, and carbohydrates. Food
    Chem. Toxicol., 20, 79-81.
Williamson, J.M., Boettcher, B. and Meister, A. (1982). Intra-
    cellular cysteine delivery system by promoting glutathione
    synthesis. Proc. Natl. Acad. Sci. USA, 79, 6246-6249.
Wright, A. and Lindroth, A. (1978). The lack of mutagenic prop-
    erties of patulin  and patulin adducts formed with cysteine
    in Salmonella test system. Mutation Res., 58, 211-215.
Yavelow, J., Finlay, T.H., Kennedy, A.R. and Troll, W. (1983).
    Bowman-Birk soybean protease inhibitor as an anticarcinogen.
    Cancer Res., 43, 2454s-2459s.

Yee, J.J. and Shipe, W.F. (1982). Effects of sulfhydryl com-
pounds on lipid oxidations catalyzed by copper and heme. J.
Dairy Sci. 65, 1414-1420.
Yoshizawa, T., Mirocha, C.J., Behrens, J.C. and Swanson, S.P.
(1981). Metabolic fate of T-2 toxin in a lactating cow.
Food Cosmet. Toxicol. 19, 31-39.
Zahnley, J.C. (1984). Stability of enzyme inhibitors and lectins
in foods and the influence of specific binding interactions.
This volume.
Zahnley, J.C. and Friedman, M. (1982). Absorption and fluores-
cence spectra of S-quinolylethylated Kunitz soybean trypsin
inhibitor. J. Protein Chemistry, 1, 225-240.

## 3

INTERACTIONS OF DIET AND IMMUNITY

David L. Brandon

U. S. Department of Agriculture
Agricultural Research Service
Western Regional Research Center
Berkeley, California  94710

INTRODUCTION

The evidence from laboratory, clinical, and epidemiological studies that diet affects immune function is compelling, although precise recommendations of dietary choices for optimum health and longevity cannot yet be stated.  In part, this situation results from complex and sometimes contradictory data.  Some of reasons for these difficulties may· be summarized as follows.

1. Results obtained in vitro must be related to conditions which apply in vivo.

2. Data from uncontrolled studies on outbred animals (i.e., humans) must be compared to those obtained from controlled studies on defined strains.

3. Direct dietary effects on immunity are difficult to separate from the indirect effects of increased infection, which often accompanies malnutrition.

4. Effects on immune response are often delayed.

5. The immune system is complex and several components may be affected by a single dietary stimulus.

Nevertheless, some interesting patterns have emerged from the data, suggesting that the nutrient content and toxicant levels of the diet may influence the immune response and, possibly,

health. It has been stated that "every nutritional insult that
promotes cancer has been shown to suppress some component of
cell-mediated immunity" (Vitale and Broitman, 1981). The purpose of this article is to summarize some of the key ideas and
formulate hypotheses, rather than to provide a comprehensive
summary of this rapidly developing field.

## KEY FEATURES OF THE IMMUNE SYSTEM

The immune system has two major functions: (1) to provide resistance against infectious disease; and (2) to defend against
developing neoplasia. However, the immune system itself is
subject to pathological conditions and to environmental influences, including diet. This article discusses the effects of
general malnutrition, the effects of specific nutrients (protein,
vitamins, trace elements), and the effects of dietary toxicants
on the immune response and concludes with recommendations for
further research.

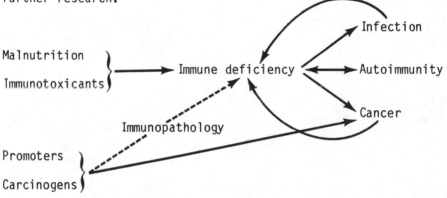

Figure 1.    Interrelations among diet, immunity, and disease.

Figure 1 illustrates a vicious circle. Malnutrition or
immunotoxic substances in the diet can result in immunodeficiency
and in greater susceptibility to infection, autoimmunity, and
neoplasia. The reason that this is a viscious circle is because
each of these three conditions can itself lead to further immunological deficiencies, by mechanisms which are only partially
understood. The link between toxicants especially carcinogens
and tumor promoters--and the immune response may have further
complexity as indicated by·the dotted line in the diagram. These
substances may cause immunopathology directly, in addition to
effecting immunosuppression indirectly as a result of neoplasia.
The phorbol esters, for example, have immunopharmacological
effects, as documented by several recent reports. (See also
also the article by Berry and Helms in this volume.) For exam-

ple, phorbol esters can impair activation of macrophages, preventing development of tumoricidal capacity (Keller et al.,
1982). On the other hand, immunostimulatory effects have been
reported on both antibody synthesis (Ralph and Kishimoto, 1982)
and on generation of cytotoxic cells (Munger and Lindquist,
1982). Phorbol esters probably affect multiple aspects of the
response by selectively activating lymphocyte subpopulations
(Andreotti, 1982). Thus, immunopharmacologically active compounds must be considered in any study of dietary influence on
immunity.

## Two major cell lineages which confer specificity on the immune response

Lymphocytes arise from stem cells in fetal liver and bone
marrow, which are processed in two different ways. Cells of
both lineages have as major functions their capabilities to
interact with antigen, to differentiate, and to proliferate.
The B cells (processed in the bone marrow) are primarily responsible for antibody production, whereas T cells (processed in
the thymus) are responsible for regulation of immune responses,
including antibody production. However, these are not completely
independent populations. For example, some cytotoxic activities,
generally classified as cellular or T-dependent immunity, depend
on antibodies. In addition, products of the B cell compartment,
can often block reactions of T cells. Pinpointing the site of
action of nutrients or toxicants is made more complex by this
situation. There are numerous subsets of each of these two major
lineages. An important subset of T cells are cytotoxic cells
which can kill other cells, usually pathogens or tumor cells.
Other T lymphocytes cooperate with B cells through surface receptors and through soluble mediators.

T cell functions. A number of parameters are usually measured to gauge the extent of T cell function. T cell levels in peripheral blood and their response to T cell mitogens such as phytohemagglutinin are commonly used quick tests, especially in clinical studies, since few cells are required and peripheral blood
provides a sufficient supply. While responsiveness to mitogens
can give a quick result, it does not adequately define the state
of the immune system. Delayed hypersensitivity assays measure T
cell function, as do quantitation of lymphokine production and
helper cell activity. Helper cell activity is necessary for both
cell-mediated immunity and humoral (antibody) immunity.

B cell functions. B cells are primarily responsible for
production of antibodies. Their function depends on recognition
of antigen presented by macrophages and on specific interactions
with helper and suppressor T cells and their products. In addi-

tion to participaton in antigen-antibody reactions and activation of the complement system, antibodies mediate some cytotoxic
reactions by killer cells. All of these functions can be readily
measured in vitro.

## Nonspecific immunity

Factors which are essential to disease resistance include
proteins of the complement system. While lacking the specificity of T and B cell products, the complement system is essential
in some cytotoxic mechanisms, in regulating lymphocyte traffic,
and in modulating a variety of lymphocyte activities (Weigle
et al., 1982). Cells which lack the specific antigen receptors
of lymphocytes are nevertheless essential for immune responses.
Macrophages are important in presenting antigens to specific
immunocytes, and natural killer (NK) cells are cytotoxic cells
which appear to recognize altered characteristics of neoplastic
cells.

## Differentiation

An important feature of the immune system is the extensive
gene rearrangement required to assemble the genetic material in
a form capable of producing functional effector molecules. Processing occurs both at the DNA and RNA levels. The extensive
shuffling of genetic information is needed to produce several
repertoires of specific recognition molecules including cell
surface receptors and antibodies.

## NUTRITION AND IMMUNITY

The interaction between nutritional status and immunity has
been discussed in recent reviews amd monographs. (Chandra and
Newberne, 1977; Suskind, 1977; Beisel, 1982). I will briefly
summarize the evidence that many aspects of immunity are sensitive to nutrition. In fact, the immune system may be one of the
most sensitive physiological systems for studying nutritional
requirements. For example, in the case of vitamin requirements,
the suggestion has been made that the requirement for optimum
function of the immune system may exceed the amount required for
normal growth (Williams et al., 1978).

## Gut-associated lymphoid system.

Food antigens are probably the most widely encountered anti-
gens--and certainly the ones encountered in the highest concentration. The gut-associated lymphoid tissue acts as a subsystem

which includes Peyer's patches--lymph nodes attached to the
intestine. These are especially important in antibody responses
involving IgA, the major immunoglobulin in seromucous secretions.
For a recent review see Dobbins (1982). IgA provides an immunological barrier against deleterious interactions of the body
with food antigens absorbed through the intestine. Antibodies
in the intestine appear to slow the absorption of specific
protein antigens (Walker et al., 1974) and, may permit more
complete digestion of protein. Impaired secretory IgA immunity
may therefore lead to less efficient uptake of protein, especially in the neonate. The secretory IgA immune response appears
sensitive to nutritional influence (McMurray et al., 1977).

Intragastric administration of antigen can induce a state of
immunologic tolerance (Vaz et al., 1977; Richman et al., 1978).
These effects were incorporated into folk medicinal practices
long before the first immunologist measured antibody titers.

A number of diseases are associated with abnormal or unusual reactions of this system and selective deficiency of IgA
is the most common immuno-deficiency syndrome in adults (Ammann
and Hong, 1971). Coeliac disease (gluten-sensitive entero-
pathy), soy intolerance, autoimmunity (Wright, 1976; Barratt,
et al., 1979), increased incidence of tumor development, and
coronary heart disease (Davies, 1976) correlate with aberrant
function of the gut-associated lymphoid system. Manifestation
of inflammatory bowel disease is associated with immunologic
reactions--including reactions with dietary antigens-although
the etiology of this disease is obscure (Bartnik and Shorter,
(1982). A deficiency in secretory IgA, caused by a defect in
synthesis of secretory component, can result in chronic intestinal infection (Strober et al., 1976).

Interactions of antigen and antibodies with the gut are especially important in the neonate. Antibodies along with cells
of the immune system are contained in the colostrum, providing
the neonate with passively acquired immunity. (For a brief review, see Faden and Ogra, 1981.) The gastrointestinal tract of
the neonate is permeable to proteins, including food antigens.
This is probably why some food antigens encountered very early in
life, before the organism has a mature immune system, can result
in food allergy. It has been suggested that apparently normal
individual variations in immune functions--such as IgA levels-
may result in vulnerability to infection and allergy in infants
(Soothill, 1976). Consequently, it is important to define nutritional requirements for optimal function of the gut-associated
lymphoid tissue.

Immunological senescence

Immunological senescence is characterized by adult onset of
a program apparently determined genetically within the major
histocompatibility complex. (See Makinodan and Kay, 1980 for a
review.) It is exemplified by the involution of the thymus
following puberty. Other changes include a general increase
in autoimmunity and a decrease in immunity to neoplasia. Immunological senescence appears to be subject to environmental--chief-
ly dietary--influence. Increased lifespan, decreased incidence
of autoimmunity and neoplasia, and slower involution of the
thymus may result from dietary restriction (discussed below).
Specific effects depend on the dietary modification and on the
age and strain of animal. The precise genes and gene products
involved have yet to be elucidated.


SPECIFIC NUTRITIONAL DEFICIENCIES AND THE IMMUNE RESPONSE

Vitamin status

There has been considerable interest in relating immunological dysfunction to specific vitamin deficiencies--especially
vitamin A, the B complex, and vitamins C and E.

Vitamin A. Investigations of the correlation between vitamin A status and immune function have not resulted in a consistent picture. In chickens, vitamin A deficiency leads to atrophy of the bursa of Fabricius, the site of B lymphocyte production (analogous to mammalian bone marrow), but not the spleen or
thymus (Panda and Combs, 1963). The response of rat lymphocytes
to both B and T cell mitogens was impaired in vitamin A-deficient
animals (Nauss et al., 1979). Vitamin A can act as an adjuvant
when administered subcutaneously or intragastrically with antigen (Falchuck et al., 1977). More extensive effects on the
immune system--hypertrophy of the thymus and thymus-dependent
areas of lymph nodes and stimulation of allograft rejection
(Medawar and Hunt, 1981)--indicate mechanisms other than simple
adjuvanticity. Vitamin A in excess (Uhr et al., 1963) or administered intraperitoneally as an emulsion (Barnett and Bryant,
1980) can result in immunosuppression. (See also Omaye, this
volume.) The results of Seifter et al. (1981) indicate that
immunopharmacological actions of carotenoids are probably determined by their activity as precursors of vitamin A or inhibitors of vitamin A biosynthesis. Thus, retinoids are good candi-
dates for nutritional and pharmacological intervention in disorders of the immune system and may be involved in development
of human cancers (Atukorala et al., 1979).

B Vitamins. Kumar and Axelrod (1978) reported that a deficiency of any of four vitamins--thiamine, riboflavin, biotin, and
folic acid--could cause a depression of humoral immunity measured
by the antibody-forming cell response to sheep erythrocytes. The
response was at least partially restored in all cases by nutritional repletion. Pyridoxine deficiency, induced experimentally
by feeding a vitamin B$_6$-deficient diet or a pyridoxine antagonist, has been implicated in abnormal antibody responses to several antigens, and in reduced cellular immunity (e.g., delayed-
type hypersensitivity, mixed lymphocyte reaction, graft vs. host
reaction) (Robson and Schwarz, 1980). Deficiencies of folic
acid and cobalamin are also associated with suppressed cell-
mediated immunity (Newberne, 1977; Williams et al., 1978). The
intake of cobalamin required for optimum immune function may
exceed the amount required for normal growth (Williams, et al.,
1978). Confirming the sensitivity of the developing immune
system, Robson and Schwarz (1975) reported that vitamin B$_6$ deficiency experienced in utero dramatically affected the immune
competence of the developing rat.

Vitamin C. The intense interest and speculation about the
role of vitamin C in defense against viral infection and cancer
has probably been responsible for the multitude of recent studies of the influence of vitamin C on the immune response. The
literature on vitamin C and the immune response has been summarized by Leibovitz and Siegel (1981). Ascorbic acid appears
to enhance chemotaxis of phagocytes, a process involved in antigen presentation to the immune system, both in vitro (Goetzl
et al., 1974) and in vivo (Boxer et al., 1976), although micro-
bicidal activity is stimulated only in immunosuppressed patients
(Olsen and Polk, 1977). There is no firm evidence that vitamin
C nutriture influences the humoral immune response, but the vitamin does influence T lymphocyte responses to mitogens (Anderson
et al., 1980; Siegel and Morton, 1977). Hence, subtle effects
on cell-mediated immunity or regulatory phenomena may occur.
Leibovitz and Siegel (1981) reported an interesting inverse relationship between spleen weight and the tissue levels of vitamin
C in the spleens of old NZB mice. The subpopulation relatively
free of autoimmune disease had higher vitamin C levels than those
with disease, possibly due to less oxidative activity associated with inflammatory reactions.

Vitamin E and selenium. Antioxidants, especially selenium
and vitamin E, have been reported to influence immune response
but documentation of direct effects is still incomplete. (For
recent reviews, see Tengerdy et al., 1981; Baumgartner, 1979;
and Spallholz, 1981.) Toxic at high concentrations, antioxidants
appear to potentiate macrophage function at lower concentrations.
Thus, they may be said to have an adjuvant effect--much like vi-

tamin A.  These substances may regulate immune responses by
influencing suppressor cell activity.  Vitamin E seems also to
suppress the secretion of corticosteroids, thus preventing the
immunosuppressive effects of this hormone, accelerating development of adult immunocompetence in young mice (Lim et al., 1981)
and stimulating the differentiation of specific T helper cell
populations (Tanaka et al., 1979).  Some of the effects of vitamin E and other antioxidants may result indirectly from blockage
of membrane lipid peroxidation (Hoffeld, 1981).  Direct action
of vitamin E on the immune system has been attributed to interaction of the vitamin with a soluble suppressor factor (Corwin
et al., 1981).  This suppressor factor appears to be an oxidation
product of spermine (Corwin and Gordon, 1982).  The hypothesized
influence of vitamin E on prostaglandin metabolism (reviewed by
Sheffy and Schultz, 1979) succinctly explains the immunostimulatory effects of vitamin E (as well as the immunosuppressive
effects of polyunsaturated fatty acids).

It is difficult to rationalize the reported antagonistic
effect of vitamins A and E when used as dietary supplements for
chicks (Tengerdy et al., 1981), since both vitamins have adjuvant-like effects in other experimental systems.  Selenium may
also have an adjuvant-like effect (Spallholz, 1981) i.e., it may
potentiate the activity of phagocytic cells involved with antigen
presentation.  A possible mechanism of selenium action is protection against free radicals generated during phagocytosis and
cytotoxic processes (Baumgartner, 1979).  Tumor-induced deficiencies of antioxidants (Baumgartner, 1979) may explain the
immunosuppression associated with malignancy.  The hypothesis
has been advanced that antioxidants may protect against genetically programmed senescence or disease-induced depression of the
immune system (Harman et al., 1977).  Thus, nutrients which
function as antioxidants may provide a means for dietary manipu-
lation of the immune response.


METAL DEFICIENCY AND THE IMMUNE RESPONSE

There is extensive documentation of the effects of zinc and
iron deficiency on the immune response (Schloen et al., 1979;
Strauss, 1978).  Iron-deficient children have impaired bacterici-
dal activity, although concentrations of serum immunoglobulin
may be even higher than normal (MacDougall et al., 1975).  Arguments have been presented which question whether host defenses
are necessarily compromised during iron deficiency (Strauss,
1978; Pearson and Robinson, 1976).  However, rats deficient in
iron fail to mount a normal primary immune response (Nalder,
et al., 1972).  Several in vitro studies have indicated impaired
responses of lymphocytes and accessory cells from iron-deficient

patients (Kulapongs et al., 1974; Sawitsky et al., 1976).

Two conditions have been recognized which convincingly im-
plicate zinc in regulation of the immune system (Brummerstedt
et al., 1977). A mutant strain of cattle is particularly sensitive to infection and has characteristic thymic aplasia. This
situation is somewhat similar to acrodermatitis enteropathica in
humans (Weisman and Flagstad, 1976). Both of these hereditable
diseases have a number of immunological sequelae, and both are
alleviated by supplementary zinc nutriture. Inadequate intake,
low bioavailability, and poor absorption can all contribute to
zinc deficiency. Pregnancy, tissue damage, disease, and malignancy are all widespread conditions which create a high demand
for zinc (Schloen et al., 1979).

Deficiencies of zinc selectively influence the T cell compartment (Fraker et al., 1977; Fernandes et al., 1979a) and also
affect the natural killer cell activity important in surveillance against tumors. Reduced level of zinc can lead to increased secretion of corticosteroids (DePasquale-Jardieu and
Fraker, 1979). Since these hormones have immunosuppressive
effects, they may be the basis some of the immunosuppression
associated with zinc deficiency and a variety of other nutritional deficiencies. Fraker et al. (1977) have identified an
apparent T cell defect in zinc-deprived A/J mice. These mice
were deficient in antibody production against sheep red blood
cells; the deficiency could be restored by transfer of thymocytes
from control animals. In an extension of the above study, Fraker
et al. (1978) reported that rapid reversal of thymus involution
and regeneration of T cell help following nutritional repletion
of zinc-deficient mice. Pekarek et al. (1977) found that zinc-
deficient unimmunized rats had increased mortality following
infections with F. tularensis. Since immunized zinc-deficient
rats actually had an increased anti-virus titer compared to controls, a defect in cell-mediated immunity is indicated (e.g.,
interferon production).

One of the sites of influence of zinc on immune response
may be at the lymphocyte level, since zinc has pharmacological
effects on lymphocytes in vitro (Berger and Skinner, 1974).
Nude mice, lacking a functional thymus, have an abnormally low
plasma level of zinc (Hattler, unpublished results quoted by
Schloen et al., 1979). However, the recent demonstration that
zinc is essential for the function of serum thymic factor, a
polypeptide hormone (Dardenne et al., 1982), supports a direct
role for zinc in the maintenance of the immune system. More
generally, zinc is an essential coenzyme in many enzyme systems
and additional functions of zinc in cell membrane function have
been postulated (Bach, 1981).

LEUCINE, LIPOTROPES AND THE IMMUNE RESPONSE

In a study of the interaction of lipotropic factors and the immune response, Williams et al. (1978) monitored the immunological consequences of a choline and cobalamin-deficient diet on C57B1/6 mice. This diet resulted in diminished response to B and T cell mitogens, decreased plaque-forming cell (PFC) response to sheep red blood cells, and an increased response to alloantigens. The results suggest a reduction of a subpopulation of T cells, including phytohemagglutinin responders and allotype suppressors. Newberne (1977) summarized the effects of lipotrope deficiency in prenatal and neonatal rats and mice on the development of the immune system. Deficiency of choline plus methionine, folic acid, or cobalamin resulted in depression of T cell-dependent cellular and humoral immunity. The same paper presented preliminary experiments indicating that folate-deficient women have impaired T cell responses to skin sensitizing agents and phytomitogens. The importance of these observations is underscored by the low levels of choline and methionine in some vegetarian diets, and the common occurrence of folic acid deficiency among pregnant women, users of oral contraceptives or anti-convulsants, and alcoholics (Newberne, 1977).

IMMUNE IMPAIRMENT ASSOCIATED WITH OVERNUTRITION

An obese strain of mice (C57B1/6Jobob) has been widely used as an animal model for obesity. Meade et al. (1979) have reported decreased cell-mediated immunity in these mice, as measured by delayed hypersensitivity, graft rejection, and cytotoxicity. The transfer of cells to a nonobese recipient resulted in normal cellular activity, implicating a defective in vivo environment. Analogous results have been reported in clinical studies of immunological function in obese patients (Hallberg, et al., 1976). High fat diets resulted in decreased resistance to infection in rats and chickens (Beisel and Fiser, 1970) and dogs (Newberne, 1966). Excess intake of leucine, when combined with a low protein diet can result in a depressed immune response (Chevalier and Aschkenasy, 1977).

The associations between overnutrition and immune function appear to be complex. Critical review reveals conflicts in the literature (Edelman, 1981), but the most convincing leads concerning mechanisms of these associations are as follows. Polyunsaturated fatty acids may be immunosuppressive because they are precursors of prostaglandins which regulate the function of immunocytes (Lichtenstein et al., 1972). The role of prostaglandins is strongly suggested by the reversal of tumor growth promotion by inhibitors of prostaglandin synthesis (Hillyard and Abraham, 1979). Some of the effects of lipid nutriture are mediated by plasma lipoproteins (reviewed by Edgington and Curtiss, 1981), thus providing a possible explanation for dis- crepancies between experiments performed in vivo and in vitro. The effect of cholesterol intake is even more confusing. Oxy- genated sterols, which inhibit cholesterol synthesis, suppress the immune response. However, nutritionally induced hypercho- lesteremia also compromises the immune system (Kos et al, 1979).


MALNUTRITION AND IMMUNITY

Protein-calorie malnutrition is a major world problem chief- ly affecting populations in the less developed nations and hospi- talized patients on total parenteral nutrition (Bistrian et al., 1975). However, there are few data from carefully controlled dietary studies of the immune response in humans. Since popula- tions which suffer from malnutrition are also subject to high morbidity of infectious and parasitic diseases, it is difficult to relate immunological effects to specific nutritional factors. The disease state generally is accompanied by immunological changes, e.g., increases in immunoglobulin levels. Plasma im- munoglobulin levels may exceed normal controls in malnourished individuals with infection.

The human population is genetically heterogeneous. The advances in immunogenetics over the last fifteen years document the pervasive genetic influence on immunity. Numerous immuno- logical disorders have been correlated with the presence of cer- tain alleles of the major histocompatibility complex. It is reasonable to hypothesize that the interaction between diet and immunity is also regulated at the genetic level.

Malnutrition is a complex disease, rarely characterized by a well defined deficit of specific nutrients. For example, the protein composition of the diet often influences the trace ele- ment content. Differences between animal models, in which pro- tein deficiency often stimulates cell-mediated immunity, and human studies, which have often indicated depressed cell-mediated immunity in protein restriction, can be rationalized by deficien- cies of zinc which occur in protein-poor human diets. Consistent with this picture, most amino acid deficiencies exert less influ- ence on the immune response than on overall growth rate (Hill, 1982).

Severe malnutrition generally affects the T cell rather than the B cell compartment. The basis for immune impairment of both children and experimental animals may well be atrophy of the thymus. The thymus is the slowest organ to recover fol- lowing nutritional repletion (McFarlane, 1977). Severe malnu-

trition experienced <u>in utero</u> or early in life has profound
effects on the development of the immune system--so severe that
the term "nutritional thymectomy" describes the resulting immunological consequences. The T cell levels in peripheral blood
may be as low as 20% of controls. Hypogammaglobulinemia, reduced
responses to mitogens, and impaired cell-mediated immunity,
may continue for months following nutritional repletion (Chandra,
1974 and 1975). It is not known whether this deficiency impairs
immunity in the adult, or whether it affects susceptibility to
immunity-linked disorders (autoimmunity, allergy, malignancy).
In animal studies, the offspring of nutritionally deprived
rats were deficient in T cell responses, but not in response to
a B cell mitogen (Gebhardt and Newberne, 1974). In this study,
progeny of lipotrope-deficient rats had a particularly low
response to phytohemagglutinin in the spleen, but not in the
thymus, suggesting a blockage in either maturation or migration
of T cells.

IgA responses are nutritionally sensitive, and it seems
likely that this is due to the strong T cell dependence of these
functions. Even moderate malnutrition impairs secretory immunity
(McMurray et al., 1977; Sirisinha et al., 1977; Faulk et al.,
1977), one of the primary barriers against infection. These
studies indicate that secretory IgA may be deficient even though
serum immunoglobulin levels are normal or above normal.

Some workers have observed an increased concentration of
serum IgE in marasmic children even in the absence of parasitic
infection (Kikkawa et al., 1973; Johansson et al., 1968). Along
with IgA, IgE production is very dependent on thymic influence.
Therefore, it seems reasonable to hypothesize the existence of
nutritionally sensitive, class-specific suppressor T cell populations.

Although fetal and neonatal malnutrition are associated
with similar immunological impairment (Chandra, 1977), the effects of malnutrition in the fetal period appear more persistent.
Nutritional influences on immune competence of the offspring
may be exerted at two levels: at the level of development of
the embryo (Hurley, 1968), or at the level of transfer of
immunoglobulin, immunocytes, and regulatory factors through the
colostrum and milk (Ogra et al., 1977). It is noteworthy that
neonates which are breast-fed have higher levels of secretory
IgA than bottlefed infants (Chandra, 1978). This may be due
to a dietary factor that favors development of IgA immunity.

EFFECTS OF MODERATE MALNUTRITION AND CONTROLLED DIETARY

RESTRICTION

Studies of moderate malnutrition probably have more relevance to the population of the developed nations of the world.
Interestingly, moderate malnutrition often stimulates immune
functions, which are suppressed in severe malnutrition. In model
systems, such effects are often seen as stimulation of the response to mitogens. As with the suppressive effects of severe
malnutrition, the effects are most dramatic when studied in
young animals, but effects can be documented in adults as well.

The relationship between controlled dietary restriction
and immune response has been discussed in a number of recent reviews (Good et al., 1982; Fernandes and Good, 1981; Weindruch
and Makinodan, 1981). Some of the consequences of dietary restriction appear straightforward, and there is a general tendency
for low calorie diets to sustain immune functions in aging mice.
For example, the increase in resistance to the development of
spontaneous and transplanted tumors in protein-deficient animals
has been ascribed to the depression in synthesis of blocking
antibodies which shield tumor cells from cytotoxic T cells (Jose
and Good, 1971). In addition, increases in suppressor cell
populations have been linked to immunological senescence (Roder
et al., 1978). It is reasonable to speculate that certain suppressor cell populations are nutritionally sensitive.

MODEL SYSTEMS FOR STUDYING DIETARY RESTRICTION AND IMMUNITY

The complexity of the immune response and the interaction
with nutrition has prompted the extensive use of model systems.
Autoimmunity and malignancy generally increase as normal animals
age. Hence, premature development of these conditions can be
used as a model for immunological senescence. The model systems
most widely used are strains of mice which succumb to autoimmune
diseases and, in some cases, malignancy, in accordance with a
genetic program. Fernandes and Good (1981) have summarized data
from four model systems, and the general pattern is that increased longevity, with delayed onset of autoimmunity, results
from restricted dietary intake (generally low calorie, low protein intake). The $F_1$ of the NZB x NZW cross--the "B/W" mouse--
has an abnormally short lifespan and is subject to a variety of
autoimmune diseases. Low calorie diets markedly increased the
longevity of these mice (Good, 1977), delaying the onset of autoimmune hemolytic anemia and immune complex glomerulonephritis.
The protein content of the diet has a minor influence on longevity. Evidence has been presented for a nutritionally sensi-

tive suppressor cell population in these mice (Fernandes, et al, 1978). In contrast, a positive effect of a protein restriction on immune function has been noted in NZB mice, which have a genetically programmed decline in T lymphocyte function and a tendency to develop a lethal autoimmune hemolytic anemia (Fernandes et al., 1979b). In C3H mice, both the total caloric intake and the fat content of the diet are crucial in determining the development of mammary adenocarcinoma. A low fat-low calorie diet offers maximal protection against tumorigenesis. The complexity of such model systems is underscored by the lack of precise correlation between autoimmunity and aging. For example, a low protein diet results in maintenance of T cell function and delayed onset of hemolytic anemia, but does not significantly extend the lifetime or or prevent the development of autoimmune renal disease in NZB mice. Although the mechanisms of the NZB syndrome are not known, it is conceivable that protein restriction inhibits the expansion of some clones of aberrant T cells responsible for autoimmunity. The balance between T and B cell immunity can be varied by manipulating the duration of protein deficiency (Good, 1977). The relevance of these models to normal animals is indicated by the observation that caloric restriction of mouse strains with normal lifespan increases the percentage which survive to old age and preserves a number of immune functions which otherwise decline with age. These studies present a picture of impaired immune response resulting from "overnutrition," which is a significant problem in the U. S. and other countries.

In some studies, caloric restriction was effective even when commenced in adulthood (Weindruch and Walford, 1982), but some immunological correlates of aging can be altered only if restriction is imposed on young animals (Nandy, 1981). Since some of the effects of controlled caloric intake or severe nutritional deficiency are most dramatic when experienced at an early age, the developmental aspects of immunity are probably most nutritionally sensitive.

IMMUNOTOXICOLOGY

Several recent reviews have covered various aspects of immunotoxicology, especially the effects of environmental contaminants on immune response (Faith et al., 1980; Sharma and Zeeman, 1980; Caren, 1981; Sharma, 1981; and Davies, 1981). Immunosuppressive effects of toxicants (Vos, 1977), food-borne immunotoxicants (Archer, 1978), and developmental aspects of immunotoxicology (Roberts and Chapman, 1981) have been discussed in more specialized review articles. Briefly, chemicals in the food supply can inhibit normal development of antibody-producing cells (B cells), and adversely affect the cellular immune response which is dependent primarily on T cells and adherent cells. Many of the toxic effects on the immune system are observed before more general pathological conditions develop (Blakley and Archer, 1981). (Another class of immunotoxic responses are allergic reactions to food components, involving IgE antibodies or delayed-type hypersensitivity, will not be discussed here).

The developing immune system is especially sensitive to toxic substances. This process has been studied by fetal or neonatal exposure to toxicants (Ways et al., 1980; Luster et al., 1979). Spyker (1975) noted that some immunotoxic effects may remain latent until the animal is old. These delayed effects may be related to immunological senescence. Holt and Keast (1977) hypothesized that toxic substances can accelerate immunological senescence. These authors noted that immunotoxicants sometimes produce biphasic responses--first stimulation and then suppression. They also noted that local immunosuppression (e.g., in the respiratory tract) could occur even in the presence of systemic stimulation of immunity. Chang et al. (1981) found that altered lymphocyte populations are found in the peripheral blood of patients who have been exposed to polychlorinated biphenyls. The range of apparently immunotoxic compounds includes steroid hormones and synthetic estrogens (Ways et al., 1980), fungal toxins (Rosenstein et al., 1979), and oxidized lipids (Humphries and McConnell, 1979). All of these compounds are found in the food supply, but their significance to immune function and human health remains to be determined.

Effects on immune response may remain unnoticed until the organism is exposed to stress such as infection. It seems likely that traditional toxicology protocols might overlook these subtle changes in the immune system. Selected subpopulations of cells may be involved, and effects may not be observable until an appropriate response is elicited. Thus, the timing of exposure and response evaluation may be critical. Model studies (Speirs and Speirs, 1979) indicate that some immune responses-- such as enhancement of IgE responses to antigenic challenge--may be characteristic of chronic immunotoxicity. Verification of this hypothesis would provide an approach to routine immunotoxicological screening of humans.

Sex hormones. Roberts and Chapman (1981) emphasize the importance of studying both male and female animals, since there appears to be an interaction between sex and immunocompetence. This interaction may explain some of the immunotoxic effects of diethylstilbesterol and other reproductive hormones administered to neonatal mice (Ways et al., 1980). Notably, the immunotoxic effects were latent for a considerable period of time and were observed only in some assays of immune response. The general immunosuppression observed in adult mice exposed neonatally to hormones was preceded by an earlier mild elevation of immune response. Most of the effects were subtle, but diethylstilbestrol caused substantially reduced responses in several assays. These authors suggest that the eventual appearance of genital tract and mammary gland lesions following neonatal exposure to reproductive hormones may result from impaired immune surveillance.

Phorbol esters. Phorbol esters are an interesting class of compounds which are synthesized by plants of the Euphorbiaceae family. (See also the article by Berry and Helms in this volume.) They promote skin tumors and also affect the immune system. They stimulate some subpopulations of human T-lymphocytes to divide (Kaplan et al., 1982) and alter the activity of natural killer cells (Kolb et al., 1981). The effect may be adjuvant-like, since the phorbol esters have a macrophage-replacing activity in some culture systems (Koretzky et al., 1982). Vitamin A derivatives appear to inhibit the mitogenic activity of phorbols (Skinnider and Giesbrecht, 1981), but it is not clear whether this is related to the immunopharmacological effects of retinoids discussed above.

Other immunotoxicants. The list of potentially immunotoxic substances is large (see reviews cited above), and includes diverse compounds--heavy metals such as lead (Blakley and Archer, 1981) and fungal alkaloids such as trichothecene alkaloids (Nasuda et al., 1982). The obvious questions raised are defining the range of compounds in our diet which are immunotoxic and the quantitation of the health risk associated with toxicant consumption. The lack of correlation between some in vivo and in vitro studies (e.g., Blakley and Archer, 1982) indicate that risk will be difficult to assess.


CONCLUSIONS

Based primarily on immunological studies in model systems, dietary recommendations, though tentative, should include moderation of caloric and lipid intake. The recommended dietary intakes of trace elements and antioxidants, for selected populations such as pregnant women, should be reexamined with reference to the relevant literature in immunology. The nutritional needs of the aged also require further research, since the elderly may experience nutritionally sensitive immunological impairment Chandra et al., 1982).

In contrast to the tentative nature of dietary recommendations, several areas for further research are well defined. Since the developing immune system appears to have enhanced sensitivity to nutritional impairment, research focussing on developmental immunology is most likely to reveal important mechanisms. The mouse provides a useful model system, since the neonatal mouse is in a state of immunological development similar to the human fetus at six months of gestation. The development of new model systems, both in vivo and in vitro, would be useful. The mechanisms of zinc and antioxidant effects on immune response and the detailed consequences of various lipid intakes are two broad areas for further study. In the case of lipids, attention should be given to the role of fats as precursors to prostaglandins and the influence of prostaglandins on immune responses. Hormones, such as corticosteroids, influence immune responses and the relationship of nutrient and toxicant intake to hormone levels and immunological senescence should be investigated. The thresholds of nutritional and toxicological influences on immune response need definition to assess the relevance of experimental findings to human populations. The use of immunological criteria in assessing nutritional status has been described (Miller, 1978), but further research is required to develop reliable immunological methods for epidemiological studies. These should rely on cells and proteins of the blood, since this tissue is relatively easy to sample.

Inflammatory reactions can result in production of mutagens (Weitzman and Stossel, 1982) and antibodies against dietary antigens are found in inflammatory bowel disease (Haeney et al., 1982). These observations indicate that food and immunity can interact in a dramatic way. The relationship between immune reactions in the gut, colon cancer, and diet may well be very significant to human health. It seems possible that the immune system could influence gut flora and their production of mutagens. Therefore, the relationships among dietary choices, gut immunity, and immunopathology should be studied further.

In summary, the interactions of diet and immunity occur on many levels. Further research in this area is required to help insure a healthful diet.

REFERENCES

Ammann, A.J. and Hong, R. (1971). Selective IgA deficiency: presentation of 30 cases and a review of the literature. Medicine 50:223-236.

Anderson, R., Oosthuizen, R., Maritz, R., Theron, A., Van Rensburg, A. J. (1980) Am. J. Clin. Nutr. 33:71-76.

Andreotti, P. E. (1982). Phorbol ester tumor promoter modulation of alloantigen specific T lymphocyte responses. J. Immunol. 129:91-96.

Archer, D.L. (1978). Immunotoxicology of foodborne substances: An overview. J. Food Protect. 41:983-988.

Atukorala, S., Basu, T., Dickerson, J., Donaldson, D. and Sakura, A. (1979). Vitamin A, zinc, and lung cancer. Br. J. Cancer 40:927-931.

Bach, J. F. (1981). The multi-faceted zinc dependency of the immune system. Immunol. Today 2:225-227.

Barnett, J. B. and Bryant, R. L., (1980). Adjuvant and immunosuppressive effects of retinol and Tween 80 on immunoglobulin G production in mice. Int. Arch. Allergy Appl. Immunol. 63:145-152.

Barratt, M.E.J., Strachan, P.J., and Porter, P. (1979). Immunologically mediated nutritional disturbances associated with soya protein antigens. Proc. Nutr. Soc. 38:143-150.

Bartnik, W. and Shorter, R.G. (1981). The immunology of inflammatory bowel disease. Materia Medica Polonia. 12:77-83.

Baumgartner, W.A. (1979). Antioxidants, cancer, and the immune response, in Kharasch, N. (ed.) "Trace Metals in Health and Disease (New York: Raven Press) 287-305.

Beisel, W.R. (1982). Single nutrients and immunity. Amer. J. Clin. Nutr. 35:417-468.

Beisel, W.R., and Fiser, R.H., Jr. (1970). Lipid metabolism during infectious illness. Am. J. Clin. Nutr. 23:1069-1970.

Berger, N.A., and Skinner, A.M. (1974). Characterization of lymphocyte transformation induced by zinc ions. J. Cell Biol. 61:45-48.

Bistrian, B.R., Blackburn, G.L., Scrimshaw, N.S., and Flatt, J.P. (1975). Cellular immunity in semi-starved states in hospitalized adults. Am. J. Clin. Nutr. 28:1148-1155.

Blakley, B.R., and Archer, D.L. (1981). Effect of lead acetate on the immune response in mice. Toxicol. Appl. Pharmacol. 61:18-26.

Blakley, B.R., and Archer, D.L. (1982). Mitogen stimulation of lymphocytes exposed to lead. Toxicol. Appl. Pharmacol. 62: 183-189.

Boxer, L.A., Watanabe, A.M., Rister, M., Besch, H.R., Allen, J., and Baehner, R.L. (1976). Correction of leukocyte function in Chediak-Higashi syndrome by ascorbate. New Engl. J. Med. 295:1041-1045.

Brummerstedt, E., Basse, A., Flagstad, T, and Andresen, E. (1977). Animal model of human disease: acrodermatitis enteropathica, zinc malabsorption. Am. J. Pathol. 87:725-728.

Caren, L.D. (1981). Environmental pollutants: effects on the immune system and resistance to infectious disease. Bio-Science 31:582-586.

Chandra, R.K. (1974). Immunocompetence in low birth weight infants after intrauterine malnutrition. Lancet ii: 1393-1394.

Chandra, R.K. (1975). Fetal malnutrition and postnatal immuno-competence. Am. J. Dis. Child. 129:450-454.

Chandra, R.K. (1977). Cell mediated immunity in fetally and post-natally malnourished children from India and Newfoundland, in R.M. Suskind (ed.) Malnutrition and the Immune Response (New York: Raven Press) 111-115.

Chandra, R.K. (1978). Immunological aspects of human milk. Nutr. Rev. 36:265-272.

Chandra, R. K., Joshi, P., Au, B., Woodford, G., and Chandra, S. (1982). Nutrition and immunocompetence of the elderly: effect of short-term nutritional supplementation on cell-mediated immunity and lymphocyte subsets. Nutr. Res. 2:223-232.

Chandra, R. K. and Newberne, P. M. (1977). Nutrition, Immunity and Infection: Mechanisms of Interactions (New York: Plenum Press) 246 pp.

Chang, K-J., Hsieh K-H., Lee, T-P., Tang, S-Y., and Tung, T-C. (1981). Immunologic evaluation of patients with polychlorinated biphenyl poisoning. Determination of lymphocyte subpopulations. Toxicol. Appl. Pharmacol. 61:58-63.

Chevalier, P. and Aschkenasy, A. (1977). Hematological and immunological effects of excess dietary leucine in the young rat. Am. J. Clin. Nutr. 30:1645-1654.

Corwin, L. M., and Gordon, R. K. (1982). Vitamin E and immune regulation. Ann. N. Y. Acad. Sci. 393:437-51.

Corwin, L. M., Gordon, R. K., and Shloss, J. (1981). Studies of the mode of action of vitamin E in stimulating T cell mitogenesis. Scand. J. Immunol. 14:565-572.

Dardenne, M., Pléau, J.-M., Nabarra, B., Lefrancier, P., Derrien, M., Choay, J., and Bach, J.-F. (1982). Contribution of zinc and other metals to the biological activity of the serum thymic factor. Proc. Natl. Acad. Sci. USA 79:5370-5373.

Davies, D.F. (1976). Immunological aspects of atherosclerosis. Proc. Nutr. Soc. 35:293-295.

Davies, G.E. (1981). Toxicology of the immune system. Histo-Chem. J. 13:879-884.

DePasquale-Jardieu, P., and Fraker, P. J. (1979). The role of corticosterone in the loss in immune function in the zinc-deficient A/J mouse. J. Nutr. 109:1847-1855.

Dobbins, W. O. (1982). Gut immunophysiology. A gastroenterolo-
    gists view with emphasis on pathophysiology. Am. J. Physiol.
    242:G1-G8.
Edelman, R. (1981). Obesity: does it modulate infectious disease
    and immunity? in Selvey, N., and White, P.L. (eds.) Nutri-
    tion in the 1980's: Constraints on Our Knowledge (New York:
    Alan R. Liss, Inc.) 327-338.
Edgington, T. S., and Curtiss, L. K. (1981). Plasma lipoproteins
    with bioregulatory properties including the capacity to regu-
    late lymphocyte function and the immune response. Cancer
    Res: 41:3786-3788.
Faden, H., and Ogra, P.L. (1981). Breast milk as an immunologic
    vehicle for transport of immunocompetence, in Lebenthal, E.
    (ed.) Textbook of Gastroenterology and Nutrition in Infancy.
    (New York: Raven Press) 355-361.
Faith, R. E., Luster, M. I., and Vos, J. G. (1980). Effects on
    immunocompetence by chemicals of environmental concern. Re-
    views in Biochem. Toxicol. 2:173-212.
Falchuk, K. R., Walker, W. A., Perrotto, J. L., and Isselbacher,
    K. J. (1977). Effect of vitamin A on the systemic and
    local antibody responses to intragastrically administered
    bovine serum albumin. Infect. Immunity, 17:361-365.
Faulk, W.P., Paes, R.P., and Marigo, C. (1977). The Immuno-
    logical system in health and malnutrition. Proc. Nutr.
    Soc. 35:253-261.
Fernandes, G., Friend, P., Yunis, E.J., and Good, R.A. (1978).
    Influence of dietary restriction on immunologic function
    and renal disease in (NZB X NZW)F$_1$ mice. Proc. Natl.
    Acad. Sci. USA 75:1500-1504.
Fernandes, G., and Good, R.A. (1981). Immunological basis for
    the prolongation of life by nutritional manipulation. In
    Boers, R.E., Jr., and Bassett, E.G. (eds.) Nutritional
    Factors: Modulating Effects on Metabolic Processes, (New
    York: Raven Press) 437-461.
Fernandes, G., Nair, M., Onoe, K., Tanaka, T., Floyd, R., and Good,
    R.A. (1979a). Impairment of cell mediated immunity functions
    by dietary zinc deficiency in mice. Proc. Natl. Acad. Sci.
    USA 76:457-461.
Fernandes, G., West, A., and Good, R. A. (1979b). Nutrition, im-
    munity and cancer - a review, Part III: Effects of diet on
    the diseases of aging. Clin. Bull. 9:91-106.
Fraker, P. J., Haas, S. M., Luecke, R. W. (1977). Effect of zinc
    deficiency on the immune response of the young adult A/J
    mouse. J. Nutr. 107:1889-1895.
Fraker, P. J., DePasquale-Jardieu, P., Zwickl, C. M., and Lueke,
    R. W. (1978). Regeneration of T-cell helper function in
    zinc-deficient adult mice. Proc. Natl. Acad. Sci. USA 75:
    5660-5664.

Gebhardt, B.M., and Newberne, P.M. (1974). Nutrition and immuno-
    logical responsiveness: T-cell function in the offspring of
    lipotrope and protein-deficient rats. Immunol. 26:489-495.
Goetzl, E.J., Wasserman, S.I., Gigli, I., and Austen, K.F. (1974).
    Enhancement of random migration and chemotactic response of
    human leukocytes by ascorbic acid. J. Clin. Invest. 53:813-
    818.
Good, R. A. (1977).  Influence of protein or protein calorie mal-
    nutrition and zinc deficiency on immunity. Clin. Bull. 9:
    3-12.
Good, R.A., Fernandez, G., and Day, N.K. (1982). The influence of
    nutrition on development of cancer immunity and resistance to
    mesenchymal Diseases in Annott, M.S., Van Eys, J., and Wang,
    Y.M. (eds.) Molecular Interrelations of Nutrition and Cancer
    (New York: Raven Press) 73-90.
Haeney, M. R., Goodwin, B. J. F., Barratt, M. E. J., Mike, N., and
    Asquith, P.  (1982).  Soy protein antibodies in man: their
    occurrence and possible relevance in celiac disease. J.
    Clin. Pathol. 35:319-322
Hallberg, D., Nilsson, B. S., and Baekman, L. (1976). Immuno-
    logical function in patients operated on with small intes-
    tinal shunts for morbid obesity. Scand. J. Gastroenterol.
    11:41-48.
Harman, D., Heidrick, M. L. and Eddy, D. E., (1977).  Free radical
    theory of aging: effects of free radical reaction inhibitors
    on the immune response. J. Am. Geriat. Soc. 9:400-407.
Hill, C. H. (1982).  Interaction of dietary amino acids with the
    immune response. Fed. Proc. 41:2818-2820.
Hillyard, L. A., and Abraham, S. (1979).  Effect of dietary
    polyunsaturated fatty acids on growth of mammary adeno-
    carcinomas in mice and rats. Cancer Res. 39:4430-4437.
Hoffeld, J. T. (1981).  Agents which block membrane lipid per-
    oxidation enhance mouse spleen cell immune activities. In
    vitro relationship to the enhancing activity of 2-mercapto-
    ethanol. Eur. J. Immunol. 11:371-376.
Holt, P. G., and Keast, D. (1977) Environmentally induced changes
    in immunological function: acute and chronic effects of
    inhalation of tobacco smoke and other atmospheric contami-
    nants in man and experimental animals. Bacteriol. Rev.
    41:205-216.
Humphries, G. M. K. and McConnell, H. M.  (1979).  Potent immuno-
    suppression by oxidized cholesterol. J. Immunol. 122: 121-
    126.
Hurley, L. (1968). Approaches to the study of nutrition in mammal-
    ian development. Fed. Proc. 27:193-198.
Johansson, S.G.O., Mellbin, T., and Vahlquist, B.O. (1968).  Im-
    munogloblin levels in Ethiopian preschool children with spe-
    cial reference to high concentrations of immunoglobulin E.
    Lancet i:118-1121.

Jose, D. G., and Good, R. A. (1971). Absence of enhancing anti-
    body in cell-mediated immunity to tumor heterografts in
    protein-deficient rats. Nature 231:323-325.
Kaplan, J. H., Lee, L.S., and Lockwood, H. (1982). Identifica-
    tion by monoclonal antibodies of T cell subpopulations
    activated by 12-0 tetradecanoylphorbol-13-acetate. Immunol.
    Lett. 4:269-274.
Keller, R., Keist R., Adolf, W., Opferkuch, H. J., Schmidt, R.,
    and Hecker, E. (1982). Tumor promoting diterpene esters
    prevent macrophage activation and suppress macrophage tumori-
    cidal capacity. Exp. Cell Biol. 50: 121-134.
Kikkawa, Y., Kamimura, K., Hamajima, T., Sakiguchi, T., and Kawai,
    T. (1973). Thymic alymphoplasia with hyper-IgE-globulinemia.
    Pediatrics 51:690-696.
Kolb, J.P.B., Senik, A., and Castagna, M. (1981). In vitro
    stimulation of mouse natural killer cell activity by phorbol
    esters. Cell. Immunol. 65: 258-271.
Koretzky, G.A., Daniele, R.P., and Nowell, P.C. (1982). A phor-
    bol ester 12-0-tetradecanoylphorbol-13-acetate can replace
    macrophages in human lymphocytes cultures stimulated with a
    mitogen but not with an antigen. J. Immunol. 128:1776-1780.
Kos, W. L., Loria, R. M., Snograss, M. J., Cohen, D., Thorpe, T.
    G., and Kaplan, A. M. (1979). Inhibition of host resistance
    by nutritional hypercholesterolemia. Infect. Immun. 26:658-
    667.
Kulapongs, P., Vithayasai, V., and Suskind, R. and Olson, R.E.
    (1974). Cell-mediated immunity and phagocytosis and killing
    function in children with severe iron deficiency anemia.
    Lancet ii, 689-691.
Kumar, M., and Axelrod, A.E. (1978). Cellular antibody synthesis
    in thiamin, riboflavin, biotin and folic acid-deficient rats.
    Proc. Soc. Exp. Biol. Med. 157:421-423.
Leibovitz, B., and Siegel, B. V. (1981). Ascorbic acid and the
    immune response, in Phillips, M. and Baetz, A. (eds.) Diet
    and Resistance to Disease (New York: Plenum Press) 1-25.
Lichtenstein, L. M., Gillespie, E., Bourne, H.R., and Henney, C.
    S. (1972). The effects of a series of prostaglandins on
    in vitro models of the allergic response and cellular im-
    munity. Prostaglandins 2:519-528.
Lim, T. S., Putt, N., Safranski, D., Chung, C., and Watson, R.
    R. (1981). Effect of vitamin E on cell mediated immune
    response and serum corticosterone in young and maturing
    mice. Immunol. 44:289-296.
Luster, M. I., Faith, R. E., McLachlan, J. A., and Clark, G. C.
    (1979). Effect of in utero exposure to diethylstilbesterol
    on the immune response in mice. Toxicol. Appl. Pharmacol.
    47:279-285.

MacDougall, L. G., Anderson, R., McNab, G. M., and Katz, J. (1975). The immune response in iron-deficient children: impaired cellular defense mechanisms with altered humoral components. J. Pediatr. 86:833-843.

Makinodan, T. and Kay, M. M. B. (1980). Age influence on the immune system. Adv. Immunol. 29:287-330.

McFarlane, H. (1977). Cell-mediated immunity in clinical and experimental protein-calorie malnutrition, in Suskind, R.M. (ed.) Malnutrition and The Immune Response (New York: Raven Press) 127-133.

McMurray, D. N., Rey, H., Casazza, L. J., and Watson, R. R., (1977). Effect of moderate malnutrition on concentrations of immunoglobulins and enzymes in tears and saliva of young Colombian children. Am. J. Clin. Nutr. 30:1944-48.

Meade, C. J., Sheena, J., and Mertin, J. (1979). Effects of the obese (ob/ob) genotype on spleen cell immune function. Int. Arch. Allergy Appl. Immunol. 58:121-127.

Medawar, P. B. and Hunt R., (1981). Anti-cancer action of retinoids. Immunol. 42:349-353.

Miller, Carol, L. (1978). Immunological assays as measurements of nutritional status: A review. J. Parent. Ent. Nutr. 2:554-566.

Munger, W. E. and Lindquist, R. R., (1982). Effect of phorbol esters on alloimmune cytolysis. Cancer Res. 42:5023-5029.

Nalder, B. N., Mahoney, A. W., Ramakrishnan, R., and Hendricks, D. G. (1972). Sensitivity of the immunological response to the nutritional status of rats. J. Nutr. 102:535-542.

Nandy, K., (1981). Effects of caloric restriction on brain-reactive antibodies in sera of old mice. Age 4:117-121.

Nasuda, E., Takemota, T., Tatsuno, T., Obara, T. (1982). Immunosuppressive effect of a trichothecene mycotoxin fusarenone x in mice. Immunol. 45:743-751.

Nauss, K. M., Mark, D. A., and Suskind, R. M., (1979). The effect of vitamin A deficiency on the in vitro cellular immune response of rats. J. Nutr. 109:1815-1823.

Newberne, P. M., (1977). Effect of folic acid, B$_{12}$, choline, and methionine on immunocompetence and cell-mediated immunity, in Suskind, R.M. (ed.) Malnutrition and the Immune Response (New York:Raven Press) 375-386.

Newberne, P. M. (1966). Effects of overnutrition on resistance of dogs to distemper virus. Fed. Proc. 25:1701-1710.

Ogra, S.S., Weintraub, D., and Ogra, P.L. (1977). Immunologic aspects of human colostrum and milk. 3. Fate and absorption of cellular and soluble components in the gastrointestinal tract of the newborn. J. Immunol. 119:245-248.

Olsen, G.E., and Polk, H.C. (1977). In vitro effect of ascorbic acid on corticosteroid-caused neutrophil dysfunction. J. Surgical Res. 22:109-112.

Panda, B. and Combs, G. F. (1963). Impaired antibody production in chicks fed diets low in vitamin A, pantothenic acid, or riboflavin. Proc. Soc. Exp. Biol. Med. 113:530-534.

Pearson, H. A., and Robinson, J. E. (1976). The role of iron in host resistance. Adv. Pediatr. 23:1-33.

Pekarek, R. S., Hoagland, A. M., and Powanda, M. C. (1977). Humoral and cellular immune responses in zinc deficient rats. Nutr. Rep. Int. 16:267-276.

Ralph, P., and Kishimoto, T. (1982). Tumor promoter phorbol myristic acetate is a T cell dependent inducer of immuno-globulin secretion in human lymphocytes. Clin. Immuno. Immunopathol. 22:340-348.

Richman, L. K., Chiller, J. M., Brown, W. R., Hansen, D. G., and Vas, N. M. (1978). Enterically induced immunologic toler-ance, I. Induction of suppressor T lymphocytes by intra-gastric administration of soluble proteins. J. Immunol. 121:2429-2434.

Roberts, D. W., and Chapman, J. R. (1981). Concepts essential to the assessment of toxicity to the developing immune system, in Kimmel, C. A. and Buelke-Sam, J. (eds.) Developmental Toxicology (New York: Raven Press) 167-190.

Robson, L.C., and Schwartz, M. R. (1975). Vitamin B$_6$ defic-iency and the lymphoid system. II. Effects of vitamin B$_6$ in utero on the immunological competence of the offspring. Cell Immunol. 16:145-152.

Robson, L. C., and Schwarz, M. R. (1980). The effects of vitamin B$_6$ deficiency on the lymphoid system and immune responses. in Tryfiates, G.P. (ed.) Vitamin B$_6$ Metabolism and Role in Growth (Westport, CT: Food and Nutr. Press, Inc.) 205-222.

Roder, J. C., Duwe, A. K., Bell, D. A., and Singhal, S. K. (1978). The role of suppressor cells. Immunol. 35:837-847.

Rosenstein, Y., LaFarge-Frayssinet, C., Lespinats, G., Loisillier, LaFont, P., and Frayssinet, C. (1979). Immunosuppressive activity of Fusarium toxins. Effects on antibody synthesis and skin grafts of crude extracts. T$_2$-toxin and diacetoxy-scirpenol. Immunol. 36:111-118.

Sawitsky, B., Kanter, R., and Sawitsky, A. (1976). Lymphocyte re-sponse to phytomitogens in iron deficiency. Am. J. Med. Sci. 272:153-162.

Schloen, L. H., Fernandes, G., Garofalo, J. A., and Good, R. A. (1979). Nutrition, Immunity, and Cancer: A review, 2. Zinc immune function and cancer. Clin. Bull. 9:63-81.

Seifter, E., Rettura, G., and Levenson, S.M., (1981). Carotenoids and cell mediated immune responses in Charalambous, G. (ed.) Quality of Foods and Beverages: Chemistry and Technology (New York: Academic Press) 335-347.

Sharma, R. P. (1981). Immunologic considerations in toxicology. V. 1 and 2. (Boca Raton: CRC Press) 174 pp. and 176 pp.

Sharma, R. P., and Zeeman, M.G. (1980). Immunologic alterations by environmental chemicals: relevance of studying mechanisms versus effects. J. Immunopharmacol. 2:285-307

Sheffy, B. E., and Schultz, R. D. (1979) Influence of vitamin E and selenium on immune response mechanisms. Fed. Proc. 38: 2139-43.

Siegel, B. V., and Morton, J. I. (1977). Vitamin C and the immune response. Experientia 33:393-395.

Sirisinha, S., Suskind, R.M., Edelman, R., Asvapaka, C., and Olson, R.E. (1977). Secretory IgA in Thai children with protein-calorie malnutrition in Suskin, R.M. (ed.) Malnutrition and the Immune Response (New York: Raven Press) 195-199.

Skinnider, L. F., Giesbrecht, K. (1981). Inhibition of phorbol myristate acetate and phytohemagglutinin stimulation of human lymphocytes by 13-cis-retinoic acid and ethyletrinoate. Experientia 37:1345-1346.

Soothill, J. F. (1976). Allergy and infant feeding, Proc. Nutr. Soc. 35:283-284.

Spallholz, J. E. (1981). Anti-inflammatory, immunologic and carcinostatic attributes of selenium in experimental animals. in Phillips, M. and Baetz, A. (eds.) Diet and Resistance to Disease (New York: Plenum Press) 43-62.

Speirs, R. S., and Speirs, E. E. (1979). An in vivo model for assessing effects of drugs and toxicants. Drug Chem. Toxicol. 2:19-33.

Spyker, J. M. (1975). Assessing the impact of low level chemicals on development: behavioral and latent effects. Fed. Proc. 34:1835-1844.

Strauss, R. G. (1978). Iron deficiency, infections, and immune function: a reassessment. Am. J. Clin. Nutr. 31:660-666.

Strober, W., Krakauer, R., Klaeveman, H., Reynolds, H. Y., and Nelson, D. L. (1976). Secretory component deficiency: a disorder of the IgA immune system. New Engl. J. Med. 294: 351-356.

Suskind, R. M. (ed.) (1977). Malnutrition and the Immune Response. (New York: Raven Press) 455 pp.

Tanaka, J., Fujiwara, H., and Torisu, M. (1979). Vitamin E and immune response: I. Enhancement of helper T cell activity by dietary supplementation of vitamin E in mice. Immunol. 38: 727-34.

Tengerdy, R.P., Mathias, M. M., and Nockels, C. F. (1981). Vitamin E immunity, and disease resistance, in Phillips, M. and Baetz, A. (eds.) Diet and Resistance to Disease (New York: Plenum Press) 27-42.

Uhr, J. W. Weissmann, G., and Thomas, L. (1963). Acute hypervitaminosis A in guinea pigs, II. Effects on delayed-type hypersensitivity Proc. Soc. Exp. Biol. Med. 112:287-291.

Vaz, N. M., Maia, L. C. S., Hanson, D. G., and Lynch, J. M. (1977). Inhibition of homocytotropic antibody responses in adult inbred mice by previous feeding of the specific antigen. J. Allergy Clin. Immunol. 60:110-115.

Vitale, J. J., and Broitman, S. A., (1979). Lipids and immune function. Cancer Res. 41:3706-3710

Vos, J. G. (1977). Immune suppression as related to toxicology. CRC Crit. Rev. Toxicol. 5:67-101.

Walker, W.A., Isselbacher, K.J., and Bloch, K.J. (1974) Immunologic control of soluble protein absorption from the small intestine. A gut surface phenomenon. Am. J. Clin. Nutr. 27:1434-1440

Ways, S. C., Blair, P. B., Bern, H. A., and Staskawicz, M. O. (1980). Immune responsiveness of adult mice exposed neonatally to diethylstilbestrol, steroid hormones, or vitamin A. J. Environ. Pathol. Toxicol. 3: 207-20.

Weigle, W. O., Morgan, E. L., Goodman, M. G., Chenoweth, D. E., Hugli, T. E. (1982). Modulation of the immune response by anaphylatoxin in the microenvironment of the interacting cells. Fed. Proc. 41:3099-3103.

Weindruch, R., and Walford, R. L. (1982). Dietary restriction in mice beginning at one year of age: effect on life span and spontaneous cancer incidence. Science 215:1415-1416.

Weindruch, R. H., and Makinodan, T. (1981). Dietary restriction and its effects on immunity and aging, in Selvey, N. and White, P.L. (eds.) Nutrition in the 1980's: Constraints on our Knowledge (New York: Alan R. Liss, Inc.) 319-326.

Weisman, K., and Flagstad, T. (1976). Hereditary zinc deficency (adema disease) in cattle. Acta Derm. Venereol. 56:151-154.

Weitzman, S. A., and Stossel, T. P. (1982). Effect of oxygen radical scavengers and antioxidants on phagocyte induced mutagenesis. J. Immunol. 128:2770-2772.

Williams, E. A. J., Gebhardt, B. M., Yee, H., Newberne, P. M. (1978). Immunological consequences of choline-vitamin $B_{12}$ deprivation. Nutr. Rep. Int. 17:279-292.

Wright, R. (1976). Immunology of food intolerance in adults. Proc. Nutr. Soc. 35:285-292.

## 4

ROLE OF EPIGENETIC FACTORS IN DIETARY CARCINOGENESIS

David L. Berry[1] and C. Tucker Helmes[2]

[1]United States Department of Agriculture, WRRC,
Toxicology & Biological Evaluation Research Unit
800 Buchanan Street, Berkeley, CA   94710

[2]Biological and Environmental Chemistry Department
Life Sciences Division, SRI International
333 Ravenswood Avenue, Menlo Park, CA   94025

Sixty to eighty percent of all human cancers are associated with our environment and environmental agents. The environment is a complex mixture of chemicals that originate from industrial and technological development and, most importantly, natural sources. Many industrial chemicals, pesticides, insecticides and food additives have exhibited carcinogenic properties in various animal model systems, and occupational and drug exposure has led to human cancers. However, estimates attribute the majority of human cancers associated with the environment not to intentional or accidental chemical exposure, but to the vast number of naturally occurring chemicals in our environment (Doll and Peto, 1981).

Carcinogenesis encompasses many different diseases, and the exact cause of a lesion is a multifactorial process including nature of exposure, duration, age, route of exposure and numerous environmental factors. Data from animal models indicate that carcinogenesis is a multi-step process that involves genetic and epigenetic events. Boutwell (1974) proposed that complete carcinogenesis could be divided into initiation and promotion. Initiation is defined as an irreversible genetic event (mutational), and promotion is a time-and-dose-dependent process which is, to a certain degree, reversible. The genotoxic events could be a direct change in genetic material through attack by radiation or chemicals. The genotoxic change could result from a direct mutational lesion, faulty error-prone DNA repair, DNA polymerase error, or infidelity of DNA synthesis. The promotion process is termed "epigenetic" since no direct interaction of the promoter with the genetic material has been detected. The process involves conversion of the damaged cells into a dormant tumor cell followed by propagation of that cell into a viable growing tumor. The epigenetic events

result in selective cell growth and loss of endogenous and exogenous cell growth controls that eventually lead to an increased
tumor yield and a decreased latency period. Compounds that are
tumor promoters are a class of cocarcinogens which are not in
themselves mutagenic or carcinogenic, but which cause tumors when
applied after an irreversible initiation event has occurred. Since
promotion requires repeated application with a promoter and is a
reversible process (Boutwell, 1974, Diamond et al., 1980), an understanding of the process is essential in order to develop methods of
controlling the incidence of cancer in humans.

Dietary factors may account for up to thirty-five percent
of the total environmentally related cancers and there is limited
information on specific dietary causative agents. Diets are a
complex mixture of plant and animal matter with geographical and
ethnic variations. Considering that the latency period for tumor
development may extend for twenty years and that we are a mobile
population, specific dietary causes and effects are difficult to
assess. The National Academy of Sciences report on Diet, Nutrition
and Cancer (1982) provides a combination of human epidemiological
investigations with experimental animal data to indicate dietary
influences on disease development. Reviews by Petering (1978),
Hirono (1979), Reddy et al. (1980), Doll and Peto (1981), Hirono
(1981), and Lower et al. (1982) have presented data both on the
role of diet in various animal tumor models and on epidemiological
human data. However, these reviews have covered all aspects of
dietary influences on disease development (i.e. complete carcinogens, mutagens, modifiers of carcinogenesis, classes of carcinogens,
and dietary trends). It is the aim of this chapter to relate
dietary constituents to their epigenetic or tumor-promoting activity
and to review the effects of these dietary constituents in animal
models and presumptive human models.

ANIMAL MODELS

Mammary and Respiratory

Enhancement of murine mammary tumorigenesis by elevated dietary
fat was first described by Tannenbaum (1942). Mice fed a diet containing 12% fat (hydrogenated cottonseed oil) at six times the control dietary fat level had a tumor incidence significantly above the
control animals. These initial investigations have been confirmed
and extended by many laboratories using both mice and rats (Welsch
and Aylsworth, 1983). In all of the investigations, the experimental protocol has been based on various levels of dietary fat and
the enhancement of a spontaneous mammary tumor or x-ray-induced or
chemically induced mammary tumor. The animals were fed high-fat
diets, which led to an increased mammary tumor incidence and/or a
reduction in the latency period of mammary tumor development when
compared to animals fed at a low or moderate fat level. The role

of high dietary fat in mammary tumor promotion appears to be applicable to most of the experimental rodent systems and does not appear to be influenced by caloric consumption (Carroll, 1981).

The nature of the fat in the diet influences the development of mammary tumors. High levels of saturated fat enhance spontaneous mammary tumor development and DMBA-initiated mammary tumors in rats (Rogers and Wetsel, 1981). The enhancement of tumor development is increased when levels of unsaturated or polyunsaturated fats are given in addition to saturated fats, indicating that unsaturated lipids play an important role in mammary tumor promotion. Investigations by Rao and Abrahams (1976) indicated that dietary lineolate enhanced transplantable tumor growth in C3H mice. These findings were extended by Hopkins et al. (1981) to rats, further demonstrating that either ethyl linoleate or fish oil (rich in linolenates) could serve as the source of unsaturated lipid. Moreover, their results indicated that as little as 3% polyunsaturated fat could enhance the tumor response when fed in a high-saturated fat diet. Carroll (1981) demonstrated that once the requirement for polyunsaturated fat was met, it did not matter whether the remaining fat was saturated or unsaturated. The exact role of linolenates in tumor development is not known, but it is perhaps related to the role of these acids in cell proliferation.

The mechanism of dietary fat enhancement of mammary tumorigenesis is controversial and is based on the question of hormonal influences in tumor development. Numerous reports have stated that serum prolactin levels have been elevated by high dietary fat (Williams et al., 1982), while other reports have indicated that no significant elevation of serum prolactin occurred under high dietary fat (Welsch and Aylsworth, 1983). Given the number of investigations and conflicting reports on serum hormone concentrations, there is as yet no clear relationship between high dietary fat levels and elevated prolactin levels. The possibility that hormone dependent tissues may become either sensitized or nonresponsive to endocrine secretion as a result of high dietary fat is also subject to conjecture and cannot be validated (Welsch and Aylsworth, 1983). Studies designed to demonstrate endocrine enhancement of mammary tumorigenesis have been inconclusive in the mouse and rat mammary tumor models, which indicates that non-endocrine or indirect mechanisms may be involved in promotion of rodent mammary tumors (Wagner et al., 1980; Welsch and Aylsworth, 1983). Investigations involving immunosuppression and peripheral lymphocyte populations (Wagner et al., 1980) and cell-to-cell communication (Welsch and Aylsworth, 1983) are limited and, although unsaturated fats alter immune suppression and metabolic cooperativity, the role of these processes in tumor promotion remains unknown. Perhaps the most important information regarding fats and promotion is that unsaturated fatty acids appear to directly stimulate cell growth and proliferation of mammary epithelium (Smith, H: Personal communication). Fatty acids can markedly change membrane composition which can result in structural changes, membrane fluidity, enzyme activity, cyclic nucleotide and prostaglandin biosynthesis, and transport processes. Many of the above events can stimulate cells to divide via a biochemical cascade. This area of research provides some of the more plausible mechanisms regarding dietary fat and its promoting mechanism.

Enhancement of respiratory tumors by dietary agents was initially described by Clapp et al., (1974) when they noted that mice treated with diethylnitrosamine followed by treatment with 0.75% butylated hydroxytoluene (BHT) for 16 months developed lung squamous cell carcinomas. Witschi and Lock (1978) extended the initial findings to the lung adenoma assay using urethan as an initiator and BHT as a promoting agent. BHT stimulated DNA synthesis in total pulmonary cells; the mechanism of the increase was due to regeneration of cells that were killed or damaged by BHT and therefore not a direct effect on cell proliferation. Recent data indicate that doses of BHT (400 mg/kg, ip) enhance adenoma formation in Swiss Webster mice and that type II alveolar cell proliferation can be dissociated from the promoting effect. The use of BHT to stimulate selective cell proliferation probably relates to a process of regenerative hyperplasia and not exclusively to cell proliferation. The doses of BHT used in the rodent lung squamous-cell carcinoma assay are many-fold in excess of normal human dietary consumption of BHT. The enhancement of tumors occurs only following treatment with a subcarcinogenic dose of a carcinogen and, therefore, is a promotional event. Much more data must be generated with this model before specific mechanisms can be determined and before other potential respiratory promoters can be identified.

## Esophagus and Forestomach

Esophageal tumors can be enhanced in rats by feeding various plant tannins (Morton, 1972). The tannins that appear to be most active are the hydrolyzed pyrogallol tannins and the condensed catechol tannins. In both cases, the tannins increase the tumor yield and decrease the latency period for tumor development. Tannins are normal secondary plant products and many are associated with insect resistance. The exact mechanism of tumor enhancement by tannins remains unknown, but tannins are irritants and allergens.

Extracts from pickled vegetables from Linxian County in China were recently described to be both mutagenic and carcinogenic to cells in culture (Cheng et al., 1980). Further investigations by the group indicated that Roussin's red, a compound isolated from pickled vegetables, was a potential promoting agent (Cheng et al., 1981). Roussin's red enhanced transformation of C3H 10T 1/2 cells in culture, was capable of inducing epidermal hyperplasia in mice, and promoted epidermal tumors in mice. Although no direct evidence was presented for esophageal promotion in animals, it was assumed by the authors that the human levels of esophageal tumors in Linxian could be partially attributed to the presence of Roussin's red in the pickled vegetables consumed by the general populous. Roussin's red is probably produced by fungi in the pickling process and is a strong irritant and hyperplastiogen.

Tumors of the forestomach can be promoted in mice, rats and hamsters. Using mice initiated with diethylnitrosamine and promotion with dietary BHT, Clapp et al. (1974) demonstrated that 0.75% BHT in the diet could enhance hyperplasia and squamous-cell carcinomas of the forestomach. The development of the disease progressed with time and was similar to that seen in the epidermis with the phorbol diesters (Boutwell, 1974). The fatty acid ester methyl oxo-octadecanoate was shown to have promoting properties in mouse forestomach initiated with 4-nitroquinoline-1-oxide (Kiaer et al., 1975). The same compound was shown to promote DMBA-initiated skin tumors in the same mouse strain. Using rats and hamsters, Goerttler et al. (1982) demonstrated that the skin tumor promoting phorbol diester 12-0-tetradecanoylphorbol-13-acetate (TPA) could promote hyperplasia, papillomas and squamous-cell carcinomas in the forestomach of initiated animals. Oral administration of TPA in sesame oil produced changes in the forestomach epithelium that were exactly parallel to changes seen in the epidermis.

The chewing of betel quid ingredients in Asia is often associated with high incidences of oral, pharyngeal and esophageal tumors in man. Ranadive et al. (1979) reported that an undefined mixture of betel nut, betel leaf, lime and catechu produced both oral and gastric tumors in Syrian hamsters. Since this mixture produced lesions, subsequent fractionation experiments demonstrated that betel quid with tobacco and betel quid alone could serve as a promoter of stomach lesions. These experiments tend to verify, in laboratory animals, the possible causal relationship of betel quid chewing and cancer in humans.

The mechanism of promotion in the esophagus and the forestomach appears to be similar to that of the skin (Boutwell, 1974; Diamond et al., 1980). BHT, TPA and betel quid are all inflammatory and cause irritation and hyperplasia. The process of sustained hyperplasia leads to increased cell turnover, possible gene activation and cell selection. The end result of these processes is the development of papillomas followed by squamous-cell carcinomas.

## Stomach

Gastric tumors in rats can be promoted by a number of naturally occurring plant products. A number of phenols (Farnsworth et al., 1976), hydrazones (Braun et al., 1981) and catechols (Hirono, 1972; Hirono, 1981) have been shown to promote stomach tumors in mice and Syrian hamsters. The antioxidant butylated hydroxyanisole (BHA) was recently shown to enhance spontaneous gastric tumors in F344 rats (Ito et al., 1983). In all reported cases, both male and female animals were responsive to promotional stimuli by phenols, catechols, hydrozones and BHA.

Limited experimental data on promotion in the stomach make it impossible to explore possible mechanisms of action of these compounds. Critical experiments on promoting dose, biochemical, and histopathological data on the progression of the lesion have not been adequately developed. Future investigations, especially involving BHA, should address the development of histopathological and biochemical data to explain the resultant long-term lesions in the stomach.

Liver

Next to the mouse epidermis, more information is known regarding promotion in the liver than in any other organ. Tumor promotion has been demonstrated in trout, mouse and rat liver with a variety of compounds including cyclopropenoid fatty acids, estrogens, phenobarbital and organohalides. Promotion has been demonstrated in both male and female rats, implying that there are no sex differences with regard to promoting sensitivity.

Initial reports by Lee et al. (1971) indicated that low levels (100 ppm) of methylsterculate fed in the diets of aflatoxin $B_1$ initiated trout resulted in enhanced hepatocellular carcinoma formation. There was a threshold no response level followed by an increase in promotional response up to a maximum of 100 ppm for the cyclopropenoid fatty acid. Peraino et al. (1971) reported that phenobarbital (PB) could decrease the initiating ability of 2-acetylaminofluorene (AAF) by presumably modifying activation, and could serve as a promoter if given after AAF treatment in the diet. Their data indicated that phenobarbital increased the hepatoma yield and decreased the time for the onset of tumors. They further demonstrated that PB could stimulate hepatocellular proliferation. Subsequent results (Peraino et al., 1973, 1975 and 1977) indicated that: (1) prolonged exposure to PB was necessary for promotion; (2) initiation by AAF was irreversible; (3) latency period was neodecreased; (4) tumor growth rate was increased; (5) number of plastic foci increased; and (6) new compounds such as BHT and dichlorodiphenyltrichloroethane (DDT) were promoters in the hepatic system. All promoting compounds were capable of stimulating DNA synthesis in the liver but not at the same level, suggesting differences in potency. As in the epidermis, initiation was irreversible and promotion was reversible up to a point. Evidence of chronic hepatic toxicity was lacking, indicating that the response was not necessarily regenerative in nature but that it could be an adaptive metabolic response to the persistant levels of phenobarbital. Promotion by structurally related compounds has been demonstrated by Kimura et al. (1976), Nishizumi (1976) and Ito et al. (1978) using the polychorinated biphenyls as promoting agents in both male and female rats. The potency of the PCBs is equivalent to DDT, and partial hepatectomy enhances the promoting potential of the PCBs. Dietary fat, especially trienes, are capable of enhancing spontaneous (C3H mice) and 1,2-dimethyl-hydrazine initiated (CD1 mice) hepatomas when fed at fat levels of 17% of the total dietary caloric level (Brown, 1981). Wanless and Medline (1982) and Toper (1978) presented data on ethinyl estradiol, estradiol-17-phenylpropionate and estradiol benzoate as promoters of rat hepatomas in diethylnitrosamine and N-nitrosomorph-oline initiated animals. The increased tumor yield and decreased latency appeared to be similar to the data of Peraino et al. (1978) using phenobarbital as a promoter.

The mechanism of action of drug induced hepatocellular tumors is quite similar to epidermal carcinogenesis. There are a series of biochemical changes such as enzyme induction that are parallel in liver and skin. Hyperplastic nodules result from promoter treat-ment and these nodules regress if the promoting stimulus is removed up to a critical point, after which the nodule is committed to develop into a hepatoma. The preneoplasic nodules contain enzyme altered foci for γ-glutamyltranspeptidase, canalicular ATPase, glu-cose-6-phosphatase and iron deficient foci (Pitot et al., 1982). It is not clear whether or not all of the enzyme altered foci re-sult in hepatomas but they are certainly direct evidence of enhanced clonal growth or cell section and proliferation of presumably initiated cells. The promotional model can be modified by hormonal and dietary factors and this model has potential use for extrapola-tion to the human population.

## Large Intestine and Bowel

Dietary fat can enhance the incidence of both chemically induced and spontaneous tumors of the lower gastrointestinal tract in laboratory animals. Reddy et al. (1978) demonstrated that rats initiated with either methylazoxymethanol or N-methyl-N-nitrosourea could be promoted by diets containing high fat levels from corn oil or beef lard. The study demonstrated that high levels of beef or soy protein (40%) and high levels of fat (20%) enhanced the tumor yield when compared to dietary protein levels of 20%. Additional investigations by Rose and Nahrwold (1982) indicated that 1,2-di-methylhydrazine initiated rats and guinea pigs could be promoted by high dietary fat levels. Investigations by Reddy et al. (1978) demonstrated that colon tumors in germ-free rats could be promoted by elevated dietary fat.

The apparent mechanism of promotion by dietary fats in the large intestine and colon relates to biliary excretion as a response to high fat levels (Reddy et al., 1981). The primary bile acids cholic and chemodeoxycholic acid enhance MNNG tumor formation in normal and germ-free rats (Narisawa et al., 1974). These acids are subject to microbial 7-α-dehydroxylation to deoxycholic acid and lithocholic acid (Reddy et al., 1977). The process of 7-α-dehydroxylation can result in the development of a higher colonic pH which also may contribute to potential enhanced promotion (Thornton, 1981). All of the known bile acids that have been tested for promoting activity have been positive. Bile acids enhance intestinal epithelial cell growth and induce cell proliferation (Reddy et al., 1976; Cohen et al., 1980; Reddy, 1981). The increased cell proliferation could result in more rapid expression of genomic lesions and increased tumor yields. It is not known if high levels of bile acids result in cell selection for a specific epithelial cell type, but increased cell turnover could progress via selective process. The mechanism of colonic promotion relates to increased cell turnover and/or a process of regenerative hyperplasia, none of which are well characterized in the large intestine and colon.

Colon tumors can be enhanced by carrageenan, a complex mixture of sulfated galactose polymers (DeCarvalho, 1980; Hopkins, 1981). These compounds are found in sea weeds and commonly used in foods as emulsifiers, stabilizers, thickeners and gelling agents. Carrageenan enhances methylnitrosourea induced colon tumors in rats when given at levels of 15% in the diet (Hopkins, 1981). Oohashi et al. (1981) reported that degraded carrageenan could induce squamous metaplasia and ulcerative lesions in the rat colon when fed for a period of up to 9 months. The serial study indicated that degraded carrageenan led to squamous metaplasia and some areas progressed to squamous cell carcinomas and enhancement of a low level spontaneous tumor in the colon.

The bile acids, which result from high fat diets, and carrageenans both act as dietary promoters in experimental animals. Mechanistically, both agents stimulate cell proliferation and cell turnover which enhances the probability for expression of malignantly transformed cells within the population of exposed epithelium. Clearly, further mechanistic investigations are necessary to determine the key factors that trigger cell proliferation and that may be responsible for specific cell selection during the promotion process.

## Bladder

The artificial sweeteners saccharin (2,3-dihydro-3-oxybenziso-sulphonazole) and the sodium and calcium salts of cyclamate (N-cyclohexylsulphamate) are effective promoters of tumors in the rat urinary bladder (Hicks et al., 1973; Hicks et al., 1975). Animals initiated with N-methyl-N-nitrosourea (NMU) at doses that did not produce tumors and subsequently fed diets containing saccharin or sodium cyclamate for the duration of the investigation, had a significant increase in urinary bladder tumors. Maximum tumor response was attained by 60 to 80 weeks of promotion. Saccharin and cyclamate were previously shown to be very weak carcinogens in mice when life-time feeding studies were performed at dosages up to 5% of the total diet (Roe et al., 1970). In a review by Reuber (1978) on the carcinogenicity of saccharin, some 10 studies in rats and 3 studies in mice indicated that saccharin was most probably a weak urinary bladder carcinogen and that it had promoting potential. Studies with Syrian golden hamsters and Rhesus monkeys did not indicate tumorigenicity associated with saccharin. Male rats were more susceptible to the promotional effects of saccharin than females, and the animals on chronic feeding studies had an increased incidence of renal disease.

Elevated levels of tryptophan and its metabolites are found in urine of a high percentage of patients with bladder carcinomas, and the role of tryptophan in bladder carcinogenesis is being investigated in laboratory animals. Recent investigations by Cohen et al. (1979) indicated that DL-tryptophan could promote N-[4-(5-nitro-2-furyl)-2-thiazolyl]formamide (FANFT) initiated F344 rat urinary bladder tumors. Tryptophan was less potent than saccharin and may require metabolism to an unknown species in order to promote since several tryptophan metabolites are found in urine. In an attempt to isolate possible promoting compounds in DL-tryptophan, L-tryptophan was used in a promotion experiment following FANFT initiation (Fukushima et al., 1981). The L isomer of tryptophan was much less active than the DL mixture but the experimental conditions were not comparable in the two studies. In an investigation by Wang et al. (1976), allopurinol was shown to enhance FANFT bladder tumors, but did not affect tryptophan or tryptophan metabolites in female F344 rats.

Promotion in the urinary bladder is similar to other models since the lesions go through the stages of hyperplasia, hyperplastic nodules and later form carcinomas. Cell proliferation in the urinary bladder is an essential part of the promotion process (Hicks and Chowaniec, 1977). Focal hyperplasia, dysplasia and a high number of mitotic figures in transitional cells and urothelium are common in animals fed dietary saccharin (Chowaniec and Hicks, 1979). Male and female rats administered saccharin in the drinking water consumed less water and voluntarily less diet than if saccharin was mixed in the diet, and they did not develop bladder tumors (Chowaniec and Hicks, 1979). While increased cell proliferation is an essential component of the promotional process, hyperplasia may not be a sufficient stimulus for promotion. Isoproterenol given to rats caused urothelia hyperplasia but did not enhance MNU initiated bladder tumors (Hicks, 1980). Clearly, other biochemical properties of saccharin have to be investigated since hyperplasia is a necessary but not sufficient property to promote tumors. The exact mechanism that triggers cell proliferation and the nature of the urothelia cells stimulated and/or selected by saccharin exposure are unknown. Future investigations should focus on these problems as well as on understanding more about reversibility of stages of promotion and the relationship of early biochemical changes to the development of focal and nodular hyperplasia.

## Pancreas

Dietary fat enhances both the incidence and number of induced pancreatic tumors in the rat and Syrian hamster (Pour et al., 1981; Roebuck et al., 1981). The progress of the disease in both the rat and the hamster models is similar to the development of the disease in humans with regard to morphology and biochemical characteristics (Pour et al., 1981; Jacobs, 1983). Animals are initiated with ip or sc injections of azaserine, N-nitrosobis(2-oxopropyl)amine (BOP) or other N-nitrosobisamines followed by dietary feeding of known levels of fats (saturated and unsaturated). Animals normally develop pancreatic tumors in a time frame of 25 to 45 weeks of promotion. Unsaturated fats promote tumors more efficaciously than saturated fats (Roebuck et al., 1981), and other compounds such as cyclopropenoid fatty acids (Scarpelli, 1975) and soy bean flour (Morgan et al., 1977) promote pancreatic ductular and acinar adenomas and adenocarcinomas. The hamster model results in the majority of lesions in ductular cell epithelium while the rat model yields lesions primarily in the acinar cell epithelium, but both types of lesions develop in the respective models.

Promotion in the rodent pancreas occurs over a prolonged period and is linked to increased cell proliferation in both ductal and acinar cell epithelium. In rats, unsaturated fats are required to promote azaserine-induced tumors (Roebuck et al., 1981), and saturated fats following unsaturated fats will promote tumors in rats (Roebuck et al., 1981). Although the data is not complete, unsaturated fats stimulate pancreatic cell turnover better than saturated fats in rats (Jacobs, 1983) and hamster (Birt et al., 1981). Cyclopropenoid fatty acids stimulate DNA synthesis in rats fed diets containing increasing doses of cyclopropenoid fatty acids as rapidly as as 30 hours after treatment (Scarpelli, 1975). Mitotic index was enhanced and the majority of the mitotic activity was in acinar cells. Ductal epithelium in hamsters proliferates, leading to focal hyperplasia (Moore et al., 1983) followed by enzyme altered foci and tumors (Moore et al., 1983). Hamsters fed a protein-free diet during initiation with N-nitrosobis(2-oxopropyl)amine and the early stages of promotion developed fewer pancreatic tumors (adenomas and carcinomas) than animals fed a purified 18% casein diet (Pour et al., 1983). Partial pancreatectomy in hamsters results in enhanced tumor yield in animals that were pancreatectomized one week week prior to carcinogen treatment (Pour et al., 1983). Prior treatment with alloxan inhibited tumor formation in the hamster pancreas, and the decrease in tumor incidence probably related to carcinogen metabolism and was not a modification of promotion (Pour et al., 1983).

In the rat and hamster models, enhanced cell proliferation by promoters leads to an increase in pancreatic tumors. The exact mechanisms involved are not known, but stimulation of cell turnover by elevated dietary fat or stimulation of regenerative hyperplasia, or by a hormonal mechanism involving cholecystokinin could result in the observed enhancement. Very limited information is available on induction of enzymes involved in cell proliferation for both ductal and acinar cells, and it is not clear which compounds are capable of promotion. Future areas of research should focus on mechanistic aspects of promotion that may lead to an understanding of pancreatic tumor promotion in humans.

## Other Tissues

Tumor promotion has been demonstrated in the epidermis (Boutwell, 1974), hemopoietic and lymph system (Slaga, 1981) and in the thyroid (Hall and Bielschowsky, 1949). The models are based on an increase in tumor yield and a decreased latency as in the previously described models. The mouse is the most widely used species although rabbit and rat have been used. Since limited information is available for the hemopoietic and lymph and thyroid models, the majority of this section will center on the epidermis.

In the experimental animal models for promotion, the mouse epidermis is the most extensively defined system where several classes of compounds exhibit promoting activity (Boutwell, 1964; Boutwell, 1974; Blumberg, 1980; Diamond et al., 1980; and Slaga, 1981). The majority of the compounds identified as having promoting potential are topically applied rather than ingested; however, many of these compounds occur in edible plants or feedstuffs. Long-chain fatty acids and fatty acid esters possess promoting activity (Holsti, 1959; Arffmann and Glavind, 1971). Citrus oils contain known promoters (Roe and Peirce, 1960; Homburger and Boger, 1968) and many plants used for edible oil production contain promoting substances (Ranadive et al., 1971). The most widely studied promoters in the epidermis are the diesters of diterpene alcohol phorbol that were isolated from croton oil (Boutwell, 1964; Blumberg, 1980).

The mechanism of promoting action in the epidermis relates to the ability of the compounds to cause irritation, inflammation and cell proliferation. The phorbol diester 12-0-tetradecanoylphorbol-13-acetate (TPA) when applied to mouse epidermis in nanomolar quantities produces dramatic histopathological changes (Boutwell, 1964) and biochemical changes in the epidermis (Diamond et al., 1980). Irritation and hyperplasia are insufficient properties alone to yield complete promotion. The phorbol diesters bind to a cellular membrane receptor (Blumberg, 1981), and binding appears to correlate with the promoting activity of the phorbol diester. However, with the wide range of plieotrophic effects of the phorbol diesters in a variety of cells (in vivo and in vitro), it is not clear that binding to this membrane receptor is promotion specific. Boutwell (1974) postulated that gene activation was a necessary part of promotion and that a hormonal mechanism (such as a receptor mechanism) could be paramount to promotion. Recent data indicating that the phorbol diester binding protein co-purifies with a calcium-dependent, phospholipid-dependent protein kinase C has created much excitement since the gene product of the SV40 virus (src gene) is also a protein kinase. The exact role of the kinase in promotion remains obscure, but of all the promotion models, this avenue of pursuit appears to provide the most well-defined series of biochemical events associated with the process of promotion.

Rapid progress in the area of kinase activation and subsequent elucidation of substrate-specificity and associated biochemical cascades may indicate a specific set of biochemical and cellular events which lead to tumor promotion in the epidermis. Once these events have been identified, it may be possible to demonstrate a similar series of biochemical and cellular events in other organ systems where promotion occurs, thus clarifying the mechanism of tumor promotion in other target organs.

PRESUMPTIVE HUMAN MODELS

Epidemiologic evidence suggests that cancers of the digestive tract, especially stomach, colon and pancreas, mammary gland, and prostate, are strongly associated with dietary factors. Inherently, epidemiological data is less structured than experimenntal animal data, but valuable information can be derived from studies on tumor incidences in various groups or populations in the world. For example, by comparing trends in tumor incidence, it is evident that cancers of the colon, breast and prostate have increased slightly over the past 40 years in the United States, indicating an environmental influence on these tumors. In Japan there is a low incidence of tumors in the colon, breast and prostate but, recently, an increasing trend in these tumors has begun and it appears associated with the Westernization of the Japanese diet (NAS, 1982). These data provide evidence that dietary pattern influences rates of tumor development in specific organs rather than a relationship to industrialization.

Mammary

Epidemiologic evidence in concert with experimental data implicates excessive fat intake as a major environmental determinant in the etiology of breast cancer. There is a positive correlation between the intake of dietary fat in various populations and mortality rates from breast cancer, with Japan having low fat intake and mortality and Denmark, the Netherlands and the United States having a high fat intake and a higher mortality rate (Wynder, 1980). Studies of Japanese who have migrated to the U.S. reveal an increase in breast cancer incidence. Successive generations eventually attain a tumor rate equivalent to the host country (Wynder, 1980; Williams et al., 1982). Recent data from Japanese populations indicated that breast tumor incidence increases with the Westernization of their diets. Hence, the combined epidemiologic data indicates a strong association between dietary fat and breast tumor development.

Elevated fat levels in experimental animals enhance mammary epithelial proliferation. There is conflicting evidence for the involvement of prolactin levels and tumor development in experimental animals, and the same is true for humans (Wynder, 1980; NAS, 1982; Weisburger, 1982). Effects on endocrine balance, especially prolactin levels, are difficult to isolate and interpret in humans but it is assumed that prolactin is a liporegulatory hormone as it is in other mammals. Additional hormones may be involved in regulating mammary gland growth and lipids may exert specific membrane effects on mammary cells. The role of high dietary fat intake in enhancing mammary tumor incidence is unknown but the evidence is strongly supportive of the concept that high levels of dietary fat act as promotors of human breast cancer.

## Stomach

Gastric tumor rates vary from nation-to-nation with countries like Japan and Iceland having high rates and the United States, Canada and New Zealand having low rates of incidence and mortality (Reddy et al., 1980). The relative rates of gastric tumors in the U.S. has declined in both men and women during the last 40 years, which may indicate a dietary rather than an industrialization relationship (Williams et al., 1982).

The major factors associated with gastric tumors appear to be the intake of cured meat products, dried salted fish, smoked fish and low vegetable intake (Williams et al., 1982; NAS, 1982). In one instance, tannins have been implicated in the etiology of gastric tumors in humans (Morton, 1972). For the most part, gastric cancer rates correlate with intake of foods containing compounds capable of being activated in the stomach to alkylnitrosourea structures (Weisburger et al., 1982). Many of these compounds are initiators rather than promoters and compounds like vitamin C and E can inhibit tumor formation in experimental animals.

In animal models, elevated sodium chloride levels promoted gastric tumor development (Williams et al., 1982). Since many of

the cured meat products contain high salt and, in some cases, high
fat it is possible that both salt and high fat levels can promote
gastric tumor formation. Clearly, many of the pickled fish products
are high in salt content and could serve as a promotion source.
However, at present there is not plausible mechanism to explain how
these compounds can promote gastric tumors in experimental animals
or humans.

## Colon

Epidemiologic data indicate that colon cancer rates are associated with dietary factors. Colon cancer has a relatively low
incidence in African countries and Japan and a rather high incidence
in the United States (Bababunmi, 1978; Reddy et al., 1980; Zaridge,
1981; NAS, 1982). Migrant studies of Japanese moving to the U.S.
indicate that the imigrant assumes the tumor rate of the host
country with the time (Williams et al., 1982; NAS, 1982). Japanese
colon tumor rates are increasing with Westernization over the past
40 years (Reddy et al., 1980). Combined, these data support a
dietary and/or lifestyle factor in the etiology of colon cancer.

Data from experimental animal model systems indicate that high
levels of dietary fat enhance or promote colon tumors in these
models (Weisburger et al., 1982). The high fat, low fiber diets
place elevated risk on human populations for colon cancer and coronary heart disease (Williams et al., 1982). In experimental animals
high fat diets stimulate cholesterol biosynthesis which stimulates
bile acid production in the lower gastrointestinal tract (Weisburger
and Horn, 1982; Williams et al., 1982). Bile acids, when given at
elevated concentrations or when synthesized to meet a high cholesterol challenge, are true promoters in the colonic epithelium
(Reddy et al., 1980; Williams and Horn, 1982). The type of fat
(saturated vs. unsaturated) does not seem to influence the tumor
incidence. Diets high in fat and low in fiber will concentrate
the promoter levels via lack of stool bulk. Lack of stool bulk not
only serves to concentrate promoter levels but initiator levels
as well. Although there is no specific human data on cholesterol
biosynthesis as a function of fat intake, there appears to be
elevated bile salts in stools from high fat diet individuals (Reddy
et al., 1982). It is assumed that the major effect of high dietary
fat appears to enhance endogenous cholesterol biosynthesis that
creates a need for increased bile acid biosynthesis.

## Pancreas

Epidemiologic evidence indicates a Westernized diet, typically
high in total fat content, is an important etiologic factor in
pancreatic tumorigenesis. The positive correlation between total
fat consumption in various countries and pancreatic tumor incidence
suggests that excessive fat ingestion is an important disease fac-

tor (Fraumeni, 1975; Kritchevsky, 1975; Wynder, 1975; Morosco and Goeringer, 1980). The correlation between fat and pancreatic cancer incidence is increased by including smoking and alcohol consumption as etiologic agents (Wynder, 1975; Morosco and Goeringer, 1980; NAS, 1982). Hence, there exists a suggestive line of evidence that elevated dietary fats are major etiologic determinants in pancreatic tumor development.

Limited data from the rat indicates that diets containing high dietary fat (20%, unsaturated) promoted azaserine initiated pancreatic tumors (Roebuck et al, 1981), and data in hamsters indicate that high fat diets promoted nitrosamine induced pancreatic tumors (Birt et al., 1981). Lipids in these diets can cause injury to acinar cells (Morosco and Goeringer, 1980), and unsaturated lipids can stimulate cell proliferation via specific membrane effects. Both ductular and acinar cell proliferation could be stimulated via either a pure proliferative process or a necrosis followed by regenerative hyperplasia. To date, there is limited metabolic epidemiology in humans to support the experimental pancreatic models (Reddy et al., 1982). However, given the existing animal data it is plausible that elevated dietary fat could promote pancreatic tumors in humans, and further epidemiologic investigations to substantiate this hypothesis would seem timely and prudent.

## Prostate

Epidemiological evidence from international populations indicate a positive relationship between prostatic cancer and environmental factors (Reddy et al., 1982; NAS, 1982). Correlations between dietary fat intake and age-adjusted mortality for prostatic cancer show a positive correlation similar to that for breast cancer (Reddy et al., 1982). Positive correlations between coffee intake, meat fats, milk products and high protein intake and human prostatic tumor incidences have been demonstrated (NAS, 1982). Japanese migrants and their descendants moving to the U.S. assume the increased tumor incidence and their dietary habits reflect elevated dietary fat intake (NAS, 1982). Within the U.S., there are great variations in rates of prostatic cancer among regions and ethnic groups (Williams et al., 1982; NAS, 1982). Curiously, American blacks have a greater risk for prostate cancer than American whites (Weisburger and Horn, 1982; Williams et al., 1982; NAS, 1982). Noteworthy, the incidence of prostatic cancer correlates with other organ sites associated with the diet such as breast and colon cancer.

Given the positive correlation between the incidence of prostatic cancer and consumption of high fat diets, a possible mechanism of enhancement could be similar to that occurring in the mammary gland. Dietary fats can modify hormonally dependent processes as well as specific membrane responses leading to increased cell proliferation (NAS, 1982; Welsch and Aylsworth, 1983). However, there is no experimental animal model available at this time to successfully address some of these questions. Further progress in understanding the process of prostatic tumor development will require additional epidemiology and validation in an experimental animal model.

Other tissues

Epidemiological data from national and international populations indicate probable dietary influences on human esophageal and bladder tumors. The data is not extensive but trends are apparent from these investigations.

Esophageal tumors in humans are associated with ingestion of pickled foods, alcohol consumption and smoking habits (Doll and Peto, 1981). Studies of tumor incidences in Curacao indicated that tannins, which are normal plant constituents, are closely associated with elevated esophageal tumor rates (Morton, 1972). Recent investigations from North China indicate that Roussin's red (a constituent in pickled cabbage) is a promoter and possibly implicated as a human promoter of esophageal tumors (Cheng et al., 1980; NAS, 1982). Additional information on other dietary influences on esophageal tumor development await further epidemiological and experimental investigations.

Most of the epidemiological data on the association of bladder tumors and the environment are centered on coffee consumption and artificial sweeteners. From case-control studies, there appears to be no causal relationship between coffee consumption and bladder tumor incidence (NAS, 1982; Omenn, 1982). Experimental animal models would predict that saccharin should promote tumors in the human. However, to this date there is no epidemiologic data that would indicate an increased tumor rate associated with populations consuming quantities of artificial sweeteners (Saccharin and Bladder Cancer, 1900, NAS, 1982). Perhaps the time frame is not correct or the dose is too low to demonstrate an effect in the human population. Further studies, especially on diabetic populations, are required to ascertain if artificial sweeteners can serve as promoters in human bladder cancer.

CONCLUSIONS

Extrapolation of risk factors from rodent to humans is difficult because of differences in longevity, mobility, and the multistage development of cancer, but correlations can be made between dietary factors and tumor incidence. Based on experimental animal models and epidemiologic investigations, the combined evidence is strongly suggestive for a causal relationship between fat intake and occurrence of cancer for the breast, stomach, and colon. Data from animal studies suggest that increased total fat increases the tumor incidence in the colon, stomach and breast and that lower fat levels lower the risk for tumor development. Although the experimental and epidemiological data is not complete, dietary fat intake affects tumor incidence in the animal pancreas and human prostate.

The hallmark of promotion is the increased tumor yield and decreased latency period. If one better understands the mechanism(s) of the promotional process, one might be able to intervene in the process and either reverse the process of tumor development or prolong the latency such that tumors would not appear in the normal life span. Experimental data indicate that reversal or inhibition of promotion is possible by administering vitamins A, C or E and the trace metals selenium and zinc. Drugs like the glucocorticoids, protease inhibitors and cyclic AMP analogs also inhibit tumor promotion. At present, the exact mechanism of action of these compounds is unknown and many anti-promoters may yet be discovered.

Further research into basic mechanisms of dietary promotion should be conducted. Promoting substances should be fractionated and purified. Once specific promoting activities can be assigned to a compound, mechanisms can be proposed and potential prophylactic measures can be developed. Additional animal models need to be developed for research on organ sites where promotion has not been previously demonstrated, and the scope of human metabolic epidemiology expanded with regards to additional dietary determinants of cancer.

Although there is sufficient data to indicate causal relationships between diet and cancer, drastic changes in dietary guidelines are premature at this time. Since many vitamins and trace elements are anti-carcinogens or anti-promoters, more research should be done on understanding their mechanisms of action. Further research on the role of dietary fiber intake on the inhibition of tumor development is warranted since the mechanism is unknown and there may be adverse effects on nutritional status in high fiber diets. As more information becomes available, it may be possible to modify dietary patterns slightly to reflect decreased consumption of deleterious compounds and increased consumption of beneficial components such that an overall decrease in the risk of diet and cancer could result.

REFERENCES

Arffmann, E. and Glavind, J., 1971, Tumor-promoting activity of fatty acid methyl esters in mice, Experientia, 27:1465-1466.
Bababunmi, E.A, 1978, Toxins and carcinogens in the environment: An observation in the tropics, J. Toxicol. Environ. Health, 4:691-699.

Birt, D.F., Salmasi, S. and Pour, P.M., 1981, Enhancement of experimental pancreatic cancer in Syrian golden hamsters by dietary
fat, J. Nat. Cancer Inst., 67:1327-1332.

Blumberg, P.M., 1980, In vitro studies on the mode of action of the
phorbol esters, potent tumor promoters: parts 1 & 2. CRC Crit.
Rev. Toxicol., 8:153-234.

Braun, R., Dittmar, W. and Greeff, U., 1981, Considerations on the
carcinogenicity of the mushroom poison gyromitrin and its
metabolites, J. Appl. Toxicol., 1:243-246.

Brown, R.R., 1981, Effects of dietary fat on incidence of spontaneous and induced cancer in mice, Cancer Res., 41:3741-3742.

Boutwell, R.K., 1964, Some biological aspects of skin carcinogenesis, Prog. Exp. Tumor Res., 4:207-250.

Boutwell, R.K., 1974, The function and mechanism of promoters of
carcinogenesis, CRC Crit. Rev. Toxicol., 2:419-443.

Carroll, K.K., 1981, Neutral fats and cancer, Cancer Res., 41:3695-
3699.

Carroll, K.K., 1982, Dietary fat and its relationship to human
cancer, in: "Carcinogens and Mutagens in the Environment, Vol.
1, Food Products," H.F. Stich, ed., CRC Press, Boca Raton,
Florida.

Cheng, S. J., Sala, M., Li, M.H., Wang, M.Y., Pot-Deprun, J. and
Chouroulinkov, I., 1980, Mutagenic, transforming and promoting
effect of pickled vegetables from Linxian County, China,
Carcinogenesis (Lond), 1:685-692.

Cheng, S.J., Sala, M., Li, M.H., Courtois, I. and Chouroulinkov, I.,
1981, Promoting effect of Roussin's red identified in pickled
vegetables from Linxian, China, Carcinogenesis (Lond), 2:313-319.

Chowaniec, J. and Hicks, R.M., 1979, Response of the rat to saccharin with particular reference to the urinary bladder, Br. J.
Cancer, 39:355-375.

Clapp, N.K., Tyndall, R.L., Cumming, R.B. and Otten, J.A., 1974,
Effects of butylated hydroxytoluene alone or with diethylnitrosamine in mice, Food Cosmet. Toxicol., 17:367-371.

Cohen, S.M., Arai, M., Jacob., J.B. and Friedell, G.H., 1979,
Promoting effect of saccharin and DL-tryptophan in urinary
bladder carcinogenesis, Cancer Res., 39:1207-1217.

Cohen, B.I., Raicht, R.F., Deschner, E.E., Takahashi, M., Sarwal,
A.N. and Fazzini, E., 1980, Effect of cholic acid feeding on
N-methyl-N-nitrosourea induced colon tumors and cell kinetics
in rats, J. Nat. Cancer Inst., 64:573-578.

Cummings, J.H., Branch, W.J., Bjerrum, L., Paerregaard, A., Helmes,
P. and Burton, R., 1982, Colon cancer and large bowel function
in Denmark and Finland, Nutr. Cancer, 4:61-66.

DeCarvalho, S., 1980, Carrageenan, a common food additive with
immunosuppressant and cocarcinogenic properties. J. Clin.
Hematol. Oncol., 10:6472.

Diamond, L., O'Brien, T.G. and Baird, W.M., 1980, Tumor promoters
and the mechanism of tumor promotion. Adv. Cancer Res., 32:1-74.

Doll, R. and Peto, R., 1981, The causes of cancer: quantitative estimates of avoidable risks of cancer in the United States today, J. Nat. Cancer Inst., 66:1193-1308.

Editorial, 1980, Saccharin and Bladder Cancer, Lancet, 855-856.

Farnsworth, N.R., Bingel, A.S., Fong, H.H.S., Saleh, A.A., Christenson, G.M. and Saufferer, S.M., 1976, Oncogenic and tumor-promoting spermatophytes and pteridophytes and their active principles, Cancer Treat. Rep., 60: 1171-1214.

Fraumeni, Jr., J.F., 1975, Cancer of the pancreas and biliary tract: epidemiological considerations. Cancer Res. 35:3437-3446.

Fukushima, S., Friedell, G.H., Jacobs, J.B. and Cohen, S.M., 1981, Effect of L-tryptophan and sodium saccharin on urinary tract carcinogenesis initiated by N-[4-(5-nitro-2-furyl)-2-thiazolyl] formamide, Cancer Res., 41:3100-3103.

Goerttler, K., Loehrke, H., Schweizer, J. and Hesse, B. 1982, Diterpene ester-mediated two-stage carcinogenesis, In: "Carcinogenesis, Vol 7, Cocarcinogens and Biological Effects of Tumor Promoters", E. Hecker, N.E. Fusenig, W. Kunz, F. Marks and H.W. Thielmann, eds., Raven Press, New York.

Gross, R.L. and Newberne, P.M., 1976, Naturally occurring substances in foods, Clin. Pharmacol. Ther., 22:680-698.

Hall, W.H. and Bielschowsky, F., 1949, The development of malignancy in experimentally-induced adenomata of the thyroid, Br. J. Cancer, 30:534-541.

Hicks, R.M., Wakefield, J. St.J and Chowaniec, J., 1973, Impurities in saccharin and bladder cancer, Nature, 243:424-426.

Hicks, R.M., Wakefield, J. St.J. and Chowaniec, J., 1975, Evaluation of a new model to detect bladder carcinogens or co-carcinogens: results obtained with saccharin, cyclamate and cyclophosphamide, Chem. Biol. Interact., 11:225-233.

Hicks, R.M. and Chowaniec, J., 1977, The importance of synergy between weak carcinogens in the induction of bladder cancer in experimental animals and humans, Cancer Res., 37:2943-2949.

Hicks, R.M., 1980, Multistage carcinogenesis in the urinary bladder, Br. Med. Bull., 36:39-46.

Hill, M.J., Taylor, A.J., Thompson, M.H. and Wait, R., 1982, Fecal steroids and uriniary volatile phenols in four Scandinavian populations, Nutr. Cancer, 4:61-66.

Hirono, I., 1979, Naturally-occurring carcinogenic substances, GANN, 24:85-102.

Hirono, I., 1981, Natural carcinogenic products of plant origin, CRC Crit. Rev. Toxicol., 8:235-277.

Holsti, P., 1959, Tumor promoting effects of some long chain fatty acids in experimental skin carcinogenesis in the mouse, Acta Pathol. Microbiol. Scand. Sect. B. Microbiol., 46:51-58.

Homburger, F. and Boger, E., 1968, The carcinogenicity of essential oils, flavors and spices: a review, Cancer Res., 28:2372-2374.

Hopkins, G.J., Kennedy, T.G. and Carroll, K.K., 1981, Polyunsaturated fatty acids as promoters of mammary carcinogenesis induced in Sprague-Dawley rats by 7,12-dimethylbenz(a)anthracene, J. Nat. Cancer Inst., 66:517-522.

Hopkins, J., 1981, Carcinogenicity of carrageenan, Food Cosmet. Toxicol., 19:779-781.

Ito, N., Tatematsu, M., Hirose, M., Nakanishi, K. and Murasaki, G. 1981, Enhancing effect of chemicals on production of hyperplastic liver nodules induced by N-2-fluorenyl-acetamide in hepatectomized rats, GANN, 69:143-144.

Ito, N., Fukushima, S., Hagiwara, A., Shibata, M. and Ogiso, T. 1983, Carcinogenicity of butylated hydroxyanisole in F344 rats, J. Nat. Cancer Inst., 70:343-352.

Kiaer, H.W., Glavind, J. and Arffmann, E., 1975, Carcinogenicity in mice of some fatty acid methyl esters, Acta Path. Microbiol. Scand., Sect. A Pathol., 83:550-558.

Kimura, N.T., Kanematsu, T. and Baba, T., 1976, Polychlorinated biphenyl(s) as a promoter in experimental hepatocarcinogenesis in rats, Krebs. Klin. Onkol., 87:257-266.

Kritchevsky, D., 1975, Formal discussion of "Cancer of the pancreas and biliary tract: epidemiological considerations"., Cancer Res. 35:3450-3451.

Lee, D.J., Wales, J.H. and Sinnhuber, R.O., 1971, Promotion of aflatoxin induced hepatoma growth in trout by methylmalvalate and sterculate, Cancer Res., 31:960-963.

Lower, G.M.,Jr. and Kanarek, M.S., 1982, Risk, susceptibility and epidemiology of proliferative neoplastic disease: descriptive vs. mechanistic approaches, Med. Hypotheses, 9:33-49.

Miller, E.C. and Miller, J.A., 1979, Overview on the relevance of naturally occurring carcinogens, promoters and modulators of carcinogenesis in human cancer, In: "Naturally occurring carcinogens-mutagens and modulators of carcinogenesis," E.C. Miller, J.A. Miller, I. Hirono, T. Sugimura, S. Takayama, eds., Japan Science Society Press, Tokyo Univ., Park Press, Baltimore.

Morgan, R.G.H., Levinson, D.A., Hopwood, D., Saunders, J.H.B. and Wormsley, K.G., 1977, Potentiation of the action of azaserine on the rat pancreas by raw soya bean flour, Cancer Lett., 3:87-90.

Morosco, G.J. and Goeringer, G.C., 1980, Lifestyle factors and cancer of the pancreas: a hypothetical mechanism, Med. Hypotheses, 6:971-975.

Morton, J.F., 1972, Further associations of plant tannins and human cancer, Q.J. Crude Drug Res., 12:1829-1841.

Narisawa, T., Magadia, N.E., Weisburger, J.H. and Wynder, E.L., 1974, Promoting effect of bile acids on colon carcinogenesis after intrarectal instillation of N-methyl-N-nitro-N-nitrosoguanidine in rats, J. Nat. Cancer Inst., 53:1093-1097.

National Academy of Sciences, 1982, Diet Nutrition and Cancer, National Academy Press, Washington, D.C.

Nishizumi, M., 1976, Enhancement of dimethylnitrosamine hepatocarci-nogenesis in rats by exposure to polychlorinated biphenyls or phenobarbital, Cancer Lett., 2:11-16.

Omenn, G.S., 1982, Predictive identification of hypersusceptible individuals, J. Occup. Med., 24:369-374.

Oohashi, Y., Ishioka, T., Wakabayashi, K. and Kuwabara, N., 1981, A study on carcinogenesis induced by degraded carrageenan arising from squamous metaplasia of the rat colorectum, Cancer Lett., 14:267-272.

Pariza, M.W., 1983, Carcinogenicity/Toxicity testing and the safety of foods, Food Tech., 37:84-86.

Peraino, C., Fry, R.J.M. and Staffeldt, E., 1971, Reduction and enhancement by phenobarbital of hepatocarcinogenesis induced in the rat by 2-acetylaminofluorene, Cancer Res., 31:1506-1512.

Peraino, C., Fry, R.J.M., Staffeldt, E. and Kisieleski, W.E., 1973, Effects of varying the exposure to phenobarbital on its enhance-ment of 2-acetylaminofluorene-induced hepatic tumorigenesis in the rat, Cancer Res., 33: 2701-2705.

Peraino, C., Fry, R.J.M., Staffeldt, E. and Christopher, J.P., 1975, Comparative enhancing effects of phenobarbital, amobarbital, diphenylhydentoin a and dichlorodiphenyltrichloroethane on 2-acetyl-aminofluorene-induced hepatic tumorigenesis in the rat, Cancer Res., 35:2884-2890.

Peraino, C., Fry, R.J.M., Staffeldt, E. and Christopher, J.P., 1977, Enhancing effects of phenobarbitone and butylated hydroxytoluene on 2-acetylaminofluorene-induced hepatic tumorigenesis in the rat, Food Cosmet. Toxicol., 153:93-96.

Peraino, C., Fry, R.J.M. and Grube, D.D., 1978, Drug-induced en-hancement of hepatic tumorigenesis, In: "Carcinogenesis, Vol. 2., Mechanisms of Tumor Promotion and Co-carcinogenesis," T.J. Slaga, A. Sivak and R.K. Boutwell, eds., Raven Press, New York.

Petering, H.G., 1978, Diet, nutrition and cancer, In: "Inorganic and Nutritional Aspects of Cancer," G.N. Schrauzer, ed., Plenum Press, New York.

Pitot, H., Goldsworthy, T., Moran, S., Sirica, A.E. and Weeks, J.E. 1982, Properties of incomplete carcinogens and promoters in hepatocarcinogenesis, In: "Carcinogenesis, Vol. 7, Cocarcino-gens and Biological Effects of Tumor Promoters," E. Hecker, N.E. Fusenig, W. Kung, F. Marks and H.W. Thielmann, eds., Raven Press, New York.

Pour, P.M., Runge, R.G., Birt, D., Gingell, R., Lawson, T., Nagel, D., Wallcare, L., and Salmasi, S.S., 1981, Current knowledge of pancreatic carcinogenesis in the hamster and its relevance to human disease, Cancer, 47:1573-1587.

Ranadive, K.J., Gothoskar, S.V. and Tezabwala, B.U., 1972, Carcino-genicity of contaminants in indigenous edible oils, Int. J. Cancer, 10:652-666.

Ranadive, K.J., Ranadive, S.N., Shiapurkar, N.M. and Gothoskar, S.V., 1979, Betel quid chewing and oral cancer: experimental studies on hamsters, Int. J. Cancer, 24:835-843.

Rao, G.A. and Abraham, S., 1976, Enhanced growth rate of transplantable mammary adinocarcinoma induced in C3H mice by dietary lineolate, J. Nat. Cancer Inst., 60:849-853.

Reddy, B.S., Narasawa, T., Weisburger, J.H. and Wynder, E.L., 1976, Promoting effects of deoxycholate in colon adenocarcinomas in germ-free rats, J. Nat. Cancer Inst., 56:441-442.

Reddy, B.S., Watanabe, K., Weisburger, J.H. and Wynder, E.L., 1977, Promoting effect of bile acids in colon carcinogenesis in germ-free and conventional F344 rats, Cancer Res., 37:3238-3242.

Reddy, B.S., Weisburger, J.H. and Wynder,E.L., 1978, Colon cancer: bile salts as tumor promoters, In: "Carcinogenesis, Vol. 2, Mechanisms of Tumor Promotion and Cocarcinogenesis," T.J. Slaga, A. Sivak and R.K. Boutwell, eds., Raven Press, New York.

Reddy, B.S., Cohen, L.A., McCoy, G.D., Hill, P., Weisburger, J.H. and Wynder, E.L., 1980, Nutrition and its relationship to cancer, Adv. Cancer Res., 32:237-345.

Reddy, B.S., 1981, Dietary fat and its relationship to large bowel cancer, Cancer Res., 41:3700-3705.

Reuber, M.D., 1978, Carcinogenicity of saccharin, Environ. Health Perspect., 25:173-200.

Roe, F.J.C. and Peirce, W.E.H., 1960, Tumor promotion by citrus oils: tumors of the skin and urethral orifice in mice, J. Nat. Cancer Inst., 24:1389-1403.

Roe, F.J.C., Levy, L.S. and Carter, R.L., 1970, Feeding studies on sodium cyclamate, saccharin and sucrose for carcinogenic and tumor-promoting activity, Food Cosmet. Toxicol., 8:135-145.

Roebuck, B.D., Yager, J.D. and Longnecker, D.S. 1981, Dietary modulation of azaserine-induced pancreatic carcinogenesis in the rat, Cancer Res., 41:888-893.

Roebuck, B.D., Yager, J.D., Jr., Longnecker, D.S. and Wilpone, S.A., 1981, Promotion by unsaturated fat of azaserine-induced pancreatic carcinogenesis in the rat, Cancer Res., 41:3961-3966.

Rogers, A.E. and Wetsel, W.C., 1981, Mammary carcinogenesis in rats fed different amounts and types of fats, Cancer Res., 41:3735-3737.

Scarpelli, D.G., 1975, Preliminary observations on the mitogenic effect or cyclopropenoid fatty acids on rat pancreas, Cancer Res., 35:2278-2283.

Schramm, V.I. and Gibel, W., 1969, Pflanzenstoffe als Karzinogene and Kokarzinogene Substanzen. II. Karzinogene Substanzen Bei Pteridophyten (Filicinae) und Spermatophyten (Lycadinae, Dicotyledonae, Monocotytledonae), Archiv. fur Geschwulstforschung, 33:169-188.

Slaga, T.J., 1981, Food additives and contaminants as modifying factors in cancer induction, In: "Nutrition and Cancer: Etiology and Treatment," G.R. Newell and N.M. Ellison, eds., Raven Press, New York.

Sugimura, T., 1979, Naturally occurring genotoxic carcinogens, In: "Naturally occurring carcinogens-mutagens and modulators of carcinogenesis," E.C. Miller, J.A. Miller and T. Sugimara, eds., Japan Science Society Press, Tokyo/Univ. Park Press, Baltimore.

Tannenbaum, A., 1942, The genesis and growth of tumors. III. Effects of a high fat diet, Cancer Res., 2:468-475.

Taper, H.S., 1978, The effect of estradiol-17-phenylpropionate and estradiol benzoate on N-nitrosomorpholine-induced liver carcinogenesis in ovariectomized female rats, Cancer, 42:462-467.

Thornton, J.R., 1981, High colonic pH promotes colorectal cancer, Lancet, 1081-1083.

Wagner, D.A., Naylor, P.H., Kim, U., Shea, W., Ip, C. and Ip, M.P. 1982, Interaction of dietary fat and the thymus in the induction of mammary tumors by 7,12-dimethylbenz(a)anthracene, Cancer Res., 42:1266-1273.

Wanless, I.R. and Medline, A., 1982, Role of estrogens as promoters of hepatic neoplasia, Lab. Invest., 46:313-320.

Wang, C.V., Hayashida, S., Pamakau, A.M. and Bryan, G.T., 1976, Enhancing effect of allopurinol on the induction of bladder cancer in rats by N[4-(5-nitro-2-furyl)-2-thiazolyl] formamide, Cancer Res., 36: 1551-1555.

Weisburger, J.M. and Horn, C., 1982, Nutrition and cancer: Mechanisms of genotoxic and epigenetic carcinogenesis in nutritional carcinogenesis, Bull. N.Y. Acad. Med., 58:296-312.

Welsh, C.W. and Aylsworth, C.F., 1983, Enhancement of murine mammary tumorigenesis by feeding high levels of dietary fat: a hormonal mechanism, J. Nat. Cancer Inst., 70:215-221.

Williams, G.M., Weisburger, J.H. and Wynder, E.L., 1982, Lifestyle and cancer etiology, In: "Carcinogens and Mutagens in the Environment, Vol. 1, Food Products," H.F. Stich, ed., CRC Press, Boca Raton, Florida.

Witschi, H. and Lock, S., 1978, Butylated hydroxytoluene: A possible promoter of adenoma formation in mouse lung, In: "Carcinogenesis, Vol. 2, Mechanisms of Tumor Promotion and Co-carcinogenesis," T.J. Slaga, A. Sivak, and R.K, Boutwell, eds., Raven Press, New York.

Witschi, H. and Kehrer, J.P., 1982, Adenoma development in mouse lung following treatment with possible promoting agents, J. Toxicol. Environ. Health, 8:171-183.

Wynder, E.L., 1975, An epidemiological evaluation of the causes of cancer of the pancreas, Cancer Res., 35:2228-2233.

Wynder, E.L., 1980, Dietary factors related to breast cancer, Cancer, 46: 899-904.

Zaridze, D.G., 1981, Diet and cancer of the large bowel, Nutr. Cancer, 2:241-249.

# 5

DEFENSES AGAINST AFLATOXIN CARCINOGENESIS IN HUMANS

M. Sid Masri
Western Regional Research Center
Agricultural Research Service
U.S. Department of Agriculture
Albany, California 94710

ABSTRACT

Aspects of work on aflatoxin mainly from the author's studies are discussed, emphasizing procedures, strategies and insights related to protection against aflatoxin carcinogenesis in man due to transmission in the food chain. Five lines of defense are discussed. First, is the interfacing of the farm animal as a biologic filter between contaminated crop (feed) and human food derived from animal products. Second, is chemical detoxification of meals with ammonia and of contaminated milk by ultraviolet light. Third is prevention of mold growth and aflatoxin elaboration with ammonium carbonate for potential application during storage and drying of agricultural products. Fourth, is the development of ultrasensitive analytical methods for aflatoxins with structural proof of identity. And, fifth are natural defenses related to the role of hydroxylation of mammalian metabolism in detoxification, especially the pathway to aflatoxin $Q_1$, a major detoxifying mechanism in primates, and the epoxidation pathway and its relation to prevention of aflatoxin residue transmission into edible tissues.

INTRODUCTION

The brilliant work of a group of British scientists investigating outbreaks of widespread deaths of turkeys, pigs, and calves in the United Kingdom in 1960, ushered in an era of new insights on the interrelationships between the agricultural feed and food supply, mycotoxins, and cancer [1-3]. Demonstration of the association of the problem with metabolites of a ubiquitous mold, the high

115

potency and carcinogenicity of the toxin and the widespread susceptibility of animal species to the toxin, all stimulated world-wide
interest and concerted research efforts. The ubiquitous nature of
the responsible Aspergillus flavus mold, which infects agricultural crops had clear implications of potential serious impacts on
national agricultural economies and public health.

In this report, results, mainly from the author's work (and
thus are not meant to be comprehensive of the field) are reviewed,
especially those which provide examples of useful strategies for
defending against exposure to aflatoxin carcinogenesis in man
through the food chain. The following topics which point to this
emphasis are covered: a) the role of the farm animal in the food
chain as a biologic filter, b) chemical detoxification and control with ammonia, c) inactivation of aflatoxin $M_1$ in milk by
ultravioltet light, d) prevention of mold growth and of aflatoxin
production with ammonium carbonate, e) comparative metabolism and
detoxification pathways in mammals and primates, and f) development
of relevant sensitive methodology. Most of the experimental work
had been published 4, 5, 8-17, 32, 33.

EXPERIMENTAL PROCEDURES

Feeding trials with pigs, steers and dairy cows were carried
out to study animal husbandry and toxicologic effects of graded
levels of aflatoxin in the feed and to study potential transmission
of aflatoxin or its metabolities into edible tissues of meat, liver,
milk, etc. The feeding experiments were with 115 hogs, 50 steers,
and six cows; other tests were with poultry, (broilers and laying
hens). The results were reported [4-7]. Results on transmission
will be discussed further here. Both chemical and bioassay methods
were used to check on transmission of residues. Tissues were
collected (after 4-5 months of feeding) from the pigs and steers
immediately after slaughter, chilled with dry ice, then ground and
freeze-dried and stored in the cold until used. The tissues
collected were muscle, blood, liver, kidney, spleen and adipose
tissue.

Chemical analysis of the freeze-dried tissues (FDT) was carried
out by extracting FDT samples with methanol in a blender. Water
was added to the methanol extract and the aqueous methanol solution
was defatted with SkellySolve F prior to partitioning with chloroform. Residue from the chloroform extracts was examined by TLC on
silica gel. In some tests, the chloroform extracts were subjected
to column cleanup before TLC examination [8].

Milk was collected from Holstein cows given known amounts of
aflatoxin $B_1$ ($B_1$). The milk was freeze-dried and stored in the
cold until used for the isolation and structure proof of aflatoxin

$M_1$ ($M_1$), and for developing analytical methods for $M_1$, as well as for the study of excretion pattern of $M_1$ in milk on known levels of $B_1$ intake, as previously published [5, 9-11].

The following experiment was carried out to test inactivation of aflatoxin $M_1$ in milk with u.v. A 40 g sample of freeze-dried experimental milk from the cow receiving a daily dose of 80 mg aflatoxin $B_1$ was used [5,9]. The dried milk had 1400 ppb aflatoxin $M_1$. The 40 g sample was reconstituted to a volume of 400 ml with distilled water in a blender and then divided equally into two Pyrex trays (base: 38 x 24 cm) giving a milk layer depth of about 2 mm. One sample was kept in the dark and the other was exposed in the tray for one hour to long and short wavelength ultraviolet from a source placed about 10" above the tray in a Chromatoview cabinet. Both samples were then worked up side by side to determine aflatoxin $M_1$ content according to our method [10].

Tests of detoxification of aflatoxin in peanut meals by treatment with ammonia were done either at ambient temperature and atmospheric pressure using 100 ml conc. ammonium hydroxide (28-30% $NH_3$) per kg of meal or at elevated temperature, ammonia pressure and moisture content [12]. The treated meals were tested for reduction of toxicity in 2-day-old Peking ducklings given the meals in the ration.

Tests of the effect of ammonium carbonate on the prevention of mold growth and aflatoxin elaboration were done in static cultures using moistened rice or whole almond kernels as substrates and were inoculated with A. parasiticus NRRL 2999. A pellet of about 1/2 g ammonium carbonate was added one day after inoculation to each flask containing about 20 g of the inoculated substrate moistened with about 10 ml $H_2O$ containing mineral supplement. Each substrate was tested in 10 incubations in 250 ml Erlenmeyer flasks. The incubated substrates were sampled one week after inoculation to check for presence of aflatoxin.

Bioassay tests with ducklings were carried out by incorporating the test substance (meals or FDT) in the duckling ration and observing mortality, growth and gross and histopathologic changes in sacrificed ducklings. Duck tissue specimens were fixed in buffered 10% formalin for histopathology. In the duckling bioassays of the FDT for the aflatoxin residue transmission studies, both a 2-week and a 6-week duration feeding tests were done with most tissues, and about 20 ducklings were used per test.

In order to test the sensitivity of the duckling bioassay, an experiment was carried out with two-day-old ducklings that were fed for four weeks rations containing known levels of aflatoxin $B_1$ provided from a spiked peanut meal. The levels tested were: 0, 25, 50, 100, 200, 300, 400, 600, and 800 ppb $B_1$. Each level was

tested with 17 ducklings. The ration consisted of Purina chick
starter plus 8% peanut meal, except with groups on 600 and 800 ppb,
for which higher percentages of the spiked meal had to be used to
provide the desired aflatoxin concentrations. Two peanut meals
were used for the rations containing 0 through 400 ppb: a high
quality meal which was free of aflatoxin, and the spiked meal which
had 5400 ppb $B_1$. The proportion of each meal was adjusted in the
rations to provide the desired aflatoxin concentrations. All
rations were also supplemented with 0.1% lysine, 0.1% methionine
and 6 g $MnSO_4$/kg. Deaths were recorded during the test. Individual
body weights were obtained weekly. At the end of the fourth
week, the surviving ducks were autopsied and liver specimens were
saved for histopathologic examination. Histopathologic changes
were ranked according to severity from 0 (lesions absent, tissues
normal) to +4 (distinct tumor-like foci throughout liver). The
mortality and growth rate data were analysed statistically.

For the _in vitro_ metabolic studies, methods of liver fractionation and incubation and carcinogenicity tests of aflatoxin $Q_1$ in
feeding trials in the trout were carried out as described [13-17].

## RESULTS AND DISCUSSION

### Role of Farm Animal as Biologic Filter - Chemical Analysis

Analysis of the methanol extracts of the FDT from the pigs
and steers on the different dietary levels of aflatoxin showed no
$B_1$ or its free metabolites by the TLC method that was available at
the time which had a detection sensitivity limit of about 5 ppb in
the FDT. In later tests, sensitivity of the method was improved
to the 1 ppb level by the use of column cleanup procedure on the
chloroform extract before TLC[8]. Selected tissue samples (especially
from the higher level groups) were reanalysed with this modified
procedure with similar results as summarized in Table 1. These
results point to rapid clearance of free aflatoxins by excretion
mainly in urine[9] and by efficient metabolism and binding of
aflatoxin to DNA and other macromolecules (discussed later). The
rapid removal is further corroborated by comparison of results in
Table 1 for the two groups of steers on 1000 ppb $B_1$ in the ration.
No $B_1$ was detected with the group that was fasted for 72 hours
before slaughter. Comparison of our results with those of Krogh et
al. [18] is also instructive. These workers found slightly higher
residues of $B_1$ + $M_1$, especially in liver and kidney of pigs on
aflatoxin regimen. The higher values may be due to the relatively
short time between last meal and slaughter in their experiments, in
which "the pigs were fed for the last time at 7 A.M. as is the
custom in Danish husbandry, and slaughtered at 11 A.M. the same
day". In our experiments, the animals to be sacrificed were fasted
the afternoon of the day previous to slaughter, and thus feed had

Table 1.    Aflatoxin  Residue  Analysis  in  Freeze-Dried  Tissues[a]

| Group, (ppb $B_1$ in ration) | Tissue | FDT ppb $B_1$ |
|---|---|---|
| Pigs, 810 | Blood, Muscle, Spleen | 2 |
| | Kidney | 4 |
| | Liver | 5 |
| Steers, 1000 | Blood, Muscle, Liver Kidney, Spleen | 1-2 |
| Steers, 1000 fasted 72h | Blood, Muscle, Liver Kidney, Spleen | 0 |

[a]Basis dried tissues; residue analyses of selected tissue samples from groups on lower levels were negative with few instances with ambiguous traces below method sensitivity and without apparent correlation to level in ration.  Aflatoxin $M_1$ was also seen in trace amounts at the higher levels and distinctly in kidney at ~5-10 ppb.  FDT = freeze-dried tissue.

been withheld for about 18 hours.  This procedure fits more closely with slaughter practice in the United States and avoids a sanitation burden at slaughter.  Furthermore, the presence of excreted aflatoxins in the urine and feces in the vicinity of slaughter and tissue collection may inadvertently result in cross contamination of tissues from urine, feces or particles of feed on the animal hide.  These remarks suggest a useful precautionary strategy against contamination of meat at slaughter (this very likely also applies to many other toxins which do not have a tendency to bioaccumulate). Our results lead us to recommend a short fast (two to three days) and adequate cleansing of animals before slaughter especially if feed is suspected of contamination.  Alternatively, the animals may be maintained for a few days before slaughter on feed known to be free of aflatoxin.  The short withdrawal allows time for the animal to complete the excretion and metabolic action, and avoids cross-contamination of meat from excreta.

## Excretion of $M_1$ in Milk

Unlike the findings with the experimental pig and steer FDT, analyses of $M_1$ in the freeze-dried milk of the cows on aflatoxin regimen[5] showed relatively high concentration of $M_1$ as summarized in Tables 2 and 3.  Although the amounts of $M_1$ (and lower amounts of

Table 2.  Aflatoxin $M_1$ Concentration[a] (ppb) and Total Content (Percentage of Aflatoxin $B_1$ Fed) in Dried Milk from Cows 1 to 6 During Aflatoxin $B_1$ Feeding.  (See footnotes next page).

| Week on B1 | Cow 1 (ppb) | Cow 1 (%) | Cow 2 (ppb) | Cow 2 (%) | Cow 3 (ppb) | Cow 3 (%) | Cow 4 (ppb) | Cow 4 (%) | Cow 5 (ppb) | Cow 5 (%) | Cow 6 (ppb) | Cow 6 (%) |
|---|---|---|---|---|---|---|---|---|---|---|---|---|
| 1 | b | --- | 70 | 2.80 | 35 | 0.69 | 85 | 1.05 | 130 | 0.80 | 240 | 0.80 |
| 2 | 70 | 2.10 | 65 | 2.92 | 30 | 0.49 | 90 | 1.03 | 200 | 1.18 | 600 | 1.37 |
| 3 | 105 | 2.29 | 75 | 3.37 | 50 | 0.79 | 85 | 1.12 | 180 | 1.04 | 1,500[c,e] | 2.84 |
| 4 | 135 | 1.91 | 90 | 3.94 | 55 | 0.79 | 90 | 0.95 | 180 | 0.77 | | |
| 5 | 130 | 1.21 | | | | | | | 175 | 0.67 | | |
| 6 | 155 | 1.06 | 85 | 3.32 | 65 | 1.01 | 100 | 0.96 | c,d | | | |
| 7 | 120[c] | 0.87 | | | 75 | 1.05 | | | | | | |
| 8 | | | 85 | 3.26 | 90 | 1.30 | 100 | 0.81 | | | | |
| 10 | | | 85 | 3.20 | 65 | 0.82 | | | | | | |
| 11 | | | | | 70 | 0.77 | 120 | 1.01 | | | | |
| 12 | | | 85 | 3.03 | 95 | 0.99 | | | | | | |
| 14 | | | 90 | 3.06 | 80[c] | 0.87 | 95[c] | 0.75 | | | | |
| 16 | | | 70[c] | 2.37 | | | | | | | | |

(continued)

Footnotes to Table 2

$^a$ Values are rounded to the nearest 5 ppb.  Cow 1 was on varying
  daily dose (shown in Table 3); cows 2 through 6 were on daily
  dosages of 5, 10, 20, 40, and 80 mg $B_1$, respectively.
$^b$ Not detected.  When values are not shown:  milk was either not
  collected or not analysed.
$^c$ Last week on $B_1$, except for Cow 2, which was continued on $B_1$ for
  38 weeks; dried milk in 38th week had 45 ppb $M_1$.
$^d$ Milk was inadvertently overheated and overconcentrated; heated
  sample had 30 ppb $M_1$.  $^e$ Dried milk also contained 190 ppb $B_1$.
  (Masri et. al., $^5$).

$B_1$) that were found, represented only a small fraction (1–2%) of
the ingested dose, the absolute amounts were substantial, in view
of the high potency of the aflatoxins, and the high susceptibility
of the young.  In retrospect, it is perhaps not surprising that the
milk contained these levels of aflatoxin $M_1$, considering the high

Table 3.   Aflatoxin $M_1$ in Milk from Cow 1 while on Weekly Varying
           Dosage of Aflatoxin $B_1$

| Week's Number on expt. | $B_1$ intake in week (mg) | Fresh milk for week (lb)$^a$ | Avg. feed intake, lb/day Hay | Concen- trate | $M_1$ in dry milk (ppb) |
|---|---|---|---|---|---|
| 1 | 0.4 | 428 | 25.1 | 24.6 | – |
| 2 | 67.0 | 423 | 22.4 | 29.4 | 70 |
| 3 | 89.0 | 419 | 25.0 | 32.0 | 107 |
| 4 | 120.0 | 380 | 23.2 | 28.1 | 133 |
| 5 | 169.0 | 350 | 24.3 | 29.7 | 129 |
| 6 | 206.0 | 313 | 23.2 | 24.0 | 154 |
| 7 | 175.0 | 282 | 22.2 | 9.0 | 118 |
| 8 | 0.0 | 290 | 22.7 | 21.2 | 18$^a$ |

$^a$ Note drop in milk production with dosage; also note drop in $M_1$ in
  milk after withdrawal of $B_1$.

degree of metabolic activity and rich blood supply of the mammary tissue and the large volume of blood circulated through this gland. In a functional sense, the mammary gland could be viewed as a filtering organ in a way analogous to the kidney; milk secretion may be looked upon as another excretory route of aflatoxins (and very likely many other toxins) from the body. Excretion of $M_1$ in milk emphasizes the importance of avoiding contaminated feed for dairy cows. Here, detoxification of meals (e.g. with ammonia) becomes especially important and applicable.

Comparison of results with cows 5 and 6 in Table 2 is noteworthy. Although daily dosage differed only by a factor of 2 (40 vs 80 mg), the concentration of $M_1$ was about eight-fold higher with cow 6 in the third week. This concentration difference is probably mainly due to sharply reduced milk production with cow 6 compared to cow 5 (Masri et al.[5]). Despite the drop in milk production, the total "excretion" of $M_1$ in milk (cow 6) as a percent of the $B_1$ dose actually rose in the third week to 2.8% from 1.4% in the previous week. The drop in milk production, the rise of $M_1$ concentration, and the higher excretion as a percent of the dose, all point to the significant role of the liver rather than the mammary tissue in the biotransformation to $M_1$, and to the effective role of the mammary gland in its "excretory" function of filtering out $M_1$ from the circulation, even when its milk fat and protein synthesis and secretory functions were failing. The $M_1$ concentration difference may also be related to nonlinear pharmacokinetics of biotransformation (hydroxylation and alternate pathways) in relation to dose, and to dose-dependent difference in excretion rates and thresholds.

Regarding the relatively high excretion of $M_1$ in milk, however, the following fortunate situations, from the point of public health, may be pointed out. Withdrawal of contaminated feed results in rapid disappearance of $M_1$ from the milk as shown in Table 4. Also, aflatoxin $M_1$ is much less carcinogenic than $B_1$ (Sinnhuber et al. [19]); it is also much less stable than $B_1$. The industry practice of milk collection and distribution to the consumer results in mixing and dilution of potential contaminated lots; pasteurization may also enhance instability of $M_1$ (see footnote to Table 2). And, as discussed later, irradiation with ultraviolet light is potentially an effective method for inactivation of aflatoxin $M_1$.

## Duckling Bioassay of Freeze-dried Tissues

In the design of the bioassay to check on transmission of toxicity into the experimental pig and steer tissues, we chose to test the whole tissue rather than extracts or fractions thereof in order to avoid potential problems that might be associated with extraction, fractionation and instability of purified fractions,

Table 4.    Aflatoxin $M_1$ in Milk from Cows 3,4, and 6 Following Withdrawal of Aflatoxin $B_1$.

| | Concentration of aflatoxin $M_1$ in dry milk (ppb) | | | | | |
| | $M_1$ the week before withdrawal | Days after withdrawal | | | | |
| Cow | | 1 | 2 | 3 | 4 | 7 |
|---|---|---|---|---|---|---|
| 3 | 82 | 44 | 17 | 5 | | 5 |
| 4 | 96 | 55 | 27 | | 3 | 1 |
| 6 | 1500[a] | 718 | 434 | 285 | 160 | |

[a] Dried milk also contained 190 ppb $B_1$; after withdrawal, the milk contained 290, 265, 245, and 175 ppb $B_1$ for days 1 through 4, respectively (Masri et. al.[5]).

and to insure testing not only for free aflatoxin itself and its known free metabolites but also for any tissue-bound known or unknown toxic derivatives. Accordingly, the whole dried tissues were incorporated in the ration of the ducklings. Lean muscle tissue, was incorporated at 30%, and liver, blood, fat, kidney, and spleen at 10%. Composition of the duck rations are shown in Tables 5-8. Kidney and spleen were tested by feeding only a few ducklings. Some tests with kidney and spleen were done by crop-tubing suitable extracts of the tissues as well as by feeding the extracted tissues.

The outstanding finding from all the bioassay tests with the tissues is that no changes could be attributed to the treatment level when compared to results with tissues from control groups (no aflatoxin). The bioassay tests thus corroborated results of the chemical residue analyses, demonstrating the effectiveness of interfacing the farm animal as a biologic filter between contaminated feed and human food.

## Method Sensitivity of Duckling Bioassay

In order to check suitability and sensitivity of the bioassay method, an experiment was carried out with groups of two-day-old ducklings that were fed known levels of aflatoxin for four weeks. Results of mortality and growth rate as criteria of toxicity were treated statistically. Ducklings livers were also examined for

Table 5.   Duck Bioassay Ration with Freeze-Dried Steer Muscle Tissue

| Muscle[a] | 30% | USP XIV salt | 5% |
|---|---|---|---|
| Casein[b] | 1-2% | Vit. mix | 4% |
| Corn Oil[c] | 1.5-3.% | $MnSO_4$ | 1g/kg |
| Methionine | 0.25% | | |

[a] Proximate analysis in Table 8.
[b] To a final 22% protein in ration.
[c] To a final 9% fat in ration.


Table 6.   Duckling Bioassay Ration Formulated with Freeze-dried
           Blood or liver of Pigs and Steers.

| | % | | % |
|---|---|---|---|
| Liver or blood[a] | 10 | Methionine | 0.5 - 0.8 |
| | | Choline chloride | 0.3 |
| Casein | 17.8 | | |
| Salt USP XIV | 5. | Alfalfa meal | 1.7 |
| Corn oil | 7. | Corn meal | 54 |
| Vitamin mix | 4. | $MnSO_4$ | 1g/kg |

[a] Provides   7% protein, 1% salt and 0.7 - 1.0% fat.


histopathologic changes.  For this experiment, the ration consisted
of Purina Starter plus about 8% peanut meal.  Results on mortality
are shown in Tables 9-12 and those on growth in Table 13.  Results
of the statistical treatment of the mortality and growth data are
discussed below.

    Mortality.  Groups on 25, 50, and 100 ppb dose levels had no
deaths in the four weeks and were not included in the potency probit
analysis.  Table 9 summarizes the deaths in the other groups and
relates the mortality by days and dose level.  Comparison of the
averaged dose level for all the groups on 200 ppb and above

Table 7.  Duck Bioassay Ration with Freeze-Dried Pig and Steer
Fat Tissue

| | | | |
|---|---|---|---|
| Fat | 10% | Choline Chloride | 0.3% |
| Casein | 35% | Alfa-cell | 5.% |
| Corn Oil | 2% | Glycine | 0.3% |
| Salt USP XIV | 5% | Corn meal | 36.4% |
| Methionine | 1% | $MnSO_4$ | 1g/kg |

(combined), with the net response (averaged % deaths), for the
various time periods, gives results shown in Table 10.  These values
indicate that over time, the average dose level decreases at a
decreasing rate.  The calculated dosage levels required for certain
LD's are shown in Table 11.  The calculated relative potency of
days 14, 21, and 28, when compared to day 7, are shown in Table 12.
The values suggest that potency (mortality response) levels off
around the third to fourth week.  A plot of the percent deaths vs.
log dosage level for the different weeks, shows a linear relationship.  Thus for day 7, the average kill for the different dose
levels is less than 10%, while the average for day 21 is over 50%.
Within limits, the regression lines for the different days can be
extrapolated to determine the percent kill for different dose
levels within these time limits (Table 11).  The results with
mortality very likely represent a complex interplay of factors.  On
the one hand, the overall measured response is enhanced with time
by an increasing rate of daily intake, absolute cumulative intake,
and length of physiologic exposure time.  On the other hand, counteracting factors with time, are decreased susceptibility with age
of ducks that survive, potential repair of physiologic damage, and
selection of resistant animals through early deaths of less resistant ducklings.

The choice of both a two-week and a six-week periods for the
bioassays of the FDT was based on several considerations.  First,
statistical potency probit analysis indicated increase in potency
after the first, and a leveling off around the fourth week (Table
12).  Potency at two weeks was intermediate, being about 2.5 times
that of the first and about one half that of the fourth.  A short
feeding period was also desirable because of the suspected higher
sensitivity and susceptibility of the young.  Discovery of histopathologic changes might be enhanced by the short test, while the
animals are still young.  (The potency comparison referred to above

Table 8.    Proximate analyses of Steer Freeze-dried Muscle and Liver
            Tissues

| Group (ppb $B_1$ in ration) | % Total Solid | % Crude fat | %N |
|---|---|---|---|
| | Muscle | | |
| 0 | 96.4 | 12.2 | 12.6 |
| 100 | 96.0 | 13.2 | 12.4 |
| 300 | 95.4 | 12.3 | 12.5 |
| 700 | 96.2 | 10.0 | 12.9 |
| 1000 | 95.5 | 13.1 | 12.4 |
| 1000 (fasted 72h) | 96.7 | 16.2 | 12.0 |
| | Liver | | |
| 0 | 96.8 | 6.8 | 11.2 |
| 100 | 96.0 | 7.0 | 10.8 |
| 300 | 96.0 | 8.6 | 10.8 |
| 700 | 95.8 | 8.5 | 11.0 |
| 1000 | 95.7 | 9.7 | 11.4 |
| 1000 (fasted 72h) | 95.6 | 8.0 | 10.9 |

was based on mortality and not histopathologic change as a criterion).  Finally, duckling feed consumption up to two weeks is
relatively small (not much FDT would be needed), and the logistics
of animal care are relatively simple; but both feed consumption and
difficulty of duckling care increase rapidly after the second week.

Growth - (Analysis of Variance and Mean Separation).  Analysis
of variance was used to determine effect of dosage level and length
of treatment on body weight gain.  Both level and length of time

Table 9.    Relation  of  Mortality  (% Deaths)  in  Ducklings  to  Dose
Level  of  Aflatoxin  $B_1$  (ppb  in  ration)  and  Time  on  Level
(days)[a]

| Days | Aflatoxin level | | | | |
|------|------|------|------|------|------|
|      | 200 | 300 | 400 | 600 | 800 |
| 7 | 00.00% | 00.00 | 00.00 | 5.88 | 11.76 |
| 14 | 5.88 | 00.00 | 29.41 | 76.47 | 82.35 |
| 21 | 23.53 | 41.18 | 47.06 | 100.00 | 100.00 |
| 28 | 35.29 | 52.94 | 58.82 | 100.00 | 100.00 |

[a] The  basal  ration  consisted  of  Purina  starter  to  which  was  added
about 8% peanut meal.   The  aflatoxin  in  the  ration  was  provided
from  aflatoxin  spiked  peanut  meal.    The  Purina  starter  had  not
less  than  18%  crude  protein  or  3%  fat,  and  not  more  than  6%
crude  fiber,  7%  ash,  or  2%  added  minerals.


Table 10.   Relationship of Averaged Aflatoxin $B_1$ Dose Levels for All
Groups  (on  200-800 ppb)  with  the  Averaged  Net % Deaths
for  the  Various  Time  Periods[a]

| Days | $B_1$ Average dosage level | Average % death |
|------|------|------|
| 7 | 696 | 8.0 |
| 14 | 471 | 44.0 |
| 21 | 354 | 58.0 |
| 28 | 328 | 63.0 |

[a] Deaths before the end of a week were considered in the calculation.

Table 11.   Calculated Aflatoxin Dosage Levels in Ration of Ducklings
(ppb $B_1$) Required for $LD_{30}$ – $LD_{90}$

| Days | LD30 | LD50 | LD70 | LD90 |
|------|--------|--------|--------|--------|
| 7 | 1050.47 | 1336.23 | 1699.71 | 2405.81 |
| 14 | 397.87 | 506.10 | 643.78 | 911.21 |
| 21 | 253.27 | 322.17 | 409.80 | 580.04 |
| 28 | 222.11 | 282.53 | 359.38 | 508.68 |

Table 12.   Relative Potency of Days 14, 21, and 28 on Aflatoxin $B_1$
Regimen, Compared to Day 7[a]

| Days | Upper Limit | Potency | Lower Limit |
|------|-------------|---------|-------------|
| 14 | 3.69 | 2.64 | 1.89 |
| 21 | 5.89 | 4.15 | 2.92 |
| 28 | 6.77 | 4.73 | 3.30 |

[a] With upper and lower 95% confidence limits. Based on mortality
response.

on treatment as well as their interaction, affected body weight
significantly (P< .001). Table 13 shows that as the $B_1$ dose
level increased, weight gain over time decreased. For dose levels
of 100 and 200 ppb, the effect of dose level on weight over time
was distinctly greater when compared to levels of 0 through 50 ppb
for all weeks. Similarly, the 200 ppb level was significantly
different (P< .05) from the 300 to 400 ppb levels after 21 days.
In other words, rate of weight increase from day 7 to 21 was greater
among ducklings on the lower levels (less than 200 ppb) compared to
the higher levels (200, 300, and 400 ppb). After 21 days however,
there was a tendency of surviving ducklings among the higher levels
to increase their rate of weight gain (tolerance or selection for
resistance through early deaths of less resistant ducklings?).
Also, up to 300 ppb level and between days 14 to 28, the decrease

Table 13.  Means of Weights of Ducklings at Different Diet Levels
          and Days

| Diet levels (ppb $B_1$) | Days 7 | 14 | 21 | 28 | Diet means (Body wts.,g) |
|---|---|---|---|---|---|
| | Body weights, g | | | | |
| 0 | 225.0 | 607.6 | 1125.9 | 1517.3 | 869.0 |
| 25 | 196.8 | 578.8 | 1065.9 | 1453.2 | 823.7 |
| 50 | 187.9 | 502.9 | 989.4 | 1377.1 | 764.3 |
| 100 | 156.2 | 384.1 | 824.1 | 1209.1 | 743.4 |
| 200 | 141.5 | 263.1 | 520.8 | 909.5 | 458.7 |
| 300 | 130.6 | 224.7 | 431.0 | 644.5 | 357.7 |
| 400 | 127.6 | 206.7 | 385.0 | 694.3 | 353.4 |
| 600 | 104.1 | 167.5 | a | a | 464.8 |
| 800 | 88.0 | 156.7 | a | a | 451.4 |
| Days Means | 150.8 | 343.6 | 729.4 | 1081.3 | 576.3 |

a All ducklings dead

in weight gain was more pronounced at the higher dose levels.  All
ducklings had died at 21 days on levels of 600 and 800 ppb.

Histopathology.  Ranking of severity of histopathologic changes
of livers of sacrificed ducklings are shown in Table 14.  The
following supplementary notes summarize gross pathology observations
at autopsy.

a) A characteristic type of hemorrhage of liver; perhaps best
   described as transverse streaky lesion was commonly found at
   lower levels; streak effect was most prominent at 25 ppb $B_1$,
   and to a much lesser extent was also observed in controls (0
   ppb).

Table 14.  Liver Bile Duct Hyperplasia in Ducklings Fed Aflatoxin $B_1$

| Diet levels (ppb $B_1$) | Ranking of hyperplasia[a] | | | | | No. deaths by 4th week |
|---|---|---|---|---|---|---|
| | 0 | +1 | +2 | +3 | +4 | |
| 0 (control) | 17 | | | | | 0 |
| 25 | 7 | 10 | | | | 0 |
| 50 | 2 | 3 | 7 | 5 | | 0 |
| 100 | | 3 | 3 | 5 | 1 | 0 |
| 200 | | | | 8 | 3 | 6 |
| 300 | | | | 3 | 5 | 9 |
| 400 | | | | 1 | 3 | 10 |

[a] 2-day old ducklings, 17 per group, were fed the levels for 4 weeks; 2 more groups on 600 and 800 ppb $B_1$ (not shown) were all dead by the 4th week; livers from dead ducklings were not includeded; data on 5 livers of group on 100 ppb level and 3 livers on 400 ppb level are missing. Increasing severity of hyperplasia from 0 (not present, not observed) to +4 (tumor-like foci).

b) Granular appearance of liver could be discerned at low levels, and became more pronounced at higher levels giving a cobblestone appearance. Cobblestone effect was distinct and frequent at 100 ppb and occasionally at lower levels. Transverse streaks and cobblestone effects were not observed concurrently. Cobblestone effect increased in severity at higher levels, i.e. 200 and 300 ppb with definite appearance of tumors throughout liver in 50% of livers at 300 ppb.

c) Fatty infiltration of liver apparently could be observed at

virtually any level becoming a more common and expected finding
at 100 ppb, perhaps even lower.

d) Variation in gross kidney appearance with increasing levels was
difficult to judge and more difficult to describe. It appeared
as a color change from deep dark red (normal) to a more bland
tan shade, and as a surface texture change from very granular
(normal) to smooth.

## Detoxification with Ammonia and  Inhibition with Ammonium Carbonate

Of the many reagents and methods tried in the literature,
aqueous or gaseous ammonia employed at ambient or elevated temperature, pressure and moisture content, remain the most practical and
convenient effective means for detoxifying contaminated meals without loss of nutritive value[12]. Effectiveness of treatment at ambient temperature and at elevated temperature and pressure is shown
in Tables 15-17. Detoxification of meals not only protects the
health of the animal, but also further insures against transmission
of residues into the food chain especially milk.

Table 15.  Ammonia Treatment of Peanut Meal Containing Aflatoxin
$B_1$--Feeding Test in Ducklings[a]

| Meal | Mean body wt (g) at end of: | | No. ducklings in group | Deaths at end |
|---|---|---|---|---|
| | 1st week | 2nd week | | |
| Treated [b] | 108 | 191 | 10 | 0 |
| Untreated | 84 | 95 | 10 | 6 |

[a] 2-day old ducklings were fed ration containing in percent:  peanut
meal, 70; casein, 14; corn oil, 12; USP XIV salt mix., 2; and vitamin mix., 2.  The original (untreated) meal had 1300 ppb $B_1$
and the following composition in percent:  total N, 7.1; crude
fiber 13.9; crude fat, 2.3; ash, 5.5; and moisture, 6.7.
[b] Using 100 ml conc. ammonium hydroxide (28-30% $NH_3$) per kg meal
for two days at ambient temperature and pressure.  The ammonia was
then purged and the meal dried in a forced draft oven at 70C
(Masri et. al.[12]).

Table 16. Detoxification of Aflatoxin in Peanut Meal with Ammonia at Ambient Temperature and Pressure

| Meals [a] | | | | Ducklings [b] | | | |
| No. | Treatment | ppb $B_1$ | No. fed | Mean body wt (g) at end of: | | Deaths at end of 2nd week | Remarks; autopsy |
| | | | | 1st week | 2nd week | | |
| 1 | Contaminated untreated | 1,300 | 10 | 102 | 140 | 9 | Fatty yellowish liver at death |
| 2 | Meal No. 1 treated with $NH_4OH$ for 7 days | 260 | 8 | 143 | 363 | 0 | Livers appear normal |
| 3 | Meal No. 1 treated with $NH_4OH$ for 14 days | c | 10 | 158 | 378 | 0 | Livers appear normal |
| 4 | Meal No. 3 (purged of ammonia) but autoclaved 1 hr at 120C | c | 10 | 170 | 431 | 0 | Livers appear normal |
| 5 | Uncontaminated high quality meal (control) | 0 | 20 | 171 | 385 | 0 | Livers healthy and normal |

a Meal No. 1 was the starting meal for Nos. 2,3, and 4; it is also the same starting meal in Table 15. Treatment was with 100 ml/kg, then meals were purged of $NH_3$ (ambient temp.) then dried in forced-draft oven εt 70C.

b 2-day old ducklings at start had average weight of 45 g. See text for ration formula; meals were mixed at about 70%.

c Not determined. (Masri et. al.[12])

Table 17. Detoxification of Aflatoxin B$_1$ in Peanut Meal with Ammonia (20 psig 95C, 1h)

| Meals | | | Ducklings | | |
|---|---|---|---|---|---|
| | | | Mean body wt (g) at end of: | | |
| No. | Treatment | ppb B$_1$ | 1st week | 2nd week | Remarks; liver histopathology |
| 6 | Contaminated, untreated | 709 | 114 | 199 | Severe bile duct epithelial hyperplasia; fatty degeneration and necrosis. |
| 7 | Meal No. 6 bone-dried, then ammoniated | 203 | 135 | 261 | Similar but less severe changes than with meal No. 6. |
| 8 | Meal No. 6 moistened to 9.6% H$_2$O then ammoniated | 25 | 154 | 282 | Epithelial proliferation absent, liver tissue normal. |
| 9 | As No. 8, except moistened to 14.6% | 17 | 145 | 322 | Epithelial proliferation absent, liver tissue normal. |
| 5 | Uncontaminated high quality meal (control, untreated) | 0 | 162 | 391 | Epithelial proliferation absent, liver tissue normal. |

(Masri et. al.[12])

In addition to the bioassays of the ammoniated meals in ducklings, a preliminary feeding test of another ammoniated meal was carried out with six pigs fed the meal at 15% in the ration for 14 weeks. The spiked meal contained originally 5400 ppb $B_1$ and after ammoniation the level had dropped to about 250 ppb. The treatment was with 100 ml conc. $NH_4OH/kg$ for four weeks at 98° F and atmospheric pressure before purging the ammonia with air. The experiment with the six pigs given the ammoniated meal was run as a side experiment of the larger main feeding trial with pigs on graded level of aflatoxin in the ration. The group in the main experiment on the highest level of 810 ppb $B_1$ (provided by 15% of the same unammoniated meal containing 5400 ppb) served as a positive control; and the group on the 5 ppb level (15% of control peanut meal containing 36 ppb), served as another (negative) control. Also, another control group received 15% soybean meal instead of peanut meal. The basal ration for the pigs consisted in percent: whole ground barley, 75; peanut (or soybean) meal,15; alfalfa, 5; bone-and-meat meal, 4; salt, 0.5; and vitamin mixture, 0.5. The control groups had ten pigs per group. Body weight data are summarized in Table 18. These show distinct improvement due to ammoniation when the average of 14 weekly means for each group and the final means at 14 weeks are compared. Also the group on the ammoniated meal appeared healthy at autopsy and the livers were similar to those of the negative control group, and were free of the histopatholigic lesions that were observed with livers of the group receiving the unammoniated meal.

Results of the incubation experiment with ammonium carbonate to control mold growth and prevent elaboration of aflatoxin in static cultures (rice and almond substrates inoculated with A. parasiticus) showed that no aflatoxin was produced after one week of incubation; under similar conditions without ammonium carbonate, usually about 1/4-1/2 mg $B_1/g$ substrate would be produced. This experiment suggests an alternate approach to the process that was successfully developed at the Northern Regional Research Center in which circulating anhydrous ammonia is used in storage bins to prevent mold and aflatoxin elaboration in corn during air drying or storage. We suggest that the use of ammonium carbonate pellets, dispersed in the stored crop in a bin or a silo may be similarly effective. The slowly released $NH_3$ by decomposition of the carbonate in the vicinity of the pellets would diffuse throughout providing a buffered ammonia atmosphere (alkaline) without the need for circulating equipment. The carbonate is a milder reagent and may be easier to handle. The $CO_2$ from the decompositon would also help exclude $O_2$ from the atmosphere. Similar reagents such as ammonium bisulfite may also be effective. Field tests are needed to check potential application.

Table 18.  Effect of Ammoniation of Aflatoxin $B_1$ Spiked Peanut Meal on Growth of Pigs[a]

| Meal No. | Treatment | No. of animals | Body weight (lb) avg. weekly means and S.E. (14 wks.) | | Body wt. (lb) Mean & S.E. at 14 wks. | |
|---|---|---|---|---|---|---|
| 1. | Spiked meal untreated (5400 ppb $B_1$) | 10 | 114.42 | 1.67 | 169.2 | 6.12 |
| 2. | Meal #1, ammoniated for 2 wks, 98F | 6 | 140.21 | 2.32 | 217.6 | 7.95 |
| 3. | Control meal (36 ppb $B_1$) | 10 | 138.10 | 1.67 | 215.2 | 6.12 |
| 4. | Soybean meal, control | 10 | 145.10 | 1.67 | 223.9 | 6.12 |

[a] 6 barrows and 4 gilts (Duroc-Jersey) for groups 1, 3, and 4; and 4 barrows and 2 gilts for group 2. Pigs were 13-14 weeks old, weighing approx. 50 lb. at start and were fed for 14 weeks. Individual body weights were recorded weekly.  The peanut and soybean meals were fed at 15% in the ration.  S. E. = standard error of the mean.

Table 19.   Conversion of Aflatoxin $B_1$ to $M_1$ and $Q_1$ by Liver
            Postmitochondrial Fraction from Rats Treated or Untreated
            with Phenobarbital

| Rat No. | Aflatoxin $B_1$ level (mg/g liver equivalent) | % Substrate to: $M_1$ | $Q_1$ | Residual $B_1$ (% of added) |
|---|---|---|---|---|
| **Untreated rats** | | | | |
| U1 | 0.2 | 0.9 | 1.9 | 48 |
|    | 0.4 | 1.0 | 1.6 | 52 |
| U2 | 0.4 | 0.9 | 1.3 | 60 |
|    | 0.8 | 0.5 | 0.7 | 75 |
| U3 | 0.8 | 0.4 | 0.7 | 85 |
|    | 2.0 | 0.2 | 0.5 | 84 |
| U4 – U8 (pooled 5 livers) | 0.2 | 0.7 | 1.3 | 56 |
|    | 0.8 | 0.5 | 1.0 | 66 |
| U9 – U14 (pooled 7 livers) | 0.2 | 0.9 | 1.2 | 53 |
|    | 0.4 | 0.6 | 0.7 | 62 |
| **Treated rats** | | | | |
| T1 | 0.8 | 3.2 | 7.6 | 19 |
| T2 | 1.0 | 2.3 | 5.3 | 29 |
|    | 2.0 | 1.9 | 4.2 | 40 |
| T3 – T12 (pooled 10 livers) | 0.4 | 1.4 | 3.5 | 21 |
|    | 1.1 | 1.2 | 3.6 | 32 |
| T13 – T20 (pooled 8 livers) | 0.4 | 1.8 | 4.1 | 11 |
|    | 1.2 | 1.0 | 3.2 | 39 |

(Masri et. al.[15]).

Table 20. Conversion of Aflatoxin $B_1$ to $M_1$ and $Q_1$ by Monkey and Chicken Liver Postmitochondrial Fractions (Footnotes on next page).

| Species | No. of Animals | $B_1$ Substrate (mg/g liver equivalent) | % Conversion to: $M_1$ mean (range) | $Q_1$ mean (range) | Residual $B_1$ (% of original) mean and range |
|---|---|---|---|---|---|
| Monkey[a] | 1 | 0.5 | 0.3 | 27.5 | 34 |
| | | 1.0 | b | 29.0 | b |
| Macaca irus | | | | | |
| Saimiri sciureus | 4[c] | 0.2 | 0.58 (0.30–0.90) | 32.3 (19–49) | 37 (15–56) |
| | | 1.0 | 0.26 (0.25–0.32) | 30.3 (28–33) | 54 (50–58) |
| Saimiri sciureus (dead 1 hr) | 1 | 0.2 | b | 7.9 | 78 |
| Chicken | | | | | |
| White Leghorn cockerels | 3[d] | 0.2 | 0.16 (0.08–0.26) | 0.14 (0.07–0.20) | 57 (47–68) |
| | | 1.0 | 0.09 (0.05–0.12) | 0.11 (0.06–0.14) | 62 (57–68) |
| White Leghorn cockerels | 1 | 0.4 | 0.08 | 0.06 | 47 |

Footnotes to Table 20

[a] High conversion of aflatoxin $B_1$ to $Q_1$ (average 32% and as high as 52%) was also obtained with <u>Macaca mulatta</u> (Rhesus) monkeys (Masri et. al.[13, 15]).

[b] Not determined.

[c] Both substrate levels were tested with livers from three monkeys; liver of fourth monkey was tested at the lower substrate concentration only.

[d] Each liver was tested at both substrate levels. (Masri et. al.[15]).

## Inactivation of Aflatoxin $M_1$ in Milk with Ultraviolet Light

In the course of isolation and structure proof of aflatoxin $M_1$ from experimental milk[9], and development of analytical methods for aflatoxin $M_1$[10,11], we noted much greater instability of aflatoxin $M_1$ relative to $B_1$, especially when it was exposed to air and light. Also, well separated streaks of aflatoxin $M_1$ on silica gel thinlayer chromatoplates, when excessively exposed to ultraviolet light during visualization of streak position, gave a multitude of new fluorescent bands upon rechromatogrpahy. These observations led us to test instability of $M_1$ in reconstituted experimental milk to ultraviolet irradation in thin film (See EXPERIMENTAL). The results showed that the sample that was kept in the dark had 26.2 µg of $M_1$ (93.6% recovery). On the other hand, the sample of reconstituted milk that was exposed to ultraviolet for 1 hr had only 2.3 µg representing only 8.2% of the original content (~92% destruction). Exposure in a very thin film may be an effective practical process with short exposure time. Potential rancidity problem could be overcome with appropriate filters to block out certain wavelengths.

## Aflatoxin $Q_1$ Pathway in Primates and Metabolic Detoxication

Indentification of aflatoxin $M_1$ in milk and in urine represents the first demonstration of a biotransformation of $B_1$ by mammalian metabolic action. The hydroxylation of $B_1$ to $M_1$ and excretion of the latter in urine and milk[9] represents a mammalilan detoxification pathway; Sinnhuber <u>et al</u>[19]. showed much reduced carcinogenicity of $M_1$ in the trout. Excretion of $M_1$ in urine of sheep was only about 5% of the $B_1$ dose. Similarly, our studies on the excretion of $M_1$ in milk of cows on aflatoxin $B_1$ regimen showed that the $M_1$ in the milk accounted for only 1–2% of the $B_1$ dose, suggesting that the pathway to $M_1$ is probably not a major detoxifying pathway. (Although $M_1$ is also excreted through other routes, and <u>de novo</u> $M_1$ may be removed <u>via</u> epoxidation and covalent binding to DNA and other macromolecules).

Comparative metabolic studies in the literature were aimed to test whether differences in susceptibility to aflatoxin carcinogenesis among species were related to differences in metabolic disposition of $B_1$. The studies were also aimed at identifying the proximate carcinogenic metabolite of $B_1$, since $B_1$ itself is not an immediate carcinogen per se, and requires activation to the proximate carcinogen (now known to be the 2,3-oxide of $B_1$ through epoxidation of the terminal furan ring) which interacts with DNA to produce the carcinogenic effect.

Our comparative metabolic studies in vitro with liver microsomal fractions resulted in the discovery of aflatoxin $Q_1$ as a hydroxylated mammalian metabolite of $B_1$, isomeric with $M_1$, with the position of the hydroxyl group at the carbon $\beta$ to the carbonyl of the cyclopentenone ring [13-15]. Aflatoxin $Q_1$ possesses the intact unsaturated terminal furan ring as aflatoxins $B_1$ and $M_1$, and is thus potentially susceptible to epoxidation through further metabolic action. Quantitative aspects of the comparative in vitro studies on the rate of conversion to $Q_1$ in different species resulted in the important discovery that aflatoxin $Q_1$, rather than $M_1$, represented a remarkably major pathway in primate liver including human liver as contrasted to other mammals. The results are summarized in Tables 19 and 20. The conversions to $Q_1$ with monkey livers from three species accounted for 20-30% of the added $B_1$ substrate compared to a conversion of only 0.1-3% with other mammalian and avian species [15]. The in vitro conversion to $M_1$

Table 21. Hepatocarcinoma Incidence in Trout Fed Aflatoxins $Q_1$ or $B_1$ [a]

| Group | Diet | % Incidence at: | | | Deaths (no. fish) |
| --- | --- | --- | --- | --- | --- |
| | | 4 mos. | 8 mos. | 12 mos. | |
| 1 | Basal alone | 0 | 0 | 0 | 2 |
| 2 | Basal + 20 ppb $Q_1$ | 0 | 0 | 0 | 4 |
| 3 | Basal + 20 ppb $B_1$ | 0 | 56 | 83 | 5 |

[a] Fish 6 mos. old when started, fed for 1 year. Each group 2 x 120 fish in two tanks. 40 fish were sampled from each group at 4 mos. and again at 8 mos., remaining sacrficed at 12 mos.

(Masri et. al.[16], and Hendricks et. al. [17]).

with all livers, was roughly the same, about 1-3%. The high
conversion of aflatoxin $B_1$ to $Q_1$ in primate liver and the potential
of epoxidation of $Q_1$ on the basis of intact unsaturated terminal
furan ring prompted us to test carcinogenicity of $Q_1$ in rainbow
trout in collaboration with Professors Sinnhuber and Hendricks at
Oregon State University[16,17]. The tests showed greatly reduced
carcinogenicity relative to $B_1$. The results summarized in Tables
21 and 22 show that aflatoxin $Q_1$ is about 100-fold less carcinogenic
than aflatoxin $B_1$. This finding provides another important insight,
namely, that the pathway to aflatoxin $Q_1$ represents a major successful detoxifying mechanism in primates. It should be added however,
that the 100-fold reduction of carcinogenicity does not remove
aflatoxin $Q_1$ from the class of potent carcinogens.

Examination of the results in Tables 19 and 20 on the _in vitro_
conversion of $B_1$ by rat and monkey livers shows that a substantial
part of the added substrate is unaccounted for (after adding $M_1$ +
$Q_1$ + unused recovered $B_1$ substrate). Also, incubation with $^{14}C$-
aflatoxin $B_1$ showed that substantial amount of radioactivity was
bound to the liver fraction. Epoxidation of the $B_1$ substrate (and

Table 22.  Hepatocarcinoma Incidence in Trout Fed Aflatoxin $B_1$ and
           $Q_1$ and Methyl Sterculate (MS)[a]

| Group | Diet | 6 mos. | 9 mos. | 12 mos. | Deaths (no. fish) |
|---|---|---|---|---|---|
| 1 | Basal alone | 0 | 0 | 0 | 0 |
| 2 | Basal + 100 ppb $Q_1$ | 0 | 0 | 10.6 | 7 |
| 3 | Basal + 50 ppb MS | 0 | 5 | 41 | 13 |
| 4 | Basal + 100 ppb $Q_1$ + 50 ppb MS | 15 | 30 | 89 | 1 |
| 5 | Basal + 4 ppb $B_1$ | 0 | 25 | 48 | 6 |

[a] Fish started at 4 mos. old fed for 1 year except group 5 which
were 2 mos. when started. Each group: 2 x 80 fish in two tanks.
$Q_1$ was omitted with groups 2 and 4 the last 2 mos. At 6 and 9
mos., 2 x 10 fish were sampled from each group, remaining were
sacrificed at 12 mos.

(Masri et. al.[16], and Hendricks et. al.[17]).

probably of $M_1$ and $Q_1$ produced during incubation) may be responsible for the missing balance. The epoxidation pathway and subsequent binding of the 2,3-oxides with DNA in vivo are probably responsible for the (milder) carcinogenicity of aflatoxins $M_1$ and $Q_1$. These aflatoxins, from a structural consideration, could potentially undergo epoxidation, although the presence of the hydroxyl group in their structure (which renders them less planar than $B_1$) may hinder their interaction and intercalation with DNA and covalent binding. Hydroxylation may also enhance their excretion in vivo through increased solubility and potential conjugation. Binding of $M_1$ to DNA at the $N^7$-guanyl residue in vivo has been tentatively identified by Croy and Wogan [20] and by Hertzog et al.[21].

## Epoxidation Pathway and Detoxifying Role in Trophic Chain

Swenson and coworkers[22-24] demonstrated activation of $B_1$ by microsomal oxygenase to $B_1$-2,3-oxide. This reactive electrophile binds covalently to DNA to initiate the carcinogenic process. The major binding to DNA (~70%) is principally at the $N^7$ position of the guanine residue [20,21,25-28]. Thus, hydrolysis of the DNA adduct gives mainly 2,3-dihydro-2-($N^7$-guanyl)-3-hydroxyaflatoxin $B_1$. This derivative ($N^7$-Gua-$B_1$) is also excreted in urine of rats given $B_1$[29]. The DNA $N$-$^7$-Gua-$B_1$ adduct is unstable in vitro and has a short half-life (hours) in liver DNA as shown by Croy and Wogan [20] and Hertzog et al.[21]. Kinetic studies showed efficient removal of this lesion from liver DNA by two main processes: a) spontaneous chemical (and very likely enzymic excision repair) depurination of the affected guanine residues; and, b) conversion of this lesion to two new more stable lesions, which appear to result from ring scission of the positively charged imidazole moiety of the affected (7,9-disubstituted) guanine residues. Two putative DNA-formamidopyrimidine-$B_1$ adducts (DNA-FAPY-$B_1$) are formed. These new adducts have relatively long half-lives and accumulate in DNA upon repeated $B_1$ dosing. Their structures as deduced from hydrolysis of liver DNA have been proposed. When extensively reacted DNA from aflatoxin $B_1$ treated rats is hydrolysed, 2,3-dihydro-2,3-dihydroxyaflatoxin $B_1$ ($B_1$-2,3-diol) is also released [20,21]. The diol itself may react with DNA[30].

The exact mechanisms in which these lesions initiate carcinogenesis are not known. They may involve the generation of apurinic sites (from excision of $N^7$-Gua-$B_1$) which may result in single strand breaks. Also, the persistent DNA-FAPY-$B_1$ lesions would have different base-pairing properties at replication from the original guanine residues from which they are derived. The unexcised DNA adducts may result in expressible genetic damage[31].

The biological effects of the free (hydrolysed) reaction products of all these DNA lesions are unknown. While the lesions are detrimental to the organism in which they are initially produced, the biological effects of ingestion of the free or DNA bound reaction products in a recipient organism are also unknown, although results of our bioassays of pig and steer FDT suggest that the products are detoxified or biologically unavailable since no toxicity was found in the ducklings consuming these tissues. The recent findings on the covalent binding of $B_1$-2,3-oxide to DNA and kinetics of formation and removal of the DNA lesions, as well as results of the bioassays of the FDT, point to the important role of the epoxidation pathway as a successful mechanism or strategy in preventing transmission of toxicity upward in the trophic chain, by containment or removal of the genotoxicity at initial contact points at the lower trophic levels. This is not surprising, in view of the recent knowledge of the chemistry and kinetics of these lesions which show that the major lesion is rapidly removed by depurination followed by excretion. The other persistent lesions (FAPY-$B_1$) are probably biologically unavailable or may give rise to products lacking the unique structural features (esp. 2,3 double bond) and reactivity of $B_1$ requisite for its genotoxicity.

The metabolism of $B_1$ to $M_1$ (which does not involve the epoxidation pathway) and subsequent excretion of $M_1$ in milk is a serious exception to the prevention of transmission of toxicity from feed to food as was discussed earlier. The situation arises due to efficient excretion of this metabolite by the mammary gland. Inefficient epoxidation of de novo $M_1$ and/or inefficient reaction of $M_1$-2,3-oxide with DNA may also be involved.

Note on Analytical Methodology

The development of sensitive analytical methods had been an indispensible tool in advancing research in aflatoxins, whether the research need was for survey of agricultural crops, transmission of residues from feed to food, following the course of detoxification, preparation of spiked meals for experimental use, or for toxicologic and metabolic studies. Methods had to be developed as the research progressed, since there were no available or established methods. The task has been difficult for other reasons also. First, there are the inherent problems of statistical sampling procedures, compounded by typically focal infections, low concentration of natural occurrence and instability after extraction. Second, the high biologic potency and carcinogenicity dictate interest in and concern with very low concentrations. Finally, there are analytical problems associated with low concentrations, such as the need for certainty and proof of structural identity in measured analyses, the presence of different interfering substances in different commodities, and the difficulty of securing authentic standards. On the other hand, three properties of aflatoxin facilitated development of analytical methodology. First, is the very intense fluorescence, high molar absorptivity, and distinctive absorption maxima of aflatoxins in the ultraviolet region. Second, is the unique elemental composition of the aflatoxins. This property allows the mass spectrometer, when used in the high resolution selected ion monitoring mode, to monitor ions having specific elemental compositions, as opposed to monitoring an integral mass signal, which may be composed of several ions of different elemental composition. Third, is the high biologic potency of aflatoxins affording sensitive bioassay methods for confirmation of identity. All these distinctive properties coupled with advances in separation and identification techniques and instruments such as TLC, HPLC, high resolution mass spectrometry and mass spectral selected ion monitoring, have contributed a wealth of powerful analytical methods. These should also be considered as yet another line of defense, as monitoring, identifying, and measuring tools against aflatoxin exposure.

Our contribution in analytical methods include column cleanup procedures and separation of $B_1$ and $M_1$ in different fractions[8], methods for aflatoxin $M_1$ in milk[10,11], and an ultrasensitive mass spectral method capable of quantitative analysis at the nanogram level of a mixture of aflatoxins (without the need for separation) with proof of identity[32,33].

In conclusion, I described some of our results of mycotoxin research, representing operational lines of defense against the potential of aflatoxin carcinogenesis in man through potential toxin transmission in the food supply. Chemical, biological, and metabolic detoxification processes and mechanisms, and appropriate actional strategies are also discussed.

Reference to a company and/or product named by the Department is only for purposes of information and does not imply approval or recommendation of the product to the exclusion of others which may also be suitable.

ACKNOWLEDGEMENTS

I thank V. Garcia Randall, J. R. Page and J. Sucy for assistance, L. A. Goldblatt, H. Vix, H. Gardner, F. Dollear, R. Sinnhuber, J. Hendricks and O. Shotwell for collaboration, D. O'Connel, L. Whitehand and B. Mackey for statistical treatments, W. Gagne for histopathologic examinations, Drs. Mendel Friedman, A. Douglas King and Charles C. Huxsoll for reading the manuscript, and Esperanza Gonzalez for assistance in preparing the manuscript.

REFERENCES

1.  R.   Allcroft,   and   R.   B.   A.   Carnaghan,   Toxic   products   in
    ground nuts,  Biologic  effects,  Chem.  and  Ind.  Jan.  12,
    1963, p. 50.

2.  K.   Sargeant,   Toxic   products   in   groundnuts,   chemistry   and
    origin, Ibid, p. 53.

3.  P.   K.   C.   Austwick,   and   G.   Ayerst,   Groundnut   microflora   and
    toxicity, Ibid. p. 55.

4.  A.  C.  Keyl,  A.  N.  Booth,  M.  S.  Masri,  M.  R.  Gumbmann and W.  E.
    Gagne,  Chronic  effects  of  aflatoxin  in  farm  animal  feeding
    studies,  In Proc.  First U.S.-Japan Conf.  Toxic Microorga-
    nisms, M. Herzber, ed. (1970).

5.  M.   S.  Masri,  V.  C.  Garcia  and  J.  R.  Page.  The  aflatoxin M
    content of milk from cows fed known amounts of aflatoxin,
    Vet. Rec., 84:146 (1969).

6.  F.   H.  Kratzer,  D.  Bandy,  M.  Wiley  and  A.  N.  Booth,  Aflatoxin
    effects in poultry,  Proc.  Soc.  Exp.  Biol.  Med.,  131:1281
    (1969).

7.  M.  R.  Gumbmann,  S.  N.  Williams,  A.  N.  Booth,  Pran Vohra,  R.  A.
    Ernst and M. Bethard,  Aflatoxin  susceptibility  in  various
    breeds of poultry, Ibid. 134:683 (1970).

8.  M.   S.  Masri,  Partition  of  aflatoxin  $B_1$,  J.  Am.  Oil  Chemists
    Soc., 47:61 (1970).

9.  M.   S.   Masri,   R.   E.   Lundin,   J.   R.   Page   and   V.   C.   Garcia
    Crystalline aflatoxin  $M_1$  from  urine  and  milk,  Nature
    215:753 (1967).

10. M.  S. Masri,  J.  R. Page and V.  C. Garcia, Analysis for aflatoxin
    M  in  milk,  J.  Assoc.  Official  Anal.  Chemists,  51:594
    (1968).

11. M.   S.,  Masri,  J.  R.  Page  and  V.  C.  Garcia,  Modification  of
    method for aflatoxins in milk, Ibid. 52:641 (1969).

12. M.   S.  Masri,  H.  L.  E.  Vix  and  L.  A.  Goldblatt,  Process  for
    detoxifying  substances  contaminated  with  aflatoxin,  U.S.
    Patent 3,429,709.

13. M.  S.  Masri,  W.  F.  Haddon,  R.  E.  Lundin  and  D.  P.  H.  Hsieh,
    Aflatoxin  $Q_1$.  A  newly  identified  major  metabolite  of

aflatoxin $B_1$ in monkey liver, J. Agric. Food Chem., 22:512 (1947).

14. D. P. H. Hsieh, J. I. Dalezios, R. I. Kreiger, M. S. Masri and W. F. Haddon, Use of monkey liver microsomes in production of aflatoxin $Q_1$, Ibid. 22:515 (1974).

15. M. S. Masri, A. N. Booth, and D. P. H. Hsieh. Comparative metabolic conversion of aflatoxin $B_1$ to $M_1$ and $Q_1$. Life Sciences 15:203 (1974).

16. M. S. Masri, D. J. Hendricks and R. O. Sinnhuber, Aflatoxin $Q_1$- major detoxifying pathway in primates, carcinogenicity tests in rainbow trout, Toxicon 17 (Suppl. 1):116 (1979).

17. J. D. Hendricks, R. O. Sinnhuber, J. E. Nixon, J. H. Wales, M. S. Masri and D. P. H. Hsieh, Carcinogenicity of $AFQ_1$ to rainbow trout, J. Natl. Cancer Inst., 4:523 (1980).

18. P. Krogh, B. Hald, E. Hasselager, A. Madsen, H. P. Mortensen, A. E. Larsen and A. D. Campbell, Aflatoxin residues in bacon pigs, Pure and Appl. Chem. 35:273 (1973).

19. R. O. Sinnhuber, D. J. Lee and J. H. Wales, Hepatic carcinogenesis of aflatoxin $M_1$ in rainbow trout, J. Natl. Cancer Inst. 53:1285 (1974).

20. R. G. Croy and G. N. Wogan, Removal of covalent aflatoxin $B_1$ adducts from rat liver DNA, Cance Res., 4:197 (1981).

21. P. J. Hertzog, J. R. L. Smith and R. C. Garner, Removal of DNA-bound aflatoxin $B_1$, in rat liver and in vitro, carcinogenesis, 1:787 (1980).

22. D. H. Swenson, J. Lin, E. C. Miller and J. A. Miller, Aflatoxin $B_1$-2,3-oxide as a probable intermediate in the covalent binding of aflatoxin $B_1$ and $B_2$ to rat liver DNA and ribosomal RNA in vivo, Cancer Res., 37:172 (1977).

23. D. H. Swenson, E. C. Miller and J. A. Miller, Aflatoxin $B_1$-2,3-oxide: evidence for its formation in rat liver in vivo and by human microsomes in vitro, Biochem. Biophys. Res. Commun., 60:1036 (1974).

24. D. H. Swenson, J. A. Miller and E. C. Miller, 2,3-Dihydro-2,3-dihydroxy-aflatoxin $B_1$: an acid hydrolysis product of an RNA-aflatoxin $B_1$ adduct formed by hamster and rat liver microsomes in vitro, Biochem. Biophys. Res. Commun., 53:1260 (1973).

25.  J. K. Lin, J. A. Miller and E. C. Miller, 2,3-Dihydro-2-(guan-
     7-yl)-3-hydroxy-aflatoxin $B_1$, a major acid hydrolysis
     product of aflatoxin $B_1$-DNA or ribosomal RNA adducts formed
     in hepatic microsome-mediated reactions and in rat liver
     in vivo, Cancer Res., 37:4430 (1977).

26.  R. G. Croy, J. M. Essigmann, V. N. Reinhold and G. N. Wogan,
     Identification of the principal aflatoxin $B_1$-DNA adduct
     formed in vivo in rat liver, Proc. Natl. Acad. Sci. (USA),
     75:1745 (1978).

27.  C. N. Martin and R. C. Garner, Aflatoxin $B_1$-oxide generated by
     chemical or enzymic oxidation of aflatoxin $B_1$ causes
     guanine substitution in nucleic acids, Nature 267:863
     (1977).

28.  R. G. Croy, J. E. Nixon, R. O. Sinnhuber and G. N. Wogan,
     Aflatoxin $B_1$-DNA adducts in rainbow trout embryos and
     liver, Carcinogenesis 1:903 (1980).

29.  R. A. Bennett, J. M. Essigmann and G. N. Wogan, Excretion of
     aflatoxin-guanine adduct in the urine of aflatoxin $B_1$-
     treated rats, Cancer Res., 41:650 (1981).

30.  B. F. Coles, A. M. Welch, P. J. Hertzog, J. R. Lindsay Smith
     and R. C. Garner, Biological and chemical studies on 8,9-
     dihydroxy-8,9-dihydro aflatoxin $B_1$ and some of its esters,
     Carcinogenesis, 1:79 (1980).

31.  L. L. Yang, V. M. Maher and M. McCormick, Error-free excision
     of the cytotoxic mutagenic $N^2$-deoxyguanosine DNA adduct
     formed in human fibroblasts by 7,8-dihydroxy-9,10-epoxy-
     7,8,9,10-tetrahydrobenzo [d]pyrene, Proc. Natl. Acad. Sci.
     USA, 77:5933 (1980).

32.  W. F. Haddon, M. S. Masri, Determination of aflatoxins by high-
     resolution mass spectroscopy, Toxicon 17 (Suppl. 1):64
     (1979).

33.  W. F. Haddon, M. S. Masri, V. Randall, R. Elsken and B.
     Menneghelli, J. Assoc. Official Anal. Chemists, Mass
     spectral analysis of aflatoxins, 60:107 (1977).

# 6

METABOLISM OF FOOD TOXICANTS:  SACCHARIN AND AFLATOXIN $B_1$,

A CONTRAST IN METABOLISM AND TOXICITY

J. L. Byard

Department of Environmental Toxicology
University of California
Davis, California 95616

ABSTRACT

The artificial sweetener, saccharin, and the secondary mold product, aflatoxin $B_1$, are present in many foods consumed by humans. Both chemicals produce cancer in rats. For this reason, they have aroused concern among scientists and the public as to the carcinogenic risk that they pose to humans. A comparison of these two chemicals reveals striking contrasts in potency, metabolism, mechanism-of-action, and experimental approach to assessing metabolism. These contrasts are examined in detail to illustrate the importance of metabolism in safety evaluation.

INTRODUCTION

The metabolism of a food chemical is often a critical determinant in evaluating toxic hazard to humans. Also, the methods for assessing metabolism in humans is usually dictated by the toxic potency of the chemical and its mechanism-of-action. These concepts can be illustrated by comparing the potency, metabolism and mechanism-of-action of two carcinogenic food chemicals, aflatoxin $B_1$ and saccharin.

DISCUSSION

Aflatoxin $B_1$ is a food contaminant produced by the molds Aspergillus flavus and Aspergillus parasiticus. Peanuts and corn are two foods, among others, that are commonly contaminated by this potent hepatocarcinogen (Heathcote and Hibbert, 1978). Assessment

147

of human hazard due to aflatoxin $B_1$ ingestion is based on epidemiology, animal toxicity studies, and animal metabolism studies.
Human epidemiology studies indicate a clear relationship of
increasing incidence of liver cancer with increasing levels of
dietary aflatoxin $B_1$ (Heathcote and Hibbert, 1978). For example,
daily ingestion of 3.5 ng/kg body weight occurred in a population
with a rate of 0.7 liver cancers/$10^5$ population/year, whereas a
population with a daily ingestion of 222 ng/kg body weight had 25
liver cancers/$10^5$ population/year. Intermediate exposures produced
intermediate rates of liver cancer. Because of confounding factors,
such as the presence of other dietary carcinogens and the occurrence
of hepatitis B virus, there is considerable uncertainty in using
epidemiological data to estimate the carcinogenic potency of
aflatoxin $B_1$.

     Laboratory rats develop a high incidence of malignant liver
cancers when fed diets containing levels as low as 50 ppb aflatoxin
$B_1$ (Wogan et al., 1974). Hyperplasia and carcinoma of the liver
were elevated at the lowest dose of 1 ppb! However, adult mice are
refractory to the hepatocarcinogenic effects of aflatoxin $B_1$ (Wogan,
1973). This striking species difference creates considerable
uncertainty as to the human susceptibility to the hepato-
carcinogenicity of aflatoxin $B_1$. Are we more like the rat or the
mouse?

     Species differences in susceptibility can often be explained by
differences in metabolism, particularly in metabolic pathways that
produce the biochemical lesion which leads to carcinogenicity.
Aflatoxin $B_1$ is metabolically activated to an electrophilic epoxide
that forms stable covalent adducts with the nitrogen and oxygen
atoms of bases in DNA (Essigman et al., 1977; Lin et al., 1977).
Replication of DNA containing such adducts results in frameshift
mutations (McCann et al., 1975). Hence, aflatoxin $B_1$ is genotoxic
by two criteria, chemical modification of DNA and mutagenesis.
These two processes are considered to be the initiating events in
the carcinogenic action of aflatoxin $B_1$. Mutagenicity is difficult
to assess in liver or in primary cultures of hepatocytes because
liver cells replicate very slowly. However, the extent of covalent
binding of aflatoxin $B_1$ to macromolecules, particularly to DNA,
corelates well with species susceptibility to carcinogenesis.
Primary hepatocyte cultures from rats as compared to mice form
twenty times as many macromolecular adducts of aflatoxin $B_1$ (Decad
et al., 1979). A similar species difference in DNA adducts was
measured in liver slices incubated in aflatoxin $B_1$ (Booth et al.,
1981). In vivo, rat liver forms fifty times as many DNA adducts as
mouse liver (Croy and Wogan, 1981). Hence, for a genotoxic
chemical, such as aflatoxin $B_1$, there appears to be a good corela-
tion between the level of DNA adducts and carcinogenicity. Using
this concept, how might we better estimate human susceptibility?

We can't intentionally feed aflatoxin $B_1$ to humans to measure the carcinogenic dose-response. Considering the genotoxicity of aflatoxin $B_1$, we can't ethically give a trace dose of radiolabeled aflatoxin $B_1$ to humans to measure DNA adducts in a liver biopsy. One possible approach is to measure metabolism and binding to DNA in primary hepatocyte cultures prepared from human liver biopsies. In a preliminary experiment, the covalent binding of aflatoxin $B_1$ to macromolecules precipitated by trichloroacetic acid was measured in primary hepatocyte cultures prepared from one human liver biopsy.[1] The level of binding was about 10% compared to about 13% in the rat and about 0.6% in the mouse. Although we can't predict human susceptibility based on one individual, this data illustrates how a cellular approach to metabolism may be useful in assessing the susceptibility of humans to genotoxic chemicals.

Another carcinogenic chemical in foods is the synthetic sweetener, saccharin. In discussing saccharin, a comparison with aflatoxin $B_1$ is helpful to contrast differences in potency, metabolism, mechanism-of-action, and approach to measuring metabolism in humans. Typical human exposure levels to saccharin range from an average of about 30 mg/day to a maximum of about 300 mg/day (Cranmer, 1980), a range of about 0.43 - 4.3 mg/kg body weight/day. Several epidemiological studies have not found a relationship between ingestion of saccharin and cancer (Cranmer, 1980). The consensus at this time is that human epidemiological studies do not support a relationship between exposure to saccharin and cancer. Much lower levels of aflatoxin $B_1$ suggest a clear relationship between dietary level and liver cancer in humans.

In the laboratory rat, saccharin produces bladder cancer when fed at levels of 5% and 7.5% of the diet (Arnold et al., 1980; Cranmer, 1980). Although carcinogenic, saccharin is about one fifty-millionth as potent as aflatoxin $B_1$. What is the basis for this great difference in potency?

Metabolism studies in laboratory animals, including the rat, indicate that saccharin is not metabolized (Byard and Golberg, 1973; Mathews et al., 1973; Minegishi et al., 1972). Saccharin is readily absorbed from the gastrointestinal tract and is rapidly excreted by the kidneys. Saccharin does not form covalently bound adducts with DNA in the bladder of the rat in an assay that would have detected one mole of saccharin per $10^8$ moles of DNA phosphate (Lutz and Schlatter, 1977). By contrast, aflatoxin $B_1$ has a covalent binding index in liver of 10,300 (Lutz, 1979) compared to a value for saccharin in the bladder of less than 0.05 (Lutz and Schlatter, 1977). Saccharin also appears to be nonmutagenic (Cranmer, 1980).

---

[1] Byard, J.L., J.A. Reese, C.E. Green, C.B. Salocks and B.H. Ruebner, 1981, Unpublished data.

Therefore, one can conclude that saccharin is nongenotoxic, and produces bladder cancer through a nongenotoxic mechanism. Changes in urinary volume and pH, as well as the occurrence of calculi in the bladder have been proposed as possible mechanisms by which saccharin chronically irritates the bladder epithelium, thereby promoting carcinogenesis (Cranmer, 1980). These changes in urinary physiology are reversible and occur only at very high levels of saccharin.

Based on the lack of irreversible genotoxicity, ethical metabolism studies of saccharin can and have been done in humans (Byard et al., 1974; Sweatman et al., 1981). Although covalent binding to bladder DNA was not measured in these studies, such measurements could probably be ethically done in human volunteers, who for some medical reason are scheduled for bladder surgery.

In summary, our knowledge of potency and mechanism can aid our decision as to how to assess potency in humans. Nongenotoxic carcinogens, such as saccharin, permit direct determination of metabolism in humans. The metabolism of genotoxic chemicals, such as aflatoxin $B_1$, cannot be measured directly in humans, but primary cultures of the target cell, derived from human tissues, may provide a good indication of in vivo human metabolism. Such metabolic data is essential for assessing human susceptibility from carcinogenicity bioassays in laboratory animals.

REFERENCES

Arnold, D.L., C.A. Moodie, H.C. Grice, S.M. Charbonneau, B. Stauric, B.T. Collins, P.F. McGuire, Z.Z. Zawidzka and I.C. Munro, 1980, Long-term toxicity of ortho-toluenesulfonamide and sodium saccharin in the rat, Toxicol. Appl. Pharmacol., 52:113.

Booth, S.C., H. Bösenberg, R.C. Garner, P.J. Hertzog and K. Norpoth, 1981, The activation of aflatoxin $B_1$ in liver slices and in bacterial mutagenicity assays using livers from different species including man, Carcinogenesis, 2:1063.

Byard, J.L., E.W. McChesney, L. Golberg and F. Coulston, 1974, Excretion and metabolism of saccharin in man: II. Studies with $[^{14}C]$ labeled and unlabeled saccharin, Fd. Cosmet. Toxicol., 12:175.

Byard, J.L. and L. Golberg, 1973, Metabolism of saccharin in laboratory animals, Fd. Cosmet. Toxicol., 11:391.

Cranmer, M.F., 1980, "Saccharin, a Report," American Drug Research Institute, U.S.A.

Croy, R.G. and G.N. Wogan, 1981, Quantitative comparison of covalent aflatoxin-DNA adducts formed in rat and mouse livers and kidneys, J. Natl. Cancer Inst., 66:761.

Decad, G.M., K.K. Dougherty, D.P.H. Hsieh and J.L. Byard, 1979,
    Metabolism of aflatoxin $B_1$ in cultured mouse hepatocytes:
    Comparison with rat and effects of cyclohexene oxide and
    diethyl maleate, Toxicol. Appl. Pharmacol., 50:429.
Essigman, J.M., R.G. Croy, A.M. Nadzan, W.F. Busby, V.N. Reinhold,
    G. Buchi and G.N. Wogan, 1977, Structural identification of the
    major DNA adduct formed by aflatoxin $B_1$ in vitro, Proc. Natl.
    Acad. Sci. U.S.A., 74:1870.
Heathcote, J.G. and J.R. Hibbert, 1978, "Aflatoxins: Chemical and
    Biological Aspects," Elsevier, New York.
Lin, J., J.A. Miller and E.C. Miller, 1977, 2,3-Dihydro-3-(guan-7-
    yl)-3-hydroxy-aflatoxin $B_1$, a major acid hydrolysis product of
    aflatoxin $B_1$-DNA of ribosomal RNA adducts formed in hepatic
    microsome-mediated reactions and rat liver in vivo, Cancer.
    Res., 37:4430.
Lutz, W.K., 1979, In vivo covalent binding of organic chemicals to
    DNA as a quantitative indicator in the process of chemical
    carcinogenesis, Mutation Research, 65:289.
Lutz, W.K. and C.H. Schlatter, 1977, Saccharin does not bind to DNA
    of liver or bladder in the rat, Chem.-Biol. Interactions,
    19:253.
Mathews, H.B., M. Fields and L. Fishbein, 1973, Saccharin
    Distribution and excretion of a limited dose in the rat, J.
    Agr. Food Chem., 21:916.
McCann, J., E. Choi, E. Yamasaki and B.N. Ames, 1975, Detection of
    carcinogens as mutagens in the Salmonella/microsome test: Assay
    of 300 chemicals, Proc. Nat. Acad. Sci. USA, 72:5135.
Minegishi, K.I., M. Asahina and T. Yamaha, 1972, The metabolism of
    saccharin and related compounds in rats and guinea pigs, Chem.
    Pharm. Bull., 20:1351.
Sweatman, T.W., A.G. Renwick and C.D. Burgess, 1981, The
    pharmacokinetics of saccharin in man, Xenobiotica, 11:531.
Wogan, G.N., 1973, Aflatoxin carcinogenesis, Methods Cancer Res.,
    7:309.
Wogan, G.N., S. Paglialunga and P.M. Newberne, 1974, Carcinogenic
    effects of low dietary levels of aflatoxin $B_1$ in rats, Fd.
    Cosmet. Toxicol., 12:681.

# EFFECT OF DIET ON T-2 TOXICOSIS[1]

T.K. Smith and M.S. Carson

Department of Nutrition
University of Guelph
Guelph, Ontario, Canada N1G 2W1

## ABSTRACT

T-2 toxin is an emetic trichothecene mycotoxin produced by
Fusarium molds.  This compound causes feed refusal, emesis and
lesions in the gastrointestinal tract of livestock, poultry and man.
Studies in our laboratory have indicated that the feeding of high
fibre diets, non-nutritive mineral additives and high fat diets
can largely overcome feed refusal caused when T-2 toxin is fed to
rats.  Subsequent experiments were designed to determine the mechanism by which such diets exert this effect.  Rats were fed for
two weeks diets containing varying levels of alfalfa meal, bentonite
or corn oil in a casein-based semi-purified diet.  Rats were then
orally dosed with [$^3$H] T-2 toxin and urine and feces were collected
for 21 hours after which all animals were killed and tissues ex-
cised.  Diet had no significant effect on the fraction dose of $^3$H
excreted in the urine.  Significant increases in fecal excretion
of $^3$H were seen, however, with all test diets.  Only high fat diets
reduced hepatic residues of $^3$H while alfalfa had a similar effect
in kidney and both alfalfa and bentonite lowered muscle residues.
It was concluded that such dietary treatments overcome T-2 toxicosis mainly by promoting fecal excretion of toxin thereby reducing
absorption and biological half-life.

## INTRODUCTION

T-2 toxin [3 -hydroxy-4β, 15-diacetoxy-8α-(3 methylbutyryloxy)-
12, 13-epoxytrichotehc-9-ene] is a trichothecene mycotoxin produced

---

[1]Supported by the Natural Sciences and Engineering Research Council
of Canada and the Ontario Ministry of Agriculture and Food.

by various species of Fusarium, Myrothecium Trichoderma, Cephalo-
sporium, Verticimonosporium and Stachybotrys (Ueno, 1977). About
fifty trichothecenes have been identified although new compounds
are still being reported (Ishii and Ueno, 1981). Trichothecene
mycotoxicoses have been described in man and many animal species
and are characterized by feed refusal, hemorrhage of the mucosal
epithelium of the stomach and intestine and emesis (Ueno, 1977).
T-2 toxin is one of the more common trichothecenes and has been associated with the death of Wisconsin dairy cattle fed Fusarium
contaminated corn (Hou et al., 1972) A more recent report identified T-2 toxin as a component of moldy barley which caused outbreaks of mycotoxicoses in Canadian geese, ducks, horses and swine
(Puls and Greenway, 1976).

Dietary fibre has been demonstrated to protect against toxicoses resulting from numerous xenobiotic compounds (Ershoff, 1974).
Ershoff reported that unrefined plant fibres can overcome the toxicity of food dyes when fed to rats (Ershoff, 1977). Such effects
have sometimes been attributed to the physical properties of fibres
(Takeda and (Kirieyama, 1979) as well as their ability to alter
intestinal transit time (Takeda et al., 1982). Chadwick et al.
(1978), however, observed that dietary fibre could also alter the
metabolism of foreign compounds.

Diets rich in fibre have been shown to overcome the toxicity
of zearalenone (Smith, 1980a), a uterotropic Fusarium mycotoxin.
Alfalfa was particularly effective and has been shown to alter
zearalenone metabolism (Smith, 1980b; James and Smith, 1982).

Investigators have recently attempted to treat toxicoses caused
by environmental agents through the use of orally-administered, non-
nutritive polymers that can reduce gastrointestinal absorption of
the toxin. Cholestyramine resins singly and in combination with
mineral oils have been shown to promote fecal excretion of brominated
biphenyls (Kozman et al., 1982) while divinylbenzene/styrene polymers
have been reported to overcome toxicity of zearalenone (Smith, 1980b;
Smith, 1982). In vitro studies with bentonite have indicated this
pellet binding agent may bind aflatoxin (Masimango et al., 1979).

Experiments were conducted, therefore, to determine the capacity
of diets containing assorted fibres, non-nutritive mineral additives
and oils to overcome T-2 toxicosis in rats.

MATERIALS AND METHODS

Animals and Diets

Male weanling Wistar rats (50-55 g, Woodlyn Labs., Guelph,
Ontario) were used in all experiments. Animals were individually
housed in stainless steel cages and were exposed to 12 hours of

light daily with water supplied ad libitum. Each diet was fed to
ten rats chosen at random with growth, feed consumption and feed
efficiency (gain/feed consumed) calculated for each rat.

Test diets were formulated to be isonitrogenous with the
casein-based semipurified control diet (Carson and Smith, 1983)
although no attempt was made to keep diets isoenergetic. In all
growth trials, diets were fed for two weeks with and without crystalline T-2 toxin (Myco-Labs, Chesterfield, MO) which was dissolved
in acetone and added dropwise to the appropriate diets to give a
final concentration of 3 µg/g diet.

Growth Trials

In the first trial, purfied fibres were added to the control
diet to determine any effect of overcoming toxin-induced feed refusal. Cellulose (United States Biochemical Corp., Cleveland, OH),
hemicellulose (Masonite Corp., Laurel, MS), pectin (United States
Chemical Corp., Cleveland, OH) and lignin (Reed Paper Co., Quebec,
Quebec) were fed at 20% of the control diet by substitution for
cellulose and a fraction of cornstarch.

The second trial was designed to test pectin and lignin at
lower levels than those used in trial 1. Pectin was supplied at
0, 2.5, 5 and 10% while lignin was included at 0, 2.5, 5 and 7.5%.

Dehydrated alfalfa meal (Martin Feed Mills, Elmira, Ontario),
a lignin-rich feedstuff, was added to the control diet at 0, 5, 10,
15, 20 and 25% in the third trial. Diets were maintained isonitrogenous by appropriate reductions in the level of casein.

In the fourth trial, several non-nutritive polymers were
tested to determine their effect on feed refusal and growth depression caused by T-2 toxin. The control diet was supplemented with
bentonite (BDH Chemicals, Toronto, Ont.), Zonolite Verxite (W.R.
Grace and Co., Ajax, Ont.; purified vermiculite-hydrobiotite composed of hydrated magnesium silicate containing variable amounts
of other exchangeable mineral ions principally potassium, iron and
calcium), or synthetic anion or cation exchange resins
(divinylbenzene/styrene polymer; Culligan Inc., Guelph, Ont.).
Additions were 5% of the diet and were made at the expense of cornstarch in all trials.

The effectiveness of different levels of anion-exchange resin
was measured in the fifth trial. The control diet was supplemented
with 2.5, 5.0, 7.5 and 10.0% resin. Bentonite was provided at
2.5, 5.0, 7.5 and 10.0% in the sixth trial.

In the seventh trial, corn oil was fed at 0.5, 6.0, 10.0,
15.0 and 20.0% while lard was supplied at 2.0, 6.0 10.0, 15.0 and
20.0% in the eighth trial.

Enzyme Assay

     Rats fed alfalfa, bentonite and corn oil supplements were
killed at the end of the growth trials and livers were removed for
measurement of non-specific hepatic esterase activity (EC 3.1.1.1;
Dabich et al. (1968)).  Specific activity was expressed as µmoles
α-napthol produced/minute/mg protein with protein determined according to Lowry et al. (1951).

Radionuclide Purification

     [$^3$H] T-2 toxin was prepared by the Wilzbach method (New
England Nuclear, Boston, MA).  The crude product was dissolved in
acetone:water (1:1) and allowed to stand for several hours.  All
labile tritium was removed with a rotary evaporator.  The residue
was dissolved in benzene and was applied to a column of silica gel
60 (14.5 x 1.5 cm, 70-230 mesh, BDH Chemicals, Toronto, Ont.) which
was developed with 100 ml benzene followed by 300 ml benzene:acetone
(3:1).  Five ml fractions were collected and a small portion of each
was applied to a TLC silica gel GF plate (Fisher Scientific Co.,
Toronto, Ont.) which was developed in benzene:acetone (63:37).
Fractions containing T-2 toxin were identified by comparison with
known standards.  The toxin was then recrystallized by the method
of Bamburg et al. (1968).  Specific activity of [$^3$H] T-2 toxin was
determined to be 0.172 mCi/mg with the concentration of T-2 toxin
measured by gas chromatography (Mirocha et al., 1976).  The in vitro
method of Chi et al. (1978) was used to test for possible $^3$H exchange under acidic conditions similar to those of the stomach.

Excretion and Residue Trials

     Rats were fed diets containing 0, 5, 12.5 or 20% alfalfa; 5,
5.7 or 10% bentonite and 1.25, 6, 10, 15 or 20% corn oil for two
weeks.  Rats were weighed on the fourteenth day and 7.5 x 10$^7$
dpm [$^3$H] T-2 toxin/kg body weight was added to 4 g of the appropriate
diet at the beginning of the dark cycle.  After about two hours feed
was again supplied ad libitum.  Urine and feces were collected for
21 hours after dosing after which time all rats were decapitated
and liver, kidneys and a muscle sample were removed for determination
of residual $^3$H.  Small intestine samples were also taken for rats
fed alfalfa and bentonite.

     A second trial was then conducted using the same protocol
except that the toxin was administered topically to rats fed the
diets containing alfalfa or bentonite.  Rats were shaved on the back
at the base of the skull and [$^3$H] T-2 toxin was applied to the skin
in 0.1 ml acetone.  All samples were collected as previously described.  Radioactivity in urine, feces and tissues was determined
according to Carson and Smith (1983).

Determination of Passage Time

The effect of additives on passage time of ingesta was determined according to Guncaga et al. (1974). Rats were fed the diets used in excretion and residue trials containing alfalfa bentonite or corn oil for one week after which time all diets were fed supplemented with 1% chromium sequioxide, an inert fecal marker. Feces were collected at hourly intervals after six hours and the initial a-pearance of chromium in feces was considered to be the passage time of the ingesta.

Statistical Analyses

Data were analyzed by one-way analysis of variance appropriate for completely randomized design (Snedecor and Cochran, 1967). The effect of diet on measured parameters was determined by individually comparing treatment groups to controls.

RESULTS

Growth Trials

Animals fed 20% lignin were removed from the experiment testing effects of purified fibres on growth because they were suffereing from severe diarrhea. This symptom was made more severe when T-2 toxin was fed. The addition of T-2 toxin depressed final body weight and feed consumption of all rats except those fed pectin. Animals fed 20% pectin were, however, already stunted. Feed efficiency was not influenced by the addition of T-2 toxin to the control diet or to diets containing pectin. Reductions were seen, however, when toxin was added to diets containing cellulose or hemicellulose.

In the second growth trial, final body weights and feed consumption were reduced when T-2 toxin was fed regardless of the level of pectin. Only rats fed 2.5% pectin did not have reduced feed efficiencies when the toxin was fed.

Dietary lignin was somewhat more beneficial. Rats exposed to T-2 toxin again had depressed final body weight and feed consumption although 2.5 or 5% lignin lessened this effect when compared to animals fed the control diet. Feed efficiency decreased in response to T-2 toxin only with the feeding of 0 or 7.5% lignin.

Alfalfa, a feedstuff rich in lignin, overcame the depressive effects of T-2 toxin on final body weights and feed consumption when fed at 20% in the third trial. Feed efficiency was unaffected by T-2 toxin regardless of the level of dietary alfalfa.

Rats were fed various non-nutritive polymers and T-2 toxin in the fourth trial and the addition of T-2 toxin to the control diet again reduced final body weights and feed consumption. Zonolite Verxite and cation exchange resin had little effect on these responses but the decrease in final body weights and feed consumption was reduced by the feeding of bentonite and anion exchange resin. Since both feed consumption and growth rates were reduced by T-2 toxin, feed efficiencies were relatively unaffected.

The fifth growth trial was conducted in order to determine if the possible protective effect of 5% anion exchange resin could be duplicated or increased when this material was fed at other levels. The growth depressing effects of T-2 toxin were minimized by the feeding of 5.0% resin while the reduction in feed consumption was overcome to the greatest degree by 7.5% resin. The feeding of 10.0% resin depressed growth and lowered feed efficiency. The increase in feed consumption noted with increased inclusion of resin was likely due to reduced dietary energy density. No attempt was made to ensure isoenergetic diets since dietary fat was shown to partially overcome feed refusal due to T-2 toxin.

The sixth experiment showed bentonite to be the most effective treatment tested. As little as 2.5% bentonite greatly increased feed consumption and final body weights of rats fed T-2 toxin. The feeding of 10% bentonite completely overcome the toxicosis.

In the seventh trial, as little as 6.0% corn oil overcame the feed refusal and reduced body weight caused by T-2 toxin. Similar results were seen with lard.

Enzyme Assay

The activity of hepatic esterase (EC 3.1.1.1.) was not significantly affected by dietary alfalfa, bentonite or corn oil. Supplementation of the control diet with 3 μg/g of T-2 toxin also failed to induce this enzyme.

Excretion and Residue Trials

The influence of dietary alfalfa, bentonite and corn oil on fecal excretion of $^3$H following oral administration of [$^3$H] T-2 toxin is given in table 1. There was no effect of diet on the fraction of dose excreted in urine but fecal excretion increased when 12.5 or 20% alfalfa was added to the basal diet. The only effect of 20% alfalfa on the relative amounts of T-2 toxin and known metabolites in feces was an increase in unmetabolized T-2 toxin. Hepatic residues of T-2 toxin and metabolites were not affected by dietary alfalfa. Residues in muscle were lowered, however, when 12.5 or 20% alfalfa was fed and renal residues declined with 20% alfalfa feeding. Intestinal contents of rats fed 12.5% alfalfa contained more $^3$H than controls while increased residues were found in the intestinal wall of rats fed 5% alfalfa. Fecal excretion of $^3$H following topical administration increased when 12.5 or 20% alfalfa was fed while 20% alfalfa reduced hepatic radioactivity. There was a slight but insignificant decrease in the fraction of the dose of [$^3$H] T-2 toxin excreted in urine when bentonite was added to the diet. As little as 5% bentonite, however, sharply increased fecal excretion of the toxin. This effect caused an increase in total toxin excreted when all levels of bentonite were fed.

Residues in liver and kidney were not significantly altered by bentonite feeding but all levels of bentonite lowered residues in muscle. Residual $^3$H in intestinal contents and the intestinal cell wall increased compared to controls when 5.0% bentonite was fed, other levels of bentonite had no effect.

Increasing bentonite caused consistently more fecal excretion of $^3$H following topical administration but this effect was not significant. No effect of diet was seen on urinary excretion of $^3$H or on tissue residues.

Supplements of dietary corn oil also had little effect on urinary excretion of $^3$H but fecal excretion was sharply increased. Residual $^3$H in liver was reduced compared to controls when 20.0% corn oil was fed but there was no effect on other tissues.

Effect of Diet On Passage Time

Passage time in rats fed the control diet was determined to be ten hours. Each increment in dietary alfalfa concentration reduced retention of feces until rats fed 20% alfalfa excreted feces in six hours. All levels of bentonite reduced the passage time to seven hours as did all levels of corn oil.

DISCUSSION

Experiments were conducted to determine the effects of purified and natural fibre sources on feed consumption and growth rate of rats fed T-2 toxin. The depressions in growth rate and feed consumption of animals consuming the toxin were similar to those observed by Marasus et al. (1969) and Vesonder et al. (1979). It has been suggested that feed refusal caused by T-2 toxin may be due to inflammation of oral cavities and intestinal tracts of intoxicated animals (Ueno, 1977). Some rats in the current experiments suffered a slight nasal hemorrhage but acute affects (Marasus et al., 1969; Kosuri et al., 1971) were not observed. No attempt was

Table 1.    Effect Of Diet On Fecal Excretion Of $^3$H Following An Oral
            Dose Of [$^3$H] T-2 Toxin

| Dietary alfalfa (%) | | | |
| --- | --- | --- | --- |
| 0 | 5.0 | 12.5 | 20.0 |
| 18.6[1] | 18.2 | 27.4 | 43.1** |
| ±3.2 | ±4.0 | ±2.6 | ±3.0 |

| Dietary bentonite (%) | | | |
| --- | --- | --- | --- |
| 0 | 5.0 | 7.5 | 10.0 |
| 18.6 | 36.4** | 46.6** | 47.9** |
| ±3.2 | ±3.3 | ±2.3 | ±3.8 |

| Dietary corn oil (%) | | | | |
| --- | --- | --- | --- | --- |
| 1.25 | 6.0 | 10.0 | 15.0 | 20.0 |
| 18.6 | 31.6** | 32.7** | 37.8** | 51.1** |
| ±3.2 | ±3.7 | ±3.7 | ±3.4 | ±3.5 |

[1]% of dose, mean ± SEM for 10 rats.

*,**Different from diet devoid of additive (P<0.05 and P<0.01,
respectively).

made to make the experimental diets isoenergetic since the feeding
of fat was shown to partially overcome feed refusal associated with
T-2 toxicosis.

Cellulose and hemicellulose were not found to overcome the
growth depression and feed refusal associated with the consumption
of T-2 toxin.  This may be due to inability of cellulose to act
as a binding agent in the gastrointestinal tract (Story and
Kritchevsky, 1976).

The lack of effect of T-2 toxin on feed consumption of rats
fed 20% pectin was likely due to the already low consumption of this
diet.  This concept was supported by the observation that lower
levels of dietary pectin had little effect on the anorectic response
to T-2 toxin.  High levels of pectin have been reported to reduce
feed consumption, weight gain and feed efficiency (Rao and Bright-
See, 1979).

Rats fed 20% lignin were removed from the experiment because they were suffering from severe diarrhea. Lignin has been described as an inhibitor of digestion (George et al., 1980) and this may have contributed to the growth depression when high levels of lignin were fed. Lower levels of lignin caused a slight decline in feed consumption only in those rats fed diets without T-2 toxin. It was concluded that 5% lignin has a slight beneficial effect at overcoming feed refusal and growth depression due to T-2 toxin.

Alfalfa was tested because it is a fibrous feedstuff relatively rich in lignin. Feed consumption and growth increased with increasing alfalfa only when T-2 toxin was included in the diet. No effect of T-2 toxin was seen when 20% alfalfa was fed. This corresponds to feeding 3.3% lignin since the alfalfa was determined to contain 16.7% lignin. The effect of feeding alfalfa at lower levels was less consistent, however, due to large individual variability, particularly in growth rates. Feeding alfalfa in excess of 20% would likely be of little benefit since high alfalfa diets have been shown to decrease feed intake of rats (Myer and Cheeke, 1975).

Anion-exchange resin has been shown to alleviate zearalenone toxicosis by binding the toxin in the digestive tract to prevent absorption (Smith, 1980b; Smith, 1982). T-2 toxin may be less anionic than zearalenone at intestinal pH since less beneficial effect of the resin was seen with T-2 toxin. Any beneficial effects of this type of resin may, therefore, be less related to ion-exchange capacity than to its capacity to physically entrap the toxin in a polymeric matrix. Some anionic behavior of T-2 toxin is likely, however, since in the fourth growth trial anion exchange resin was clearly more effective at overcoming the toxicosis than cation exchange resin.

The ability of bentonite to effectively overcome feed refusal due to T-2 toxin is not likely due to ion exchange since this product is a colloidal hydrated aluminum silicate. It has the property, however, of forming highly viscous gels when mixed with water. Absorption of toxin may therefore be reduced as bentonite traps molecules. This property likely accounts for the observation that bentonite can bind aflatoxin in vitro (Masimango et al., 1979).

An alternative hypothesis to account for the beneficial effects of bentonite could be that this treatment was promoting metabolism of T-2 to less active higher metabolites. Chadwick et al. (1978) have reported that dietary fibre can promote the metabolism of pesticides even though the polymeric fibres cannot be absorbed from the gastrointestinal tract.

It is not so clear how supplements of dietary lipid could overcome T-2 toxicosis. The lipids fed in the current studies are readily absorbed and metabolized in contrast to mineral oil which

was found by Rozman et al. (1982) to promote fecal excretion of
brominated biphenyls. The oils may have a soothing effect on the
irritated gut mucosa; they may act as a laxative and reduce absorption; or they may promote tissue breakdown of the toxin.

T-2 toxin is believed to undergo selective C-4 deacetylation
by hepatic microsomal carboxyesterase to yield HT-2 toxin (Ohta
et al., 1977; Otta et al., 1978). Further metabolism by the same
enzyme leads to the stepwise conversion of HT-2 toxin to the more
polar metabolites 4-deactylneosolaniol and T-2 tetraol (Yoshizawa
et al., 1980a). The lack of effect of T-2 toxin on the activity
of heaptic esterase may be due to the non-specific nature of this
enzyme. The low dose of toxin may not have been enough to cause
substrate induction of the enzyme. The lack of effect of additives
on the activity of hepatic esterase adds support to the concept
that they do not alleviate T-2 toxicosis by promoting catabolism of
the toxin.

[$^3$H] T-2 toxin was excreted in the feces and this is in agreement with literature reports (Matsumoto et al., 1978; Yoshizawa
et al., 1981). Increasing dietary alfalfa markedly increased fecal
excretion of $^3$H but little effect of diet on urinary losses was
noticed. The latter observation may have been due to some exchange
of $^3$H between T-2 toxin and water in vivo. This would cause a small
but constant loss of $^3$H which might mask slight dietary effects.
No evidence of exchange was seen, however, using the in vitro test
of Chi et al. (1978).

The increased fecal losses of $^3$H when alfalfa was fed coupled
with the lack of effect on urinary excretion indicate that alfalfa
is likely binding T-2 toxin in the intestinal lumen and blocking
absorption of the toxin. Part of these fecal losses can also be
attributed, however, to increased biliary excretion as indicated by
the topical administration experiment.

Chi et al. (1978) have shown that T-2 toxin and metabolites are
excreted through the bile. It has also been reported that alfalfa
promotes biliary excretion (Malinow, 1980). Increased biliary losses
of $^3$H seen when alfalfa was fed following topical dosing may have
caused the reduced hepatic residues of $^3$H. The fraction of topical
dose lost in biliary excretion could not be determined, however,
because absorption through the skin was not complete at 21 hours.

The liver has been reported to be a major site of residual
[$^3$H] T-2 toxin (Matsumoto et al., 1978). It is also an important
organ for metabolism (Yoshizawa et al., 1980a) and excretion
(Kosuri et al., 1971) of the compound. The slight but insignificant
decrease in residual hepatic $^3$H when 20% alfalfa was fed may have
been due to increased biliary excretion. Kidney is also thought to
be a site of T-2 toxin elimination (Kosuri et al., 1971). The

feeding of diets high in alfalfa likely blocked intestinal absorption of the toxin which caused reduced renal residues and had a slight effect on urinary excretion of $^3$H. This is supported by the reduction in muscle residues indicating that high alfalfa diets were likely reducing the $^3$H available for distribution of peripheral tissues. The rapid absorption and tissue distribution of ingested [$^3$H] T-2 toxin has been reported in the cow (Yoshizawa et al., 1981), mouse (Matsumoto et al., 1978) and chicken (Chi et al., 1978). The digesta of control rats in the current study contained the least $^3$H which indicates that the toxin had likely already been absorbed. The increasing $^3$H in the digesta of rats fed 5% and 12.5% alfalfa was likely due to the binding of [$^3$H] T-2 toxin in the gut lumen and delayed absorption. The levels of $^3$H in the digesta of rats fed 20% alfalfa were as low as control rats and this was likely due to the reduced transit time of digesta in rats fed this diet. This would cause the toxin to be lost into the large intestine. The effect of alfalfa on reduction of digesta transit time as shown in the chromic oxide market study is in agreement with literature reports which have shown that high fibre diets decrease transit time (George et al., 1980; Stasse-Walthuis et al., 1980).

The small contribution of T-2 toxin and known metabolites HT-2, 4-diactylneosolaniol and T-2 tetraol to excreted $^3$H in feces is in agreement with earlier workers (Matsumoto et al., 1978; Yoshizawa et al., 1980b). The major metabolites of T-2 toxin have not been identified. The lack of effect of dietary alfalfa on the relative proportions of the known metabolites, however, indicates that the protective effects of alfalfa against T-2 toxicosis are not likely mediated through changes in tissue metabolism. It can be concluded that diets high in alfalfa protect against T-2 toxicosis in rats by reducing intestinal absorption of the toxin and increasing fecal excretion.

Similar conclusions can be made for the effect of dietary bentonite. The increased fecal excretion of orally-administered [$^3$H] T-2 toxin when bentonite was fed coupled with lack of effect of diet following topical administration of the toxin prompts the conclusion that bentonite is blocking intestinal absorption of T-2. Intestinal transit time is also reduced as was seen with alfalfa feeding. The reduction in muscle residues is likely due to reduced intestinal absorption and less distribution to peripheral tissues.

Bentonite did not cause increased fecal losses of $^3$H following topical administration and this is in contrast to alfalfa treatments. There are no literature reports on the effect of bentonite on bile flow and it must be assumed that bentonite has no effect on this parameter.

The lack of effect of bentonite on the relative proportions of known metabolites of T-2 toxin coupled with the lack of effect of

bentonite on hepatic esterase activity further supports the concept
of bentonite as an intestinal blocking agent.

The role of dietary fats in alleviation of T-2 toxicosis is
not easily explained.  No evidence was gathered to show that dietary
fats altered T-2 metabolism.  Hepatic esterase activities were
unchanged and the degree of unsaturation of lipid had no effect on
feed refusal.  The fatty acid composition of dietary oils has been
shown to alter mixed frunction oxidase activity in rats (Gefferth
and Blakovits, 1977) and it was thought that this might influence
T-2 toxin metabolism.

The excretion and residue studies conducted with diets of
varying corn oil content differed from the diets used in growth
trials in that all diets were made isoenergetic through the addition
of cellulose.  It was the presence of cellulose that caused increased
fecal losses of $^3$H and reduced transit time.  Subsequent experiments
using diets not made isoenergetic still showed significant increases
in fecal losses with supplemental lipid but the magnitude of the
differences was reduced.  Transit time was markedly longer when
diets without cellulose were fed.

It is unlikely that the reduced transit time seen when isoenergetic high fat diets were fed is of physiological significance
since cellulose fed alone did not overcome T-2 induced feed refusal.

Diets rich in fat promote bile flow (Mayes, 1977).  It is not
known if the high fat diets fed in these studies promote biliary
losses of T-2 since no experiments have been conducted measuring
fecal losses of $^3$H following topical administration of the toxin.
Since high fat diets without cellulose have little effect on transit
time, it is unlikely that these diets reduce T-2 toxicosis due to
a laxative effect.  It is suggested that high fat diets likely
promote feed consumption by reducing intestinal absorption of the
toxin.

## REFERENCES

Bamburg, J.R., Riggs, N.V. and Strong, F.M. (1968).  The structures
      of toxins from two strains of Fusarium tricinctum.  Tetrahedron
      24, 3329-3336.
Carson, M.S. and Smith, T.K. (1983).  Effect of feeding alfalfa
      and refined plant fibres on the toxicity and metabolism of
      T-2 toxin in rats.  J. Nutr. 113: in press.
Chadwick, R.W., Copeland, M.R. and Chadwick, C.J. (1978).  Enhanced
      pesticide metabolism - A previously unreported effect of dietary
      fibre in mammals.  Food Cosmet. Toxicol. 16, 217-225.
Chi, M.S., Robison, T.S., Mirocha, C.J., Swanson, S.P. and Shimoda,
      W. (1978).  Excretion and tissue distribution of radioactivity
      from tritium-labelled T-2 toxin in chicks.  Toxicol. Appl.

Pharmacol. 45 391-402.

Dabich, D., Chakrapani, B. and Syner, F.N. (1968). Purification and properties of esterases characterisitic of adult rat brian. Biochem. J. 110, 713-719.

Ershoff, B.H. (1974). Anti-toxic effects of plant fiber. Am. J. Clin. Nutr. 27, 1395-1398.

Ershoff, B.H. (1977). Effects of diet on growth and survival of rats fed toxic levels of tartrazine (FD & C Yellow No. 6). J. Nutr. 107, 822-828.

Gefferth, G. and Blaskovits, A. (1977). The effect of different oil contents of the diet on the mixed function oxidase in the liver of rats. Nutr. Met. 21 (Suppl. 1), 246-248.

George, J.R., Harbers, L.H. and Reeves, R.D. (1980). Digestive response of rats to fibre type, level and particle size. Nutr. Rep. Inter. 21, 313-322.

Guncaga, J., Lentner, C. and Haas, H.G. (1974). Determination of chromium in feces by atomic absorption spectrophotometry. Clin. Chim. Acta 57, 77-81.

Hou, I.-C., Smalley, E.B., Stron, F.M. and Ribelin, W.E. (1972). Identification of T-2 toxin in moldy corn associated with a lethal toxicosis in dairy cattle. Appl. Microbiol. 24, 684-690.

Ishii, K. and Ueno, Y. (1981). Isolation and characterization of two new trichothecenes from Fusarium Sporotrichoides strain M-1-1. Appl. Environ. Microbiol. 42, 541-543.

James, L.J. and Smith, T.K. (1982). Effect of alfalfa on zearalenone toxicity and metabolism in rats and swine. J. Anim. Sci. 55, 110-118.

Kosuri, N.R., Smalley, E.B. and Nichols, R.E. (1971). Toxicological studies of F. Tricinctum (cordal) Synder et Hansen from moldy corn. Am. J. Vet. Res. 32, 1843-1850.

Lowry, O.H., Rosenbrough, N.J., Farr, A.L. and Randall, R.J. (1951). Protein measurement with folin phenol reagent. J. Biol. Chem. 193, 265-275.

Malinow, M.R., McLaughlin, P. and Stafford, C. (1980). Alfalfa seeds: effects on cholesterol metabolism. Experientia 36, 562-564.

Marasas, W.F.O., Bamburg, J.M., Smalley, E.B., Strong, F.M., Rogland, W.L. and Degurse, P.E. (1969). Toxic effects on trout, rats and mice of T-2 toxin produced by the fungus Fusarium tricinctum. Toxicol. Appl. Pharmacol. 15, 471-482.

Masimango, N., Remacle, J. and Ramout, J. (1979). Elimination par des argiles conflantes de l'aflatoxine $B_1$ des milieux contamines. Ann. Nutr. Aliment. 33, 137-147.

Matsumoto, H., Ito, T. and Ueno, Y. (1978) Toxicological approaches to the metabolites of Fusaria. XII. Fate and distribution of T-2 toxin in mice. Jpn. J. Exp. Med. 48, 393-399.

Mayes, P.A. (1977). Digestion and absorption from the gastrointestinal tract. In: Review Of Physiological Chemistry (Harper, H.A., Rodwell, V.W. and Mayes, P.A., ed.), p.p.

202-217, Lange Medical Publications, Los Altos, CA.

Mirocha, C.J., Pathre, S.V., Schuerhamer, B. and Christensen, C.M.
    (1976).  Natural occurrence of Fusarium toxins in feedsuffs.
    Appl. Environ. Microbiol. 32, 553-556.

Myer, R.O. and Cheeke, P.R. (1975).  Utilization of alfalfa meal
    and alfalfa protein concentrate by rats.  J. Anim. Sci.
    40, 500-508.

Ohta, M., Ishii, K. and Ueno, Y. (1977).  Metabolism of tricho-
    thecene mycotoxins.  I. Microsomal deacetylation of T-2
    toxin in animal tissues.  J. Biochem. 82, 1591-1598.

Ohta, M., Matsumoto, H., Ishii, K. and Ueno, Y. (1978).  Metabolism
    of trichothecene mycotoxins.  II. Substrate specificity of
    microsomal deacetylation of trichothecene.  J. Biochem. 84,
    697-706.

Puls, R. and Greenway, J.A. (1976).  Fusario toxicosis from barley
    in B.C. II. Analysis and toxicity of suspected barley.  Can.
    J. Comp. Med. 40, 16-19.

Rao, A.V. and Bright-See, E. (1979).  Effect of graded amount of
    dietary pectin on growth parameters of rats.  Nutr. Rep. Inter.
    19, 411-417.

Rozman, K.K., Rozman, T.A., Williams, J. and Greim, H.A. (1982).
    Effect of mineral oil and/or cholestyramine in the diet on
    biliary and intestinal elimination of 2, 4, 5, 2', 4',
    5'-hexabromobiphenyl in the rhesus monkey.  J. Toxicol. Environ.
    Health 9, 611-618.

Smith, T.K. (1980a).  Influence of dietary fiber, protein and
    zeolite on zearalenone toxicosis in rats and swine.  J. Anim.
    Sci. 50, 278-285.

Smith, T.K. (1980b).  Effect of dietary protein, alfalfa and zeolite
    on excretory patterns of 5', 5', 7', 7'-[$^3$H] zearalenone in
    rats.  Can. J. Physiol. Pharmacol. 58, 1251-1255.

Smith, T.K. (1982).  Dietary influences on excretory pathways and
    tissue residues of zearalenone and zearalenols in the rat.
    Can. J. Physiol. Pharmacol. 60, in press.

Snedecor, G.W. and Cochran, W.G. (1967)  Statiscial Methods.
    The Iowa State University Press, Ames, IA.

Stasse-Walthuis, M., Albers, H.F.F., Van Jeversen, J.G.C., Wil de
    Jong, J., Hautvast, J.G.A.J., Hermus, R.J.J., Katan, M.B.,
    Bryden, W.G. and Eastwood, M.A. (1980).  Influence of dietary
    fibre from vegetables and fruits, bran or citrus pectin on
    serum lipids, fecal lipids, and colonic function.  Am. J.
    Clin. Nutr. 33, 1745-1756.

Story, J.A. and Kritchevsky, D. (1976).  Comparison of the binding
    of various bile salts and bile acids in vitro by several types
    of fibres.  J. Nutr. 106, 1291-1294.

Takeda, H. and Kiriyama, S. (1979).  Correlation between the
    physical properties of dietary fibres and their protective
    activity against amaranth toxicity in rats.  J. Nutr. 109,
    388-396.

Takeda, H., Tsujita, J. and Emoto, T. (1982). Nutritional significance of dietary fiber in counteracting the amaranth-toxicity in rats: A possible explanation of the mechanism. Nutr. Rep. Inter. 25, 169–187.

Ueno, Y. (1977). Trichothecenes: Overview Address. In: Mycotoxins in Human and Animal Health (Rodricks, J,V., Hesseltine, C.W. and Mehlman, M.A., ed.), pp. 189–207, Pathotox Publishers, Inc., Park Forest South, IL.

Vesonder, R.F., Ciegler, A., Burmeister, H.R. and Jensen, A.H. (1979). Acceptance by swine and rats of corn amended with trichothecene. Appl. Environ. Microbiol. 38, 344–346.

Yoshizawa, T., Swanson, S.P. and Mirocha, C.J. (1980a). In vitro metabolism of T-2 toxin in rats. Appl. Environ. Microbiol. 40, 901–906.

Yoshizawa, T., Swanson, S.P. and Mirocha, C.J. (1980b). T-2 metabolites in the excreta of broiler chickens administered $^3$H-labelled T-2 toxin. Appl. Environ. Microbiol. 39, 1172–1177.

Yoshizawa, T., Mirocha, C.J., Behrens, J.C. and Swanson, S.P. (1981). Metabolic fate of T-2 toxin in a lactating cow. Food Cosmet. Toxicol. 19, 31–39.

# 8

SAFETY OF MEGAVITAMIN THERAPY

Stanley T. Omaye

Nutrients Research Unit
Western Regional Research Center, USDA, ARS
Berkeley, California

ABSTRACT

Carbon compounds that are needed in small amounts in the diet because they are not made in the body of vertebrates are defined as vitamins. Excluded from this definition are vitamins D, K, and niacin which can be synthesized by the organism or, as in the case of vitamin K, by the host's intestinal bacteria. Lack of such vitamins can result in characteristic deficiency diseases. The therapeutic use of such compounds (megavitamin intake) is based on the spectacular effect of vitamins on deficiency diseases; however, evidence that the ingestion of large amounts of vitamins beyond the "Recommended Daily Allowances" (RDA) is beneficial is not within the basic concept of nutrition. Vitamins, like many substances, may be toxic when taken in large quantities, especially the fat-soluble vitamins, and the concept of "more is better" is a common misconception. Vitamin supplements can be suggested only in the unusual cases of patients having inadequate intake, disturbed absorption (genetic or otherwise), or increased tissue requirements. A well-balanced diet that includes a wide variety of foods from each of the four food groups is adequate for the supply of vitamins, as well as other nutrients, in healthy people. This paper will review some of the recent findings regarding vitamin toxicity and the mechanisms of toxicity.

INTRODUCTION

Definition of a vitamin

A vitamin is a chemical compound, an organic molecule not made

in the organism, that is required in small amounts in food to
sustain the normal metabolic functions of life.  The key to this
definition is that this chemical compound must be supplied to the
organism because the animal cannot synthesize vitamins.  Lack of
it produces a specific deficiency syndrome and supplying it cures
that deficiency.  An exception to this definition is vitamin D,
which can be made in the skin upon adequate exposure to sunlight.
However, without adequate exposure, the animal is dependent on a
dietary source.  Biotin, panthothenic acid, and vitamin K are made
by bacteria in the human intestine, based on a symbiotic relationship and, thus, are not required by the human.  Niacin can also be
synthesized in humans from the amino acid tryptophane.

The definition of a vitamin has been extended by some (Spector,
1980) to include those chemical compounds required by a specific
tissue but not synthesized by that tissue.  For example, in certain
species, vitamin C can be synthesized from glucose in the liver but
not in the Central Nervous System; therefore, vitamin C is not
considered a vitamin for these animals.  Vitamin C for the brain
must be drawn from the blood, from the vitamin C that entered the
blood from the diet, or was synthesized in the liver.  Subsequently,
vitamin C could be considered a vitamin for the brain since it must
be obtained from outside the brain.  Another example is the vitamin
niacin, which cannot be synthesized from tryptophane in mammalian
brain; however, the synthesis of niacin from tryptophane occurs in
mammalian liver (Spector and Kelly, 1979; Spector, 1979).

Evolution of the vitamin concept

There are 13 vitamins known to be important in human nutrition.
Included in this class of essential nutrients are vitamin A, B-
complex (actually composed of a group of 8 vitamins), vitamin C,
vitamin D, vitamin E, and vitamin K.  We can subdivide vitamins
into: 1) the lipid (fat)-soluble compounds (vitamins A, D, E, and
K), meaning they dissolve readily in fat and 2) the water-soluble
compounds, vitamin B-complex (thiamin, riboflavin, niacin, vitamin $B_6$, panthothenic acid, vitamin $B_{12}$, biotin, and folic acid)
and vitamin C, which dissolves readily in aqueous solutions.  It
has been three decades since the last vitamin ($B_{12}$) was discovered
in 1948; and as stated by Herbert (1980), it is probable that no
new vitamins will be found.  There are substances which are growth
factors for other forms of life, such as bacteria, but not for
humans. Para-aminobenzoic acid (PABA), bioflavonoids (vitamin P),
choline, inositol, lipoic acid, and ubiquinone are just a few of
these examples.

The concept of, and need for, vitamins developed through many
feeding experiments with animals and some with humans.  These
experiments pointed to the fact that substances other than minerals,
water, carbohydrate, protein, and essential fats were required for

good nutrition. Vitamins became established by the deficiency
diseases that result when they are absent from the diet. These
disorders produce severe, distinctive symptoms and signs. The
discovery of the vitamins resulted in the lessened incidence of
these diseases, i.e., pellagra, rickets, scurvy and others; so
these diseases are now rare in the United States and in most of
the Western World.

## Intake:  Nutritional, therapeutic, and toxic

Vitamins function in two basic ways, either as a nutrient
or vitamin or as a chemical (Herbert, 1980). When the function
is known, fat-soluble vitamins function as regulators of specific
metabolic activity, and the water-soluble vitamins function as
coenzymes. Although rather exact roles for some of the vitamins
in the chain of metabolic events are understood, it is safe to say
that the complete function of any one of the vitamins in the body
is unknown. What we do know about specific coenzyme functions is
that the encouragement of doses of at least tenfold above the
recommended dietary allowance (RDA) serves no nutritional function
(Herbert, 1977). Vitamins enter the body as a component of food,
travel to the tissues/cells that need them, are taken into the
cells, and converted into a coenzyme form. In some cases the
vitamin enters the cell as the coenzyme form already. A protein
within the cell, called apoenzyme, combines with the vitamin coenzyme to form a holoenzyme. The holoenzyme or enzyme then serves
the vitamin function of catalyzing certain specific metabolic-
biochemical reactions. It appears that only when combined with
its apoenzyme within the cell can a vitamin function as a vitamin.
Since the quantity of protein, as well as the quantity of apoenzyme,
any cell can make per unit time is limited (Schimke and Doyle,
1970), the amount of enzyme formation through the combination of
apoenzyme with vitamin-coenzyme is also limited. Therefore, when
the cell is producing apoenzyme at maximum capacity, extra vitamin-
coenzyme coming into the cell cannot possibly perform its vitamin
function because there is no apoenzyme available to combine with.
Coincidentally, the apoenzymes for vitamins are generally found to
be saturated at the levels approximating the RDA (1980) or less
(Herbert, 1980). Therefore, if vitamins produce a physiological
response beyond nutritional requirements, we should label those as
pharmacological or, at extreme, toxicological. In contrast there
is the suggestion that certain apoenzymes in certain tissues in
various individuals may have varying affinities for the vitamin-
coenzymes (Pauling, 1974). The concentration of vitamin-coenzyme
needed to produce the amount of active enzyme necessary for optimal
health varies with the individual. Megavitamin therapy could increase the concentration of coenzyme and shift the equilibrium of
the reactions of apoenzyme and coenzyme toward more of the active
enzyme and, therefore, could produce more of the required product.

The term megavitamin is misleading and deceptive. The term is used originally by non-scientific people to refer to very large doses of vitamins. By definition the prefix "mega" means large or great. However, it is often used in the layman literature with overtones of magic and power. The intent is to encourage the healthy person to consume concentrations of vitamins several times beyond the doses suggested by the RDA. This is not to say that there is no role in medicine for massive-dose vitamin therapy. The unequivocal use for vitamin therapy is vitamin deficiency (Herbert, 1980). As Herbert (1980) has indicated, there are situations in which the use of megavitamin therapy is rationally based and efficacious: genetic diseases where the patient was born with a defect in the machinery for utilizing vitamins; diseases where there is a defective transport of vitamins across cell membranes; and situations where an antidote is needed for the toxicity of anti-vitamins. In regard to the two major groups of disorders, the malabsorption syndromes and inborn errors of metabolism, Hodges (1980) has tabulated the situations where megavitamin therapy appears useful. Malabsorption of nutrients can result from a variety of disorders, including abnormality of the mucosa or excessively rapid motility. Malabsorption can result from any form of steatorrhea, whether it's from pancreatic disease, hepatobiliary disease, or a small bowel disorder, the consequence being substantial decreases of lipid-soluble nutrients that are essential. A number of genetic abnormalities involve the need for the metabolism of essential nutrients. A disproportionate number of genetic abnormalities involve vitamin $B_6$, perhaps because it plays an important role in a great many enzyme systems (Hodges, 1980; Rosenberg, 1970).

Vitamins are by far one of the world's most widely used pharmaceutical products (Cumming et al, 1981). Over $1.5 billion are expended each year on food supplements, including vitamins, in the United States (Scala, 1980). Many of our population are lacking in sound knowledge of nutrition to make sensible use of vitamin supplementation. The old concept that if a little is good, then a lot or more should be better prevails. This concept ignores the fact that the single factor that determines the degree of harmfulness of a compound is the dose of that compound (Loomis, 1978). Some chemicals which produce death in microgram doses are commonly thought of as being extremely toxic; other chemicals may be relatively harmless following doses in excess of several grams. Because of the great range of concentrations of chemicals for the potential of producing harm, categories of toxicity have been devised (Doull et al, 1980). The most innocuous of substances when taken into the body in sufficient amount may lead to undesirable, if not distinctly harmful, effects. If a sufficient dosage is taken into the body, a harmful effect may be the consequence, i.e., the ability of that biologic agent to carry on a function is seriously impaired or destroyed. On the other hand, the most harmful of all chemical agents can be taken into the body in sufficiently small

amounts so that there may be no ill effect from such chemicals.
Therefore, the harmfulness or safeness of any chemical is primarily
related to the concentration of that chemical in the organism or
the dose received by that animal.  Subsequently, there is a danger
of harm from megavitamin consumption which comes from the simple
faith many have in using colossal amounts of vitamins for a variety
of minor and meager problems, or for serious disorders for which
there is no remedy.  In the past, most concern for toxicity problems
resulting from the ingestion of vitamins was centered around the
fat-soluble vitamins.  Interest in health problems associated with
massive doses of fat-soluble vitamins has gained new impetus because
of a surge in the practice of self-dosing with massive amounts of
vitamins and controversies regarding the legal aspects of unlimited
access by people to high potency preparations.  Fat-soluble vitamins
are retained efficiently in the body, especially vitamins A and D,
of which single large doses may be stored in the body for months.
Therefore, it is not unreasonable to expect that the presence of a
vitamin and its metabolic products might produce pharmacological,
if not toxicological, responses.

Generally, excesses in water-soluble vitamins are rapidly
excreted.  Until recently little concern was directed to water-
soluble vitamin toxicity; however, as this publication will show,
there is substantial reason for concern and research.  Even now
research is beginning to associate some side effects with excess
intakes of the water-soluble vitamins.

LIPID-SOLUBLE VITAMINS

The four lipid-soluble vitamins A, D, E, and K are often
grouped together because of their solubility in fats and their
insolubility in aqueous solutions.  They share some common physical
properties, but each has a distinct biochemical function in the
organism.  In general: vitamin D serves as the precursor of a
hormone which in turn is necessary for the regulation of calcium
and phosphorus metabolism; vitamin A can be seen as a light-
sensitive, protein-bound pigment necessary for the visual process
and also required for specific cellular differentiation; vitamin
E plays a role as an important biological antioxidant; and vitamin
K, as a coenzyme precursor, is involved in the synthesis of blood
clotting factors.

Toxicity can be induced by high dietary intakes of lipid-
soluble vitamins, especially vitamin A and D, because these vitamins
are readily stored in the body.  Severe tissue reactions occur when
large doses of vitamins A and D are consumed (DeLuca, 1982; National
Nutrition Consortium, 1978).  For the most part, toxicity or hypervitaminosis is the result of indiscriminate use of, as well as self-
medication with, dietary supplements and not from the consumption

of usual diets (Jukes, 1979; Macupinlac and Olson, 1981; National
Nutrition Consortium, 1978). With the exception of patients with
identified deficiency syndromes where vitamin supplements are warranted, there does not appear to be a demonstrated health value
in consuming fat-soluble vitamins in excess of the RDA (RDA, 1980).

## Vitamin A

Therapy. Deficiency states respond well to treatments of high
doses of vitamin A, at a daily dose of 30,000–50,000 IU (Table 1).
The recent suggested recommended allowance for the adult male of
5000 IU has not changed since the previous edition of the RDA (1980)
in 1974. Night blindness and milder conjunctival changes seem to
reverse themselves within a few days of 30,000 IU per day (Roels
and Lui, 1973). Caution is advised in the dosing of children. In
most patients, oral treatment is recommended and decidedly more
effective than parenteral. However, the oral route is counter-
indicated in conditions of gastrointestinal tract lesions where
aqueous dispersion of the vitamin should be used.

Toxicity. In rats large doses of vitamin A produced loss of
hair, emaciation, paralysis, and death following within a few days
(Jenkins, 1978). However, there seems to be a wide range of tolerance to high concentrations of vitamin intake by various species.
Polar bears and seals can tolerate extremely large daily doses (up
to 60,000 µg/kg body weight) before being clinically affected (Hayes
and Hegsted, 1973), whereas the calf and pig are in a range of
danger comparable to man (1,000–3,000 µg/kg) (Hayes and Hegsted,
1973). Common symptoms of toxicity which vary from species to
species include anemia, loss of hair, skin lesions, and raised
intercranial pressure together with degenerative atrophy of various
organs (Cumming et al, 1980; DiPalma and Ritchie, 1977; Jenkins,
1978; Hathcock, 1976).

Humans have been exposed to large doses of vitamin mostly from
supplements (Jenkins, 1978; Cumming et al, 1980). Toxicity resulting from the ingestion of natural foodstuffs is rare, the
exceptions being the result of the consumption of polar-bear liver
or the ingestion of fish livers from those large fishes, such as
cod or halibut, capable of high stores of vitamin A (Nater and
Douglas, 1970; Hayes and Hegsted, 1973; Arnrich, 1978; Knudson and
Rothman, 1953). A review of the past literature (Korner and Vollm,
1975) suggests that 75% of the cases of hypervitaminosis A reported
over the past four decades have been acute, with young children the
principal victims (Woodard et al, 1961). Fortunately, acute adverse
reactions to vitamin A are for the most part transitory (Olson,
1972; Arnrich, 1978). Marie and See (1954) have described the symptoms of acute vitamin poisoning in children, vomiting, drowsiness,
and bulging of the fontanelle. Table 1 documents that adults tend
to complain of severe headaches, vertigo, abdominal pain, vomiting,

Table 1.  Vitamin A

| Recommended Intake | Therapeutic Dose | Toxic Concentration |
|---|---|---|
| 5000 IU | 30,000 – 50,000 IU | Acute:<br>Infants  75,000 – 300,000 IU<br>Adults   2 – 5 million IU<br>Chronic:<br>Infants  18,000 – 60,000 IU/day<br>Adults  100,000 IU/day |

| Therapeutic Uses | Toxic Symptoms |
|---|---|
| Vitamin A deficiency.<br>Steatorrhea or other deficiency<br>Hyperkeratosis, Xerophthalmia,<br>  Night blindness. | Acute:<br>Infants – anorexia, bulging fontanelles, hyper irritability, vomiting.<br>Adults – headache, dizziness, drowsiness, nausea, vomiting.<br>Chronic:<br>anorexia, headache, blurred vision, sore muscles, loss of hair, bleeding lips, cracking and peeling skin, nose bleed, liver and spleen enlargement, anemia.<br>Teratogenic. |

and diarrhea as the characteristic symptoms of acute hypervitaminosis A (Knudson and Rothman, 1953).

Chronic hypervitaminosis most often was due to medical over-prescription and self-overdose.  The chronic poisoning of children was often due to poor judgment of adults, e.g., if a small amount of a preparation is good for a child, a larger amount should be better

(Ostwald and Briggs, 1966). Young children suffering from chronic
hypervitaminosis A exhibit premature epiphyseal bone closing, cor-
tical regional thickening and long bone growth retardation (Cumming
et al, 1981). Investigations by Ruby and Mital (1974) indicate
that permanent bone malformations seem relatively rare. Others
have suggested that bone changes induced by vitamin A toxicity are
among the most long lasting effects (Hayes and Hegsted, 1973).
In adults, chronic hypervitaminosis has resulted most often as the
result of high vitamin A taken for the treatment of acne or other
skin disorders (Ostwald and Briggs, 1966). Chronic toxicity (Table
1) in man results in anorexia, headaches, blurred vision, muscle
soreness after exercise, hair loss, general drying and flaking of
the skin with pruritis, and nose bleeds. Anemia may be present and
the liver and spleen enlarged (Bartolozzi et al, 1967; Pease, 1962;
Hayes and Hegsted, 1973). There have not been any reports of
deaths unequivocally ascribed to vitamin A excess, but most serious
is the implied sequelae to toxicity being irreparable damage to the
liver, fibrosis, and cirrhosis (Muenter et al, 1971; Rubin et al,
1970). It has been suggested that in hypervitaminosis A, the
retinoids interfere with cellular membrane integrity (Dingle, 1964);
and it has been shown that vitamin A enhances the activation and
release of proteolytic enzymes by cell organelles, especially
lysosomes (Fell, 1970). Small amounts of vitamin within the range
of the RDA (1980) stabilize the membrane by providing cross-link
between lipids and protein of membrane. Large amounts of vitamin A
when combined with cell membrane lipoprotein and exogenous protein
play some important protective role in vitamin A toxicity. Because
the cell capacity to synthesize this protein is limited, the cell
membrane is exposed to unbound vitamin A when the intake of the
lipid-soluble vitamin is exceedingly high leaving the membrane
unprotected (Cumming et al, 1981; Jenkins, 1978).

There is no evidence that the carotenoids, as such, exhibit
toxic effects. Loading the patient with carotenoids does not
elevate serum vitamin A, it may even lower vitamin A to some ex
tent. Excess carotenoid ingestion does cause yellow pigmentation
(xanthosis) in children.

Recently, significant emphasis has been placed on the tera-
tological effects of high vitamin A consumption. The developing
mesenchymal cell seems particularly prone to toxicity (Marin-
Padilla, 1960). The main malformations noted in the early studies
include resorption of the fetus, cleft palate, spina bifida,
anencephaly, and others (Cohlan, 1953; Vorhees, 1974; Nolen, 1969;
Kochhar, 1973; Robens, 1970; Geelen, 1973).

There has been a renewed interest in hypervitaminosis A and
its prevention because of the suggested efficacy and the synthesis
of new analogs for use in three physiological disorders. The
three disorders include acne vulgaris, enhancement of the immune
system function, and cancer.

Ostwald and Briggs (1966) questioned the efficacy of massive doses of vitamin A in the treatment of acne or similar skin disorders.  Vitamin A has been used for many years in the treatment of various skin disorders characterized by abnormal keratinization (Pochi, 1982), but not without problems (Youmans, 1960).  There is some evidence that synthetic derivatives of vitamin A, retinoic acid forms, are beneficial in treating severe acne patients; however, caution is suggested, since complete evaluation of the side effects is not available (Pochi, 1982).

Vitamin A does appear to influence cell—mediated immune mechanisms.  In vitamin A deficiency, antibody production after immunization is suppressed whereas large doses of vitamin A stimulate production (Beisel, 1982).  A similar response is also found for cell—mediated (nonhumoral) immunity.  It does seem clear that vitamin A deficiency increases the host susceptibility to infections (Beisel et al, 1981) and conversely, at least in animals supplemented with vitamin A, host susceptibility to infection is diminished.

During the last few years, research regarding possible preventive and therapeutic uses of vitamin A, or more specifically, its synthetic analogs in cancer has increased.  Epidemiological studies support the hypothesis that vitamin A deficiency is related to an increased susceptibility to epithelial carcinoma (Rivlin, 1982) and, to an extent, animal experiments seem to support this hypothesis (Spron et al, 1976; Spron and Newton, 1979; Basu, 1979).  Non-synthetic vitamin A compounds are of limited usefulness as chemopreventive agents owing to their intrinsic toxicity; however, excess amounts of retinoids exhibit some ability to retard the development of tumors (Wolf, 1980).  A number of new synthetic retinoid analogs that have diminished toxicity are under investigation, and the data suggest better targeting to specific sites and better potent antitumor activity than natural vitamin A (Sporn, 1979, Peter et al, 1980).

## Vitamin D

Therapy.  Vitamin D occurs in two forms:  1) vitamin $D_2$ or ergocalciferol, which is produced by ultraviolet irradiation of ergosterol, a plant steroid and 2) vitamin $D_3$ or cholecalciferol, the naturally occurring form of the vitamin in animal tissues.  Both are equally effective in man, and the latter is often placed in the class of hormones since calciferol is formed by the action of sunlight on 7-dehydrocholesterol in the skin (Frieden and Lipner, 1971).  Vitamin D increases bone sensitivity to parathyroid hormones and promotes intestinal absorption of calcium and phosphate.

Chronic deficiency produces rickets in children and osteomalacia in adults.  The exact requirement for vitamin D has not yet been established because the amount is dependent on a number of variables

such as length and intensity of light exposure and skin color.
Chronically bed-ridden patients, not exposed to sunlight and/or with
some degree of steatorrhea, should have a daily intake of 400 IU
vitamin D supplied exogenously.  Daily dose of 1,000-4,000 is often
prescribed for rapid recovery from fully developed rickets (Weiss,
1970); and sometimes, depending on the response, a dose of 30,000 IU
has been used (Table 2).

The most active form of vitamin D is the 1, 25-dihydroxy-
cholecalciferol or calcitriol (DeLuca, 1974).  Vitamin $D_3$ is first
hydroxylated in the 25-position in the liver and then hydroxylated
again in the alpha-position in the kidneys.  It is understandable
that for satisfactory vitamin D therapy, functional kidneys are
important.

Toxicity.  Most cases of hypervitaminosis D found in the
literature are the consequences of overdosing with either cholecalciferol or ergocalciferol.  Toxic concentrations of vitamin D

Table 2.   Vitamin D

| Recommended Intake | Therapeutic Dose | Toxic Concentration |
|---|---|---|
| 400 IU | Infants<br>  3000 IU/day<br>Adults<br>  150,000 IU/day | Acute:  1000 - 3000<br>          IU/kg/day<br>Chronic: 10,000 -<br>          500,000 IU/day |

| Therapeutic Uses | Toxic Symptoms |
|---|---|
| Prophylaxis of rickets<br>Hypoparathyroidism<br>Adult osteomalacia | Acute:<br>  Anorexia, nausea,<br>  vomiting, diarrhea,<br>  headache, polyuria,<br>  polydipsia.<br>Chronic:<br>  Weight loss, pallor<br>  constipation, fever,<br>  hypercalcemia,<br>  calcium deposits in<br>  soft tissues. |

in man have not resulted from unlimited exposure to sunshine.  Skin tanning or the aggregation of the pigment melanin in the epithelium creates a filter for ultraviolet lights and prevents conversion of the 7-dehydrocholesterol to cholecalciferol.  The deep melanin pigmentation of indigenous tropical people is an evolutionary adaptation for protection from vitamin D toxicity (Seelig, 1969; Loomis, 1970).  The concentrations of vitamin D in certain fish oils, rich in the vitamin, are not so high that toxicity could result from ingestion.  Manifestations of vitamin D toxicity in laboratory animals are similar to those in humans overdosed with the vitamin.  A major response of animals to vitamin D overdose include hypercalcemia, producing extensive calcification of extra skeletal tissues like blood vessel walls and kidneys (Hayes and Hegsted, 1973; Ostwald and Briggs, 1966).  The symptoms (Table 2) of vitamin D toxicity in man are anorexia, nausea, vomiting, diarrhea, headache, polyuria, and polydipsia (Ostwald and Briggs, 1966).  Calcium levels are increased in serum and urine, and calcium is deposited in soft tissue.  The primary target for deposit of calcium is the kidneys which can lead to hypertension, renal failure, cardiac insufficiency, and anemia (Stults, 1981).  In some cases the effects of hypervitaminosis D can be reversed by withdrawal of vitamin D or treatment with glucocorticoids (Anderson et al, 1954).

Symptoms of acute toxicity can take several days to appear. Dose range for toxicity in children varies from 10,000 IU/day for 4 months to 200,000 IU/day for 2 weeks (Hayes and Hegsted, 1973).  The vitamin D tolerance of human adults appears to vary from 25,000 IU per day for a few weeks to over 1 million IU per day for 4 years (Ostwald and Briggs, 1966).  Some vitamin D toxicity situations result from the mistaken use of vitamin D concentrates as a result of the false idea that large amounts of vitamins promote growth and vigor (Harrison, 1978).  At one time, large concentrations of vitamin D were given to patients with rheumatoid arthritis.  Pharmacologic use of vitamin D is prescribed for hypoparathyroidism, pseudohypoparathyroidism and chronic renal diseases because of the deficiency in the renal 1-hydroxylation of 25-hydroxyvitamin D.  Pharmacologic doses of vitamin D are also used for primary hypophosphatemic vitamin D-resistant rickets in order to maintain intestinal calcium absorption in the presence of large phosphate loads.

The toxicity of these high concentrations of vitamin D is most likely due to the pharmacologic effects of 25-OH vitamin D in high concentration.  Brumbaugh and Hanssler (1973) found that 15-OH vitamin D could stimulate the 1,25di-(OH) vitamin D hormone at the receptor when present in excessive concentrations.  Circulating concentrations of 1, 25-di-(OH) vitamin D, unlike that of 25-OH vitamin D, are not greatly increased in vitamin D-intoxicated patients (Hughes et al, 1976; Haussler, 1978).

Vitamin E

   Therapy. Vitamin E activity is found in eight naturally occurring tocopherols made by plants.  α-Tocopherol (RRR-α-tocoperol)
has the greatest biological activity in biological assays such as
the prevention of infertility in deficient rats.  The requirement,
at least in animal studies, is altered by oxidizing agents (Reddy
et al, 1978; Omaye et al, 1978) increasing consumption of polyunsaturated fatty acids (Tappel, 1972).  On the other hand, the
trace element selenium and sulfur amino acid reduce the need for
this vitamin (Chow et al, 1973; Tappel, 1974; Omaye and Tappel,
1974).

   For most of the claims publicized for pharmacological doses
of vitamin E, there is no substantial scientific evidence of efficacy (RDA, 1980).  Claims have been made that high doses of
vitamin E affect muscular dystrophy, leg cramps, thrombophlebitis,
burns, hypertension, rheumatic fever, improved immunity, and retarding the aging process (Horwitt, 1980; Tappel, 1973).  There is,
however, some evidence (Table 4) to suggest the potential beneficial
effects of high dosing vitamin E in certain anemias in premature
or low birth weight infants, in patients with malabsorption syndromes, and in genetic diseases where vitamin E status is impaired
(cystic fibrosis, sickle cell anemia) (Machlin et al, 1980; Chiu
and Lubin, 1979).

Table 3.  Animal Toxicology of Vitamin E

| Species | Effects of High Doses |
| --- | --- |
| Mouse | Lung and subcutaneous tumors increased in dibenzanthracene treated animals receiving supplementary tocopherol. |
| Rat | Liver fatty infiltration with aortic intimal sclerosis. Development of ethanol enhanced fatty liver. Coagulopathy and increased mortality. |
| Chick | Increased clotting time, depressed thyroid function, decreased growth. |

Cumming, Briggs, and Briggs (1981).

Toxicity. For the most part, reviewers agree that megadoses of vitamin E are essentially nontoxic (Bieri, 1975; Horwitt, 1980) while at the same time express that indiscriminate use is unwise. Tocopherols appear to lack any inherent carcinogenicity; however, there is some evidence that they may have some procarcinogenic activity (Telford, 1949). One study suggested the potential teratogenic effects of α-tocopherol (Hook et al, 1974). They found one malformed animal out of a total of 91 offspring from 7 litters of mice who received 591 IU D-α-tocopherol (PO) on Days 7 to 11 of pregnancy. Other animal studies demonstrated fatty infiltration of the liver with deposition of cholesterol (Marxs et al, 1947). The adverse effects of high tocopherol consumption are listed in Table 4.

Table 4.   Vitamin E

| Recommended Intake | Therapeutic Dose | Toxic Concentration |
|---|---|---|
| 10 mg/day | No unequivocal evidence for therapeutic use. Low birth weight babies. | 300 mg/day or 2-12 g/day |

| Therapeutic Uses | Toxic Symptoms |
|---|---|
| Low birth weight babies. | Headache, nausea, fatigue, giddiness, blurred vision. |
|  | Suspected disruption of gonadal function. |
|  | Slight creatinuria, chapping, cheilosis, angular stomatitis, gastroinintestinal disturbances, muscle weakness. |
|  | Altered coagulation factors. |
|  | Elevated plasma lipids. |
|  | Reduced response to iron therapy. |

In 1973 the first reports appeared describing severe weakness, fatigue, and other side effects induced in healthy adults by daily doses of 800 IU of α-tocopherol (Cohen, 1973). Similar results were experienced by others (Briggs, 1974). The symptoms were associated with increased serum creatine kinase activity and with a large increase in 24-hour urinary excretion of creatine. A summary of disturbing side effects due to high vitamin E consumption is shown in Table 4.

Vitamin E has been used in cosmetic sprays and creams, and a few have been reported to produce sensitivity reaction in particular individuals to topically applied forms (Brodkin and Bleiberg, 1965; Minkin et al, 1973; Aeling et al, 1973).

Effects of excess vitamin E on hematology suggest interference with normal coagulation processes, leading to an increased vitamin K requirement (March et al, 1973). It has been suggested that α-tocophenylquinone or hydroquinone, the metabolites of α-tocopherol formed in the body, may act as antimetabolites. Melhorn and Gross (1969) showed that the hematological response of children with iron deficiency anemia to treatment with iron dextram is impaired if vitamin E supplements are given simultaneously.

Vitamin K

Therapy. The therapeutic importance of vitamin K has been extensively reviewed (Mount et al, 1982). Included in this class of chemicals is a group of naphthoquinone derivatives with anti-hemorrhagic characteristics. Natural vitamin K is composed of a series of phylloquinones and menaquinone derivatives of which vitamin $K_1$ and vitamin $K_2$ are respective representatives. Vitamin $K_3$, or menadione, is the synthetic water-soluble form which can be converted to the vitamin $K_2$ derivatives in animals (Mount et al, 1982). The requirement for vitamin K (Table 5) is met by a combination of dietary intake (vitamin $K_1$) and microbiolgical biosynthesis in the gut (vitamin $K_2$). Vitamin $K_1$, which is stored in the liver, is the preferred therapeutic agent rather than the poorly stored menadione.

Vitamin K's role is an essential co-factor for enzymatic con-version of specific glutamic acid residues of endogenous precursor proteins to alphacarboxyglutamate residues in completed protein (Suttie, 1980).

Vitamin $K_1$ is the form that is administered therapeutically. Poisoning with warfarin-related compound may be treated with vita-min $K_1$ (Mount et al, 1982). The form of treatment, parenteral or oral route, is available; however, the parenteral route is not recommended because of the potential for anaphylaxis.

Table 5. Vitamin K

| Recommended Intake | Therapeutic Dose | Toxic Concentration |
|---|---|---|
| Easy to meet needs by diet and gut bacteria synthesis. | 1-10 mg | Infants: 5-10 mg/day (ip) Premature Infants: 10 mg/day |
| Suggested intake of 70-140 µg/day. | | Adults: Essentially nontoxic. |

Water-soluble Menadione is toxic at 5-10 mg dose level.

| Therapeutic Uses | Toxic Symptoms |
|---|---|
| Bleeding associated with Vitamin K deficiency. Biliary obstruction or fistula. Malabsorption syndromes. Hepatocellular disease. Drug induced hypoprothrombinemia. | Infants: Hemolytic anemia, hyperbilirubinemia, kernicterus. Premature Infants: Kernicterus. |

Water-soluble menadione can produce anemia, skin irritation, polycythemia, splenomegaly, renal and hepatic damage, and death. Also implicated in hemolytic anemia, hyperbilirubinemia, and kernicterus. Menadione can induce erythrocyte hemolysis in individuals with genetic defect in glucose-6-phosphate dehydrogenase.

Toxicity. The major problem (Table 5) with therapeutic vitamin K administration is anaphylactoid reactions (Barash, 1978). Menadione and vitamin K in large doses have resulted in hematologic and circulatory effects. Hemolytic anemia, methemoglobinuria, polycythemia, and organ dysfunction have been noted. The effect on the hemopoietic system seems directly related to damage of the circulating red cells and by action of the quinone radical (Richards and Shapiro, 1945).

Interactions between lipid-soluble vitamins

Morgan et al (1937) showed that vitamin A in large excess protected young rats from the pathogenesis of hypervitaminosis. In general, large doses of vitamin A restored growth rate to near

normal and reduced calcium in kidney, heart, and lung tissue.  The
extent of the reversal of toxicity symptoms of vitamin D varied with
the source of vitamin A, especially with high concentrations and,
since that original study, has been confirmed by others (Belanger
and Clark, 1967; Taylor et al, 1968).  It has been suggested
(Clark and Smith, 1964) that the protective effect of vitamin A is
by increased mucopolysaccharide and collagen turnover and that
vitamin A has no influence on the enhanced intestinal calcium
absorption that occurs with high intakes of vitamin D.

Vitamin E stabilizes erythrocytes against lysis initiated by
vitamin A in vitro (Lucy and Dingle, 1964) and in vivo (Soliman,
1972).  Other reports demonstrated the conservation of hepatic
stores of vitamin A by vitamin E (Davies and Moore, 1941) and
confirmed by Cawthorne et al (1968).  However, utilizations of
β-carotene in the presence of high intakes of tocopherol were
blocked (Arnrich, 1978).  The latter study suggested that tocopherol
interacts with β-carotene in the biochemical steps that occur just
before the formation of retinol.

As discussed previously, excess tocopherol may interfere with
the normal coagulation process, producing a need for more vitamin K
(March et al, 1973).


WATER-SOLUBLE VITAMINS

Vitamin C (ascorbic acid)

Therapy.  Vitamin C has been widely used for the treatment
of a variety of conditions, including dental caries, pyorrhea, gum
infections, anemia, hemorrhagic states, and various forms of infections.  Some of these conditions might be associated with scurvy;
but only when final diagnosis based upon roentgenological evidence
in the long bones, and chemical evidence of a low ascorbic acid in
blood, is there assurance that therapy with ascorbic acid will
lead to beneficial effects.  Despite the vast clinical literature
on the various therapeutic indications for ascorbic acid, there is
no conclusive evidence that ascorbic acid is of any value except
in the relief of the symptoms associated with scurvy.  In the
treatment of fully developed scurvy, ascorbic acid may be given in
1 g a day doses until the body stores have been replenished (Burns,
1970).  Vitamin C is usually administered (Table 6) by the oral
route, although solutions of the sodium salt may be given by intramuscular or intravenous injection.  Common vehicle of administration
to children by oral route is in orange juice; however, when an
occasional infant is allergic to orange juice, ascorbic acid must
be freshly dissolved and administered soon thereafter because of
possible degradations with time (Omaye et al, 1979).

Table 6.   Ascorbic Acid (Vitamin C)

| Recommended Intake | Therapeutic Dose | Toxic Concentration |
|---|---|---|
| 60 mg/Day | 100 mg − 1 g/day | 500 mg − 10 g |

| Therapeutic Uses | Toxic Symptoms |
|---|---|
| Scurvy | Nausea, diarrhea, gastro-intestinal disturbances, flatus. Increased oxalates − stone formation. Decreased copper absorption. Increased dependency. |

Of significant interest is the investigation by Cook and Monsen (1977), who have found that the iron absorption in 63 men treated with ascorbic acid supplements in the range of 25 mg to 1 g daily resulted in a significant increase in iron absorption with vitamin supplement in a dose-related manner.  As the authors pointed out, while the effect may be of value in normal individuals, patients with idiopathic hemochromatosis, sideroblastic anemia, or other conditions in which iron absorption regulation is abnormal may be adversely affected from higher dose vitamin C supplements. Also, recent work suggests that high levels of ascorbic acid intake inversely influence copper and copper-containing proteins of blood and tissues in small animals (Milne and Omaye, 1980; Milne et al, 1981; Omaye and Turnbull, 1980).

Toxicity.  While no gross tissue changes have been reported at high doses of ascorbic acid, there is evidence of undesirable toxic effects in small animals and undesirable side effects in man (Briggs, 1978; Cumming et al, 1980).  In the rat the administration of ascorbic acid at 2.5 g/kg daily leads to damaged liver lysosomes (Lippi et al, 1981).  Thyroid activity is depressed in rats and guinea pigs receiving 0.4 g/kg ascorbic acid daily (Mallick and Deb, 1975; Deb and Mallick, 1974).

Vitamin status may be adversely influenced by excess ascorbic acid ingestion (Table 6).  β-Carotene utilization in rats is impaired when given excess ascorbic acid (Mayfield and Roehm, 1956).

Adversed nutritional effects of high doses of ascorbic acid have
been reported to lower vitamin $B_{12}$ levels (Hines, 1975; Herbert
and Jacob, 1974; Herbert, 1980). In contrast, others (Newmark and
Scheiner, 1976) have suggested that methodology problems explain
the vitamin $B_{12}$ to ascorbic acid interaction. Ascorbic acid has
also been reported to elevate the excretion of urinary 4-pyridoxic
acid, the major metabolite of vitamin $B_6$ (Sclivanova, 1960). Other
interactions between ascorbic acid and vitamin $B_6$ have also been
reported (Baker et al, 1971; Baker et al, 1964). More recently,
Shultz and Leklem (1982) concluded that short-term ascorbic acid
supplementation did not alter vitamin $B_6$ metabolism.

High doses of ascorbic acid can also influence ascorbic acid
metabolism. Guinea pigs pretreated with vitamin C become scorbutic
sooner when transferred to a deficient diet in ascorbic acid
(Gordonoft, 1960). Similar results have been reported for humans.
Frank cases of scurvy have been reported in otherwise healthy
adults who abandoned vitamin C supplements and returned to normal
dietary intake (Rhead and Schrauzer, 1971; Schrauzer and Rhead,
1973). There are no reports with convincing data that large doses
of ascorbic acid during human pregnancy are entirely safe. One
report demonstrated 16 pregnant women aborted after receiving
ascorbic acid at 6 g daily for 3 days (Briggs, 1978). High doses
of ascorbic acid given to pregnant guinea pigs shorten the gestation period and increased the incidence of stillbirths (Neuweiler,
1951; Samborckaia, 1966). There is also a report of four women
who failed to conceive while receiving vitamin C supplements
(Briggs, 1973). The suggestion from that study is that ascorbic
acid secreted into the mucus of the uterine cervix may disrupt
disulfide bonding of glycoprotein fibrils and, therefore, impede
sperm penetration.

Some individuals within a population excrete massive amounts
of urinary oxalate after the ingestion of large doses of ascorbic
acid (Driggs, 1976; Kallner, 1977), of which a few are troubled by
urinary stones (Briggs, 1973). They concluded that in Australia the
incidence of kidney stones has risen during the period approximately
similar to the mean individual consumption of vitamin C. This
suggests that individuals at risk of urinary stone formation related
to high intakes of ascorbic acid are those consuming several grams
daily, or with various metabolic disorders, which enhance susceptibility to oxalate formation.

Andrews and Wilson (1973) described a study wherein a 200 mg
daily dose of vitamin C produced a high incidence of thrombotic
episodes associated with ascorbic acid compared with the placebo.
There has also been described adverse in vitro effects of a high
concentration of ascorbic acid on human platelet structure and
function (Cowen et al, 1975). Other hematological findings include:
1) leukocyte counts of subjects who consumed large doses of vitamin

C were reduced (Robinson et al, 1975) and 2) a patient with genetic glucose-6-phosphate dehydrogenase deficiency had a fatal hemolytic crisis following infusion (iv) of 80 mg of ascorbic acid daily for two days. Mengel and Greene (1976) have found that 9 of 14 subjects given 5,000 mg of ascorbic acid daily for 3 days had substantial increases of lysis of red cell by $H_2O_2$ and in vitro index of chemical fragility of erythrocytes. The implication of such findings to clinical significance is uncertain.

No carcinogenicity has been reported. Nevertheless, there is a report showing mutagenicity following treatment of human fibroblasts with ascorbic acid in vitro (Stich et al, 1976), and several reports have shown disruption of DNA in bacteria with high amounts of ascorbic acid (Murata and Kitagawa, 1973; Richter and Loewen, 1982; Wong et al, 1974; Murata et al, 1975). They attributed this to a free radical intermediate produced by auto-oxidation of ascorbic acid and suggested that large amounts of ascorbic acid can have a damaging effect on genetic material.

Several reports illustrate drug interactions with excess ascorbic acid, such as salicyclic acid, amphetamines, warfarin, tricyclic anti-depressants, and salicylates (Briggs, 1978; Cumming et al, 1981). Others (Cumming et al, 1981) show that clinical chemical tests can be altered by high ascorbic acid intakes such as a false negative test for occult blood (Jaffe et al, 1975), a false positive Clinitest for sugar, and a false negative Testape for renal problems (Mayson, 1973; Singh et al, 1972). Invalid common tests for glycosuria (Goldner, 1971) could have serious consequences for diabetics running the risk of either diabetic ketosis or insulin shock (Herbert, 1980).

## Vitamin $B_1$ (Thiamin)

Therapy. Thiamin is necessary for decarboxylation of alpha keto acid; therefore, the requirement is proportional to the percentage of carbohydrate in the diet, approximately 0.5 mg/1,000 Kcal (RDA, 1980). The average adult daily intake is close to 0.8 mg, which suggests that some of the population might be marginal in this nutrient. Fortunately, fats and protein have a thiamin-sparing effect. Thiamin is not stored; and various outside "stresses" such as fever, hyperthyrodism, and trauma seem to influence the need for thiamin. Alcoholics with inadequate diets are particularly susceptible to thiamin deficiency.

Thiamin is specific in the treatment of conditions associated with vitamin $B_1$ deficiency or beri-beri. The etiology of peripheral neuritis frequently involves vitamin $B_1$ deficiency, such as the neuritis accompanying alcholism and pregnancy (Spector, 1980). All seem to respond well to treatment. In mild deficiency states, a daily dose of 10 mg is usually sufficient (Marks, 1975).

Table 7.   Thiamin (Vitamin $B_1$)

| Recommended Intake | Therapeutic Dose | Toxic Concentration |
|---|---|---|
| 1.5 mg/day | 10-30 mg/day | No toxic effects (PO) humans. 10-100 mg $LD_{50}$ (iv) for rats >40x (PO) |

| Therapeutic Uses | Toxic Symptoms |
|---|---|
| Beri-beri (thiamin deficiencies). Chronic alcoholism. Malnutrition (economic poor). | (iv) Anaphylactic shock, vasodilation, nausea, tachycardia, and dyspnea from repeated injections |

Doses up to 30 mg might be called upon for severe cases; and in the case of delirium tremens, a large dose of thiamin combined with other vitamins given by injection might be needed (Table 7).

Toxicity.   Thiamin from the diet is absorbed in the intestine by a carrier-mediated process (Rindi and Ventura, 1972).   The process is saturable and is sodium and energy-dependent.   Therefore, because of this active carrier-mediated transport system, an oral intake of thiamin greater than 10 mg does not generally increase blood levels of the vitamin significantly.   The oral lethal dose for thiamin is above 2000 mg/kg for rodents.   Parenteral administration of the vitamin is toxic ($LD_{50}$), in the ranges from 80-120 mg/kg (iv) and 300-500 mg/kg (ip) for rodents (Cumming et al, 1980).

Because of its low oral toxicity, serious clinical consequences have only been occasionally reported following large parenteral doses (Table 7).   Several cases of death by anaphylactic shock have been noted (Itokawa, 1978).   The symptoms of thiamin overdosage include (at 400 mg, iv) acute mental alertness, lethargy, solemnness, mild ataxia, heaviness in the limbs, and a diminution of gut tone (Gould, 1954).   Other manifestations of thiamin overdose in humans include acute hypotension and the development of hyperthyroid-like symptoms.

Klopp et al (1943) reported an interaction between thiamin and riboflavin.   There were no clinical signs of riboflavin deficiency;

but after the administration of a high thiamin dose, there was a
transitory increase in urinary riboflavin excretion.   Similar results were observed by others (Fujiwava, 1954).   One study by Inoue
et al (1956) demonstrated that massive parenteral doses of thiamin
propyl disulfide in patients with neurological disorders induced
clinical signs of riboflavin deficiency that were reversed by riboflavin administration (Nayakama, 1959a; Nayakama, 1959b).

Another possible interaction between thiamin and a vitamin has
been suggested.   Vitamin $B_6$ may have a protective effect against
excessive thiamin in rats (Morrison and Sarett, 1959; Schumacher
et al, 1965).

### Vitamin $B_2$ (Riboflavin)

Therapy.  Riboflavin is employed for the treatment of riboflavin deficiency (Marks, 1975).   Deficiencies are encountered in
patients who have undergone total or subtotal gastrectomy.   Recommended supplementation (Table 8) for recovery is in the range of
10 to 60 mg daily and by injection, if necessary.

Toxicity.   In general riboflavin is very low in toxicity.
Most of the reported problems are due to solvent toxicity which are
required to dissolve the vitamin.   Riboflavin exhibits slow solubility in water at body temperature (Rivlin, 1978).

No toxicity has been reported from the oral treatment of riboflavin.   This is likely due to the limited capacity of intestinal

Table 8.   Riboflavin (Vitamin $B_2$)

| Recommended Intake | Therapeutic Dose | Toxic Concentration |
|---|---|---|
| 1.7 mg/day | 2-15 mg/day<br>>10 g/kg (PO) | No cases |

| Therapeutic Uses | Toxic Symptoms |
|---|---|
| Ariboflavinosis | None known for humans. Anuric and vitamin cystals found in the renal tubules of injected (iv) animals. |

absorption mechanisms (Christensen, 1973). One report (Table 8) in animals found vitamin crystals in the renal tubules of anuric rats that received 0.6 g/kg (ip) (Unna and Greslin, 1942).

## Niacin (Nicotinic acid)

Therapy. Nicotinic acid is an important component of hydrogen-carrying coenzymes. The deficiency of nicotinic acid is charac-terized by severe metabolic disorders in the skin and digestive organs. This vitamin is one of the major factors missing in the development of pellagra which has symptoms of glossitis, stomatitis, anorexia, abdominal discomfort, dermatitis and diarrhea. As shown in Table 9, the daily dose for therapy is up to 500 mg for severe pellagra (Marks, 1975). Nicotinamide is useful in the treatment of glossitis, dermatitis, and mental symptoms; but because the cause of pellagra is probably not just a simple nicotinic acid deficiency, other vitamins such as vitamin $B_1$ and riboflavin may be useful (RDA, 1980).

Toxicity. Toxic effects of nicotinamide include (Table 9) flushing, pruritus, skin rash, heartburn, nausea, vomiting, diarr-hea, ulcer activation, abnormal liver function, hypertension, and hyperglycemia (Ban, 1974; Berge, 1961; Berge et al, 1961; DiPalma and Ritchie, 1977). The doses used are often greater in excess of

Table 9.  Niacin (Nicotinic Acid)

| Recommended Intake | Therapeutic Dose | Toxic Concentration |
|---|---|---|
| 18 mg/day | 100–500 mg/day | 100–300 mg (PO) 20 mg (iv) |

| Therapeutic Uses | Toxic Symptoms |
|---|---|
| Pellagra, vasodilation, antilipemia, fibrinolytic agent? | Flushing, headache, cramps, nausea/vomiting. Anorexia, abnormal glucose metabolism. Abnormal plasma uric acid, liver function tests, gastric ulceration, anaphylaxis. |

recommended allowances.  The acute $LD_{50}$ in rodents is 4.5 to 7 g/kg by oral route and about 3 g/kg parenterally (Cummings et al). Liver changes are usually mild and reversible after withdrawing niacin (Bagenstoss et al, 1967).  Severe hepatic dysfunction occurs but is rare (Kohn and Montes, 1969).  There is also a reversible altered glucose tolerance in nondiabetic subjects during niacin treatment (Gurian and Alderberg, 1959), and diabetic patients have an increased insulin requirement following niacin supplements (Molnar et al, 1964).

## Folic acid

Therapy.  Folic acid deficiency in humans results in macrocytic anemia accompanied by leucopenia (RDA, 1980).  Folic acid in daily doses of about 5 mg orally is adequate therapy for deficiency cases (Table 10).

Toxicity.  Intravenous doses of 45–120 mg folic acid to young rats has been reported to induce epilepsy (Hommes et al, 1973; Obbens and Hommes, 1973).  Such research points to some metabolic product of folic acid as the active agent.

There are also marked changes in blood pressure and cardiac contracture (Jenkins and Spector, 1973).  A major concern is folic acid-induced toxicity of the kidneys.  Large doses of folic acid in animals cause immediate increase in DNA synthesis, total protein with hypertrophy, and hyperplasia of the kidney epithelial cells (Threlfall, 1968; Threlfall et al, 1967; Tilson, 1970).

Table 10.   Folacin (Folic Acid or Pteroylglutamic Acid)

| Recommended Intake | Therapeutic Dose | Toxic Concentration |
|---|---|---|
| 400 μg/day | 5 mg/day | 1 mg/day can mask $B_{12}$ deficiency symptoms |

| Therapeutic Uses | Toxic Symptoms |
|---|---|
| Folate deficiency | Allergic reactions? Renal toxicity in rats. Seizure in patients on antiepileptic agents. |

The most common side effect of oral treatment with 15 mg folic acid in humans is mild gastrointestinal disturbance, irritability malaise, hyperexcitability, disturbed sleep, and vivid dreams (Hunter et al, 1970). Hypersensitivity reactions do occur on occasion with certain patients (Mitchell et al, 1949).

## Vitamin B₆

Therapy. A true deficiency of vitamin $B_6$ is rare. Neurological symptoms associated with pellagra and beri-beri seem to be improved with pyridoxine. Doses of pyridoxine (40-150 mg daily) are used for the peripheral neuritis due to treatment with isoniazid and related compounds. Pyridoxine has also been used for nausea and vomiting due to a variety of conditions. It has been recommended for the treatment of agranulocytosis and seborrhoeic dermatitis (Marks, 1975).

Toxicity. Pyridoxal hydrochloride exhibits oral toxicity (Table 11) at about 2,000 mg/kg body weight (Cumming et al, 1978). No teratogenic activity has been found (Frimpter et al, 1969; Khera, 1975) for any species studied. There is one report that very large amounts of vitamin $B_6$ inhibit prolactin secretion (Canales et al, 1976) and, subsequently, could be detrimental to the lactating women. Pyridoxine at high doses does cause convulsive disorders, especially the pyridoxal hydrozones (Dubnick et al, 1960; Dipalama and Ritchie, 1977).

## Vitamin B₁₂

Therapy. Vitamin $B_{12}$ at the rate of 15-20 µg/day has been recommended (im) for the relief of pernicious anemia. Vitamin $B_{12}$ given by mouth with intrinsic factor has had variable effects (Marks, 1975).

Toxicity. The most common side effect (Table 11) due to vitamin $B_{12}$ administration is allergic reactions in sensitive individuals (Cumming, 1978). There appear to be a few reports of fatalities following injection of 3 mg/kg in mice and guinea pigs (Lipton and Steigman, 1963). There is a suggestion that vitamin $B_{12}$ might increase nucleic acid degradation and, subsequently, produce a sudden rise in serum uric acid and urinary excretion, which would be bad for the patient susceptible to gout (Cumming et al, 1978).

## Biotin and Pantothetic Acid

Therapy. Biotin at 5 mg daily injection improves infants with cutaneous lesion produced by deficiency (Marks, 1975). Pantothetic acid in the form of an alcohol has been useful for deficiency of pantothenic acid.

Table 11.   Pantothenic Acid, Pyridoxine, Biotin, and
            Vitamin $B_{12}$

| Recommended Intake | Therapeutic Dose | Toxic Concentration |
|---|---|---|
| Pantothenic Acid | | |
| None | None | 10–20 g |
| Pyridoxine | | |
| 2.2 mg/day | None | 3–4 g/kg death in animals |
| Biotin | | |
| None | None | None |
| Vitamin $B_{12}$ | | |
| 3 µg | 100 mg | None |

| Compound | Toxic Symptoms |
|---|---|
| Pantothenic Acid | Occasional diarrhea |
| Pyridoxine | Convulsions/Death in animals |
| Biotin | Delayed growth and infertility in animals |
| Vitamin $B_{12}$ | Allergic reactions and lung congestion in animals |

Toxicity.   Both of these members of the B complex vitamins are
similar in their exceedingly low toxicity and adverse reactions.
Biotin may have some influence on the inhibition of pentose phosphate pathway metabolism at extremely large overdoses (Marks, 1975).


CONCLUSION

The promotion of megadoses of vitamins in amounts exceeding
the daily Recommended Dietary Allowances (RDA) is a billion dollar
business in the United States (Jukes, 1979;  Herbert, 1980).
Advocates of megadoses of vitamins claim that vitamins in megadoses

will prevent or cure disease not of a nutritional origin and produce
optimal health beyond that needed from a nutritionally adequate
diet, i.e., if a little is good, then more is better (Jukes, 1979;
Herbert, 1980).  They also state that individual variability, or
biochemical variability, justifies high intake of vitamins.  All
these claims ignore the research and thought that went into establishing the RDA (1980), and that the RDA's do take into consideration
the range of normal variability in vitamin requirements.

Pharmacological doses of vitamins are justified only for the
reversal of deficiency states that might be the consequence of
deprivation, certain vitamin-dependent genetic diseases, problems
associated with defective vitamin transport across cell membranes
and as an antidote to anti-vitamin ingestion (Herbert, 1980; Hodges,
1981).  Intakes of large doses of vitamins provide little benefit,
are costly, and potentially harmful leading to numerous undesirable
effects, minor to serious.  Almost any vitamin taken in high concentration can be harmful, directly or indirectly, by interactions
with other nutrients.  The newly emerging data described here
clearly points to an area of enormous research potential for the
fields of nutrition, pharmacology, and toxicology.

REFERENCES

Aeling, J. L., Panagotacos, P. J., and Andreozzi, R. J., 1973,
    Allergic contact dermatitis to vitamin E aerosol deodorant,
    Arch. Dermatol., 108:579-580.
Anderson, J., Dent, C. E., Harper, C., and Philpot, G. R., 1954,
    Effect of cortisone on calcium metabolism in sacroidosis with
    hypercalcaemia.  Possibly antagonistic action of cortisone and
    vitamin D, Lancet, 267:720-724.
Andrews, C. T. and Wilson, T. S., 1973, Vitamin C and thrombotic
    episodes, Lancet, 2:39.
Arnrich, L., 1970, Toxic effects of megadoses of fat-soluble
    vitamins, in: "Nutrition and Drug Interrelations," J. H.
    Hathcock and J. Coon, eds., Acadmic Press Inc., New York,
    pp. 751-771.
Baker, E. M., Canham, J. E., Nunes, W. T., Sauberlich, H. E., and
    McDowell, M. E., 1964, Vitamin $B_6$ metabolism for adult man,
    Am. J. Clin. Nutr., 15:59-66.
Baker, E. M., Hodges, R. E., Hood, J., Sauberlich, H. E., March,
    S. C., and Canham, J. E., 1971, Metabolism of $^{14}$C- and $^{3}$H-
    labeled L-ascorbic acid in human scurvy, Am. J. Clin. Nutr.,
    24:444-454.
Ban, T. A., 1974, Negative findings with nicotinic acid in the
    treatment of schizophrenias, Int. Pharmacopsychiatry, 9:172-187.
Barash, P. G., 1978, Nutrient toxicities of vitamin K, in: "Handbook
    in Nutrition and Food," M. Rechcigl, Jr., ed., CRC Press,
    Boca Raton, pp. 97-100.

Bartolozzi, G., Bernini, G., Marianelli, L., and Corvaglia, E., 1967, Chronic vitamin A excess in infants and children, Riv. Clin. Pediatr., 80:231-290.

Basu, T. K., 1979, Vitamin A and cancer of epithelial origin, J. Hum. Nutr., 33:24-31.

Beisel, W. R., Edelman, R., Nauss, K., and Suskind, R. M., 1981, Single-nutrient effects on immunologic function, J. Am. Med. Assoc., 245:53-58.

Beisel, W. R., 1982, Single nutrient and immunity, Am. J. Clin. Nutr. (Supplement), 35:417-468.

Belanger, L. F. and Clark, I., 1967, Alpharadiographic and histological observations on the skeletal effects of hypervitaminoses A and D in the rat, Anat. Rec., 158:443-452.

Berge, K. G., 1961, Side effect of nicotinic acid in treatment of hypercholesteremia, Geriatrics, 6:416-422.

Berge, K. G., Achor, R. W. P., Christensen, N. A., Mason, H. L., and Barker, N. W., 1961, Hypercholestermia and nicotinic acid, Am. J. Med., 31:25-36.

Bieri, J. G., 1975, Vitamin E, Nutr. Rev., 33:161-167.

Briggs, M. H., 1973, Fertility and high-dose vitamin C, Lancet, 2:1083.

Briggs, M. H., 1973, Side effects of vitamin C, Lancet, 2:1429.

Briggs, M. H., 1974, Vitamin E supplements and fatigue, N. Engl. J. Med., 290:579-580.

Briggs, M. H., 1976, Vitamin C induced hyperoxaluria, Lancet, 1:154.

Brodkin, R. H. and Bleiberg, J., 1965, Sensitivity to topically applied vitamin E, Arch. Dermatol., 92:76-77.

Brumbaugh, P. F. and Haussler, M. R., 1973, 1α, 25-dihydroxy-vitamin $D_3$ receptor: Competitive binding of vitamin D analogs, Life Sci., 13:1737-1746.

Burns, J. J., 1970, Ascorbic acid (vitamin C), in: "The Pharmacological Basis of Therapeutics," L. S. Goodman and A. Gilman, eds., The Macmillian Co., London, pp. 1665-1669.

Canales, E. S., Soria, J., Zarate, A., Mason, M., and Molina, M., 1976, The influence of pyridoxine on prolactin secretion and milk production in women, Br. J. Obstet. Gynaecol., 83:387-388.

Casarett, L. J. C. and Doll, J., 1980, "Toxicology, The Basic Science of Poisons," Macmillan Publishing Co., Inc., New York.

Cawthorne, M. A., Bunyan, J., Diplock, A. T., Murrell, E. A., and Green, J., 1968, On the relationship between vitamin A and vitamin E in the rat, Br. J. Nutr., 22:133-143.

Chiu, D. and Lubin, B., 1979, Abnormal vitamin E and glutathione peroxidase levels in sickle cell anemia, J. Lab. Clin. Med., 94:542-548.

Chow, C. K., Reddy, K., and Tappel A. L., 1973, Effect of dietary vitamin E on the activities of the glutathione peroxidase system in rat tissues, J. Nutr., 103:618-624.

Christensen, S., 1973, The biological fate of riboflavin in mammals, Acta Pharmacol. Toxicol., 32(supplement 11):1-72.

Clark, I. and Smith, M. R., 1964, Effects of hypervitaminosis A and D on skeletal metabolism, J. Biol Chem., 239:1266-1271.

Cohen, H. M., 1973, Effects of vitamin E: good and bad, N. Engl. J. Med., 289:980.

Cohlan, S. Q., 1953, Excessive intake of vitamin A as a cause of congenital anomalies in the rat, Science, 117:535-536.

Cook, J. D. and Monsen, E. R., 1977, Vitamin C, the common cold, and iron absorption, Am. J. Clin. Nutr., 30:235-241.

Cowan, D. H., Graham, R. C., Shook, P., and Griffith, R., 1975, Influence of ascorbic acid on platelet structure and function, Thromb. Diath. Haemorrh., 34:50-62.

Cumming, F., Briggs, M., and Briggs, M., 1981, Clinical toxicology of vitamin supplements, in: "Vitamins in Human Biology and Medicine," M. H. Briggs, ed., CRC Press, Inc., Boca Raton, pp. 187-243.

Davies, A. W. And Moore, T., 1941, Interaction of vitamin A and E, Nature (London), 147:794-796.

Deb, C. and Mallick, N., 1974, Effect of chronic graded doses of ascorbic acid on thyroid activity, protein bound iodine of blood and deiodinase enzyme of periperhal tissues of guinea pigs, Endocrinol., 63:231-238.

DeLuca, H. F., 1974, Vitamin D: the vitamin and the hormone, Fed. Proc., 33:2211-2219.

Dingle, J. T., 1963, Action of vitamin A on the stability of lysosomes in vivo and in vitro, in: "Lysosomes," A.V.S. DeReuck and M. P. Cameron, eds., Little, Brown and Company, Boston, pp. 384-404.

DiPalma, J. R. and Ritchie, D. M., 1977, Vitamin toxicity, Ann. Rev. Pharmacol. Toxicol., 17:133-148.

Dubnick, B., Leeson, G. A., and Scott, C. C., 1960, Effect of forms of vitamin $B_6$ on acute toxicity of hydrazine, Toxicol. Appl. Pharmacol., 2:403-409.

Fell, H. B., 1970, The direct action of vitamin A on skeletal tissue in vitro, in: "The Fat Soluble Vitamins," H. F. DeLuca and J. W. Suttie, eds., Univ. of Wisconsin Press, Madison, pp. 187-202.

Frieden, E. and Lipner, H., 1971, "Biochemical Endocrinology of the Vertebrates," Prentice-Hall Inc., Englewood Cliffs.

Frimpter, G. W., Andelman, R. J., and George, W. F., 1969, Vitamin $B_6$ dependency syndrome, Am J. Clin Nutr., 22:794-805.

Fujiwara, M., 1954, Influence of thiamine on riboflavin metabolism, Vitamin, 7:206-208.

Geelen, J. A. G., 1973, Skullbase malformations in rat fetuses with hypervitaminosis A-induced exencephaly, Teratology, 7: 49-56.

Goldner, M. G., 1971, Large dose of vitamin C harmful to diabetics?, Diabetes Outlook, 6:4.

Gordonoff, T., 1960, Can water-soluble vitamins be overdosed? Research on vitamin C, Schweiz Med Wochenschr., 90:726-729.

Gould, J., 1954, The use of vitamins in psychiatric practice, Proc.
    R. Soc. Med., 47:215-220.
Gurian, H. and Aldersberg, D., 1959, The effect of large doses of
    nicotinic acid on evaluating lipids and carbohydrate tolerance,
    Am. J. Med. Sci., 237:12-22.
Harrison, H. E., 1978, Effect of nutrient toxicities in animals and
    man: vitamin D, in: "Effect of Nutrient Excesses and Toxicities
    in Animals and Man," M. Rechcigl, Jr., ed., CRC Press, Inc.,
    Boca Raton, pp. 87-90.
Hathcock, J. N., 1976, Nutrition: toxicology and pharmacology,
    Nutr. Rev., 34:65-70.
Haussler, M. R., 1978, Vitamin D: metabolism, drug interactions,
    and therapeutic application in humans, in: "Nutrition and Drug
    Interrelations," J. N. Hathcock and J. Coon, eds., Academic
    Press, Inc., New York, pp. 717-750.
Hayes, K. C. and Hegsted, D. M., 1973, Toxicity of the vitamins,
    in: "Toxicants Occurring Naturally in Foods," National Academy
    of Science, Washington, pp. 235-253.
Herbert, V. and Jacob, E., 1974, Destruction of vitamin $B_{12}$ by
    ascorbic acid, J. Am. Med. Assoc., 230:241-242.
Herbert, V., 1977, Megavitamin therapy, Contemporary Nutr., 2:#10,
    General Mills, Minneapolis.
Herbert, V., 1980, "Nutrition Cultism," George F. Stickley Co.,
    Philadelphia, pp. 141-150.
Hines, J. D., 1975, Ascorbic acid and vitamin $B_{12}$ deficiency, J.
    Am. Med. Assoc., 234:24.
Hodges, R. E., 1980, Food fads and "megavitamins", in: "Nutrition
    in Medical Practice," R. E. Hodges, ed., W. B. Saunders Co.,
    Philadelphia, pp. 288-327.
Hommes, O. R., Obbens, E. A. M. T., and Wijffels, C. C. B., 1973,
    Epileptogonic activity of sodium folate and the blood-brain
    barrier in the rat, J. Neurol. Sci., 19:63-71.
Hook, E. B., Healy, K. M., Niles, A. M. and Skalko, R. C., 1974,
    Vitamin E: teratogen or anti-teratogen?, Lancet, 1:809.
Horwitt, M. K., 1980, Therapeutic uses of vitamin E in medicine,
    Nutr. Rev., 38:105-127.
Hughes, M. R., Baylink, D. J., Jones, P. G., and Haussler, M. R.,
    1976, Radioligand receptor assay for 25-hydroxyvitamin $D_2/D_3$
    and 1,25dihydroxyvitamin $D_2/D_3$: application of hypervitamino-
    sis, D. J. Clin. Invest., 58:61-70.
Hunter, R., Barnes, J., Oakeley, H. F., and Matthews, D. M., 1970,
    Toxicity of folic acid given in pharmacological doses to
    healthy volunteers, Lancet, 1:61-63.
Inoue, K., Katsura, E., and Kariyone, S., 1956, Secondary ribo-
    flavin deficiency, Vitamin, 10:69.
Jaffe, R. M., Kasten, B., Young, D. S., and MacLowry, J. D., 1975,
    False negative stool occult blood tests caused by ingestion
    of ascorbic acid (vitamin C), Ann. Intern. Med., 83:824-826.
Jenkins, D. and Spector, R. G., 1973, The action of folate and
    phenytoin on the rat heart in vivo and in vitro, Biochem.

Pharmacol., 22:1813–1816.

Jenkins, M. Y., 1978, Effect of nutrient toxicities (excess) in animals and man: vitamin A, in: "Effect of Nutrient Excesses and Toxicities in Animals and Man," M. Rechcigl, Jr., ed., CRC Press, Inc., Boca Raton, pp. 73–85.

Jukes, T. H., 1979, Megavitamins and food fads, in: "Nutritional and Clinical Applications," R. E. Hodges, ed., Plenum Pub. Corp., New York, pp. 257–292.

Kallner, A., 1977, Serum bile acids in man during vitamin C supplementation and restriction, Acta. Med. Scand., 202:283–287.

Khera, K. S., 1975, Teratogenicity study in rats given high doses of pyridoxine (vitamin $B_6$) during organogenesis, Experientia, 31:469–470.

Klassen, C. D. and Doull, J., 1980, Evaluation of safety: toxicologic evaluation, in: "Casarett and Doull's Toxicology: The Basic Science of Poisons," J. Doull, C. D. Klausson, and M. O. Amdur, eds., Macmillan Inc., New York, pp. 11–17.

Klopp, C. T., Abels, J. C., and Rhods, C. P., 1943, The relationship between riboflavin intake and thiamin excretion in man, Am. J. Med. S., 205:852–857.

Knudson, A. G. and Rothman, P. E., 1953, Hypervitaminosis A: A review with a discussion of vitamin A, Am. J. Diseases Childdren, 85:316–334.

Kochhar, D. M., 1973, Limb development in mouse embryos. I. Analysis of teratogenic effects of retinoic acid, Teratology, 7:289–298.

Kohn, R. M. and Montes, M., 1969, Hepatic fibrosis following long acting nicotinic acid therapy: a case report, Am. J. Med. Sci., 250:94–99.

Korner, W. F. and Vollm, J., 1975, New aspects of the tolerance of retinol in humans, Int. J. Vitam. Nutr. Res., 45:363–372.

Lippi, U., Pulido, E., and Guidi, G., 1966, Lesioni lisosomiali da dosi elevate di acido ascorbico, Acta. Vitaminol, 5:177–180.

Lipton, M. M. and Steigman, A. J., 1963, Tuberculin type of hypersensitization to crystalline vitamin $B_{12}$ (cyanocobalamin) induced in guinea pig, J. Allergy, 34:362–367.

Loomis, T. A., 1978, "Essential of Toxicology," Lea and Febiger, Philadelphia.

Loomis, W. F., 1970, Rickets, Sci. Amer., 223:76–91.

Lucy, J. A. and Dingle, J. T., 1964, Fat-soluble vitamin and biological membranes, Nature (London), 204:156–160.

Machlin, L. J., 1980, "Vitamin E: A Comprehensive Treatise," Marcel Dekker, Inc., New York.

Mallick, N. and Deb, C., 1975, Effect of different doses of ascorbic acid on thyroid activity in rats at different levels of protein, Endocrinol., 65:333–339.

March, B. E., Wong, E., Seier, L., Sim, J., and Biely, J., 1973, Hypervitaminosis in the chick, J. Nutr., 103:371–377.

Marie, J. and See, G., 1954, Acute hypervitaminosis of the infant, Am. J. Diseases Children, 87:731-736.

Marin-Padilla, M., 1960, Mesodermal alterations induced by hypervitaminosis A, J. Embryol. Exp. Morphol., 15:261-269.

Marks, J., 1975, "A Guide to the Vitamins: Their Role in Health and Diseases," University Park Press, Baltimore.

Marxs, W., Marks, L., Meserve, E. R., Shimoda, F., and Deuel, H. J., 1947, Effects of the administration of a vitamin E concentrate and of cholesterol and bile salts on the aorta of the rat, Arch. Pathol., 47:440-445.

Mayfield, H. L. and Roehm, R. R., 1956, Influence of ascorbic acid and the source of B vitamins on the utilizations of carotene, J. Nutr., 58:203-217.

Mayson, J. S., Schumaker, O., and Kakamura, R. M., 1973, False-negative tests for urine glucose, Lancet, 1:780-781.

Melhorn, D. K. and Gross, S., 1969, Relationships between iron-dextran and vitamin E in iron-deficiency anemia in children, J. Lab. Clin. Med., 74:789-802.

Mengel, C. H. and Green, H. L., Jr., 1976, Ascorbic acid effects on erythrocytes, Ann. Intern. Med., 84:490-499.

Milne, D. D. and Omaye, S. T., 1981, Effect of vitamin C on copper and iron metabolism in the guinea pig. Int. J. Vit. Nutr. Res. 50: 301-308.

Milne, D. B., Omaye, S. T. and Amos, W. H., 1981, Effects of ascorbic acid on copper and cholesterol in cynomoglus monkey fed a diet marginal in copper. Am. J. Clin. Nutr., 34: 2389-2393.

Minkin, W., Cohen, H. J., and Frank, S. B., 1973, Contact dermatitis from deodorants, Arch. Dermatol., 107:774-775.

Mitchell, D. C., Vilter, R. W., and Vilter, C. F., 1949, Hypersensitivity to folic acid, Ann. Intern. Med., 31:1102-1105.

Molnar, G. D., Berge, K. G., Rosevear, J. W., McGuckin, W. F., and Achor, R. P., 1964, The effect of nicotinic acid in diabetes mellitus, Metabolism, 13:181-190.

Morgan, A. F., Kimmul, L., and Hawkins, N. C., 1937, A comparison of the hypervitaminosis induced by irradiated ergosterol and fish liver oil concentrates, J. Biol. Chem., 120:85-102.

Morrison, A. B. and Sarett, H. P., 1959, Effect of excess thiamin and pyridoxine on growth and reproduction in rats, J. Nutr., 69:111-116.

Morrison, A. B. and Sarett, H. P., 1959, Studies on B vitamin interrelationship in growing rats, J. Nutr., 68:473-484.

Mount, M. E., Feldman, B. F., and Buffington, T., 1982, Vitamin K and its therapeutic importance, J. Am. Vet. Med. Assoc., 180:1354-1356.

Muenter, M. D., Perry, H. O., and Ludwig, J., 1971, Chronic vitamin A intoxication in adults, Am. J. Med., 50:129-136.

Murata, A., and Kitagawa, K., 1973, Mechanism of inactivation of bacteriophage J1 by ascorbic acid, Agr. Biol. Chem., 37:1145-1151.

Murata, A., Oyadomari, R., Ohashi, T., and Kitagawa, K., 1975, Mechanism of inactivation of bacteriophage δ containing single-stranded DNA by ascorbic acid, J. Nutr. Sci. Vitaminol., 21: 261-269.

Nater, J. P. and Doeglas, H. M. G., 1970, Halibut liver poisoning in 11 fisherman, Acta. Derm., 50:109-113.

"National Nutritional Consortium," 1978.

Nayakama, Y., 1959, Studies on the relationship between thiamine propyl disulfide and riboflavin on riboflavin content in the urine of normal children and in urine as well as blood of healthy infants under artificial feeding, Vitamins, 16:294-300.

Nayakama, Y., 1959, Studies on the relationship between thiamine propyl disulphide and riboflavin. II. Effect of parenteral administration of TPD and esters of riboflavin content and its form in healthy infants under artificial feeding, Vitamins, 16:294-300.

Neuweiler, W., 1951, Hypervitaminosis and its relation to pregnancy, Int. Z. Vitamin Forsch., 22:392-396.

Newmark, H. L. and Scheiner, J., 1976, Destruction of vitamin B-12 by vitamin C, Am. J. Clin. Nutr., 30:299.

Nolen, G. A., 1969, Variations in teratogenic response in hyper-vitaminosis A in three strains of the albino rat, Food Cosmet. Toxicol., 7:209-214.

Obbens, E. A. M. T. and Hommes, O. R., 1973, The epileptogenic effects of folate derivatives in the rat, J. Neurol. Sci., 20:223-229.

Omaye, S. T., Reddy, K. A., and Cross, C. E., 1978, Enhanced lung toxicity of paraquat in selenium-deficient rats, Toxicol Appl. Pharmacol., 43:237-247.

Omaye, S. T. and Tappel, A. L., 1974, Effect of dietary selenium on glutathione peroxidase in the chick, J. Nutr., 104:747-753.

Omaye, S. T. and Turnbull, J. D. 1980, Effects of ascorbic acid on heme metabolism in hepatic microsomes. Life Science, 27: 441-449.

Omaye, S. T., Turnbull, J. D., and Sauberlich, H. E., 1979, Selected methods for the determination of ascorbic acid in animal cells, tissues, and fluids, in: "Methods in Enzymology," D. B. McCormick and L. D. Wright, eds., Academic Press, New York, Vol. 62, Part D, pp. 3-11.

Pauling, L., 1974, On the orthomolecular environment of the mind: orthomolecular theory, Am. J. Psychiatry, 131:1251-1267.

Pease, C. N., 1962, Focal retardation and arrestment of growth of bones due to vitamin A intoxication, J. Am. Med. Assoc., 182:980-985.

Peter, L., Bollag, W., and Mayer, H., 1980, Arotinoids, a new class of highly active retinoids, Eur. J. Med. Chem-Chim Ther., 15: 9-15.

Pochi, P. E., 1982, Oral retinoids in dermatology, Arch. Dermatol., 118:57-61.

Reddy, K. A., Omaye, S. T., Hasegawa, G. K., and Cross, C. E., 1978, Enhanced lung toxicity of intratracheally instilled cadmium chloride in selenium-deficient rats, Toxicol. Appl. Pharmcol., 43:249-257.

"Recommended Dietary Allowance (1980)," National Academy of Sciences, Washington, DC.

Rhead, W. J. and Schrauzer, G. N., 1971, Risks of long term ascorbic acid overdosage, Nutr. Rev., 11:262-263.

Richards, R. K. and Shapiro, S., 1945, Experimental and clinical studies on the action of high doses of hykinone and other menadione derivatives, J. Pharmacol. Exp. Ther., 84:93-104.

Richter, H. E. and Loewen, P. C., 1982, Rapid inactivation of bacteriophage T7 by ascorbic acid is repairable, Biochem. Biophys. Acta., 697:25-30.

Rivlin, R. S., 1978, Effect of nutrient toxicities (excess) in animals and man: riboflavin, in: "Handbook in Nutrition and Food," M. Rechcigl, Jr., ed., CRC Press, Inc., Boca Raton, pp. 25-27.

Rivlin, R. S., 1982, Nutrition and cancer: state of the art relationship of several nutrients to the development of cancer, J. Am. College Nutr., 1:75-88.

Roben, J. F., 1970, Teratogenic effects of hypervitaminosis A in the hamster and the guinea pig, Toxicol. Appl. Pharmacol., 16:88-99.

Robinson A. B., Catchpool, J. F., and Pauling, L., 1975, Decreased white blood cell count in people who supplement their diet with L-ascorbic acid, (abstract), ICRS Med. Sci. Clin. Pharma. Therapeu. Hematol. Immunol. Allgy. Metab. Nutr., 3:259.

Roels, O. A. and Lui, N. S. T., 1973, The vitamins: vitamin A and carotene, in: "Modern Nutrition in Health and Disease," R. S. Goodhart and M. E. Shils, eds., Lea and Febiger, Philadelphia, pp. 142-157.

Rosenberg, L. E., 1970, Vitamin-dependent genetic disease, Hospital Practice, 5:59-66.

Rubin, E., Florman, A. L., Degnan, T., and Diaz, J., 1970, Hepatic injury in chronic hypervitaminosis A, Am. J. Dis. Child., 119:132-138.

Samborskaya, E. P., 1964, The effect of high of ascorbic acid doses on the course of pregnancy in the guinea pig and on the progency, Bul. Exp. Biol. Med. (Moskva), 57:483-489.

Scala, J., 1980, Are food supplements necessary?, Nutr. Today, 15:25-26.

Schimke, R. T. and Doyle, D., 1970, Control of enzyme levels in animal tissues, Ann. Rev. Biochem., 39:929-969.

Schrauzer, G. N. and Rhead, W. J., 1973, Ascorbic acid abuse: effects of long-term ingestion of excessive amounts on blood levels and urinary excretion, Int. J. Vit. Nutr. Res., 43:201-211.

Schumacher, M. R., Williams, M., and Lyman, R. L., 1965, Effect
    of high intakes of thiamine, riboflavin, and pyridoxine on
    reproduction in rats and vitamin requirements of offspring,
    J. Nutr., 86:343-349.
Seelig, M. S., 1969, Vitamin D and cardiovascular, renal, and
    brain damage in infancy and childhood. Ann. N. Y. Acad. Sci.,
    147:539-582.
Selivanova, V. M., 1960, Excretion of vitamin B-6 in the urine
    of a healthy individual, Bul. Exp. Biol. Med., 50:37-39.
Shultz, T. D. and Leklem, J. E., 1982, Effect of high dose ascorbic
    acid on vitamin B6 metabolism, Am. J. Clin. Nutr., 35:1400-
    1407.
Singh, H. P., Herbert, M. A., and Gault, M. H., 1972, Effect of
    some drugs on clinical laboratory values as determined by the
    Technicon SMA-12/60, Clin. Chem., 18:137.
Soliman, M. K., 1972, Vitamin-A-Ubedosierung. II. Zytologische
    and biochemische Veranderungen im Blut von mit hohen dosen
    vitamin abzw. αtocopherol behardelten ratten, Int. J. Vitam.
    Nutr. Res., 42:576-582.
Spector, R., 1979, Niacin and niacinamide transport in the central
    nervous system, J. Neurochem., 33:1285.
Spector, R. and Kelley, P., 1979, Niacin and niacinamide accumu-
    lation by brain slicies and choroid plexus in vitro, J.
    Neurochem., 33:291.
Spector, R., 1980, Megavitamin therapy and the central nervous
    system, in: "Vitamins in Human Biology and Medicine," M. H.
    Briggs, ed., CRC Press, Inc., Boca Raton, pp. 137-156.
Sporn, M. B., Dunlop, N. M., Newton, D. L., and Smith, J. M., 1976,
    Prevention of chemical carcinogenesis by vitamin A and its
    synthetic analogs (retinoids), Fed. Proc., 35:1332-1338.
Sporn, M. B. and Newton, D. L., 1979, Chemoprevention of cancer
    with retinoids, Fed. Proc., 38:2528-2534.
Stich, H. F., Karim, J., Koropatnik, J., and Lo, L., 1976, Muta-
    genic action of ascorbic acid, Nature, 260:722.
Stults, V. J., 1981, Nutritional hazards, in: "Food Safety," H. R.
    Roberts, ed., Wiley-Interscience, New York, pp. 67-139.
Suttie, J. W., 1980, The metabolic role of vitamin K, Fed. Proc.,
    39:2730-2735.
Tappel, A. L., 1972, Vitamin E and free radical peroxidation of
    lipids, Ann. N. Y. Acad. Sci., 203:12-28.
Tappel, A. L., 1973, Vitamin E, Nutr. Today, 8:4-12.
Tappel, A. L., 1974, Selenium-glutathione peroxidase and vitamin
    E, Am. J. Nutr., 27:960-965.
Taylor, T. G., Morris, K. M. L., and Kirley, J., 1968, Effect of
    dietary excesses of vitamin A and D on some constituents of
    the blood of chicks, Br. J. Nutr., 22:713-721.
Telford, I. R., 1949, The effects of hypo and hyper-vitaminosis
    E on lung tumor growth in mice, Ann. N.Y. Acad. Sci., 52:
    132-134.

Threlfall, G., Taylor, D. M., and Buck, A. T., 1967, Studies of
    the changes in growth and DNA synthesis in the rat kidney
    during experimentally induced renal hypertrophy, Am. J.
    Pathol., 50:1-14.
Threlfall, G., 1968, Cell proliferation in the rat kidney induced
    by folic acid, Cell Tissue Kinet, 1:383-392.
Tilson, M. O., 1970, A dissimilar effect of folic acid upon growth
    of the rat kidney and small bowel, Proc. Soc. Exp. Biol. Med.,
    134:95-97.
Unna, K. and Greslin, J. G., 1942, Studies on toxicity and pharma-
    cology of riboflavin, J. Pharmacol. Exp. Ther., 76:75-80.
Vorhees, C. V., 1974, Some behavioral effects of maternal hyper-
    vitaminosis A in rats, Teratology, 10:269-274.
Weiss, W. P., 1970, Fat-soluble vitamins II vitamin D, in: "The
    Pharmacological Basis of Therapeutics," L. S. Goodman and
    A. Gilman, eds., The Macmillan Co., London, pp. 1680-1689.
Wolf, G., 1980, Vitamin A, in: "Nutrition and the Adult: Micro-
    nutrients," R. B. Alfin-Slater and D. Kritchevsky, eds., Vol.
    3B, Plenum Press, New York, pp. 97-203.
Woodard, W. K., Miller, L. J., and Legart, O., 1961, Acute and
    chronic hypervitaminosis A in a 4-month old infant, J.
    Pediatr., 9:260-264.
Youmans, J. B., 1960, Vitamin A nutrition and the skin, Am. J.
    Clin. Nutr., 8:789-792.

## 9

# PRENATAL AND DEVELOPMENTAL TOXICOLOGY OF ARSENICALS[1]

Calvin C. Willhite and Vergil H. Ferm

Toxicology Research Unit, Western Regional Research Center, United States Department of Agriculture, Berkeley, CA 94710 and Department of Anatomy and Cytology, Darmouth Medical School, Hanover, NH 03756

ABSTRACT

A variety of species, including the human, have been shown to be susceptible to the embryotoxic effects of inorganic arsenic. Malformations of the axial skeleton, neurocranium, viscerocranium, eyes, and genitourinary systems as well as prenatal death followed a bolus dose of trivalent or pentavalent inorganic arsenic. Trivalent arsenic was more teratogenic than pentavalent arsenic; in contrast, the methylated metabolites of arsenic possessed only limited teratogenic activity. Administration of inorganic arsenic to mammals results in concentration of arsenic within the placenta and small amounts are deposited within the embryo. Studies concerning the pathogenesis of arsenic-induced axial skeletal lesions revealed early failure of neural fold elevation and a subsequent, persistent failure of closure of the neural tube. Physical factors, drugs and heavy metals may modify the response to a teratogenic dose of inorganic arsenic. Medical problems associated with industrial or agricultural arsenicalism are most often typified by chronic exposure; future studies should emphasize those routes of administration and types of exposure that are characteristic of arsenic intoxication.

---

[1]Mention of trade name, proprietary product or specific equipment does not constitute a guarantee or warranty by the United States Department of Agriculture nor does it imply its approval to the exclusion of other products that may be suitable.

Arsenic has been known from antiquity as both a therapeutic agent and infamous poison. Arsenikon or arrhenikon, from which the etymology derives, refer to the Greek denoting orpiment yellow, and the historical linguistic syntheses of the Syrian zarnig and Sanskrit hiranga (gold), hari (yellow), underlie the word (Webster, 1941). The chemical and physical properties of arsenic-containing compounds have been summarized (Doak et al., 1978) and reviews of the medical use, toxicology, and environmental import of arsenic compounds have been compiled (Buchanan, 1962; Frost, 1967; Lisella et al., 1972; Schroeder, 1966; Vallee et al., 1960; NIOSH, 1975; Harvey, 1975).

The inorganic arsenites have agricultural significance as insecticides, herbicides, rodenticides, and fungicides. The lead, calcium, and sodium arsenates have been employed primarily as insecticides whereas arsenic pentoxide has found use as a defoliant and herbicide. Organic arsenicals, including the derivatives of benzene arsonic acid, have served as hog and chicken feed additives. Domestic consumption of arsenic acid in 1971 utilized as an agricultural defoliant and dessicant for cotton, citrus, and field crops was 6,073,000 lbs. and the annual consumption of all organic arsenicals used by farmers was 7,981,000 lbs. The primary agricultural use of the organic arsenicals was in pest control for cotton, sorghum, soybeans, rice, fruits, and nuts; 4,269,000 acres were treated (Andrilenas, 1971). While arsenicals are used in a variety of manufacturing processes in the pigment, glass, petrochemical, and paint industries, occupational arsenicalism has been particularly pronounced in those copper, lead and zinc smelters which process arsenic-rich ores and during the production or agricultural application of arsenic-based sheep dips and other economic poisons (Bljer, 1975; Hine et al., 1975). Arsenic is not only an industrially and agriculturally-produced contaminant, it occurs naturally as organic complexes in rather high concentrations in crustaceans. The inorganic forms of arsenic predominate in drinking water and wine (Crecelius, 1977).

Inorganic forms arsenic are more acutely toxic than the organic arsenicals and the toxicity depends upon the oxidation state of arsenic. Conversion between the oxidation states of inorganic arsenic have been shown to occur both in the environment and in the animal body. Recent studies (Vahter and Norin, 1980; Vahter, 1981) have suggested that the differences between the toxicities of trivalent and pentavalent arsenic depend upon the extent of in vivo biotransformation to methylated metabolites. Environmental oxidation favors the conversion of trivalent arsenic to pentavalent arsenic. Pentavalent arsenic may be partially reduced in vivo to trivalent arsenoxide (R·As:0), but it is the in vivo reduction of arsenate to arsenite which accounts for both toxicity and antimicrobial actions (Harvey, 1975). Rumen microflora obtained from Holstein cows reduced arsenate to arsenite (Forsberg, 1978). Ginsberg (1965) demonstrated that arsenate may be resorbed by the proximal tubule of the dog kidney and excreted as arsenite. Mealey et al. (1959) observed the urinary excretion of arsenate by humans following intravenous injection of arsenite.

A variety of microorganisms have been observed to convert inorganic arsenic to methylated derivatives. Dimethyl arsine, dimethylarsinic acid $(CH_3)_2AsOH$, and methylarsonic acid $CH_3AsO(OH)$ have been employed as agricultural herbicides and occur as microbially-formed contaminants. Cox and Alexander (1973) studied the biotransformation reactions as carried out by fungi isolated from arsenic-containing sludge. Anaerobic bacteria may also reduce arsenate As-(+5) to arsenite As(+3) and methylate the latter to form methylarsonic acid As(+3). Methylarsonic acid was subsequently reduced and methylated to yield dimethylarsinic acid As(+1) and it was ultimately reduced to dimethylarsine As(-3) (McBride and Wolfe, 1971; Braman and Foreback, 1973). Others (Braman and Foreback, 1973; Charbonneau et al., 1978; 1979; 1980; Crecelius, 1977b; Lasko and Peoples, 1975; Odanaka et al., 1980; Penrose, 1975; Smith et al., 1977; Tam et al., 1978; 1979) have documented the in vivo methylation of inorganic arsenic in humans, cows, dogs, cats, rats, mice, hamsters, and trout after oral, intramuscular, or intravenous dosing or after inhalation of various forms of inorganic arsenic. The initial reduction of arsenate to arsenite was followed by the slower synthesis of methylarsonic acid in vivo (Charbonneau et al., 1979) or in vitro as catalyzed by rat caecal microflora (Rowland and Davies, 1981). The subsequent production of dimethylarsinic acid indicated that methylarsonic acid was an intermediate in an arsenic methylation sequence by which inorganic arsenic was ultimately converted into dimethylarsinic acid. The major form of arsenic in tissue of rats following oral administration of sodium arsenate (Rowland and Davies, 1981) or in urine of mice given a single oral dose of sodium arsenate or arsenite (Vahter, 1981) was dimethylarsinic acid. Whole body retention of arsenic in mice was two-three times greater after treatment with As(+3) as compared to treatment with As(+5) (Vahter and Norin, 1980). Arsenite was methylated to a greater extent in both rats and mice than was arsenate, but the methylation of arsenic in rats was less than that in mice. Vahter (1981) suggested that the differences in extent of biotransformation by rats and mice may account for the fact that whole body retention of arsenic was twenty times greater in rats as in mice. Others (Tam et al., 1978; Charbonneau et al., 1979) suggested that hepatic mechanisms play a role in the in situ biotransformation of inorganic arsenic. Cox and Alexander (1973) and Jernelov (1975) have proposed chemical mechanisms for the methylation reactions of inorganic arsenic to trimethylarsine via dimethylarsinic acid or arsenious acid.

TERATOGENESIS AND PRENATAL MORTALITY

The deleterious influence of arsenic upon embryonic development was first described by Ancel (1946-47) who observed anomalous tail buds in chicks exposed to the metalloid. The embryopathic activity of inorganic arsenic in mammals was first reported in hamsters where arsenic administered intravenously to the dam early in gestation as sodium arsenate produced a dose-dependent syndrome of malformations including those of the cranium, genitourinary system, and ribs (Ferm and Carpenter, 1968; Ferm et al., 1971). A single intravenous injection of 20 mg/kg dibasic sodium arsenate caused cranioschisis occulta with encephalocele or cranioschisis aperta with encephaloschisis (exencephaly) in the offspring and it commonly affected all of the survivors (Fig. 1). Malformations of the neurocranium and viscerocranium resulted (Fig. 2). Dissection of fetuses taken from dams treated with a teratogenic dose of sodium arsenate revealed a variety of genitourinary disorders (Figs. 3-5). Unilateral or bilateral renal agenesis was most common when the pregnant animals were treated on the evening of the 8th day of gestation (Ferm et al., 1971).

Analysis of sodium arsenate-induced encephaloceles in fetal hamsters found a mid-line herniation of the mesencephalic roof through a defective cranium (Marin-Padilla, 1980). Within the central zone of the encephalocele, the neural tissues thinned to only a single layer of cells. Fenestrations between the neural cells were commonplace and masses of glial processes with small dendrites and axons entered the ventricular cavity through gaps between a layer of ependymal cells. Microscopic meningoceles permitted contact of neuronal and glial processes with collagen and other tissues derived from the mesoderm (Marin-Padilla, 1980). The base of the fetal hamster skull in those animals afflicted with arsenate-induced encephaloceles or exencephaly was 7-9 and 12-15% shorter than control, respectively. The reduction in longitudinal diameter of the skull base was a consequence of decreased growth of the basioccipital bone. The base of the skull was also abnormally lordotic to the vertebral axis and the lordosis was more pronounced in the case of exencephaly than in the encephalocele. The notochord of the affected offspring followed a normal intracranial course, but it was shortened and showed terminal bends and accordion-like foldings (Marin-Padilla, 1979).

The oxidation state of arsenic was important since a single intravenous injection of 20 mg/kg sodium arsenate on the morning of day 8 of gestation in the hamster killed 44% of all embryos, but an injection of 10 mg/kg sodium arsenite killed 90% (Willhite, 1981). Oral intubation of 25 mg/kg or intraperitoneal injection of 5 mg/kg sodium arsenite on days 9 or 10 of gestation in hamsters resulted in the complete resorption of all implantation sites. Injection of 5 mg/kg arsenite on days 11 or 12 resulted in a significant reduc-

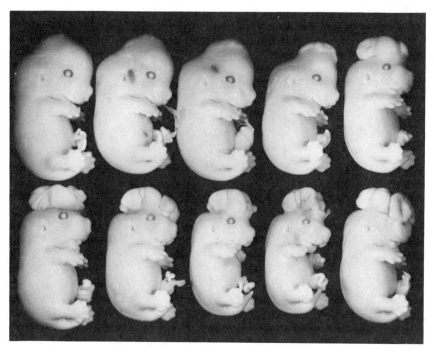

Figure 1.   Fifteen–day–old fetal littermates from a Golden hamster injected intravenously with 20 mg/kg sodium arsenate on the 8th day of pregnancy.   The animals at the upper left and center exhibit encephaloceles.   The remainder show progressive degrees of exencephaly.   (About x 1.5).

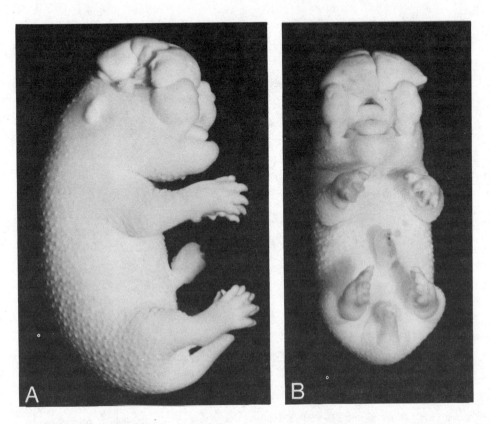

Figure 2.   A. Lateral view of a 15-day-old hamster fetus from a dam given an intravenous injection of 20 mg/kg sodium arsenate on the 8th day of gestation.   Note the exencephalia as characterized by everted, degenerative neural tissues with hemorrhagic necrosis.   B. Ventral view of the fetus shown in A.   Note the severe malformation of the craniofacial structures with bilateral anophthalmia.   (About x 3.5).

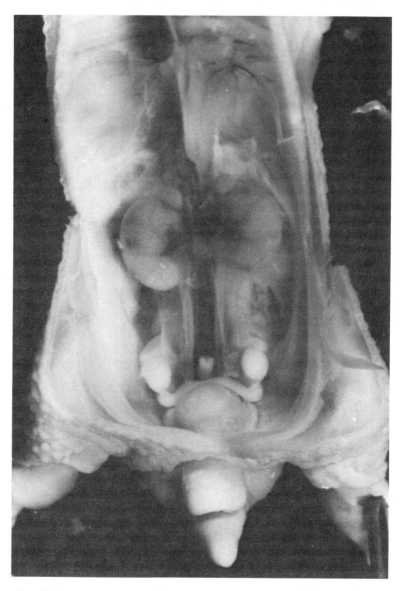

Figure 3. Urinary and reproductive tract of a male fetal hamster taken from a dam given an intravenous dose of distilled water on the 8th day of pregnancy and recovered on day 15 of pregnancy. (About x 6).

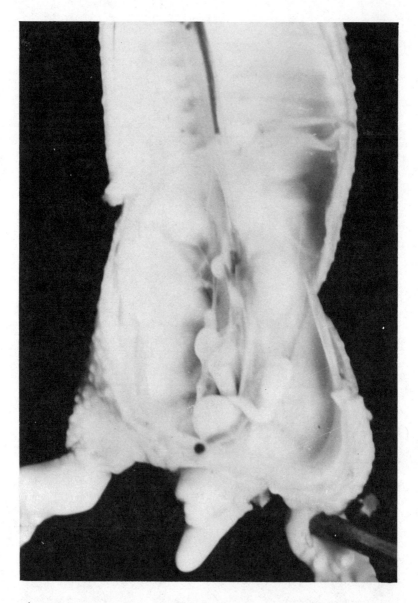

Figure 4. Urinary and reproductive tract of a male fetal hamster taken from a dam given an intravenous dose of 20 mg/kg sodium arsenate on the 8th day of pregnancy and recovered on day 15. Note the undescended testis and the uilateral renal agenesis. The adrenals are present. (About x 6).

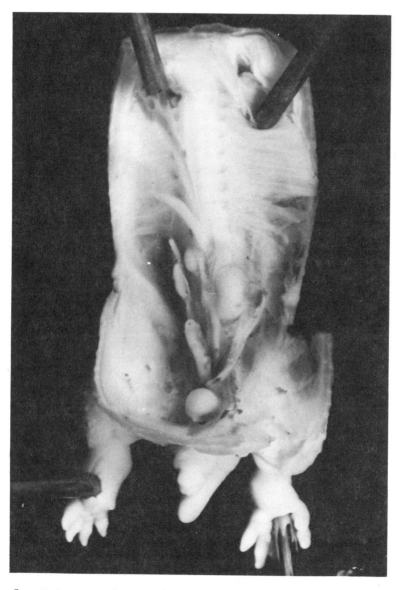

Figure 5. Urinary and reproductive tract of a female fetal hamster taken from a dam given an intravenous dose of 20 mg/kg sodium arsenate on the 8th day of pregnancy and recovered on day 15. Note the unilateral renal agenesis. The ovaries and adrenals are present. (About x 6).

tion in the average fetal body weight, but no gross or visceral abnormalities were noted (Harrison and Hood, 1981).  It is noteworthy
that arsenate injection produced not only renal and axial skeletal
malformations in hamsters, but produced a significant, concomitant
increase in the volume of the amniotic fluid surrounding those
fetuses afflicted with exencephalia (Ferm and Saxon, 1971).

The embryolethal and teratogenic activity of inorganic arsenic
in mouse and rat have been assessed.  An intraperitoneal injection
of 25 mg sodium arsenate/kg failed to induce teratogenic effects in
mice, but a similar injection of 10 mg sodium arsenite/kg showed
both teratogenic and embryolethal activity (Hood and Bishop, 1972).
Injection of 10-12 mg sodium arsenite/kg on one of days 7-12 of
gestation in mice increased the number of embryonic deaths and decreased the average fetal body weight.  Injection on days 7-10 of
pregnancy also produced exencephaly, micrognathia, abnormal tails,
open eye, rib fusions and fused or missing vertebral ossification
centers (Hood, 1972).  Prenatal mortality was increased after an
oral dose of 40 mg/kg sodium arsenite in pregnant CD-1 mice on one
of days 10-15 of gestation and malformed fetuses were recovered
following treatment on days 9 or 10 (Baxley et al., 1981).  Intraperitoneal injection of 30 or 40 mg/kg sodium arsenate on one of days
8, 9, or 10 of gestation in the rat caused congenital anophthalmia,
microphthalmia, exencephaly, renal dysgenesis, gonadal agenesis,
rib fusions, vertebral defects, subcutaneous hemorrhage and absence
of the atlas.  Injection of 20 mg/kg sodium arsenate in pregnant
rats on days 9 to 10 resulted in similar malformations, but a dose
of 50 mg/kg was lethal to all embryos regardless of the day of
treatment (Beaudoin, 1974).  Umpierre (1981) found that an intraperitoneal injection of 11 mg/kg sodium arsenite on day 10 of gestation in the rat induced a variety of skeletal, brain and eye
abnormalities.

The ability of inorganic arsenic to cross the placenta has
been described by a number of authors.  Morris and associates
(1938) observed that arsenic administered as dietary arsenic trioxide may cross the placenta of the rat.  Lugo et al. (1969) described a case of acute human inorganic arsenic intoxication which
occurred during the third trimester of pregnancy.  Placental transfer of the metalloid was followed by the death of the fetus.  Hanlon
and Ferm (1977) measured the arsenic content in hamster embryos 24
hr after an intravenous injection on the 8th day of gestation of
8.37 mg $AsO_4$/kg containing a trace of $Na_2H$ $^{74}AsO_4$.  Up to 100 ng
$AsO_4$/g wet weight appeared in the embryonic tissues.  The chorio-
allantoic placenta contained upwards of 250 ng $AsO_4$/g whereas maternal blood contained approximately 110 ng $AsO_4$/g.  Arsenic concentrations decreased by 84% over the following 72 hr in the maternal
blood, uterus and placenta, but the concentrations of arsenic decreased by 77% and 91% in maternal liver and embryonic tissues,
respectively.  Hood and colleagues (1981) studied the arsenic con-

centrations in mouse maternal, near-term fetal and placental tissues after an intraperitoneal injection or oral dose of 20 or 40 mg/kg sodium arsenate, respectively. Although the intraperitoneal dose was only one-half that given orally, the maximum fetal arsenic concentrations were 8-fold greater following parenteral administration. Intraperitoneal injection of 8 mg/kg or oral intubation of 25 mg/kg sodium arsenite in mice on day 18 resulted in peak total fetal arsenic concentrations at 4 and 6 hr post-treatment, respectively, and ranged as high as 1.27 µg/g. Placental concentrations likewise peaked at 1 and 4 hr (Hood et al., 1982).

Investigations concerning the pathogenesis of inorganic arsenic-induced terata have been carried out in hamsters, rats and mice. Carpenter and Ferm (1977) provided a preliminary account of the development of arsenic-induced exencephaly in the hamster. Histopathologic analyses of hamster embryos recovered 2 or 6 hr after an intravenous injection of a teratogenic dose of sodium arsenate failed to reveal marked differences when compared to control embryos of identical gestational age, but some delay in overall development was apparent as early as 6 hr after injection (Willhite, 1981). At 10 hr after injection of distilled water, the anterior and posterior neuropores remained open and the neural folds were elevated in preparation for fusion (Fig. 6 A), but closure of the neural tube folds had started in the mid-region where the somites were visible (Fig. 7 A). Within 10 hr after treatment with a teratogenic dose of sodium arsenate, the neural folds of treated embryos were more widely separated at both the cephalic (Fig. 6 B) and thoracic (Fig. 7 B) areas. Evidence of retardation in neural tube closure, exposure and flattening of the neuroectoderm, lack of somitic mesodermal considation, and apparent overgrowth of the neuroepithelium with collapse of the neural folds were conspicuous (Fig. 8). The appearance of arsenate-treated embryos suggested that alterations in the paraxial mesoderm and/or axial chordomesodermal system contributed to a paucity of cephalic mesoderm, a delay in approximation of the neural folds and a primary failure of closure of the neural tube (Willhite, 1981). Hamster embryos recovered 24 hr after an intravenous injection of distilled water in the dam at day 9 of gestation were approximately 4 mm long, possessed a functioning heart and had four small limb buds. Scanning electron microscopic observations of the control embryos (Fig. 9) revealed the completion of closure of the cephalic neural folds. Arsenic-induced exencephaly is characterized by non-closure of the cephalic part of the neural tube (Fig. 10). The malformation is a result of failure of the neuroepithelium to meet and fuse during the process of neurulation at a restricted area of the anterior neuropore accompanied by collapse of the neural folds (Fig. 11, 12). It is of interest of point out that the arsenic-induced changes in the cephalic region were not evident in caudal regions; in addition, we have never observed a case of arsenic-induced rachischisis in the fetal hamster which suggests a localized and specific action of arsenic in organogenesis.

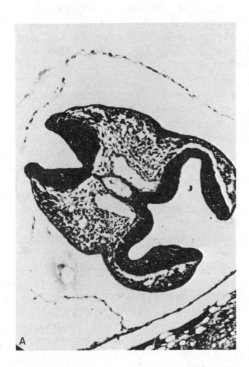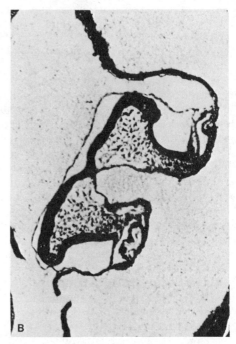

Figure 6. <u>A</u>. Cross section of the cephalic region of a hamster embryo recovered at 6:00 p.m. on day 8 of gestation at 10 hr after an intravenous dose of distilled water in the mother. The section shows the anterior (lower right) and posterior (upper left) neuro-ectoderm and approximation of the neural folds. (x 350). <u>B</u>. Trans-verse section of the cephalic region of a hamster embryo recovered 10 hr after an intravenous teratogenic dose of sodium arsenate in the dam. The embryo shows the unclosed neuroectoderm at the supe-rior edge (upper left) and the anterior level of the heart (lower right). There is a greater distance between the tips of the neural groove in treated embryos as compared to control embryos of iden-tical gestational age. The mesoderm shows enlarged intercellular spaces and is more loosely woven than that in control embryos (x 350).

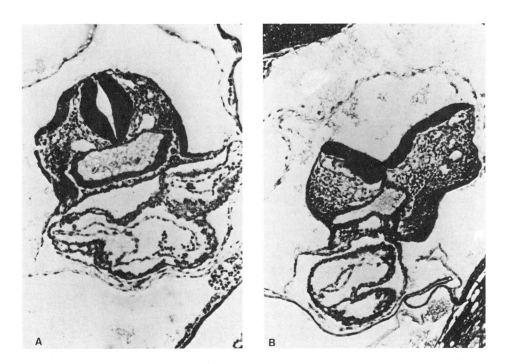

Fig. 7. A. Cross section of the middle thoracic region of a control embryo showing the closed neural tube (top) with the paired dorsal aortae. The foregut overlies the pericardial coelom and heart. B. Cross section of the middle thoracic region of a hamster embryo recovered 10 hr after a teratogenic dose of sodium arsenate. The region shows retardation in approximation and fusion of the neural folds and shows flattening and exposure of the neuroectoderm. (x 350).

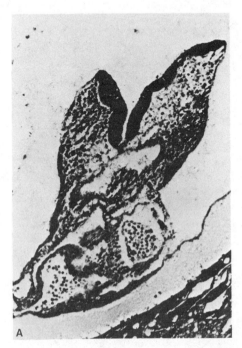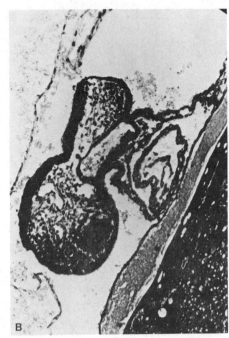

Figure 8. <u>A</u>. Transverse section of the upper thoracic region of a sodium arsenate-treated hamster embryo. The embryo shows an obvious schisis (dysraphic) defect of the neural walls with apparent overgrowth of the neuroepithelium. (x 350). <u>B</u>. Transverse section of the upper thoracic region of a sodium arsenate-treated hamster embryo. The section shows a localized schisis of the anterior region with collapse of the neural folds. Mesodermal consolidation has not occurred as evidenced by poorly formed somites. (x 350).

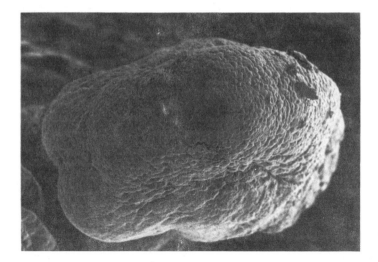

Figure 9.  Scanning electron micrograph of the anterior cephalic region of a control hamster embryo on the morning of day 9 of pregnancy at 24 hr after an intravenous dose of distilled water in the dam.  Note the closed neural tube at the site of the future brain and neurocranium.  (About x 300).

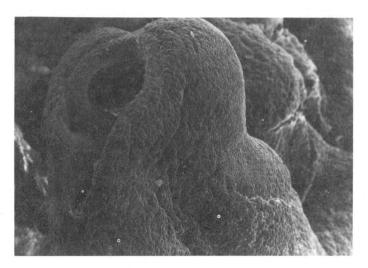

Figure 10.  Scanning electron micrograph of the anterior cephalic region of a hamster embryo on the morning of day 9 of gestation at 24 hr after an intravenous dose of 20 mg/kg sodium arsenate in the dam.  The anterior neuropore is shown at the upper portion of the specimen.  The embryo shows a persistent and obvious schisis defect. (About x 300).

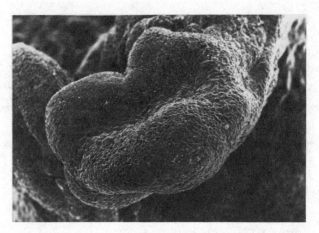

Figure 11.  Scanning electron micrograph of the anterior cephalic region of a hamster embryo on the morning of day 9 of gestation at 24 hr after an intravenous dose of 20 mg/kg sodium arsenate in the dam.  The anterior neuropore is shown at the upper portion of the specimen.  Although the neural folds have met and fused in the midline, collapse of the neural tube is evident.  (About x 300).

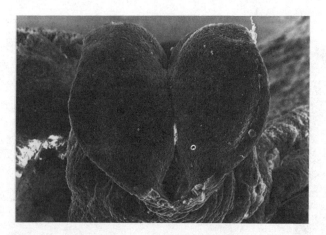

Figure 12.  Scanning electron micrograph of the anterior cephalic region of a hamster embryo on the morning of day 9 of gestation at 24 hr after an intravenous dose of 20 mg/kg sodium arsenate in the dam.  The anterior neuropore is at the center of the figure.  The embryo shows an obvious schisis defect.  The neural folds are bent laterally and show a greater distance between the edges of the neural groove.  Flattening and exposure of the neuroepithelium accompany the retardation in closure of the anterior neuropore.  These changes are typical of those embryos which will become exencephalic fetuses.  (About x 300).

Burk and Beaudoin (1977) traced the embryonic origin of sodium arsenate-induced renal dysgenesis in the rat. In all fetuses missing a kidney, the corresponding ureter was also absent. The adrenal glands, gonads and bladder were present. A substantial proportion of the male rat fetuses exhibited nondescent of the testes following a maternal dose of 50 mg sodium arsenate/kg. Males lacked the ductus deferens and seminal vesicle as well as portions of the epididymis whereas females showed only a short, blind-ending oviduct and complete absence of the uterus. Hydroureter and hypoplastic kidney were also observed. The kidney malformations were a result of failure of the ureteric bud to arise from the mesonephric duct in concert with the lack of induction of the metanephric blastema. Retarded growth of the mesonephric duct led to anomalies of the ductus deferens, seminal vesicles and epididymis. The short mesonephric duct and shortened paramesonephric duct resulted in the lack of uterine horns.

Ultrastructural studies of sections of embryonic Balb/c mice at 7 hr after an intraperitoneal injection of 45 or 60 mg/kg sodium arsenite in the dam revealed dense-staining inclusions within the neuroepithelium. The inclusions had disappeared at 21 hr, but extracellular dilatations and intracellular vesicles – said to be portions of the endoplasmic reticulum – were present and remained at up to 44 hr after arsenate treatment (Morrissey and Mottet, 1981).

Amelioration and enhancement of inorganic arsenic-induced teratogenesis by drugs and environmental agents have been reported. The simultaneous administration of selenium as sodium selenite along with a teratogenic dose of sodium arsenate protected the hamster embryo against the embryopathic activity of arsenic (Holmberg and Ferm, 1969). Thus, the absolute concentration of arsenic within the embryonic tissues may be less important than its concentration relative to the antagonist metal. Vedel et al. (1980) demonstrated that a subcutaneous injection of dimercaprol (2,3-dimercaptopropanol, BAL) at 4-8 hr prior to a teratogenic dose of sodium arsenite in CD-1 mice resulted in a reduction in both embryonic and fetal mortality, but an identical injection of dimercaprol 4-8 hr after arsenite administration enhanced the teratism. A combination of maternal hyperpyrexia and a teratogenic dose of sodium arsenate resulted in a marked increase in the embryolethal effects of arsenic and increased the frequency and severity of arsenic-induced malformations (Ferm and Kilham, 1977).

The organic forms of arsenic constitute the largest class of compounds and many have been synthesized in an effort to discover agents with therapeutic properties. Arsenicals have been superseded in the treatment of syphilis by penicillins; however, they still have value in the therapy of African trypanosomiasis and amoebic dysentery (Harvey, 1975). A limited number of studies suggest that that the organic arsenicals either failed to reach the mammalian fetus or lacked intrinsic embryotoxic activity. Traces of arsenic were found in the tissues of fetal rabbits and cats after an intravenous injection of neoarsphenamine (sodium 3-amino-3'-sulfinomethyl-amino-4,4'-dihydroxyarsenobenzene, Neosalvarsan[R]), a trivalent arsenical (Underhill and Amatruda, 1923). The fetal arsenic concentrations were independent of the total dose of drug, but the maternal liver and placental arsenic concentrations were proportionate to the total dose of neoarsphenamine. Lebedewa (1934) measured the arsenic accumulation in the mouse maternal liver and placenta after an intravenous injection of arsphenamine (3,3'-diamino-4, 4'-dihydroxyarsenobenzene dihydrochloride, Salvarsan[R]). Negligible concentrations were present in the mouse fetus. In a study of gestational age in relation to the placental transmission of arsenic, Snyder and Speert (1938) found that arsenic was concentrated in the fetal portion of the rabbit placenta after an intravenous injection of neoarsphenamine during the final trimester of pregnancy. The total concentration of arsenic in the fetus increased as parturition approached. Eastman (1931) observed the accumulation of arsenic in the human placenta after intravenous administration of arsphenamine or neoarsphenamine and found that the fetal portion of the placenta contained the greatest amounts of arsenic. The concentrations of arsenic in fetal cord blood, liver and kidney were below the limit of detection.

A number of studies have considered the teratogenic potential of the methylated metabolites of inorganic arsenic in contrast to the parent forms. Intravenous injection of arsenite was more teratogenic for the hamster embryo than an injection of arsenate, but an equivalent injection of methylarsonic acid or dimethylarsinic acid failed to show the marked teratologic activity that was characteristic of inorganic arsenic (Willhite, 1981). Hood et al. (1982b) found that intraperitoneal injection of 600 mg/kg methylarsonic acid or 1,000 mg/kg dimethylarsinic acid on days 9 or 10 of gestation in the hamster resulted in an increase in embryonic mortality and an increase in the incidence of congenital malformations as compared to control animals. Harrison and associates (1980) observed that an intraperitoneal injection of 1,500 mg/kg dimethylarsinic acid into pregnant CD-1 mice on days 8, 9 or 10 of gestation induced a variety of anomalies in 1.5, 3.7 and 1.0% of all litters, respectively. Peak total arsenic concentrations in near-term CD-1 mouse fetuses occurred at 4 hr after an intraperitoneal dose of 8 mg/kg sodium arsenite (Hood et al., 1982). Placental arsenic concentrations peaked at 1 hr post-injection. Maternal blood contained up to 2.36 µg/ml total arsenic and the near-term fetuses contained up to 1.27 µg/g. Methylated forms of arsenic accounted for 82% of the total arsenic at 12 hr after injection. The in vivo conversion of inorganic arsenic to methylated forms represents a detoxification pathway in the adult animal since the organic forms are much less toxic than the inorganic forms. The limited teratogenic activity of the methylated arsenic derivatives suggests that the metabolites may have also represented a detoxification pathway for the pregnant adult and the developing embryo.

The biochemical mechanisms by which inorganic arsenic alter normal mammalian embryogenesis are unknown. The possibility that an extraembryonic site of action in the mother is responsible for the teratogenic effects can be entertained, but the fact that arsenic readily crosses the placenta of both rodents (Hanlon and Ferm, 1977; Hood et al., 1981; 1982; Morris et al., 1938) and humans (Lugo et al., 1969) suggests that the critical sites are within the embryonic tissues themselves or that arsenic elicits disturbances in placental function. Since the placenta and associated extra-embryonic membranes are active in maintaining homeostasis for the developing embryo, disruption of those membranes and the closely regulated microenvironment could prove teratogenic for the embryo. The effect of arsenic on the rapid cell multiplication that is characteristic of embryonic growth may be related to its effects on cell division. Arsenic disrupted cell division in mice by acting on the chromatin during stages of nuclear condensation (Piton, 1929) and induced fragmentation of pyknotic nuclei and chromatin (Limarzi, 1943); Petres and associates (1970) documented chromosomal abnormalities in cultured lymphocytes obtained from humans chronically exposed to arsenic. However, it is clear that not all of the rapidly dividing and differentiating cells of the embryo are affected to the same degree - the spectrum of malformations induced by inorganic arsenic in the hamster is reasonably specific in structure and location (Figs. 1-5).

Hanlon and Ferm (1974) suggested several possible biochemical modes by which heavy metal ions may elicit embryotoxicity. Rather than a generalized effect on cell division, heavy metals or metalloids may interfere with a sensitive biochemical event in particular tissues leading to abnormal morphogenesis: a delay in normal migratory patterns, interruption of reciprocal inductive functions, and loss of contact between cells. In the case of arsenite, loss of enzyme activity may result from a specific attack upon sulfhydryl groups (-SH) contained at the active site. The presence of free sulfhydryl groups is essential for the full activity of a wide variety of enzymes including glutamic-oxaloacetic acid transaminase, pyruvate oxidase, monoamine oxidase, choline oxidase, glucose oxidase, urease, oxidoreductases (dehydrogenases) and kinases. Arsenate, however, does not interact directly with sulfhydryl groups (Johnstone, 1963), but it may compete with inorganic phosphate to form unstable arsenate esters and uncouple oxidative phosphorylation (Doudoroff et al., 1947; Crane and Lipman, 1953; Chan et al., 1969). Arsenate inhibition of phosphate-requiring reactions showed competitive kinetics and was of a nonhydrolytic nature (Mitchell et al., 1971).

DIRECTIONS FOR FUTURE INVESTIGATIONS

Future studies on the prenatal and developmental toxicology of arsenicals should consider deleterious effects in non-rodent species. Elucidation of chronic, low-level arsenicalism in relation to prenatal mortality and teratogenicity is an important step towards a more precise assessment of the possible hazard to humans and food-producing animals. The compounds should be given to the test animals by the same routes by which the target species are exposed, primarily by inhalation and in the drinking water. Attention should be paid to not only the concentration of total arsenic in maternal and embryonic tissues associated with the induction of prenatal toxicity, but to the concentrations and toxicokinetics of the various chemical species.

ACKNOWLEDGMENTS

This work was supported in part by USPHS Grant ES-00697 (V.H.F.).

REFERENCES

Andrilenas, P. A., 1971, "Farmers Use of Pesticides in 1971 - Quantities," Economic Research Service, U.S.D.A., Agricultural Economic Report No. 252.

Angel, P., 1946-1947, Recherche experimentale sur le spina bifida. Arch. Anat. Microsc. Morphol. Exp., 36: No. 1, 45-68.

Baxley, M. N., Vedel, G. C., Harrison, W. P., and Hood, R. D., 1981, Effects of oral sodium arsenite on mouse development. Teratology, 21: 27A. (Abstr).

Beaudoin, A. R., 1974, Teratogenicity of sodium arsenate in rats. Teratology, 10: 153-158.

Blejer, H. P., 1975, Arsenic: Occupational sentinel of community disease, In. "Symposium Proceedings, Int. Conf. Heavy Metals in the Environment", Toronto, Canada, Oct. 27-31, Vol. III, p. 305.

Braman, R. S. and Foreback, C. C., 1973, Methylated forms of arsenic in the environment. Science, 182: 1247-1249.

Buchanan, W. D., 1962, "Toxicity of Arsenic Compounds". Elsvier, Amsterdam.

Burk, D. and Beaudoin, A. R., 1977, Arsenate-induced renal agenesis in rats. Teratology, 16: 247-260.

Carpenter, S. J. and Ferm, V. H., 1977, Early stages in the development of exencephaly in arsenic-treated hamster embryos. Teratology, 15: 23A-24A. (Abstr).

Chan, T-L., Thomas, B. R., and Wadkins, C. L., 1973, Formation and isolation of an arsenylated component of rat liver mitochondria. J. Biol. Chem., 244: 2883-2890.

Charbonneau, S. M., Tam, G. H. K., Bryce, F., and Collins, B., 1978, in: "Trace Substances in Environmental Health. XII. A Symposium." D. D. Hemphill, ed., p. 276. University of Missouri, Columbia.

Charbonneau, S. M., Tam, G. H. K., Bryce, F., Zawidzka, Z., and Sandi, E., 1979, Metabolism of orally administered inorganic arsenic in the dog. Toxicol. Lett., 3: 107–113.

Charbonneau, S. M., Hollins, J. G., Tam, G. H. K., Bryce, F., Ridgeway, J. M., and Willes, R. F., 1980, Whole body retention, excretion, and metabolism of [$^{74}$As] arsenic acid in the hamster. Toxicol. Lett., 5: 175–182.

Cox, D. P. and Alexander, M. T., 1973, Production of trimethylarsine gas from various arsenic compounds by three sewage fungi. Bull. Environ. Contam. Toxicol., 9: 84–88.

Crane, R. K. and Lipmann, F., 1953, The effect of arsenate on aerobic phosphorylation. J. Biol. Chem., 201: 235–243.

Crecelius, E. A., 1977, Arsenite and arsenate levels in wine. Bull. Environ. Contam. Toxicol., 18: 227–230.

Crecelius, E. A., 1977b, Changes in the chemical speciation of arsenic following ingestion by man. Env. Hlth. Pers., 19: 147–150.

Doak, G. O., Long, G. G., and Freedman, L. D., 1978, Arsenic compounds. in: "Kirk-Othmer Encyclopedia of Chemical Technology", M. Grayson and D. Eckroth, eds., Vol. 3, pp. 251–266. John Wiley & Sons, Inc., New York.

Doudoroff, M., Barker, H. A., and Hassid, W. Z., 1947, Studies with bacterial sucrose phosphorylase. III. Arsenylated decomposition of sucrose and glucose-1-phosphate. J. Biol. Chem., 170: 147–150.

Eastman, N. J., 1931, The arsenic content of the human placenta following arsphenamine therapy. Amer. J. Obstet. Gynecol., 21: 60–64.

Ferm, V. H. and Carpenter, S. J., 1968, Malformations induced by sodium arsenate. J. Reprod. Fert., 17: 199–201.

Ferm, V. H. and Kilham, L., 1977, Synergistic teratogenic effects of arsenic and hyperthermia in hamsters. Env. Res., 14: 483–486.

Ferm, V. H. and Saxon, A., 1971, Amniotic fluid volume in experimentally induced renal agenesis and anencephaly. Experientia, 27: 1066–1068.

Ferm, V. H. Saxon, A., and Smith, B. M., 1971, The teratogenic profile of sodium arsenate in the golden hamster. Arch. Environ. Hlth., 22: 557–560.

Forsberg, C. W., 1978, Some effects of arsenic on the rumen microflora: An in vitro study. Can. J. Microbiol., 24: 36–44.

Frost, D. V., 1967, Arsenicals in biology. Fedn. Proc. Fedn. Am. Soc. Exp. Biol., 26: 194–208.

Ginsberg, J. M., 1965, Renal mechansism for excretion and transformation of arsenic in the dog. Amer. J. Physiol., 208: 832–840.

Hanlon, D. P. and Ferm, V. H., 1974, Possible mechansisms in metal-induced teratogenesis. Teratology, 9: 18-19.

Hanlon, D. P. and Ferm, V. H., 1977, Placental permeability of arsenate ion during early embryogenesis in the hamster. Experientia, 33: 1221-1222.

Harrison, W. P. and Hood, R. D., 1981, Prenatal effects following exposure of hamsters to sodium arsenite by oral or intraperitoneal routes. Teratology, 23: 40A. (Abstr.).

Harrison, W. P., Frazier, J. C. Mazzanti, E. M., and Hood, R. D., 1980, Teratogenicity of disodium methanarsonate and sodium dimethylarsinate (sodium cacodylate) in mice. Teratology, 21: 43A. (Abstr.).

Harvey, S. C., 1975, Heavy metals. in: "The Pharmacological Basis of Therapeutics", L. S. Goodmann and A. Gilman, eds. 5th ed., p. 924. Macmillan, New York.

Hine, C. H., Pinto, S. S., and Nelson, K. W., 1975, Medical problems associated with occupational exposure to arsenic. "Symposium Proceedings, Int. Conf. Heavy Metals in the E viro ment", Toronto, Canada, Oct. 27-31, Vol. III, p. 316.

Holmberg, R. E. and Ferm, V. H., 1969, Interrelationships of selenium, cadmium and arsenic in mammalian teratogenesis. Arch. Environ. Hlth., 18: 873-877.

Hood, R. D., 1972, Effects of sodium arsenite on fetal development. Bull. Environ. Contam. Toxicol., 7: 216-222.

Hood, R. D. and Bishop, S. L., 1972, Teratogenic effects of sodium arsenate in mice. Arch. Environ. Hlth. 24: 62-65.

Hood, R. D., Harrison, W. P., and Vedel, G. C., 1982b, Prenatal effects of arsenic metabolites in the hamster. Teratology, 25: 50A. (Abstr.).

Hood, R. D., Vedel, G. C., Zaworotko, M. J., Tatum, F. M., and Fisher, J. G., 1981, Arsenate distribution in po or ip treated pregnant mice and their fetuses. Teratology, 23: 42A. (Abstr.).

Hood, R. D., Vedel, G. C., Zaworotko, M. J., and Tatum, F. M., 1982, Distribution of arsenite and methylated metabolites in pregnant mice. Teratology, 25: 50A. (Abstr.).

Jernelov, A., 1975, Microbial alkylation of metals. in: "Symposium Proceedings, Int. Conf. Heavy Metals in the Environment", Toronto, Canada, Oct. 27-31, Vol. II, p. 845.

Johnstone, R. M., 1963, in: "Metabolic Inhibitors. A Comprehensive Treatise", R. M. Hochster and J. H. Quastel, eds., p. 99. Academic Press, New York.

Lasko, J. U. and Peoples, S. A., 1975, Methylation of inorganic arsenic by mammals. J. Agr. Food Chem., 23: 674-676.

Lebedewa, M. N., 1934, Uber den Ubergang Chemotherapeutischer Verbindungen aus dem Organismus der Mutter in den Fetus. Zentl. Gynak., 58: 1449-1456.

Limarzi, L. R., 1943, The effect of arsenic (Fowler's solution) on erythropoiesis. Amer. J. Med. Sci., 206: 339-347.

Lisella, F., Long, K. R., and Scott, H. G., 1972, Health aspects of arsenicals in the environment. J. Environ. Hlth., 34: 511-518.

Lugo, G., Cassady, G., and Palmisano, P., 1969, Acute maternal arsenic intoxication with neonatal death. Amer. J. Dis. Child, 117: 328-330.

Marin-Padilla, M., 1979, Notochordal-basichondrocranium relationships: Abnormalities in experimental axial skeletal (dysraphic) disorders. J. Embryol. Exp. Morph., 53: 15-38.

Marin-Padilla, M., 1980, Morphogenesis of experimental encephalocele (cranioschisis occulta). J. Neurol. Sci., 46: 83-99.

McBride, B. C. and Wolfe, R. S., 1971, Biosynthesis of dimethylarsine by methanobacterium. Biochemistry, 10: 4312-4317.

Mealey, J., Brownell, G. L., and Sweet, W. H., 1959, Radioarsenic in plasma, urine, normal tissues and intracranial neoplasms. Arch. Neurol. Psychiat., 81: 310-320.

Mitchell, R. A., Chang, B. F., Huang, C. H., and DeMaster, E. G., 1971, Inhibition of mitochondrial energy-linked functions by arsenate. Evidence for a nonhydrolytic mode of inhibitor action. Biochemistry, 10: 2049-2054.

Morris, H. P., Laug, E. P., Morris, H. J., and Grant, R. L., 1938, The growth and reproduction of rats fed diets containing lead acetate and arsenic trioxide and the lead and arsenic content of newborn and suckling rats. J. Pharmacol. Exp. Therap., 64: 420-445.

Morrissey, R. E. and Mottet, N. K., 1981, An ultrastructural study of the effects of arsenic on mouse neuroepithelium. Teratology, 23: 53-54A. (Abstr.).

NIOSH, National Institute of Occupational Safety and Health (1975)., "Occupational Exposure to Inorganic Arsenic. Criteria for a Recommended Standard," U. S. Dept. Health, Education and Welfare, Public Health Service, Center for Disease Control, Publ. No. 75-149., U. S. Govt. Printing Office, Washington, D.C.

Odanaka, Y., Matano, O., and Goto, S., 1980, Biomethylation of inorganic arsenic by the rat and some laboratory animals. Bull. Environ. Contam. Toxicol., 24: 452-459.

Penrose, W. R., 1975, Biosynthesis of organic arsenic compounds in brown trout (Salmo trutta). J. Fish. Res. Board Canada, 32: 2385-2390.

Petres, J., Schmid-Ulrich, K., and Wolf, U. (1970). Chromosomal anomalies in human lymphocytes in cases of chronic arsenic poisoning. Deutsch. Med. Wschr., 95: 79-80.

Piton, R., 1929, Recherches sur les actions caryoclasiques et caryocinetiques des composes arsenicaux. Arch. Int. Med. Exp., 5: 355-411.

Rowland, I. R. and Davies, M. J., 1981, In vitro metabolism of inorganic arsenic by the gastro-intestinal microflora of the rat. J. Appl. Toxicol., 1: 278-283.

Schroeder, H., 1966, Abnormal trace metals in man: Arsenic. J. Chron. Dis., 19: 85-106.

Smith, T. J., Crecelius, E. A., and Reading, T. D. (1977). Airborne arsenic exposure and excretion of methylated arsenic compound. Environ. Hlth. Pers., 19: 89-95.

Snyder, F. F. and Speert, H., 1938, The placental transmission of neoarsphenamine in relation to the stage of pregnancy. Amer. J. Obstet. Gynecol., 36: 579–586.

Tam, G. H. K., Charbonneau, S. M., Lacroix, G., and Bryce, F., 1978, Separation of arsenic metabolites in dog plasma and urine following intravenous injection of $^{74}$As. Anal. Biochem., 86: 505–511.

Tam, G. H. K., Charbonneau, S. M., Bryce, F., Pomroy, C., and Sandi, E., (1979), Metabolism of inorganic arsenic [$^{74}$As] in humans following oral ingestion. Toxicol. Appl. Pharmacol., 50: 319–322.

Umpierre, C. C., 1981, Embryolethal and teratogenic effects of sodium arsenite in rats. Teratology, 23: 66A (Abstr.).

Underhill, F. P. and Amatruda, F. G., 1923, The transmission of arsenic from mother to fetus. J. Amer. Med. Assoc., 81: 2009–2012.

Vahter, M., 1981, Biotransformation of trivalent and pentavalent inorganic arsenic in mice and rats. Env. Res., 25: 286–293.

Vahter, M. and Norin, H., 1980, Metabolism of $^{74}$As-labeled trivalent and pentavalent arsenic in mice. Env. Res., 21: 446–457.

Vallee, B. L., Ulmer, D. D., and Wacher, W. E. C., 1960, Arsenic toxicology and biochemistry. A.M.A. Arch. Ind. Hlth., 21: 132–151.

Vedel, G. C., Harrison, W. P., Baxley, M. N., and Hood, R. D., 1980, Influence of an arsenic chelating agent (BAL) on the outcome of prenatal arsenite exposure in mice. Teratology, 21: 72A–73A (Abstr.).

Webster's New International Dictionary of the English Language. 1941, W. A. Neilson, T. A. Knott, and P. W. Carhart, eds., G&C Merrian, Co., Springfield, Mass.

Willhite, C. C., 1981, Arsenic-induced axial skeletal (dysraphic) disorders. Exp. Mol. Pathol., 34: 145–158.

**10**

DIFFERENTIAL EFFECTS OF SODIUM SELENITE AND METHYLMERCURY(II) ON
MEMBRANE PERMEABILITY AND DNA REPLICATION IN HeLa S3 CARCINOMA
CELLS: A PRELIMINARY REPORT REGARDING THE MODIFICATION OF
ORGANOMERCURIAL TOXICITY BY SELENIUM COMPOUNDS

D. W. Gruenwedel

Department of Food Science and Technology
University of California, Davis, CA 95616

ABSTRACT

When viewed in terms of their concentration in the growth
medium, sodium selenite and methylmercuric hydroxide – administered individually to HeLa S3 cells – are of equal efficacy in
inhibiting DNA synthesis: the dose-response curves overlap and 50%
residual DNA synthesis occurs at 6.13 µM of either chemical. A
different picture, however, emerges if replication is expressed as
a function of the actual amounts of toxicant bound per cell. Now,
the dose-response curves do not overlap and $Na_2SeO_3$ is much more
toxic than $CH_3HgOH$: 50% inhibition of DNA replication exists at
$5.37 \times 10^{-17}$ moles of Se bound per cell and at $3.63 \times 10^{-15}$ moles
of Hg bound per cell. Further, selenite is taken up by the cells
more slowly than methylmercury and its (limiting) cellular concentration is below that of the organomercurial. Lastly, much higher
levels of selenite in the growth medium are required to bring
about the same degree of membrane damage as the one caused by
methylmercury. These differential effects may have a bearing on
the observation, well-known but thus far unexplained, that
selenite and methylmercury are strikingly less toxic to animals
when administered simultaneously than they are when administered
individually: selenium may counteract the membrane-destabilizing
characteristics of methylmercury and it may retard its binding to
the cells. Data on the inhibition of DNA synthesis have been
obtained when selenite and methylmercury are administered simultaneously to HeLa S3 cells in varied molar ratios. Best mutual
protection appears to exist when the two chemicals are present in
equimolar amounts or when there is a slight excess of selenite.

INTRODUCTION

Although more than fifteen years have passed since Parizek
and Ostadalova (1967) first reported on the protection offered by
selenium against mercury intoxication, and although numerous
investigations have been undertaken in the meantime to clarify the
metabolic interdependence existing between the two elements (cf.,
Skerfving, 1978 and the literature cited therein), there is little
agreement among toxicologists as to the biochemical basis of the
protective effect.  As is well known, both elements in the form of
their compounds, are strikingly less toxic to animals when
administered simultaneously than they are when administered
individually.  Selenium, usually employed in the form of $Na_2SeO_3$,
protects against inorganic mercury and organic mercury, e.g.,
$CH_3Hg(II)$, as well.

A number of workers (cf., Eybl et al., 1969; Burke et al.,
1974; Ganther, 1979; Iwata et al., 1981, Masukawa et al., 1982),
on the basis of results obtained from in vivo experiments, have
advanced the hypothesis that selenium counteracts the toxicity of
mercury and other heavy metals simply by forming chemically and
biologically inert metal selenides.  Since HgSe, for instance, has
a solubility product near $10^{-59}$ M, the formation of highly stable
and thus inactive -Hg-Se- bonds as the toxicity-limiting principle
appears reasonable.  On the other hand, as reported by Taylor
et al. (1978), $HgSeO_3$, an insoluble compound formed when equimolar
solutions of $Na_2SeO_3$ and $HgCl_2$ are mixed, displays in mice an $LD_{50}$
of 35.5 µmol/kg, which is only slightly below the value valid for
$Na_2SeO_3$ alone (40 µmol/kg).  In fact, HgSe by itself, when injected subcutaneously as suspension, shows an $LD_{50}$ of 400 µmol/kg in
spite of its insolubility.

Other workers have therefore suggested that selenium exerts
its detoxifying influence indirectly, viz., through metabolic
changes in the target tissues, thereby decreasing the susceptibility of an organism to the detrimental effects of mercury (cf.,
Parizek, 1978; Alexander et al., 1979).  Still others favor the
explanations that selenium accelerates the demethylation of
methylmercury (Yamane et al., 1977), yielding less toxic inorganic
mercury, or that it suppresses the formation of free radicals
generated, for instance, by homolytic fission of the carbon-metal
bond in methylmercury (Ganther, 1978).  Whatever the ultimate
explanation of the protective effect may turn out to be, the few
examples presented here show clearly that much more work needs to
be done before the metabolic antagonism of the two elements can be
understood.

Our laboratory has for some time been engaged in studying the
effects of methylmercury(II) on living systems using HeLa S3

suspension-culture cells as biological probes (Gruenwedel and Fordan, 1978; Gruenwedel and Cruikshank, 1979a; Gruenwedel et al., 1979; Gruenwedel and Friend, 1980; Gruenwedel et al., 1981; Gruenwedel, 1981) and we have also investigated the effects of $Na_2SeO_3$ on HeLa S3 cells (Gruenwedel and Cruikshank, 1979b). The purpose of the present communication is two-fold: (1) to draw attention to the fact that $CH_3Hg(II)$ and $Na_2SeO_3$ show dramatic differences in their effect on membrane permeability and on inhibition of intracellular DNA synthesis, particularly, when viewed in terms of the quantities of toxic material actually bound to a cell, and (2) to present some initial data on suppression of DNA synthesis when the two compounds are administered simultaneously to HeLa S3 cells.

METHODS

Chemicals. $CH_3HgOH$ (97+%) was purchased from Alfa Products (Ventron Corp., San Leandro, CA). Its handling has been described elsewhere (Gruenwedel and Davidson, 1966; Gruenwedel and Fordan, 1978). Radioactive-labeled $CH_3{}^{203}HgCl$, dissolved in 0.02 M sodium carbonate at a specific activity of 1.181 Ci/g, was a product of New England Nuclear (Boston, MA). $Na_2SeO_3$ (99+%) was obtained from Gallard-Schlessinger Chemical Corp. (Carle Place, NY) while radioactive-labeled $H_2{}^{75}SeO_3$, dissolved in 0.5 M HCl at the specific activity of 12.2 Ci/g, was also a product of New England Nuclear. [Methyl-$^3$H]thymidine (47 Ci/mmol) was purchased from Amersham/Searle (Arlington Heights, IL). All other chemicals were of reagent grade.

Bio-Assays

Stock Cell Culture. HeLa S3 cells were obtained as monolayer cultures from the Cell Culture Laboratory of the School of Public Health, University of California, Berkeley, CA. The cells were adapted to and maintained in suspension (asynchronous) by passaging them routinely in Eagle's minimum essential medium (MEM-Joklik modified, Grand Island Biological Co. (GIBCO), Grand Island, NY). The medium was supplemented with 10% fetal calf serum (GIBCO). The cells were kept in logarithmic growth phase by maintaining the cell density near $0.8 \times 10^6$ cells/ml. Cell growth took place in water-saturated 95% air-5% $CO_2$ at 37°C.

Intoxication and Precursor Labeling Studies. Cells were removed from a stock cell culture by centrifugation. They were incubated anew with fresh medium at a density of $0.7 \times 10^6$ cells/ml and, after gassing with 5% $CO_2$, kept for 18 hr at 37°C. They were then diluted with fresh medium to a density of $0.3 \times 10^6$ cells/ml and, after a further adaptation period of 20 min, divided into a number of 7 ml portions, each portion containing the same number of cells. Six test series were formed: one, receiving only varied concentrations of $CH_3HgOH$ and another one, receiving only varied concentrations of $Na_2SeO_3$. The four remaining series received mixtures of $CH_3HgOH$ and $Na_2SeO_3$, containing the elements Hg and Se in the molar ratios of Hg/Se = 10/1, Hg/Se = 1/10, Hg/Se = 1/3, and Hg/Se = 1/1, respectively. Stock solutions of $CH_3HgOH$ and $Na_2SeO_3$, or of the mixtures, were prepared in such a way that the addition of 70 µl of stock to 7 ml of cell suspension yielded the desired final concentration. Final toxicant concentrations in the medium amounted to [µM]: 0, 0.1, 1, 2, 5, 8, 10, 31.6, 100, and 1,000 in the case of $CH_3HgOH$ (alone) and, in the case of $Na_2SeO_3$ (alone), to [µM]: 0, 0.01, 0.1, 1, 3.16, 10, 31.6, 100, and 1,000. The equimolar mixtures contained each toxicant at levels of [µM]: 0, 0.1, 1, 10, 31.6, 100, and 1,000, respectively. The toxicant concentrations employed in the other mixtures can be gathered from Table 1.

Each series, after gassing with 5% $CO_2$, was incubated for 6 hr. DNA-precursor labeling took place during the last 30 min of the incubation period. To this end, 70 µl of [$^3$H]thymidine, dissolved in water at levels of 2 µCi, was added to each portion of a given series and incubation and labeling stopped 30 min later by immersing each vial into ice-slush and by removing the medium and all subsequent washes via vacuum filtration, making sure that only gentle suction was employed. Washing (two times) consisted of rinsing the cells with 2 ml each of ice-cold phosphate-buffered saline and, finally, by adding 2 ml each of ice-cold 5% (w/v) trichloroacetic acid. The cells were collected on Millipore filters, washed three times with 5% ice-cold trichloroacetic acid, and dried for about 30 min on a foil-covered hot plate. Radio-nuclide counting took place in a Beckman model 250 liquid scintil-lation counter.

The binding of selenite [$^{75}$Se] and methylmercury [$^{203}$Hg] to the cells has been described elsewhere (Gruenwedel and Cruikshank, 1979b; Gruenwedel et al., 1981) and need not be repeated here. Cells were counted by using the Coulter counter model $Z_B$ (Gruenwedel, 1981). The methylmercury and selenite concentrations used in the binding studies were the same used in the DNA inhibi-tion studies with the exception that with [$^{75}$Se] and [$^{203}$Hg] the lowest concentration in the medium was 1 µM and the highest 100 µM. Each concentration level contained the radionuclide of interest at the 1 µM level.

Membrane permeability was monitored via trypan blue staining (cf., Gruenwedel and Fordan, 1978).

RESULTS AND DISCUSSION

Fig. 1 contains information regarding the effects of sodium selenite (closed circles) and methylmercury(II) (open circles) on DNA synthesis in HeLa S3 cells.  In terms of concentration in the growth medium, both chemicals are of equal efficacy in suppressing replication: control cells ($pX_a$ ∞, whereby $pX_a \equiv -$ log $[X]_{added}$) as well as cells that had been in contact with either toxicant at concentrations up to 1 µM ($pX_a$ 6) do not experience inhibition, but dramatic alterations in replication take place at toxicant levels in the medium between $pX_a$ 6–5 (1–10 µM).  A residual DNA synthesis of 50% exists at 6.13 µM in the medium ($pX_a$ 5.20) for either compound.

A vastly different picture, however, is obtained if replication is not expressed as a function of the medium concentration of the toxicants but as a function of their actual amounts bound by a cell.  This can be seen from Fig. 2 where the degree of $[^3H]$thymidine incorporation into acid-insoluble cellular material has been plotted _versus_ $pX_b$.  $pX_b \equiv -$ log [moles of X bound per cell].  It

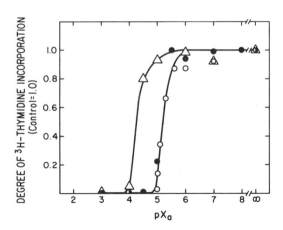

Fig. 1.  Dependence of DNA synthesis of HeLa S3 cells (normalized with respect to control) on toxicant concentration in the medium.  ( ● ) $Na_2SeO_3$ administered alone; ( ○ ) $CH_3HgOH$ administered alone; ( △ ) $Na_2SeO_3$ and $CH_3HgOH$ administered simultaneously as equimolar mixture.  $pX_a \equiv -$ log [toxicant $X]_{added}$.  In the case of the equimolar mixture, concentrations were plotted in such a way that a given $pX_a$ value refers to the concentration of eiher component in the mixture.  Thus, $pX_a$ 6 means that the mixture was 1 µM in $Na_2SeO_3$ and 1 µM in $CH_3HgOH$.  Time of incubation: 6 hr.  (( ◐ ) overlap of data points).

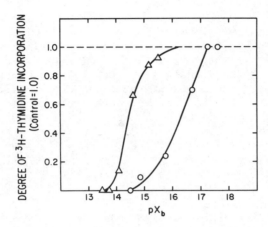

Fig. 2.  Dependence of DNA synthesis of HeLa S3 cells (normalized
with respect to control) on the amounts of the toxicants
actually boud by a cell.  ( $\circ$ ) $Na_2SeO_3$ administered
alone; ( $\triangle$ ) $CH_3HgOH$ administered alone.  $pX_b \equiv - \log$
[moles toxicant X bound per cell].  Time of incubation: 6
hr.  (dashed line refers to position of control).

is readily seen that in reality $Na_2SeO_3$ (open circles) is much
more toxic than $CH_3Hg(II)$ (open triangles).  50% inhibition of DNA
synthesis occurs at $pX_b$ 16.27 in the case of selenite and at $pX_b$
14.44 in the case of methylmercury.  Thus, when administered
individually, methylmercury needs to be present at an approximately seventy-fold higher level than $Na_2SeO_3$ in a cell in order
to produce the same degree of DNA inhibition, say 50%.

Differential effects are noted further if one looks at the
rate of toxicant binding by the cells.  This is shown in Fig. 3.
Methylmercury (dashed lines) reaches limiting intracellular levels
much more rapidly then selenite (solid lines).  Thus, it takes
$Na_2SeO_3$, at 5 μM in the growth medium, about 11 hr to reach the
$pX_b$ value at which 50% inhibition of DNA synthesis exists, and
about 4 or 1.5 hr if its medium concentration is 10 or 31.6 μM.
With $CH_3Hg(II)$, the corresponding incubation periods are 6 hr, < 1
hr, and < 1 hr, respectively.  Lastly, differential effects are
observed if one looks at the staining characteristics of the cells
in presence of methylmercury or sodium selenite.  This is shown in
Fig. 4.  Cell viability (trypan blue dye staining) has been
plotted against toxicant concentration in the growth medium.
Taking the cell viability of 50% as reference, it is readily seen
that much more $Na_2SeO_3$ is needed (by a factor of about 100) to
generate the degree of membrane damage caused by $CH_3HgOH$.  In
fact, in the case of methylmercury, either cell viability (cf.,

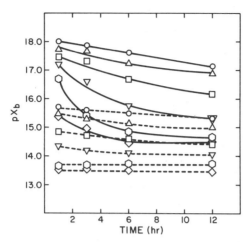

Fig. 3. Rate of toxicant binding by HeLa S3 cells. Solid lines: $Na_2SeO_3$. Dashed lines: $CH_3HgOH$. Toxicant levels in the growth medium: ( ○ ) 1 µM; ( △ ) 2 µM; ( □ ) 5 µM; ( ▽ ) 10 µM; ( ◯ ) 31.6 µM; ( ◇ ) 100 µM. Data taken from Gruenwedel and Cruikshank (1979b) and Gruenwedel et al. (1981).

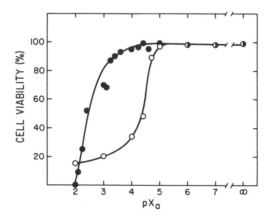

Fig. 4. Permeability of the outer plasma membrane of HeLa S3 cells with respect to trypan blue dye in the presence of $CH_3HgOH$ ( ○ ) or $Na_2SeO_3$ ( ● ). $pX_a \equiv - \log [X]_{added}$. Cell viability (trypan blue dye staining): % live cells. Time of incubation: 5-6 hr (data taken from Gruenwedel and Fordan, 1978, and Gruenwedel and Cruikshank, 1979b). [( ◑ ) overlap of data points].

Fig. 4) or inhibition of DNA synthesis (cf., Fig. 1) can serve
equally well as indicator of toxicity, for the 50% suppression
points nearly coincide. With $Na_2SeO_3$, however, substantial errors
would be introduced: 50% residual DNA synthesis takes place at $pX_a$
5.20 while 50% cell viability exists at $pX_a$ 2.45, amounting to a
concentration ratio of 560:1.

Simultaneous administration of $Na_2SeO_3$ and $CH_3HgOH$ affects
intracellular DNA synthesis as indicated in Fig. 1 (Hg/Se = 1/1,
open triangles) and Table 1. There can be no doubt that optimal
protection exists if the two chemicals are given as an equimolar
mixture. For instance, either toxicant, if administered individually at 10 µM in the medium ($pX_a$ 5), permits residual DNA synthesis near 5% after 6 hr of incubation (cf., Fig. 1). However, if
administered at the same level as equimolar mixture, residual DNA
synthesis is at 92% (cf., Fig. 1). Keeping the level of methylmercury in the medium constant at 10 µM but raising the concentration of selenite to 30 µM yields a synthesis activity of 40%, and
increasing the level of selenite further to 100 µM reduces DNA
synthesis to 10%. On the other hand, while $Na_2SeO_3$ alone at 30 µM
($pX_a$ 4.5) has abolished DNA synthesis completely, the presence of
10 µM $CH_3HgOH$ has obviously a beneficial effect since, as pointed
out already, DNA synthesis proceeds at 40% of the control (cf.,
Fig. 1). But then again, administering the two compounds at 30 µM
as equimolar mixture is still better since DNA synthesis is permitted to take place with about 80% activity of the control.

Summarizing the data on hand, it can be stated that $Na_2SeO_3$,
present either in equimolar concentration or in a not too large
excess, protects HeLa S3 cells against intoxication by methylmercury. This is of course equivalent to saying that small
amounts of methylmercury offer protection against selenite intoxication. It remains to be seen whether the opposite relation
holds, i.e., whether small amounts of $Na_2SeO_3$ reduce methylmercury
toxicity in HeLa S3 cells. The one example given in Table 1,
viz., having methylmercury present in a ten-fold excess over
selenite, does not indicate modification. For instance, the DNA
synthesis remaining in HeLa S3 cells after their exposure to 20 µM
$CH_3HgOH$ in presence of 2 µM $Na_2SeO_3$ is the same as the one found
with cells that had been exposed to 20 µM $CH_3HgOH$ alone. Since
methylmercury enters HeLa S3 cells much more rapidly than
selenite (cf., Fig.3), it appears logical that excess $CH_3HgOH$
should cause cessation of replication, provided it is present in
sufficiently large quantities to begin with. In vivo experiments
(e.g., Ganther et al., 1972; Iwata et al., 1973; Johnson and Pond,
1974; see also Skerfving, 1978) indicate that small amounts of
selenite suffice to alleviate the toxic symptoms caused by excess
$CH_3Hg(II)$ in animals and, thus, the one example presented above
would be at variance with those findings. However, the data given

Table 1. Residual DNA Synthesis in HeLa S3 Cells After Their
         Exposure to Mixtures of $Na_2SeO_3$ and $CH_3HgOH$ at the
         Indicated Levels

| $CH_3HgOH$ [μM] | $Na_2SeO_3$ [μM] | Degree of [3H] Thymidine Incorpora-tion After 6 hr of Incubation [Control = 1.00] |
|---|---|---|
| 0 | 0 | 1.00 |
| 2 | 6 | 0.71 |
| 5 | 15 | 0.52 |
| 10 | 30 | 0.40 |
| 20 | 60 | 0.03 |
| 30 | 90 | 0 |
| 2 | 20 | 0.22 |
| 5 | 50 | 0.16 |
| 10 | 100 | 0.11 |
| 15 | 150 | 0.05 |
| 20 | 200 | 0.02 |
| 30 | 300 | 0 |
| 20 | 2 | 0.02 |
| 50 | 5 | 0.01 |
| 100 | 10 | 0 |
| 200 | 20 | 0 |
| 300 | 30 | 0 |

in Table 1 are in excellent agreement with the results obtained by
Potter and Matrone (1977) concerning the growth inhibiting effects
of mixtures of $Na_2SeO_3$ and $CH_3HgCl$ on Chang's liver cells and 3T3
cells kept in culture.  They, too, find that excess $Na_2SeO_3$ offers
protection against growth inhibition brought about by the organic
mercury.  Undoubtedly, much more work needs to be done before firm
conclusions can be drawn regarding the antagonistic characteristics of mercury and selenium.

Current work in this laboratory is concerned with expanding
the studies reported here.  We are presently determining the rate
of [203]Hg and [75]Se incorporation into HeLa S3 cells when $Na_2SeO_3$
and $CH_3Hg(II)$ are administered simultaneously in varied molar
ratios and we have begun looking at their combined effects on
membrane permeability.  To us, the findings that

(1)  $Na_2SeO_3$ preserves the functional integrity of the outer
     plasma membrane of HeLa S3 cells up to much higher concentra-
     tions in the medium than $CH_3HgOH$ (cf., Fig. 4) and

(2) that $Na_2SeO_3$ enters HeLa S3 cells more slowly than $CH_3HgOH$
and, in fact, is present in cells at much lower (limiting)
levels than methylmercury at identical concentrations in the
medium (cf., Fig. 3)

are of considerable importance, for they may be the first clues to
the combined mode of action of the two compounds: it is conceivable that selenium counteracts the membrane-destroying power of
methylmercury (cf., Gruenwedel et al., 1979; Gruenwedel, 1981)
through membrane stabilization and also has a retarding influence
on methylmercury binding by the cells.

ACKNOWLEDGEMENTS

This research has been supported by funds of the University
of California and, in part, by Grant ES01115 from the U. S. Public
Health Service.

REFERENCES

Alexander, J., Hostmark, A. T., Forre, O., and von Kraemer Bryn,
M., 1979, The influence of selenium on methyl mercury toxicity in rat hepatoma cells, human embryonic fibroblasts and
human lymphocytes in culture, Acta pharmacol. et toxicol. 45:
379.

Burke, R. F., Foster, K. A., Greenfield, P. M., and Kiker, K. W.,
1974, Binding of simultaneously administered inorganic
selenium and mercury to rat plasma protein, Proc. Soc. Exp.
Biol. Med. 145: 782.

Eybl, V., Sykora, J., and Mertl, F., 1969, Einfluss von Natrium-
selenit, Natriumtellurit und Natriumsulfid auf Retention und
Verteilung von Quecksilber bei Mäusen, Arch. Toxicol. 25:
296.

Ganther, H. E., Goudie, C., Sunde, M. L., Kopecky, M. J., Wagner,
P., Oh, S.-H., and Hoekstra, W. G., 1972, Selenium: relation
to decreased toxicity of methylmercury added to diets containing tuna, Science 175: 1122.

Ganther, H. E., 1978, Modification of methylmercury toxicity and
metabolism by selenium and vitamin E: possible mechanisms,
Environ. Health Perspect. 25: 71.

Ganther, H. E., 1979, Metabolism of hydrogen selenide and methylated selenides, Adv. Nutr. Res. 2: 107.

Gruenwedel, D. W., and Davidson, N., 1966, Complexing and denaturation of DNA by methylmercuric hydroxide, J. Mol. Biol. 21:
126.

Gruenwedel, D. W., and Fordan, B. L., 1978, Effects of methylmer-
cury(II) on the viability of HeLa S3 cells, Toxicol. Appl.
Pharmacol. 46: 249.

Gruenwedel, D. W., and Cruikshank, M. K., 1979a, Effect of methyl-mercury(II) on the synthesis of deoxyribonucleic acid, ribonucleic acid and protein in HeLa S3 cells, Biochem. Pharmacol. 28: 651.

Gruenwedel, D. W., and Cruikshank, M. K., 1979b, The influence of sodium selenite on the viability and intracellular synthetic activity (DNA, RNA, and protein synthesis) of HeLa S3 cells, Toxicol. Appl. Pharmacol. 50: 1.

Gruenwedel, D. W., Glaser, J. F., and Falk, R. H., 1979, A scanning electron microscope study of the surface features of HeLa S3 suspension-culture cells treated with methylmercury(II), J. Ultrastruct. Res. 68: 296.

Gruenwedel, D. W. and Friend, D., 1980, Long-term effects of methylmercury(II) on the viability of HeLa S3 cells, Bull. Environm. Contam. Toxicol. 25: 441.

Gruenwedel, D. W., 1981, Effect of methylmercury(II) on the size of HeLa S3 carcinoma cells, Virchows Arch. [Cell Pathol.] 37: 153.

Gruenwedel, D. W., Glaser, J. F., and Cruikshank, M. K., 1981, Binding of methylmercury(II) by HeLa S3 suspension-culture cells: intracellular methylmercury levels and their effect on DNA replication and protein synthesis, Chem.-Biol. Interact. 36: 259.

Iwata, H., Okamota, H., and Ohsawa, Y., 1973, Effect of selenium on methylmercury poisoning, Res. Commun. Chem. Path. Pharm. 5: 673.

Iwata, H., Masukawa, T., Kito, H., and Hayashi, M., 1981, Involvement of tissue sulfhydryls in the formation of a complex of methylmercury with selenium, Biochem. Pharmacol. 30: 3159.

Johnson, S. L., and Pond, W. G., 1974, Inorganic vs. organic mercury toxicity in growing rats: protection by dietary selenium but not zinc, Nutr. Repts. Internat. 9, 135.

Masukawa, T., Kito, H., Hayashi, M., and Iwata, H., 1982, Formation and possible role of bis(methylmercuric)selenide in rats treated with methylmercury and selenite, Biochem. Pharmacol. 31: 75.

Parizek, J., and Ostadalova, I., 1967, The protective effect of small amounts of selenite in sublimate intoxication, Eperientia 23: 142.

Parizek, J., 1978, Interactions between selenium compounds and those of mercury or cadmium, Environ. Health Perspect. 25: 53.

Potter, S. D., and Matrone, G., 1977, A tissue culture model for mercury-selenium interactions, Toxicol. Appl. Pharmacol. 40: 201.

Skerfving, S., 1978, Interaction between selenium and methylmercury, Environ. Health Perspect. 25: 57.

Taylor, T. J., Rieders, F., and Kocsis, J. J., 1978, Toxicological
    interactions of mercury and selenium, Toxicol. Appl.
    Pharmacol 45: 347.
Yamane, Y., Fukino, H., Aida, Y., and Imagawa, M., 1977, Studies
    on the mechanism of protective effects of selenium against
    the toxicity of methylmercury, Chem. Pharm. Bull. 25: 2831.

## 11

# STRUCTURE AND STEREOCHEMISTRY OF STEROIDAL AMINE TERATOGENS

William Gaffield* and Richard F. Keeler**

*Western Regional Research Center, ARS, USDA, Berkeley
CA, 94710 and **Poisonous Plant Research Laboratory
ARS, USDA, Logan, UT 84321

ABSTRACT

Studies of teratogenic steroidal alkaloids from Veratrum and Solanum have shown that those bearing a basic nitrogen atom in ring F, shared or unshared with ring E, with bonding capabilities α to the steroid plane may be suspect as teratogens. Examples of steroidal alkaloids which produce terata but, until recently, have been of uncertain structure include muldamine and the isomeric 3, N-diformylsolasodines. The correlation of their structures with the structure-terata relationship developed by Keeler and Brown is discussed. A brief introduction to teratogenicity is presented.

## INTRODUCTION

Errors in development and abnormalities in the newborn have evoked curiosity and human sympathy from ancient times but such birth defects were accepted as hazards of procreation until the twentieth century when scientific investigations were initiated to determine their causes. Early theories which were proposed for the affliction of man with unusual physical features at birth included the views that defective children were messages from the gods foretelling events in the affairs of man, were models for creatures from Greek mythology such as sirens and cyclops, resulted from the mental impressions and emotions of pregnant women (for example, a mother frightened by a rabbit having a child with a harelip), originated from the interbreeding of man with animals or occurred as a result of cohabitation of humans with demons and witches (Warkany, 1963, Miller, 1978).

Prior to this century, important discoveries in birth development included the awareness of Hippocrates and Aristotle that
direct injury to the embryo or increased pressure on or within
the uterus could cause birth defects (Warkany, 1977), the statement of William Harvey concerning the theory of developmental
arrest (Harvey, 1651) and the first experiments performed in
animal teratology by Saint-Hilaire and his son during the 1820's
(Wilson, 1973). Around the turn of the century errors in human
development were generally attributed to genetic causes but
Warkany and Nelson (1940) reported that maternal dietary deficiencies in rats could result in predictable malformations in the
offspring. The realization that environmental factors must be
considered as potential risks to the unborn child was underscored
in 1961 with the establishment of an association between the use
by pregnant women of thalidomide, a promising sedative which was
presumed to be nontoxic, and the birth of 7-8 thousand severely
deformed children in 28 different countries (McBride, 1961, Lenz,
1962, Lenz, 1966). Due to this unfortunate incident worldwide
attention was focused on the potential teratogenic hazards of
drugs and dietary constituents.

Currently in the United States, approximately 3-7% of all
humans are born with congenital malformations, known as terata,
which may handicap, disable or result in the death of the individual (Anon., 1975, Anon., 1981). About 20% of birth defects are
thought to result from drugs, chromosomal aberrations, viral
infections and exposure to ionizing radiation. The large percentage remaining are unaccounted for at present but are presumed to
result from exposure to food additives, pesticide residues or
other environmental contaminants. The teratogenic nature of thalidomide might have gone undetected if the terata had not been so
obvious in the infants which were affected. Despite thalidomide's
potent teratogenic effect in humans, only occasional terata have
been induced in numerous strains of a wide variety of laboratory
animals (Schardein, 1976). In fact, effects similar to the limb
defects observed in man have been consistently produced in only
rabbits and primates.

At present, several drugs have been identified as human teratogens, a number of other drugs are suspected teratogens and certain additional drugs are considered to be possible teratogens
(Schardein, 1976, Shepard, 1973). However, little is known concerning the relationship of molecular structure to teratogenic
activity.

## STRUCTURE-ACTIVITY RELATIONS

An exception to the latter statement is the group of teratogenic steroidal alkaloids from toxic range plants, such as

Veratrum californicum. Their teratogenicity has been related to a particular structural feature, namely the direction of projection of the amino group and its unshared electron pair relative to the plane of the steroid (Keeler, 1983, Brown, 1978). An understanding of the teratogenic hazard of these compounds arose from studies on a double or single globe cyclopic deformity occurring in sheep known as monkey face lamb disease (Binns, et al., 1959, Binns, et al., 1960, Binns et al., 1962). Incidence of the disease often afflicted up to 30% of the lamb crop. Initial studies ruled out a genetic basis for the disease (Binns, et al., 1959) and suggested a plant teratogen as the cause since certain plants, such as Veratrum californicum, were common to ranges grazed by pregnant ewes that later produced malformed lambs. Furthur experiments by researchers at the Poisonous Plant Research Laboratory in Logan, Utah identified Veratrum as the plant responsible for birth defects and showed the precise insult period of the teratogen to be the fourteenth day of gestation (Binns, et al., 1965). These observations have enabled ranchers to reduce the incidence of monkey face lamb disease to less than 1% by keeping their pregnant ewes away from the plant until over two weeks have elapsed after rams are removed.

## Cyclopamine and Related Teratogens

Examination of over a dozen Veratrum alkaloids showed that three were capable of inducing monkey face lamb disease; jervine (1), cyclopamine (11-deoxojervine) (2) and cycloposine (3-glucosyl-11-deoxojervine) (3) all of which project the amino nitrogen α or below the plane of the steroid (Keeler and Binns, 1968).

R₁=H,  R₂=O    JERVINE    (1)

R₁=H,  R₂=H    CYCLOPAMINE (2)

R₁=GLU,  R₂=H    CYCLOPOSINE (3)

Additional experiments have demonstrated that the Veratrum steroidal alkaloid teratogens possess a broad species susceptibility and that their action is expressed as a result of a direct effect upon the embryo by a compound unaltered by maternal rumen or hepatic microsomal enzyme metabolism (Keeler, 1983, Keeler, 1979).

Evidence suggests that Veratrum teratogens may inhibit hormone
function in the developing embryo by interacting with hormone
receptor sites (Keeler, 1983), or by other competitive interference with structurally related hormones.

## Spirosolanes

Studies in rats and hamsters have indicated that solasodine
(Solanum) (4) was teratogenic in contrast to tomatidine (Lyco-
persicon) (5) and the steroidal sapogenin diosgenin (6) (Keeler
et al., 1976, Seifulla and Ryzhova, 1972). The lack of teratogenicity of diosgenin and tomatidine showed that both the presence
of the nitrogen in the F-ring and its configurational position,
with respect to the plane of the steroid, were essential for
activity. If the amino group was in the α position the compounds
were teratogenic but positioning the nitrogen on the β side of the
steroid plane resulted in inactivity. The tenfold greater teratogenicity exhibited by cyclopamine in comparison to solasodine may
be due either to the extent of α nitrogen projection or to the

SOLASODINE                                                          (4)

TOMATIDINE                                                          (5)

DIOSGENIN                                                           (6)

greater basicity (Brown and Keeler, 1978a) of cyclopamine. Since
the amino group must be favorably located in order to interact
with the active site in the embryo, the furan group present in
certain alkaloids may aid in positioning the nitrogen α with
respect to the steroid plane.

## Solanidanines

In the early 1970's, it was proposed that spina bifida and
anencephaly in humans resulted from a teratogen present in bligh-
ted potatoes (Renwick, 1972). This suggestion raised suspicion
that potato alkaloids such as solanidine, which contains a C-27
cholestane skeleton with a tertiary amino group shared by rings E

and F, might pose a teratogenic hazard for humans. Although terata were not observed upon treatment of laboratory mammals with naturally occurring demissidine (7), solanidine (8) or their glycosides (of 22R, 25S configuration), terata were induced in

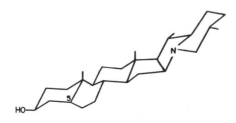

5H ▪ DEMISSIDINE  (7)
Δ5 ▪ SOLANIDINE  (8)

hamsters upon administration of a synthetic 22S, 25R solanidan epimer (Brown and Keeler, 1978c). This teratogen appears, on the basis of conformational analysis, to have the nitrogen lone pair of electrons projecting on the α side of the steroid (cf. 9) in accordance with other teratogenic Veratrum and Solanum alkaloids. On the other hand, the naturally occurring 22R, 25S solanidanes project the lone pair of electrons toward the β side of the steroidal plane (cf. 10) (Brown and Keeler, 1978c).

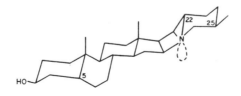

(22 S, 25 R)-5α-SOLANIDAN-3β-ol   (9)

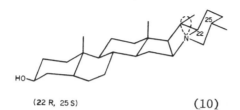

(22 R, 25 S)                        (10)

OTHER STEROIDAL ALKALOIDS

    Additional examples of steroidal alkaloids which produce
terata but, until recently, have lacked a rigorous proof of structure include muldamine and the isomeric 3,N-diformylsolasodines.

Muldamine and its Deacetyl Derivative

    Muldamine, one of three principal benzene extractable alka-
loids isolated from Veratrum californicum in 1968 (Keeler, 1968),
was initially found to be non-teratogenic in sheep (Keeler, 1975)
although later testing showed it to be marginally active in hamsters (Brown and Keeler, 1978a). The deacetyl derivative of muldamine exhibited somewhat greater teratogenicity in hamsters than
the parent alkaloid (Brown and Keeler, 1978a). Originally, muldamine was assigned a veratramine type structure (11) (Keeler,
1971) and deacetylmuldamine a 22,26-epiminocholestene structure
(12) (Kaneko et al., 1977). Structural reinvestigation of these

(11)                                              (12)

two alkaloids has shown that both of the earlier proposals are
erroneous and has firmly ootablished the structure of deacetylmuldamine as (22S,25S)-22,26-epiminocholest-5-ene-3β,16α-diol(13)
and that of muldamine as the 16-acetate (14) (Gaffield et al.,
1982). X-ray crystallographic studies in the solid state and

(13)                                              (14)

[13]C n.m.r. data in solution are consistent with the piperidine ring of deacetylmuldamine (13) positioned so that the amino group may intramolecularly hydrogen-bond to the C-16 hydroxy group (Gaffield et al., 1982). Conversely, the [13]C n.m.r. spectrum of muldamine (14) suggests a preferred conformation in which the amino group is oriented away from the C-16 acetyl group, a moiety which does not permit hydrogen-bonding (Gaffield et al., 1982).

These observations allow evaluation of the structure-terata correlation proposed by Keeler and Brown with regard to alkaloids lacking the rigidity imposed by a spirofuran ring system. General agreement is observed with the postulate since the intramolecularly hydrogen-bonded amino group of deacetylmuldamine (13) is directed below the α-face of the steroid whereas the amino nitrogen of the weaker teratogen, muldamine (14), is positioned on or above the β side of the steroid.

## Isomeric N-Formylsolasodines

The relatively high teratogenicity of N-substituted jervines, such as the N-formyl derivative, has been proposed to result either from more efficient absorption of the N-substituted alkaloid with subsequent hydrolysis producing the active species (parent alkaloid) or from a different binding site where the oxygen of the formyl group functions as the anionic center (15) or (16) (Brown and Keeler, 1978b).

(15)                    (16)

It has been known for some time that formylation of solasodine affords two isomers whereas formylation of the isomeric 22β-N-alkaloid tomatidine gives only a single derivative (Toldy and Radics, 1966a). Brown observed[1] that an unfractionated mixture of isomeric N-formylsolasodines was weakly teratogenic in hamsters

---

[1] D. Brown, unpublished observations

and his suggestion that the activity resided in one of the isomers prompted a reinvestigation of the structure of these compounds. Earlier, these isomers had been assigned structures due to either restricted rotation about the C≕N partial double bond (17) and (18) (Toldy and Radics, 1966a, 1966b) or to inversion about the

(17)                                        (18)

(19)                                        (20)

nitrogen atom (19) and (20) (nitrogen invertomers) (Kusano et al., 1975). Mass spectra, $^1$H and $^{13}$C n.m.r. spectra have shown that the isomerism is due to neither, but rather to a difference in stereochemistry at C-22. The N-formylsolasodine isomers possess 22R, 25R (major isomer) (21) and 22S, 25R (minor isomer) (22) stereochemistry and have non-chair ring-F conformations (Gaffield et al., 1983).

(21)                                        (22)

Reexamination of the biological activity of the isomeric N-formylsolasodines could shed light on the alternative mechanisms proposed by Brown and Keeler (1978b). Deformylation of the isomeric derivatives, after absorption by the animal, would produce steroidal amines having the amino group on opposite sides of the steroid plane which should exhibit differing degrees of teratogenicity. If, on the other hand, the alkaloid bonds to the active site in the embryo through the formyl group, the biological activity of the two isomers might be the same or different. A factor complicating the interpretation of biological results of these compounds involves their ability to undergo facile isomerization under acidic conditions by means of a ring-opened intermediate (Gaffield et al., 1983). Similar susceptibility to electrophilic attack resulted in lessened teratogenicity in the rabbit of cyclopamine, relative to jervine, due to opening of ring-E by stomach acid with subsequent formation of the non-teratogenic alkaloid, veratramine (Keeler, 1970). Parenteral routes of administration might be utilized to rule out ring opening from electrophilic attack.

## CONCLUDING REMARKS

Further structural or configurational revisions of alkaloids presently known to be teratogenic in addition to the discovery of additional teratogenic alkaloids may require modification or refinement of currently accepted structure-terata relationships. However, it is apparent that steroidal alkaloids bearing a basic nitrogen atom in ring-F, shared or unshared with ring-E, having bonding capabilities α to the steroid plane may be suspect as teratogens.

## REFERENCES

Anon. (1975). in: "National Foundation/March of Dimes: Facts," National Foundation, New York.

Anon. (1981). in: "Chemical Hazards to Human Reproduction", Council on Environmental Quality, II-5 and III-24.

Binns, W., Thacker, E. J., James, L. F. and Huffman, W. T. (1959). A congenital cyclopian-type malformation in lambs, J. Am. Vet. Med. Assoc., 134:180-183.

Binns, W., Anderson, W. A. and Sullivan, D. J. (1960). Further observations on a congenital cyclopian-type malformation in lambs, J. Am. Vet. Med. Assoc., 137:515-521.

Binns, W., James, L. F., Shupe, J. L. and Thacker, E. J. (1962). Cyclopian-type malformation in lambs, Arch. Environ. Health, 5:106-108.

Binns, W., Shupe, J. L., Keeler, R. F. and James, L. F. (1965). Chronologic evaluation of teratogenicity in sheep fed Veratrum californicum, J. Am. Vet. Med. Assoc., 147:839-842.

Brown, D. (1978). Structure-activity relation of steroidal amine teratogens, in: "Effects of Poisonous Plants on Livestock", R. F. Keeler, K. R. Van Kampen and L. F. James, eds., Academic Press, New York.

Brown, D. and Keeler, R. F. (1978a). Structure-activity relation of steroid teratogens. 1. Jervine ring system, J. Agric. Food Chem., 26:561-563.

Brown, D. and Keeler, R. F. (1978b). Structure-activity relation of steroid teratogens. 2. N-Substituted jervines, J. Agric. Food Chem., 26:564-566.

Brown, D. and Keeler, R. F. (1978c). Structure-activity relation of steroid teratogens. 3. Solanidan epimers, J. Agric. Food Chem., 26:566-569.

Gaffield, W., Wong, R. Y., Lundin, R. E. and Keeler, R. F. (1982). Structure of the steroidal alkaloid muldamine and its deacetyl derivative, Phytochemistry, 21:2397-2400.

Gaffield, W., Benson, M., Haddon, W. F. and Lundin, R. E. (1983). Evidence for the structure, and for a non-chair conformation in ring-F of two isomeric N-formylsolasodines, Aust. J. Chem., 36:325-338.

Harvey, W. (1651). "Exercitationes de Generatione Anamalium, Typis Du Gardianus, Impensis Octaviani, Pulleyn in Coemetario", Paulino, London.

Kaneko, K., Tanaka, M. W., Takahashi, E. and Mitsuhashi, H. (1977). Teinemine and isoteinemine, two new alkaloids from Veratrum grandiflorum, Phytochemistry, 16:1620-1622.

Keeler, R. F. (1968). Teratogenic compounds of Veratrum californicum (Durand) IV. First isolation of veratramine and alkaloid Q and a reliable method for isolation of cyclopamaine, Phytochemistry, 7:303-306.

Keeler, R. F. (1970). Teratogenic compounds of Veratrum californicum. X. Cyclopia in rabbits produced by cyclopamine, Teratology, 3:175-180.

Keeler, R. F. (1971). Teratogenic compounds of Veratrum californicum (Durand). XIII Structure of muldamine. Steroids, 18:741-752.

Keeler, R. F. (1975). Toxins and teratogens of higher plants, Lloydia, 38:56-86.

Keeler, R. F. (1978). Cyclopamine and related steroidal alkaloid teratogens: Their occurrence, structural relationship and biologic effects, Lipids, 13:708-715.

Keeler, R. F. (1983). Mammalian teratogenicity of steroidal alkaloids, in: "Biochemistry and Function of Isopentenoids in Plants", W. D. Nes, G. Fuller and L.-S. Tsai, eds., Marcel Dekker, New York.

Keeler, R. F. and Binns, W. (1968). Teratogenic compounds of Veratrum californicum. V. Comparison of cyclopian effects of steroidal alkaloids from the plant and structurally related compounds from other sources, Teratology, 1:5-10.

Keeler, R. F., Young, S. and Brown, D. (1976). Spina bifida, exencephaly, and cranial bleb produced in hamsters by the solanum alkaloid solasodine, Res. Commun. Chem. Path. Pharm., 13:723-730.

Kusano, G., Takemoto, T., Aimi, N., Yeh, H. J. C. and Johnson, D. F. (1975). Nitrogen inversion in solasodine N-methyl and N-formyl derivatives, Heterocycles, 3:697-701.

Lenz, W. (1962). Thalidomide and congenital abnormalities, Lancet, 1:271-272.

Lenz, W. (1966). Malformations caused by drugs in pregnancy, Am. J. Dis. Child., 112:99-106.

McBride, W. G. (1961). Thalidomide and congenital abonormalities, Lancet, 2:1358.

Miller, R. W. (1978). in: "Chemical Mutagens. Principles and Methods for Their Detection", A. Hollaender and F. J. de Serres, eds., Vol. 5, Plenum Press, New York.

Renwick, J. H. (1972). Hypothesis: Anencephaly and spina bifida are usually preventable by avoidance of a specific but un-identified substance present in certain potato tubers, J. Prev. Soc. Med., 26:67-68.

Schardein, J. L. (1976). "Drugs as Teratogens", CRC Press, Boca Raton, FL.

Seifulla, Kh. I. and Ryzhova, K. E. (1972). Effect of hydrocorti-sone, solasodine and solanine on the growth and develop-ment of fetuses of pregnant rats, Mater. Vses Konf. Issled. Lek. Rast. Perspekt. Ikh Ispolz. Proizvod. Lek. Prep. 1970, 1972:160-162; Chem. Abs. (1975). 83:22383t.

Shephard, T. H. (1973). "Catalog of Teratogenic Agents", The Johns Hopkins Univ. Press, Baltimore.

Toldy, L. and Radics, L. (1966a). Isolation of the rotational isomers of an amine, Tetrahedron Lett., 4753-4757.

Toldy, L. and Radics, L. (1966b). Steric structure of tomatidine and solasodanol by N.M.R. spectroscopy, Kem. Kozlem., 26: 247-259.

Warkany, J. (1963). Birth defects through the ages, in: "Birth Defects", M. Fishbein, ed., Lippincott, Philadelphia.

Warkany, J. (1977). History of teratology, in: "Handbook of Teratology", J. G. Wilson and F. C. Fraser, eds., Vol. 1, Plenum Press, New York.

Warkany, J. and Nelson, R. C. (1940). Appearance of skeletal ab-normalities in the offspring of rats reared on a deficient diet, Science, 92:383-384.

Wilson, J. G. (1973). "Environment and Birth Defects", Academic Press, New York.

## 12

MINIMIZING THE SAPONIN CONTENT OF ALFALFA SPROUTS AND LEAF PROTEIN

CONCENTRATES

A. L. Livingston, B. E. Knuckles, L. R. Teuber,
O. B. Hesterman, and L. S. Tsai

Western Regional Research Center, USDA, ARS, Berkeley
California 94710; and Department of Agronomy, University of California, Davis, California 95616

## ABSTRACT

Biologically active saponins were found to be concentrated
in the white and green alfalfa leaf protein fractions at levels
higher than in the original alfalfa. Coagulation and washing of
the leaf protein at pH 8.5 resulted in a fourfold decrease in
saponin compared to the protein coagulated at pH 6.0 and washed
at pH 4.5. The press cakes from high and low saponin alfalfa
retained 65 and 87% of the saponin. Protein fractions prepared
from a low saponin alfalfa contained saponin levels <0.07%,
compared to a saponin level of 1.33% in leaf protein prepared
form a high saponin alfalfa. The saponin contents of three
varieties of alfalfa sprouts ranged from 1.55 to 7.27% depending
upon the maturity of the sprouts. The saponin content rapidly
increased after sprouting and reached a maximum after eight days'
growth. Both starch and total sugar decreased with the age of
the sprouts while fiber increased from 8 to 18.7%.

## INTRODUCTION

Alfalfa leaf protein concentrates (LPC) have been shown to
be excellent sources of amino acids (Bickoff et al., 1975) and
other nutrients for swine (Cheeke, 1974; Cheeke et al., 1977a;
Cheeke and Myer, 1973) and poultry (Kuzmicky and Kohler, 1977)
and potentially for human consumption. However, the possible
concentration of naturally occurring biologically active compounds
in alfalfa sprouts and leafy protein concentrates makes it essential to develop procedures for minimizing such compounds during

253

alfalfa sprouting and processing. It has previously been shown by Knuckles et al. (1976) that LPC's could be prepared so as to contain lower levels of the forage plant estrogen, coumestrol, than were present in either the whole fresh or the pressed alfalfa. In addition to the forage estrogens, alfalfa plants and sprouts as well as other legumes have been found to be sources of biologically active saponins (Peterson, 1950; Walter et al., 1950; Livingston, 1959, 1979; Birk et al., 1963) which may inhibit chick growth (Heywang and Bird, 1953) or decrease egg production in laying hens (Anderson, 1957).

Cheeke et al., (1977b) evaluated the effects of high and low saponin alfalfa leaf protein concentrates in the diets of rats, rabbits and swine. He concluded that low saponin leaf protein concentrates support a higher growth rate of rats and swine than the high saponin protein concentrates. In the case of rats and rabbits he attributed the lower growth rate to decreased palatability of the high saponin concentrates.

The biologically active saponins in alfalfa are composed of long-chain carbohydrates attached principally to the sapogenin, medicagenic acid (Djerassi et al., 1957; Gestetner, 1971) with smaller amounts of the sapogenins, hederagenin and lucernic acid, present. (Figure 1, Shaney et al., 1972, Livingston, 1959). All three of these saponins contain carboxyl groups and therefore tend to be more soluble in alkaline than in acid solution.

Although there are also substantial amounts of the soya – saponins present in alfalfa, these do not apparently contribute to chick growth or egg production inhibition and are not precipitable with cholesterol. (Reshef, et al., 1976) The ability to complex with cholesterol also apparently accounts for the fungistatic activity of the saponins of both medicagenic acid and hederagenin.

Recently Malinow et al. (1982, 1977a, 1977b, 1977c) in a cooperative study with our laboratory, the Western Regional Research Center of the United States Department of Agriculture, demonstrated that antiatherosclerotic effects of alfalfa meal in monkeys was due to the inhibition of cholesterol absorption by alfalfa saponins.

In the course of our study on saponin in leaf protein concentrates we had occasion to examine a number of common vegetable food products regarding their saponin contents. We were impressed by the very high saponin contents of commercially grown alfalfa sprouts found in the market place and accordingly have made further study on possible methods to minimize the saponin in alfalfa

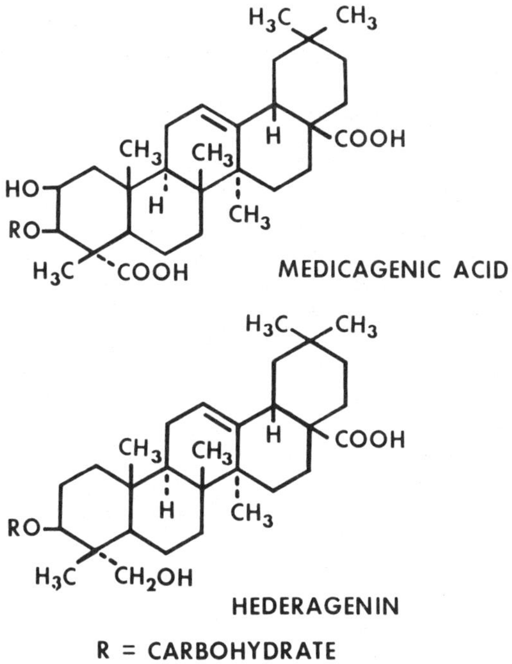

Figure 1.   Chemical structures of two saponins.

sprouts. These results are also included in the manuscript. Previously we defined sprouting conditions which produced sprouts of both high nutritional quality and fresh weights (Hesterman, 1981). A similar environment was employed for producing the sprouts in this study.

EXPERIMENTAL

At our laboratory we have studied the distribution of the biologically active saponins during the preparaton of LPC by two different processes. In one process (Pro-Xan process) the soluble and insoluble proteins are precipitated together (whole LPC) to provide a chlorophyll-pigmented protein concentrate. In the second process (the Pro-Xan II Process) both a chlorophyll-pigmented (green LPC) and a white protein (white LPC) concentrate are prepared separately (de Fremery et al, 1973; Edward et al. 1975).

For our leaf protein study high and low saponin cultivars of the Ranger variety developed by Pederson and Wang (1971) and Pederson et al. (1967) were grown at the University of California (Davis) farm. The fresh alfalfa was ground at its natural pH (pH 5.8-6.1) and pressed with a twin-screw press to provide a press cake and a whole green juice. The juice was later adjusted to pH 8.5 by the addition of concentrated $NH_4OH$. The whole LPC was prepared by heating 1500 gram portions of the green juice to 80-85°C by direct steam injection, holding for 5 minutes, then cooling to room temperature. The coagulated protein curd was transferred to a nylon filter bag, pressed to 35 to 40% solids, and freeze-dried (Edward et al., 1977). The brown juice from the pressed green curd was freeze-dried for saponin analyses.

Fractionated LPC was prepared by heating the whole green juice to 60°C for 20 sec. by steam injection and cooling immediately (deFremery et al., 1973). A weighed portion of the heated juice was centrifuged at 0°C and 27000g for 20 minutes to separate the green LPC from the white LPC. The clear liquid was decanted from the green solids and heated on a steam bath to 85°C for 5 min., then rapidly cooled in an ice bath. The coagulated white LPC was cooled, collected by centrifugation, washed with three portions of water and then freeze-dried.

In some studies, the pH of coagulation and of the washing of the white LPC was varied to determine the effects upon saponin distribution. The alfalfa solubles from the white LPC preparation were also freeze-dried.

The finding of a high level of saponin in a commercial alfalfa sprout sample prompted us to examine other alfalfa sprout
samples and to investigate some growing conditions to minimize
the saponin content of sprouts. Accordingly: three alfalfa
cultivars representing a range in fall dormancy were used in
this study. Ranger was the dormant cultivar, Calverde was the
semidormant cultivar, and Moapa was the nondormant cultivar.
Conditions for growing the sprouts were chosen as conditions
that are known to be frequently used by commercial alfalfa sprout
growers. Constant values were 12 hours light 21C, 8 liters of
water applied in 24 hours per 500 grams of seed, water applied
at 2 hour intervals, and harvest at 2, 4, 6 and 8 days. The
seeds were also analyzed biologically for saponin content.

Seed for each experiment was placed in a glass jar and soak-
ed for 6 hours, after which excess water was decanted. Seed was
left in the jar for an additional 24 hours and once during this
period the germinating seed was watered and excess water decanted.
The partial germinating seed was then placed in pots with drainage
holes in the bottom, and the pots were placed in a growth chamber.
The equivalent of 11g of dry seed was placed in each pot. Water
was applied by an overhead mist system controlled by a timer.
In each trial containers were placed in the growth chamber in
randomized design. Cultivars were replicated four times in each
trial.

Saponin analysis of these products and of the alfalfa sprouts
was by the procedure of Livingston et al. (1977). The samples
were assayed in duplicate at three levels using a Trichodernia
bioassay, comparing the results with a standard saponin, calcula-
ting saponin by means of a slope ratio analysis.


RESULTS AND DISCUSSION

Table 1 presents the saponin content of alfalfa products
prepared from high and low saponin cultivars. Saponin contents
of the whole LPCs from either low or high saponin alfalfa were
higher than the respective green LPCs, but in both instances the
saponin contents were highest in the white LPCs coagulated and
washed at pH 6.0. By coagulation and washing at pH 8.5 the
saponin contents of the white LPCs were greatly reduced. The
saponin content of the white LPC prepared at pH 8.5 from high or
low saponin alfalfa was actually lower than that of the initial
whole alfalfa. The apparent problem of the saponin complexing
with protein can be largely solved by coagulation and washing of
the protein at an alkaline pH (8.5).

Table 1.    SAPONIN CONTENT (DRY BASIS) OF PRODUCTS FROM THE

PREPARATION OF FRACTIONATED LPC FROM LOW AND HIGH SAPONIN ALFALFA

| Product | Low saponin alfalfa saponin content, % | High saponin alfalfa saponin content, % |
|---|---|---|
| Whole alfalfa | .14 | 1.71 |
| Pressed alfalfa | .13 | .66 |
| Whole LPC (Pro-Xan)[a] | .10 | 1.90 |
| Alfalfa solubles[a] | .06 | .65 |
| Green LPC (Pro-Xan II)[a] | .08 | 1.54 |
| White LPC[a] | .21 | 2.88 |
| Alfalfa solubles[a] | .06 | .47 |
| White LPC[b] | .12 | .97 |
| Alfalfa solubles[b] | .20 | 1.17 |

[a]From coagulation and washing pH 6.0.

[b]From coagulation and washing pH 8.5.

    As shown in Table 2 most of the dry matter and saponin re-
mained in the whole alfalfa.  Although the white LPCs coagualted
at pH 6.0, contained only 1.5 and 1.6% of the dry matter, these
concentrates did contain higher proportions of recovered saponin
than the other protein products.  Preparation of the white LPC
from high saponin alfalfa by coagulation and washing at pH 8.5
greatly reduced the proportion of saponin in the protein, leaving
most of the saponin in the alfalfa solubles.

Table 2.  DISTRIBUTION OF SAPONIN AND DRY MATTER (DRY BASIS)

DURING THE PREPARATION OF FRACTIONATED LPC FROM LOW AND HIGH

SAPONIN ALFALFA

| Product | Low saponin alfalfa | | High saponin alfalfa | |
| --- | --- | --- | --- | --- |
| | Dry matter Yield, wt. % | Saponin % of recov. | Dry matter Yield, wt. | Saponin % of recov. |
| Press cake | 80.3 | 86.8 | 76.0 | 65.0 |
| Green LPC[a] | 8.7 | 5.8 | 11.1 | 22.1 |
| White LPC[a] | 1.5 | 2.7 | 1.6 | 6.0 |
| Alfalfa solubles[a] | 9.5 | 4.7 | 11.3 | 6.9 |
| White LPC[b] | | | 1.9 | 2.2 |
| Alfalfa solubles[b] | | | 10.5 | 15.0 |

[a]From coagulation of pH 6.0.

[b]From coagulation of pH 8.5.


The effects of varying the pH during coagulation and washing
upon the saponin content of white LPC is presented in Table 3.
As would be expected, due to the carboxyl groups on the biologically active saponins and their increased solubility in alkaline
solutions, the white LPC prepared by coagulating  and washing at
pH 8.5 contained less saponin than the products prepared at
lower pHs.  It would seem that the pH of coagulation is the most
important since the white LPC coagulated at pH 8.5 contained
less than half the saponin of the white LPC coagulated at pH 6
and then washed at pH 8.5.

Tung et al. (1977) prepared alfalfa whole dried juice from
high and low saponin alfalfas for feeding experiments using day-
old chicks.  Unlike most LPCs which are prepared by coagulating
and separating the protein from the juice prior to drying these
preparations contained all of the soluble components of the
leaves as well as the proteins.

Table 3.  EFFECT OF PH OF COAGULATION AND WASH UPON THE SAPONIN
CONTENT OF WHITE LPC[a]

| pH of coagulation | pH of wash | Saponin content, %[b] |
|---|---|---|
| 6.0 | 4.5 | 1.09 (.92 - 1.28)[c] |
| 6.0 | 6.0 | .55 (.29 -  .81) |
| 6.0 | 8.5 | .57 (.40 -  .75) |
| 8.5 | 4.5 | .46 (.29 -  .63) |
| 8.5 | 6.0 | .47 (.30 -  .64) |
| 8.5 | 8.5 | .24 (.20 -  .29) |

[a]Saponin content of the whole alfalfa was 0.25%, 95% confidence
limits, 0.20 - .29%.

[b]Dry basis.

[c]Numbers in parenthesis are 95% limits.

In this study there was more than a 50% reduction in weight gain
of the chicks fed the dried alfalfa juice prepared from the
first cutting of high saponin alfalfa and nearly the same reduction in the gain of the chicks using the second cutting study.

Preliminary poultry feeding studies at our laboratory and
cooperative swine feeding trials demonstrated that the growth
inhibition effects of saponins in alfalfa LPC is not serious and
most of the soluble solids are removed during processing. Kuzmicky et al. (1971) (Table 4) found that there were no significant
differences in chick weight gain from any of the rations containing corn-soy only, 27% LPC or 54% LPC. Although the 54% LPC
produced wet excreta, the addition of cholesterol did not increase
the gain.

Myer, et. al. (1975) replaced soybean meal by LPC in the
diets of growing finishing swine. In one study LPC was included
at levels of 0, 5, 10, 15, 20 and 24% in a barley-based diet.
Growth at all levels was excellent, indicating that alfalfa leaf

Table 4.   CHICK WEIGHT GAIN, AND FEED EFFICIENCY (GAIN/FEED)[1]

| Ration | Average Gain/Chick[2] | Feed Efficiency[3] |
|---|---|---|
| | Grams | |
| Corn-soy | 578[a] | 0.701[a] |
| 5% Pro-Xan | 577[a] | .701[a] |
| 10% Pro-Xan | 567[a] | .684[b] |
| 20% Pro-Xan | 569[a] | .669[c] [d] |
| 20% Pro-Xan + Lysine | 557[a] | .68[b] [c] |
| 20% Pro-Xan + Lysine + Cholesterol | 581[a] | .67[b] [c] [d] |
| 20% Pro-Xan + K + Mg | 578[a] | .668[d] |

[1]From Kuzmicky et al. 1971.

[2]Day-old white-rock cockerels were fed a corn-soy ration for 7 days followed by the experimental ration for 19 days.
There were 4 replicates of 7 chicks per replicate for each ration. Average chick starting weight, 126 grams.

[3]Means not followed by the same letter are significantly different.

protein can be used to replace soy-bean meal in swine rations.

The saponin contents of a number of vegetable food products are presented in table 5. Most of the food products contained levels of saponin similar to that in the white LPC. However, the commercial sample of alfalfa sprouts contained a level of saponin far exceeding any of the food products or any of high saponin alfalfa meal samples analyzed at the laboratory. This apparant high saponin content of the alfalfa sprouts may be due in part to the shortened carbohydrate side chain attached to the medicagenic acid aglycone. The alfalfa sprouts bearing a resemblance to alfalfa roots in this respect. Further studies will be needed to confirm this possibility.

Table 5.   SAPONIN CONTENT OF PLANT FOOD PRODUCTS

AND WHITE LPC[a]

| Product[b] | Saponin[c] found, % | |
|---|---|---|
| White LPC | 0.07 | (0.05 - 0.09)[d] |
| Alfalfa sprouts | 7.93 | (6.58 - 9.49) |
| Spinach leaves (frozen) | 0.07 | (0.06 - 0.08) |
| Green beans (frozen) | 0.05 | (0.04 - 0.06) |
| Red beans | 0.02 | (0.01 - 0.03) |
| Brussel sprouts (frozen) | 0.03 | (0.02 - 0.04) |
| Peas (dry) | 0.03 | (0.02 - 0.04) |
| Black eyed peas (dry) | 0.03 | (0.02 - 0.04) |
| Soyabeans (dry) | 0.01 | |
| Bean sprouts | 0.01 | |
| Grean peas (frozen) | 0.01 | |
| Snow peas (frozen) | 0.01 | |

[a]From Livingston et. al. 1979.

[b]Fresh and frozen samples were freeze-dried prior to analysis.

[c]Dry weight basis.

[d]Numbers in parentheses are 95% confidence limits

Table 6. SAPONIN CONTENT OF ALFALFA MEALS

| Variety and Source | Drying Method | Saponin Found, % | 95% Conf. Limits, % |
|---|---|---|---|
| Lahontan-Kansas | dehydrated | 0.19 | 0.16-0.25 |
| Northrupt King Thor-Kansas | dehydrated | 0.67 | 0.54-0.86 |
| California Common | dehydrated | 0.21 | 0.09-0.34 |
| Kansas Common | dehydrated | 0.58 | 0.46-0.75 |
| Ranger-Nebraska | dehydrated | 0.55 | 0.42-0.73 |
| Lahontan-Nevada | suncured | 0.37 | 0.25-0.52 |
| Lahontan-California | dehydrated | 0.16 | 0.12-0.20 |
| Lahontan-California | Freeze-dried | 0.11 | 0.08-0.14 |
| Ranger-Low Saponin | dehydrated | 0.14 | 0.12-0.16 |
| Ranger-High Saponin | dehydrated | 1.71 | 1.33-2.11 |

In addition to minimizing the possible saponin content by coagulation and washing the LPC at an alkaline pH of 8.5, it is also desirable to use alfalfa varieties known to be low in saponin content. The saponin analyses of a number of commercially grown alfalfa varieties form various parts of the United States is presented in Table 6. The wide range of saponin contents emphasizes the importance of conducting saponin analyses on alfalfa plants intended for use in preparing LPC and utilizing only those found to be low in the biologically active saponins.

Table 7 presents the saponin contents of alfalfa sprouts grown from three varieties of alfalfa seeds. The saponin contents of all three varieties increased rapidly after sprouting. Moapa contained the highest level of more than 7% reached after 6 days of sprouting. In order to avoid the highest level it seems that sprouts should not be consumed after more than 4 days of sprouting. Varietal differences in saponin content of the sprouts would seem to be only slight with both Caliverde and Ranger having similar saponin contents after a similar time of sprouting.

Table 7.  SAPONIN CONTENT OF ALFALFA SPROUTS

| Variety | Maturity | Saponin | 95% Confidence Limits |
|---------|----------|---------|------------------------|
|         | Days     | %       | %                      |
| CALIVERDE | 0 | 0.14 | 0.09 - 0.18 |
|           | 2 | 1.78 | 1.48 - 2.09 |
|           | 4 | 2.11 | 1.80 - 2.43 |
|           | 6 | 3.00 | 2.67 - 3.36 |
|           | 8 | 3.59 | 3.23 - 3.99 |
| MOAPA     | 0 | 0.12 | 0.08 - 0.15 |
|           | 2 | 2.01 | 1.69 - 2.37 |
|           | 4 | 3.24 | 2.74 - 3.62 |
|           | 6 | 7.27 | 5.81 - 8.85 |
|           | 8 | 7.24 | 5.70 - 8.82 |
| RANGER    | 0 | 0.11 | 0.08 - 0.15 |
|           | 2 | 2.25 | 1.84 - 2.76 |
|           | 4 | 2.24 | 1.83 - 2.75 |
|           | 6 | 2.78 | 2.02 - 3.61 |
|           | 8 | 3.14 | 2.38 - 4.00 |

In addition to saponin contents we also analyzed the alfalfa sprouts for proximate contents. These data are presented in Tables 8 and 9. Starch and total sugar decreased while fiber increased due to photosynthesis during sprouting. The largest decrease in total sugar occurred between 4 and 6 days after sprouting began.

There was an increase in percentage protein after sprouting 6-8 days (Table 9). There was also a decrease in fat and an

increase in ash.  This decrease in ash may be due to the leaching
of soluble solids during watering.

Although alfalfa saponins were found to be non-toxic in Cyno-
mologus Macaques by Malinow et al. (1982a), and were very effective in lowering plasma cholesterol levels and decreasing athero-
genesis, this laboratory did subsequently report that both alfalfa
seeds and sprouts induced systemic lupers erythematosus (anemia)
in non-human primates (Malinow, 1982b; Bardana, 1982).  This
syndrome in monkeys was shown to be due to the presence of L-canavanine sulfate in the alfalfa seeds and sprouts.  This compound
produced a similar deleterious effect upon a human consuming
high levels of alfalfa seeds (Malinow, 1981).  Accordingly,
while the saponin contents of alfalfa sprouts may be minimized
by growing the sprouts to only 4 days maturity, the presence of
a proven deleterious compound such as L-canavanine sulfate in
the sprouts suggest a hazard in humans consuming large quantities
of sprouts as a health food.

Table 8.  PROXIMATE COMPOSITION OF ALFALFA SPROUTS

| Variety | Maturity Days | Wt. g | Total Sugar | Starch % | Fiber |
|---|---|---|---|---|---|
| CALIVERDE | 2 | 9.00 | 7.60 | 3.13 | 10.97 |
|  | 4 | 8.24 | 6.38 | 2.81 | 13.47 |
|  | 6 | 7.85 | 4.10 | 2.16 | 15.27 |
|  | 8 | 7.99 | 2.25 | 2.16 | 17.97 |
| MOAPA | 2 | 9.31 | 8.30 | 4.32 | 10.89 |
|  | 4 | 8.99 | 7.28 | 2.66 | 13.41 |
|  | 6 | 8.07 | 4.40 | 2.20 | 15.27 |
|  | 8 | 7.88 | 2.20 | 1.94 | 16.21 |
| RANGER | 2 | 9.05 | 6.75 | 3.91 | 10.97 |
|  | 4 | 8.10 | 5.80 | 2.38 | 14.83 |
|  | 6 | 8.30 | 3.60 | 2.16 | 14.63 |
|  | 8 | 8.00 | 1.50 | 1.73 | 17.19 |

Table 9.   PROXIMATE COMPOSITION OF ALFALFA SPROUTS

| Variety | Maturity Days | Protein | Fat % | Ash | CA | P |
|---------|---------------|---------|-------|-----|-----|-----|
| CALIVERDE | 2 | 41.41 | 9.97 | 3.65 | .27 | .67 |
| | 4 | 41.80 | 7.03 | 4.15 | .32 | .65 |
| | 6 | 42.33 | 5.69 | 5.30 | .43 | .78 |
| | 8 | 43.57 | 4.39 | 6.29 | .53 | .68 |
| MOAPA | 2 | 41.94 | 7.78 | 3.92 | .27 | .72 |
| | 4 | 41.78 | 6.57 | 4.62 | .32 | .78 |
| | 6 | 40.93 | 4.44 | 4.94 | .38 | .76 |
| | 8 | 43.81 | 2.75 | 6.42 | .47 | .84 |
| RANGER | 2 | 38.93 | 8.04 | 3.90 | .28 | .69 |
| | 4 | 38.31 | 5.35 | 5.04 | .38 | .82 |
| | 6 | 41.50 | 3.28 | 5.55 | .42 | .84 |
| | 8 | 41.21 | 2.75 | 5.78 | .44 | .81 |

REFERENCES

Anderson, J. O., Poultry Sci., 36:873 (1957).
Berdana, E. J., Malinow, M. R., Houghton, D. C., McNulty, W.P.,
        Kirk, D., Wvepper, K. D., Parker, F., Piroysky, B.,
        Amer. J. Kidney Diseases, 1:345 (1982).
Bickoff, E. M., Miller, R. E., Saunders, R. M., Kohler, G. O.,
        "Protein Nutritional Quality, of Foods and Feeds",
        vol. 1, part 2, Friedman, M. Ed., Marcel Dekker, New
        York, N.Y., 1975, p. 319.
Birk, Y., Bondi, A., Gestetner, B., Ishea, I., Nature (London)
        197:1089 (1963).
Cheeke, P. R., Nutr. Rep. Int., 9:267 (1974).
Cheeke, P. R., Kinzell, J. H., deFremery, D., Kohler, G. O.,
        J. Animal Sci., 64:476 (1977a).

Cheeke, P. R., Kinzell, J. H., Pederson, M.W., J. Anim. Sci., 64:476 (1977b).

Cheeke, P. R., Myer, R.O., Feedstuffs, 45:(51),24 (1973).

deFremery, D., Miller, R. E., Edwards, R. H., Knuckles, B. E., Bickoff, E. M., Kohler, G. O., J. Agric. Food Chem., 21:886 (1973).

Djerassi, C., Thomas, D. B., Livingston, A. L., Thompson, C. R., J. Am. Chem. Soc., 79:5292 (1957).

Edwards, R. H., Miller, R. E., deFremery, D., Knuckles, B. E., Bickoff, E. M., Kohler, G. O., J. Agric. Food Chem., 23:620 (1975).

Gestetner, B., Phytochemistry, 10:2221 (1971).

Hesterman, O. O. B., Teuber, L. R., Livingston, A. L., Crop Sci., 21:720 (1981).

Heywang, B. W., and Bird, H. R., Poultry Sci., 38:968 (1954).

Knuckles, B. E., deFremery, D., Kohler, G. O., J. Agri. Food Chem., 24:1177 (1976).

Kohler, G. O., Bickoff, E. M., Spencer, R. R., Witt, S. C., Knuckles, B. E., Tech. Alfalfa Conf. Proc., 10th ARS-74-46, 1968, p. 71.

Kohler, G. O., and Knuckles, B. E., Food Tech., 1977, p. 191.

Kuzmicky, D. D., and Kohler, G. O., Poultry Sci., 56:1510 (1977).

Kuzmicky, D. D., Livingston, A. L., Knowles, R. E., Kohler, G. O., Guenther, E., Olson, E. E., Carlson, C. W., Poultry Sci., 56:1504 (1977).

Kuzmicky, D. D., Kohler, G. O., Bickoff, E. M., 11th Tech. Alfalfa Conf. Proceedings, 1971, p. 58.

Livingston, A. L., J. Org. Chem., 24:1567 (1959).

Livingston, A. L., Knuckles, B. E., Edwards, R. H, deFremery, D., Miller, R. E., Kohler, G. O., J. Ag. Food Chem., 27:362. (1979).

Livingston, A. L., Whitehand, L. C., Kohler, G. O., J. Assoc. off. Anal. Chem., 60, 957 (1977).

Malinow, M. R., Bardana, E. J., Jr., Pirofsky, B., Craig, S., and McLaughlin, P., Science, 216:415 (1982a).

Malinow, M. R., McLaughlin, P., Kohler, G. O., and Livingston, A. L., Steroids, 29:105 (1977a).

Malinow, M. R., McLaughlin, P., Kohler, G. O., and Livingston, A. L., Clin. Res., 25:A97 (1977b).

Malinow, M. R., McLaughlin, P. Stafford, C., Kohler, G. O., and Livingston, A. L., Int. Conf. on Atherosclerosis, Carlson, L. A., Ed., Raven Press, N.Y., 1978, p. 183.

Malinow, M. R., McNulty, W. P., Houghton, D. C., Kessler, S., Stenzel, P., Goodnight, S. H., Jr., Bandana, E. J., Jr., Palotay, J. L, McLaughlin, P., and Livingston, A. L., J. Med. Primatol., Primatol., 11:106 (1982b).

Myer, R. O., Cheeke, P. R., and Kennick, W. H., J. Anim. Sci., 40:885 (1975).

Pederson, M. W., Zimmer, D. E., McAllister, D. R., Anderson, J. O., Wilding M. D., Taylor, G. A., and McGuire, C. F., Crop Sci., 7:349 (1967).

Pederson, M. W., and Wang, L. C., Crop Sci., 11:833 (1971).

Peterson, D. W., J. Biol. Chem., 188:647 (1950).

Reshef, G., Gestetner, B., Birk, A., and Bondi, A., J. Sci. Food Agric., 27:63 (1976).

Shaney, S., Gestetner, B., Birk, Y., Bondi, A., and Kirson, I., Isreal J. Chem., 10:881 (1972).

Tung, J. Y., Straub, R. J., Scholl, J. M., and Sunde, M. L, paper no. 77-1010, presented at Annual Meeting of the American Society of Agric. Engineers, North Carolina State Univ. (1977).

Walter, E. D., Van Atta, G. R., Thompson, C. R., and Maclay, W. D., J. Am. Chem. Soc., 76:2271 (1954).

13

THE INFLUENCE OF 1-(N-L-TRYPTOPHAN)-1-DEOXY-D-FRUCTOSE [FRU-TRP]
AND ITS N-NITROSATED ANALOGUE [NO-FRU-TRP] ON THE VIABILITY AND
INTRACELLULAR SYNTHETIC ACTIVITY (DNA, RNA, AND PROTEIN SYNTHESIS)
OF HeLa S3-CARCINOMA CELLS

D. W. Gruenwedel, S. C. Lynch and G. F. Russell

Department of Food Science and Technology
University of California, Davis, CA 95616

ABSTRACT

Exposing HeLa S3 cells at 37°C to varied concentrations of,
respectively, Fru-Trp (0.1 µM – 1 mM), NO-Fru-Trp (0.1 µM – 1 mM),
and $NaNO_2$ (0.6 µM – 6 mM) for varied periods of time (1 – 36 hr)
does neither affect their viability (trypan blue dye exclusion
test) nor capability to synthesize RNA or protein but is of considerable influence on DNA synthesis in the case of NO-Fru-Trp and
$NaNO_2$, but not in the case of Fru-Trp which continues to be inef-
fective.  None of the three compounds tested is of significant
influence on cell number.  Both NO-Fru-Trp and $NaNO_2$ stimulate DNA
synthesis: a maximum of activity ([$^3$H]thymidine incorporation)
exists at the 24 hr time point of incubation, with NO-Fru-Trp, for
instance, generating a 2.5-fold increase (over control) at 1 mM
concentration in the medium while $NaNO_2$, at comparable concentration, increases DNA synthesis by a factor of 1.6 over control.
The increase in DNA synthesis is not due to stimulatory influences
on (semi-conservative) DNA replication but represents DNA
repair.  This was verified by keeping the cells under conditions
that prevent normal (semi-conservative) replication but permit
repair ("unscheduled DNA synthesis").  Two major routes are suggested by which NO-Fru-Trp could impart DNA damage and, thus,
assume mutagenic properties.

INTRODUCTION

Amino acids react readily with aldoses and ketoses upon
heating to yield 1-(N-carboxyalkyl(aryl)-amino)-1-deoxy-2-keto-
D-hexoses (Amadori compounds) and 2-(N-carboxyalkyl(aryl)-amino)-
2-deoxy-aldo-D-hexoses (Heyns compounds), respectively (Paulsen

and Pflughaupt, 1980). They are key intermediates of the so-
called Maillard reaction (non-enzymatic browning), a reaction that
contributes to the aroma, flavor, and color of many foods. Although Borsook and collaborators postulated already thirty years
ago (Borsook et al., 1952) that Amadori compounds ("fructose-amino
acids") are involved in the incorporation of amino acids into
proteins, and both Amadori-type and Heyns-type rearrangements have
been proposed to partake in the biosynthesis of certain amino
acids (cf., Lingens and Lueck, 1963; Davidson, 1966), up to now
the consensus has been that neither Amadori nor Heyns compounds
are of any true biochemical significance and that they are of
interest chiefly to the food scientist and nutritional scientist
concerned with the wholesomeness of foods. As of late, however,
there exists increasing evidence to suggest that Amadori
compounds, for instance, do play a pivotal role in vivo: the minor
hemoglobin component $HbA_{1c}$, a post-translational modification of
hemoglobin $HbA_o$ formed via an Amadori-rearranged addition of
glucose to the $\beta$ chain, is a case in point (Rhabar, 1968;
Flueckiger and Winterhalter, 1976; Shapiro et al., 1979), and the
non-enzymatic glycosylation of lens proteins in the development of
the sequelae of diabetes mellitus (Cerami et al., 1979), or the
involvement of Maillard reactions in aging (Monnier et al., 1979),
can be cited as further examples. Mester and coworkers have
directed attention to the possibility of Amadori-type reactions
taking place in hemostasis and tryptaminergic neural activity
(Mester et al., 1981).

Amadori or Heyns compounds are secondary amines and they are
therefore susceptible to nitrosation. As shown by Coughlin (1979)
and by Heyns et al. (1979), they undergo indeed a surprisingly
easy and quantitative conversion to their N-nitroso derivatives
when incubated with sodium nitrite under mildly acidic conditions
and at temperatures of physiological interest. The facility of
the conversion leads of course immediately to the question of the
health risks posed by the reaction: as is well known (cf., Committee on Nitrite and Alternative Curing Agents in Food, 1981;
Scanlon and Tannenbaum, 1981), many N-nitroso compounds display
mutagenic and/or carcinogenic characteristics and it is therefore
important also to evaluate nitrosated Amadori or Heyns compounds
in this regard.

Coughlin and coworkers (Coughlin, 1979; Coughlin et al.,
1979) found that the Amadori compound nitrosated 1-(N-L-tryptophan)-1-deoxy-D-fructose, NO-Fru-Trp[*], exhibits mutagenic

---

*Abbreviations: Fru-Trp ≡ 1-(N-L-tryptophan)-1-deoxy-D-fructose;
NO-Fru-Trp ≡ nitrosated Fru-Trp. This designation deliberately
avoids the identification of any of the three possible N-nitroso
derivatives of Fru-Trp.

activity, particularly in absence of metabolic activation, when
incubated with the S. typhimurium strains TA98 and TA100 (Ames
test). We decided to extend their study by looking also at the
mutagenic potential of NO-Fru-Trp in a short-term test employing
cultured human cells. We have used the established cell line HeLa
S3 as test organism and we have looked at the effects of NO-Fru-
Trp, Fru-Trp, and NaNO$_2$ on the cells' viability and intracellular
synthetic activity (DNA, RNA, and protein synthesis) as well as
number. The results of the two divergent short-term assays -
prokaryotic (Ames test) versus eukaryotic - should reveal the
existence of a false positive substance.


METHODS

Chemicals. Sodium nitrite was purchased from J. T. Baker Chemical
Co. (Phillipsburg, NJ). L-tryptophan was obtained from Sigma
Chemical Co. (St. Louis, MO) and anhydrous D-glucose from
Mallinckrodt, Inc. (St. Louis, MO). All chemicals were of analytical grade. L[4,5-$^3$H]leucine (85 Ci/mmole), [methyl-$^3$H]thymi-
dine (47 Ci/mmole), and [5-$^3$H]uridine (29 Ci/mmole) were purchased
from Amersham/Searl (Arlington Heights, IL).

Preparation of 1-(N-L-tryptophan)-1-deoxy-D-fructose (Fru-Trp).
The method of Sgarbieri et al. (1973) was employed in preparing
the compound. 40 g of D-glucose (0.2 mole) and 5 g of L-tryptophan (0.02 mole) were dissolved in 650 ml of absolute methanol and
refluxed under stirring for 4.25 hr. The resulting brownish
solution was concentrated under reduced pressure in a rotatory
evaporator and the syrup (about 100 ml) subjected to column
chromatography on Cellex 410 cellulose powder (Bio-Rad Laboratories, Richmond, CA). Water-saturated n-butanol was used as
eluent. Pale yellow crystals were collected from fractions 43-47
of a 65-fraction long run; they were purified further by crystallization from water-saturated n-butanol followed, lastly, by
crystallization from methanol. The final product (white crystals)
produced a single ninhydrin-positive spot in thin-layer chromatography, with $R_f$ = 0.48 (n-butanol:acetic acid:water = 4:1:1,v/v)
and $R_f$ = 0.68 (n-propanol:water = 7:3, v/v), void of any free
L-tryptophan. It emerged further as a single, tryptophan-free
symmetrical peak from an amino acid autoanalyzer, migrating
between the amino acid standards L-leucine and L-tyrosine.
Lastly, application of 3% triphenyltetrazolium chloride to a thin-
layer chromatography run revealed the absence of free fructose
from the Amadori compound.

Preparation of Nitrosated 1-(N-L-tryptophan)-1-deoxy-D-fructose
(NO-Fru-Trp). The protocol of Coughlin (1979) was used. In
brief, this involved dissolving a given amount of Fru-Trp in
water, adjusting the solution pH to 4 with 1 N HCl, and adding to

it the acidified (pH 4, 1 N HCl) solution of $NaNO_2$, whereby nitrite was always present in six molar excess over Fru-Trp. The mixture was then kept for 24 hr at 37°C, making sure that nitrosation took place in complete darkness (reaction vessel was wrapped with aluminum foil). After that, the solution pH was adjusted to 7 with 1 N NaOH and the mixture used as such in the bio-assays. The purpose of the six-fold excess of nitrite was to prevent dissociation of NO-Fru-Trp into Fru-Trp and $NaNO_2$ or $HNO_2$.

## Bio-Assays

Stock Cell Culture. HeLa S3 cells were obtained as monolayer cultures from the Cell Culture Laboratory of the School of Public Health, University of California, Berkeley, CA. The cells were adapted to and maintained in suspension (asynchronous) by passaging them routinely in Eagle's minimum essential medium (MEM-Joklik modified, Grand Island Biological Co. (GIBCO), Grand Island, NY). The medium was supplemented with 10% fetal bovine serum (ReHeis Chemical Co., Phoenix, AR). The cells were kept in logarithmic growth phase by maintaining the cell density at 0.3-1.2 X $10^6$ cells/ml. Cell growth took place in water-saturated 95% air-5% $CO_2$ at 37°C. Further details can be found elsewhere (Gruenwedel and Cruikshank, 1979).

Intoxication and Precursor Labeling Studies. Cells were removed from a stock cell culture by centrifugation. They were incubated anew with fresh medium at a density of 0.7 X $10^6$ cells/ml and, after gassing with 5% $CO_2$, kept for 18 hr at 37°C. They were then diluted with fresh medium to a density of 0.3 X $10^6$ cells/ml and divided into a number of 7 ml aliquots, each aliquot containing the same number of cells. Four test series were formed: three receiving varied concentrations of NO-Fru-Trp (always in presence of the six-fold excess of $NaNO_2$), Fru-Trp, and $NaNO_2$, respectively, and the fourth receiving medium as control. Stock solutions of NO-Fru-Trp, Fru-Trp, and $NaNO_2$ were prepared in such a way that the addition of 70 µl of stock to 7 ml of cell suspension yielded the desired final concentration. Final test concentrations ranged from pX 7 to pX 3 in the case of NO-Fru-Trp and Fru-Trp whereby pX $\equiv$ - log [substance]$_{added}$. Concentrations were varied in 0.5 pX increments. With $NaNO_2$, pX values ranged from 6.22 to 2.22 (also in 0.5 pX increments) in view of the six-fold excess of $NaNO_2$ present at each NO-Fru-Trp level (vide supra). Each series, after gassing with 5% $CO_2$, was incubated for periods lasting for 1, 3, 6, 12, 24, and 36 hr.

For cell viability determinations, a 1 ml portion of cell suspension was removed at the end of each incubation period from each aliquot and the cells subjected to trypan blue staining as described elsewhere (Gruenwedel and Fordan, 1978). Viability assays were performed in duplicate.

For precursor labeling, each test series was prepared in sets of three so as to permit the simultaneous monitoring of DNA, RNA, and protein syntheses. Precursor labeling always took place during the last 30 minutes of a given incubation period. Labeling was performed in triplicate. Further details can be found elsewhere (Gruenwedel and Cruikshank, 1979).

The effect of the three test substances on cell number was studied by maintaining cells for a total of 96 hr in the presence of 1 mM Fru-Trp, 1 mM NO-Fru-Trp plus 6 mM $NaNO_2$, and 6 mM $NaNO_2$, respectively. One-ml aliquots were removed after 1, 6, 12, 24, 36, 48, 60, 72, 84, 96 hr of incubation and assayed for the number of live cells. Cells were fed in 48 and 72 hr intervals by resuspending them in fresh medium, containing the test substances at the concentrations listed above, at the densities they had reached just prior to feeding.

"Unscheduled DNA Synthesis" (DNA Repair). The procedure of Trosko and Yager (1974) was used with modifications. Cells were removed from a stock cell culture as described above and resuspended in an arginine-deficient medium. This medium was prepared by using the GIBCO MEM amino acid kit (100X, lyophilized), leaving out L-arginine, the MEM vitamins kit (100X, lyophilized), and the K-14 spinner salt solution. Fetal bovine serum was used at a level of 2.5%. The cells were adapted to the deficient medium over a period of 66 hr, with medium change taking place at 48 hr. Incubation occurred at 37°C in 5% $CO_2$. Cell density was kept near 0.7 X $10^6$ cells/ml. At the beginning of the repair assay, the cells were placed into fresh deficient medium (fortified with 2.5% fetal bovine serum) at a density of 0.3 X $10^6$ cells/ml, divided into 5 ml aliquots, in sets of three, and the test substances NO-Fru-Trp, Fru-Trp, and $NaNO_2$ added in 50 µl portions to the respective aliquots. Each set, after renewed gassing with 5% $CO_2$, was incubated for 1, 12, 24, and 36 hr, respectively. One hour before the lapse of a given incubation period, a 1 ml portion of hydroxyurea, dissolved in water, was added to each vial. The final hydroxyurea concentration amounted to 10 mM. Incubation was continued for the remaining 60 minutes, after which [$^3$H]thymidine was added at levels of 1 µCi/vial and incubation continued for another 30 minutes in an atmosphere of 5% $CO_2$. Further cell treatment followed the protocol of Gruenwedel and Cruikshank (1979). Repair synthesis was assayed in triplicate.

Results

Studies in Complete Medium. As can be seen in Fig. 1, none of the substances tested exhibits cytotoxic effects and thus, at first glance, they do not appear to affect cell metabolism. Plotted in Fig. 1 is the viability of the cells, determined after 24 hr of incubation, as a function of test substance concentration pX. The

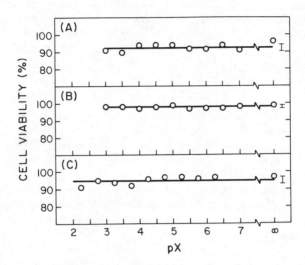

Fig. 1.    Dependence of cell viability (% live cells) of HeLa S3
cells on toxicant concentration pX at 24 hr of incubation.   pX ≡ - log [substance]$_{added}$ whereby the brackets
signal concentration in molar units.   (A) Cells kept in
presence of Fru-Trp.   (B)  Cells kept in presence of
NO-Fru-Trp, including a six-fold excess of $NaNO_2$.   Only
the NO-Fru-Trp concentrations are plotted.   (C)  Cells
kept in presence of $NaNO_2$ at concentrations identical to
those used in (B).   Solid lines represent mean values;
they are listed numerically in the text.   The bars on the
right-hand side indicate their standard deviations
(SD).   pX ∞ refers to (control) cells grown in absence of
the test substances.

percentage of live cells does not vary with toxicant concentration
over the range studied: it averages 92.5 ± 2.2, 97.7 ± 0.7, and
95.0 ± 2.1 (mean ± SD) with Fru-Trp (panel A), NO-Fru-Trp (panel
B), and $NaNO_2$ (panel C), respectively.   Similarly invariant are
plots of cell viability versus incubation time (not shown): for
instance, cells that had been kept in presence of 1 mM Fru-Trp, 1
mM NO-Fru-Trp plus 6 mM $NaNO_2$, and 6 mM $NaNO_2$ (the limiting toxicant levels employed in this study) display viability values that

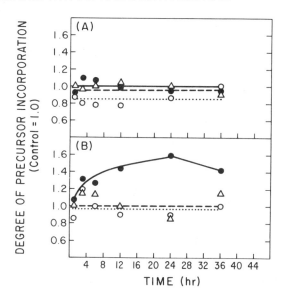

Fig. 2. Dependence of DNA (●), RNA (Δ), and protein (○) synthesis
of HeLa S3 cells (normalized with respect to control) on
treatment time (hr.) (A) Cells kept in presence of 1 mM
Fru-Trp. (——), degree of [$^3$H]thymidine incorporation =
1.00 ± 0.08 (mean ± SD). (---), degree of [$^3$H]uridine
incorporation = 0.97 ± 0.04 (mean ± SD). (•••), degree
of [$^3$H]leucine incorporation = 0.85 ± 0.10 (mean ± SD).
(B) Cells kept in presence of 6 mM NaNO$_2$. (——), degree
of [$^3$H]thymidine incorporation (best-fit curve) = 1.09 +
0.144 X ln(time), R$^2$ = 0.90, valid for 1-24 hr. (---),
degree of [$^3$H]uridine incorporation = 1.05 ± 0.14 (mean ±
SD). (•••), degree of [$^3$H]leucine incorporation = 0.97
±0.13 (mean ± SD). Label incorporation of control cells
ranges from about 5,000 cpm at 1 hr to 20,000 cps at 36
hr (± 1% SD), with background counts subtracted.

meander about 92.8 ± 2.5, 96.3 ± 2.4, and 93.8 ± 2.3% (mean ± SD),
respectively, when plotted against incubation periods of 1, 3, 6,
12, 24, and 36 hr. These data then are in close proximity to the
viability exhibited by control cells; it amounts to 97.0 ± 1.6%
when plotted against the same incubation span.

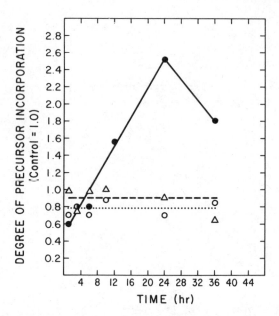

Fig. 3.    Dependence of DNA (●), RNA (Δ), and protein (○) synthesis
of HeLa S3 cells (normalized with respect to control) on
treatment time (hr).   Cells were kept in presence of 1 mM
NO-Fru-Trp plus 6 mM NaNO$_2$.   (——), degree of [$^3$H]thymi-
dine incorporation (best-fit curve) = 0.47 + 0.09 X
(time), $R^2$ = 0.98, valid for 1-24 hr.   (---), degree of
[$^3$H]uridine incorporation = 0.90 ± 0.15 (mean ± SD).
(···), degree of [$^3$H]leucine incorporation = 0.78 ± 0.07
(mean ± SD).   For actual label incorporation (cpm) see
legend to Fig. 2.

The influence of the three test substances on the synthesis
of DNA, RNA, and protein in HeLa S3 cells is shown as a function
of treatment time at limiting toxicant levels in Figs. 2 and 3
and, in Figs. 4 and 5, as a function of pX at three different
incubation periods.   It should be noted that the latter two
figures contain only information on DNA replication.   It is readily seen that the Amadori compound Fru-Trp continues to be without
influence on cell metabolism: neither DNA/RNA nor protein synthesis are significantly inhibited, or stimulated for that matter

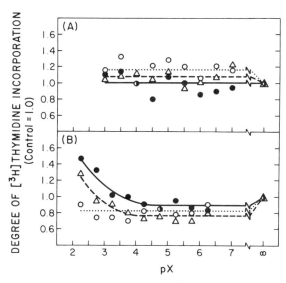

Fig. 4.  Dependence of DNA synthesis of HeLa S3 cells (normalized
with respect to control) on toxicant concentration pX
after 12 (o), 24 (●), and 36 (Δ) hr of incubation.  (A)
Cells kept in presence of 1 mM Fru-Trp.  (•••), degree of
incorporation after 12 hr = 1.16 ± 0.12 (mean ± SD).
(——), degree of incorporation after 24 hr = 1.00 ± 0.11
(mean ± SD).  (---), degree of incorporation after 36
hr = 1.08 ± 0.08 (mean ± SD).  (B) Cells kept in presence
of 6 mM NaNO$_2$.  (•••), degree of incorporation after 12
hr = 0.82 ± 0.09 (mean ± SD).  (——), degree of incorporation after 24 hr = 0.90 ± 0.07 (mean ± SD; valid from
pX 6.22 to pX 4.22).  At NaNO$_2$ concentrations < pX 4.22,
the best-fit curve yields: degree of incorporation = 2.76
X exp(-0.29pX) with R$^2$ = 0.91.  (---), degree of incorporation after 36 hr = 0.77 ± 0.12 (mean ± SD; valid from
pX 6.22 to pX 4.22).  At NaNO$_2$ concentrations < pX 4.22,
the best-fit curve yields: degree of incorporation = 2.23
X exp(-0.28pX) with R$^2$ = 0.87.  For actual label incorporation (cpm) see legend to Fig. 2.

(cf., Fig. 2, panel A; Fig. 4, panel A), and precursor incorporation fluctuates more or less randomly about the level displayed by
the controls.  Information on the various values of the mean and
the corresponding standard deviations can be found in the legends
to Figs. 2 and 4.

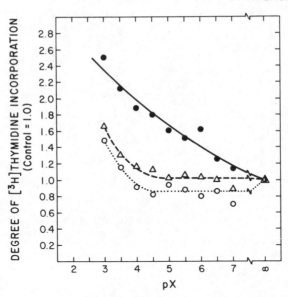

Fig. 5.  Dependence of DNA synthesis of HeLa S3 cells (normalized
with respect to control) on toxicant concentration pX
after 12 (○), 24 (●), and 36 (△) hr of incubation for
cells that had been incubated in the presence of 1 mM
NO-Fru-Trp plus 6 mM $NaNO_2$. (•••), degree of incorporation after 12 hr = 0.86 ± 0.09 (mean ± SD; valid from pX
7.00 to pX 4.00).  At concentrations < pX 4.00, the best-
fit curve yields: degree of incorporation = 4.78 X
$exp(-0.40pX)$ with $R^2$ = 0.97.  (——), degree of incorporation after 24 hr = 3.95 X $exp(-0.17pX)$ with $R^2$ = 0.96
(best-fit curve).  (---), degree of incorporation after
36 hr = 1.02 ± 0.07 (mean ± SD; valid from pX 7.00 to pX
4.5).  At concentrations < pX 4.5, the best-fit curve
yields = 3.43 X $exp(-0.26pX)$ with $R^2$ = 0.88.  All pX data
refer to NO-Fru-Trp concentrations only.  For actual
label incorporation (cpm of control) the legend to Fig. 2
should be consulted.

    Substance-related influences on cell metabolism, however,
became noticeable with sodium nitrate (Fig. 2, panel B; Fig. 4,
panel B) and, in particular, with the nitrosated Amadori compound
NO-Fru-Trp (Figs. 3 and 5).  Interestingly enough, the two compounds affect only DNA synthesis and not RNA or protein synthesis.  DNA synthesis increases with either substance with time at

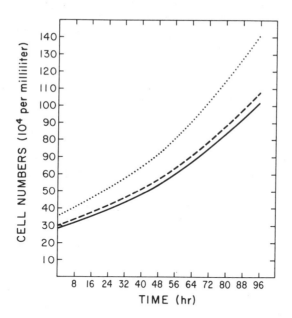

Fig. 6.  Growth of HeLa S3 cells. ($\bullet\bullet\bullet$), control cells. $N = 36.3$ X $10^4$ X exp[0.0143 X (time)]. (---), cells kept in presence of 6 mM NaNo$_2$. $N = 30.2$ X $10^4$ X exp[0.0132 X (time)]. (——), cells kept in presence of 1 mM NO-Fru-Trp plus 6 mM NaNo$_2$. $N = 28.0$ X $10^4$ X exp[0.0134 X (time)]. The coefficients of determination, $R^2$, are 0.93, 0.88, and 0.95, respectively. Not shown is the growth curve for cells kept in presence of 1 mM Fru-Trp; it runs between the curve shown for control cells and the one shown for cells kept in 6 mM NaNO$_2$. For further details see text.

given medium concentrations (Fig. 2, panel B; Fig. 3), reaching a maximum of [$^3$H]thymidine incorporation at 24 hr of incubation, and it increases with substance concentration at given treatment periods (Fig. 4, panel B; Fig. 5), whereby the maximum of synthetic activity exists again at the 24 hr time point. To see whether this increase has a bearing on cell number, we determined also the growth pattern of the cells as a function of treatment time at limiting toxicant concentrations. The results are shown

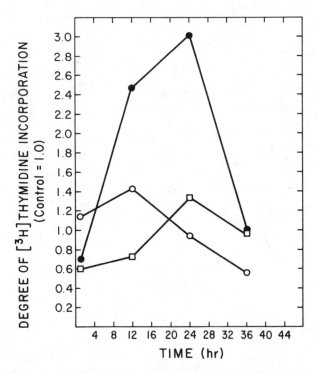

Fig. 7.   Dependence of DNA repair synthesis in HeLa S3 cells on
          treatment time. (●), cells kept in presence of 1 mM
          NO-Fru-Trp plus 6 mM $NaNO_2$. (○), cells kept in presence
          of 1 mM Fru-Trp. (□), cells kept in presence of 6 mM
          $NaNO_2$. Label incorporation has been mormalized with
          respect to control. Solid lines do not represent best-
          fit curves but merely interconnect data points. Label
          incorporation by contol cells ranges from about 400 cpm
          at 1 hr to 1,000 cpm at 36 hr (± 5% SD) (background
          counts subtracted).

in Fig. 6.   The growth curves are best-fit curves; their numerical
expressions are listed in the legend to the figure.   It is readily
seen that there is little impact, if any at all, on the growth
characteristics of the cells: if cell multiplication is approxi-
mated by the relation $N = N_o \times \exp[b \times (time)]$, with $N_o$ being the
number of cells at the start of incubation and N the number of
cells during some time of the 96 hr test period, the closeness of
the various regression coefficients b listed in the legend shows

that the excess DNA synthesis is of no consequence in regard to
cell number.

Studies in Depleted Medium. It is logical to conclude from the
evidence collected thus far that the observed excess DNA synthesis
is not so much an indication of stimulation of replication as it
is a sign of DNA repair. We tested this assumption by monitoring
the incorporation of [$^3$H]thymidine under conditions that more or
less abolish DNA replication but permit excision repair of damaged
sites ("unscheduled DNA synthesis"). The results are shown in
Fig. 7. While there is some ambiguity regarding the effects of
Fru-Trp and NaNO$_2$ on DNA repair - without doubt due to the low
rate of label incorporation by the controls - those caused by
NO-Fru-Trp are unambiguous. Nitrosated fructos-tryptophan initiates repair synthesis in HeLa S3 cells, and the maximum of repair
also exists at the 24 hr incubation time point.

DISCUSSION

The results of this study show that NO-Fru-Trp causes a
tremendous increase in DNA synthesis in HeLa S3 cells. The increase is not due to the sodium nitrite in the incubation medium,
present always as a six-fold excess over the original Fru-Trp,
since NaNO$_2$ alone, at identical concentrations, initiates excess
DNA synthesis only at concentrations above 1 mM (cf., Fig. 4,
panel B, 24 hr curve) whereas the NO-Fru-Trp/NaNO$_2$ mixture is
active already at concentrations below 1 μM (cf., Fig. 5, 24 hr
curve). Fru-Trp alone is totally inactive as can be seen from
Figs. 2 and 4 (panels A). There can be no doubt that the stimulation of DNA synthesis represents DNA repair. This follows indirectly from Figs. 3 and 6 - there is no concomitant rise in RNA
and protein synthesis as well as number when the cells are in
contact with NO-Fru-Trp - and directly from Fig. 7.

None of the compounds tested displays cytotoxicity in the
sense commonly understood: there is no change in the staining
characteristics of the cells, i.e., the outer plasma membrane
remains functional as a permeability barrier. The lack of cytotoxicity is shown further by the compounds' inability to inhibit
any of the three major macromolecular synthetic processes. If
there should be inhibition (cf., Fig. 2, Fru-Trp and NaNO$_2$, protein synthesis, 1-36 hr; Fig. 3, NO-Fru-Trp, RNA and protein
synthesis, 1-36 hr; see also Figs. 4 and 5) it is certainly marginal and presumably transient.

DNA damage in HeLa S3 cells, necessitating the observed
repair, could be due to either one of the following two possibilities: (1) as a typical N-nitrosamine, NO-Trp-Fru, following the
route of α-hydroxylation (or β or ω-oxidation), ultimately yields

alkyl or arylalkyl carbonium ions that interact with the $O^4$ position of thymidine (T), the $O^6$ position of guanosine (G), or the $N^7$ position of G. The resulting mispairing of the type ($O^4$-R)T=G, ($N^7$-R)G=T, or ($O^6$-R)G=C, representing the mutagenic damage caused by NO-Fru-Trp, would necessitate the excision of the damaged sites (cf., Margison and O'Connor, 1980). (2) the indole ring of tryptophan is known to intercalate with DNA (Helene and Dimicoli, 1972; Gabbay et al., 1972; Toulme et al., 1974) and to engage in photochemical reactions with pyrimidine bases when irradiated at wavelengths above 260 nm (Reeve and Hopkins, 1979; 1980). The two principal photoproducts are dihydro-pyrimidines (Reeve and Hopkins, 1979) and tryptophan-pyrimidine photoadducts (Reeve and Hopkins, 1980), compounds that certainly abolish Watson-Crick hydrogen bonding and again require excision of the damaged sites. As claimed by Bonnett and Holleyhead (1974), the $N^1$(indole)-NO bond in nitrosated tryptophan is extremely photolabile so that NO-Fru-Trp could conceivably engage in photochemical reactions with DNA already at wavelengths in the visible light region. It is for this reason that we performed nitrosation of Fru-Trp in the dark. Tryptophan radical cations, generated, for instance, via flash photolysis, are known to exist (cf., Adams and Wardman, 1977). On the other hand, neither Nakai et al. (1978) nor Kurosky and Hofmann (1972) mention any light-induced degradation of NO-Trp when they reacted tryptophan with nitrous acid. Clearly, there is a need for investigating the photochemical behavior of NO-Trp, NO-Fru-Trp, NO-Glu-Trp, and other related compounds, in greater detail, and our laboratory has begun such a study.

Our results are in complete harmony with the findings obtained by Coughlin (1979) employing the Ames test. He finds that NO-Fru-Trp acts on the S.typhimurium strains TA98 and TA100 much better when the assays are performed in absence of the S-9 mix. This would show NO-Fru-Trp to behave as a direct-acting mutagen; certainly no surprise when considered in light of the photo-induced activation possibilities outlined above. Further, the fact that NO-Fru-Trp interacts with TA98 to begin with – a strain especially sensitive to chemicals causing frame-shift mutations – is understandable in view of what has been mentioned above.

We have avoided identifying the sites of nitrosation in the Amadori compound and have thus resorted to denotions such as NO-Fru-Trp or "nitrosated" fructose-tryptophan. NO-Fru-Trp can exist in three nitrosated forms: as the mono-nitroso $N^1$-NO derivative, the mono-nitroso $N^\alpha$-NO derivative, and the di-nitroso $N^1$-NO,$N^\alpha$-NO derivative. To our knowledge, attempts to isolate the derivatives have been futile (Coughlin, 1979; Nakai et al., 1978). In fact, it appears that indole nitrosation is a readily reversible reaction and that the $N^1$-NO form is in existence only

when nitrite is present in excess (Nakai et al., 1978). It is for
this reason that we have performed the tests involving NO-Fru-Trp
always in presence of a six molar excess level of $NaNO_2$. We
assume that all three possible N-nitroso derivatives were present,
but of course we do not know in which proportions.

We find DNA synthesis to have reached a maximum at 24 hr of
incubation (cf., Fig. 2, panel B; Fig. 3). We find this reasonable in view of the fact that the synthesis is nothing but repair
and that there must come a point in time when most of the repair
has taken place. This seems to be the case around 24 hr of incubation. HeLa S3 cells, by the way, have a generation time of 21
hr, at least under our experimental conditions (Gruenwedel and
Cruikshank, 1979).

In conclusion, there can be no doubt that NO-Fru-Trp has
mutagenic properties. Its mutagenic action is unknown at present,
yet there are reasons to believe that it can bind covalently to
pyrimidines via the indole ring. Whether it has also carcinogenic
properties remains to be seen. We would expect nitrosated tryptophan to behave like NO-Fru-Trp, but this test still needs to be
done. Lastly, we are in the process of studying the mutagenic
potential of nitrosated 1(N-5-hydroxytryptamino)-1-deoxy-D-
fructose. Fructosyl-serotonin is closely related to Fru-Trp
structurally and it is suspected to be involved in hemostasis and
tryptaminergic neural activity (Mester et al., 1981).

## ACKNOWLEDGEMENTS

This communication contains material taken from the thesis of
Susan C. Lynch, submitted to the Graduate Division of the University of California in partial fulfillment of the M.S. degree in
Food Science. It was presented by the senior author at the 1982
ACS/SAS Pacific Conference on Chemistry and Spectroscopy, October
27-29, San Francisco, CA. This research has been supported by
funds of the University of California.

## REFERENCES

Adams, G. E., and Wardman, P., 1977, Free radicals in biology: the
    pulse radiolysis approach, in: "Free Radicals in Biology",
    Vol. III, W. A. Pryor, ed., p. 53, Academic Press, New York.
Bonnett, R., and Holleyhead, R., 1974, Reaction of tryptophan
    derivatives with nitrite, J. Chem. Soc., Perkin I: 962.
Borsook, H., Deasy, C. L., Haagen-Smit, A. J., Keighley, G., and
    Lowy, P. H., 1952, Incorporation in vitro of labeled amino
    acids into protein of rabbit reticulocytes, J. Biol. Chem.
    196: 669.

Cerami, A., Stevens, V. J., and Monnier, V. M., 1979, Role of nonenzymatic glycosylation in the development of the sequelae of diabetes mellitus, Metabolism 28 (4), Suppl. 1: 431.

Committee on Nitrite and Alternative Curing Agents in Food, 1981, "The Health Effect of Nitrate, Nitrite, and N-Nitroso Compounds", chaired by M. McCarty, chapters 1-10, National Academy Press, Washington, D.C.

Coughlin, J. R., 1979, "Formation of N-Nitrosamines from Maillard Browning Reaction Products in the Presence of Nitrite", Ph.D. Dissertation, University of California, Davis.

Coughlin, J. R., Wei, C. I., Hsieh, D. P. H., and Russell, G. F., 1979, Synthesis, mutagenicity and human health implications of N-nitroso Amadori compounds from Maillard browning reactions in the presence of nitrite, presented at Am. Chem. Soc./Chem. Soc. Japan. Chemical Congress, Honolulu, Hawaii, April 2-6.

Davidson, E. A., 1966, Metabolism of amino sugars, in: "The Amino Sugars", Vol. IIB, E. A. Balazs and R. W. Jeanloz, ed., p. 10, Academic Press, New York.

Flueckiger, R., and Winterhalter, K. H., 1976, In vitro synthesis of hemoglobin $A_{1c}$, FEBS Lett. 71: 356.

Gabbay, E. J., Sanford, K., and Baxter, C. S., 1972, Specific interaction of peptides with nucleic acids, Biochemistry 11: 3429.

Gruenwedel, D. W., and Fordan, B. L., 1978, Effects of methylmercury(II) on the viability of HeLa S3 cells, Toxicol. Appl. Pharmacol. 46: 249.

Gruenwedel, D. W., and Cruikshank, M. K., 1979, Effect of methylmercury(II) on the synthesis of deoxyribonucleic acid, ribonucleic acid and protein in HeLa S3 cells, Biochem. Pharmac. 28: 651.

Helene, C., and Dimicoli, J. C., 1972, Interaction of oligopeptides containing aromatic amino acids with nucleic acids. Fluorescence and proton magnetic resonance studies, FEBS Lett. 26: 6.

Heyns, K., Roeper, S., Roeper, H., and Meyer, B., 1979, N-nitroso sugar amino acids, Angew. Chem. Int. Ed. Engl. 18: 878.

Kurosky, A., and Hofmann, T., 1972, Kinetics of the reaction of nitrous acid with model compounds and proteins, and the conformational state of N-terminal groups in the chymotrypsin family, Canad. J. Biochem. 50: 1282.

Lingens, F., and Lueck, W., 1963, Ueber die Biosynthese des Tryptophans in Saccharomyces cerevisiae, Hoppe-Seyler's Z. Physiol. Chem. 333: 190.

Margison, G. P., and O'Connor, P. J., 1980, Nucleic acid modification by N-nitroso compounds, in: "Chemical Carcinogens and DNA", Vol. I, P. L. Grover, ed., p. 111, CRC Press, Inc., Boca Raton.

Mester, L., Szabados, L., Mester, M., and Yadav, N., 1981,
    Maillard type carbonyl-amine reactions in vivo and their
    physiological effects, in: "Maillard Reactions in Food",
    Progress in Food and Nutritional Science, Vol. 5, Nos. 1-6,
    C. Eriksson, ed., p. 295, Pergamon Press, Oxford.
Monnier, V. M., Stevens, V. J., and Cerami, A., 1981, Maillard
    reactions involving proteins and carbohydrates in vivo:
    relevance to diabetes mellitus and aging, in: "Maillard
    Reactions in Food", Progress in Food and Nutritional Science,
    Vol. 5, Nos. 1-6, C. Eriksson, ed., p. 315, Pergamon Press,
    Oxford.
Nakai, H., Cassens, R. G., Greaser, M. L., and Woolford, G., 1978,
    Significance of the reaction of nitrite with tryptophan, J.
    Food Sci. 43: 1857.
Paulsen, H., and Pflughaupt, K. W., 1980, Glycosylamines, in: "The
    Carbohydrates", Vol. IB, W. Pigman, D. Horton, and J. D.
    Wander, ed., p. 899, Academic Press, New York.
Reeve, A. E., and Hopkins, T. R., 1979, Photosensitized reduction
    of pyrimidine bases by tryptophan, Photochem. Photobiol. 30:
    677.
Reeve, A. E., and Hopkins, T. R., 1980, Photochemical reaction
    between protein and nucleic acids: photoaddition of trypto-
    phan to pyrimidine bases, Photochem. Photobiol. 31: 297.
Rhabar, S., 1968, An abnormal hemoglobin in red cells of diabet-
    ics, Clin. Chim. Acta 22: 296.
Scanlon, R. A., and Tannenbaum, S. R., ed., 1981, "N-Nitroso
    Compounds", ACS Symposium Series 174, American Chemical
    Society, Washington, D.C.
Sgarbieri, V. C., Amaya, J., Tanaka, M., and Chichester, C. O.,
    1973, Nutritional consequences of the Maillard reaction.
    Amino acid availability from fructose-leucine and fructose-
    tryptophan in the rat, J. Nutr. 103: 657.
Shapiro, R., McManus, M., Garrick, L., McDonald, M. J., and Bunn,
    H. F., 1979, Nonenzymatic glycosylation of human hemoglobin
    at multiple sites, Metabolism 28 (4), Suppl. 1: 427.
Toulme, J., Charlier, M., and Helene, C., 1974, Specific recogni-
    tion of single-stranded regions in ultraviolet-irradiated and
    heat-denatured DNA by tryptophan-containing peptides, Proc.
    Natl. Acad. Sci. USA 71: 3185.
Trosko, J. E., and Yager, J. D., 1974, A sensitive method to
    measure physical and chemical carcinogen-induced "unsched-
    uled" DNA synthesis in rapidly dividing eukaryotic cells,
    Exptl. Cell Res. 88: 47.

## 14

# SOURCES OF N-NITROSAMINE

# CONTAMINATION IN FOODS

Joseph H. Hotchkiss

Department of Food Science
Cornell University
Ithaca, New York  14853

## ABSTRACT

It has been well established that human foods may contain trace amounts of carcinogenic N-nitrosamines. Originally, it was thought that the use of nitrite as a curing agent for flesh foods was the major source of these trace compounds in the diet. Subsequent research has clearly shown that other processing and packaging procedures can also introduce trace amounts of these carcinogens into foods. These procedures include drying foods in direct flame heated air, migration from food contact surfaces and direct addition as contaminants. In addition, other reports of N-nitrosamines in foods have less well defined routes of contamination. These sources of N-nitrosamines in foods will each be briefly reviewed in this paper and recent data from our laboratory concerning N-nitrosamines in products which directly contact foods presented. We also are reporting the N-nitrosothiazolidine content of fried-out bacon fat.

## INTRODUCTION

Compounds which contain the N-nitroso functional group have raised concerns for over two decades for two reasons: First, a large majority of the over 200 individual N-nitroso compounds tested are capable of producing tumors in laboratory animals (Wishnok, 1981); and secondly, humans are exposed to small quantities of these compounds through the environment, commercial products and foods

(Fishbein, 1979). Occupational exposures may result in the highest individual exposures (Fine and Rounbehler, 1981) but the largest numbers of people have been exposed to exogenously formed N-nitroso compounds through the diet.

Several groups have demonstrated that a number of foods can contain trace quantities of VNA[1] (volatile being those compounds amenable to gas chromatography without derivatization). Analytical methodology for NVNA has not been adequately developed. Methodology for the analysis of VNA and NVNA in foods has been recently reviewed (Hotchkiss, 1981).

To date nearly all types of foods have been analyzed for VNA and, hence, some important generalizations can be made. Most importantly is that the use of nitrite as a curing agent is not solely responsible for the VNA content of foods. Several foods to which nitrite has not been intentionally added have now been shown to contain trace levels of VNAs. Equally significant is that the N-nitrosamine content of foods has decreased as a result of research in this area. In this review I will not list the VNA content of specific foods but will generalize and classify the routes and mechanisms by which foods can become contaminated. The routes of contamination can be divided into 5 groups: Additives; drying processes; migration from contact surfaces; addition of performed NA; and those for which the route is not clearly defined. Recent reviews of other aspects of N-nitroso compounds including biological actions (Digenis and Issidorides, 1979), chemical properties (Fridman et al., 1971) and food content (Gray, 1981) have been published.

ADDITIVES

The suspicion that the use of nitrite in foods might result in the formation of NA stems from an incident in which animals fed nitrite preserved fish meal developed liver necrosis. The causal agent was determined to be NDMA and it was shown that the compound resulted from the nitrosation of the amines in the fish by nitrous acid formed from nitrite (Ender et al., 1964). Nitrite is an economically and technically important food additive in the curing process in order to fix color, develop flavor and inhibit toxigenesis by Cl. botulinum.

---

[1]Abbreviations: NA, N-nitrosamines; VNA volatile N-nitrosamines; NVNA, nonvolatile N-nitrosamines; NDMA, N-nitrosodimethylamine; NPYR, N-nitrosopyrrolidine; NPRO, N-nitrosoproline; NTHZ, N-nitrosothiazolidine; NPIP, N-nitrosopiperidine; NDEA, N-nitrosodiethylamine; GC-TEA, gas chromatography - thermal energy analyzer; NMOR, N-nitrosomorpholine; NDPA, N-nitrosodipropylamine.

Since the late 1960s a substantial research effort has resulted in a body of information concerning the occurrence and formation of VNA in cured meats. This has resulted in the knowledge that the addition of nitrite to meat is not, in most cases, sufficient to routinely cause the formation of VNA. In order for cured meats to consistently contain more than 1 μg/kg VNA, the product must be subjected to temperatures greater than 100°C in a low moisture environment. The only cured product which meets these criteria is bacon. Other cured products only sporadically contain VNA in excess of 0.1 μg/kg (Gray and Randall, 1979). In a recent large survey, only 6 of 152 cooked sausage products had a VNA content greater than 5 μg/kg and only 4 of 91 dry sausages had similar VNA contents. In the same study, however, 11 of 12 dry-cured fried bacons contained VNA, some as high as 280 μg/kg . The fact that fried cured bacon consistently contains detectable VNA has been observed by numerous workers (Scanlan, 1975).

Efforts have been directed at determining the chemical mechanism and precursors to the major VNA, NPYR, found in fried bacon. While several potential precursors to NPYR have been identified, including collagen, ornithine, hydroxyproline, citrulline, putrasine and arginine, it is generally accepted that the major precursor is proline (Gray, 1976). While the free-radical mechanism proposed by Bharucha et al. (1979) is often cited as the mechanism which best fits observations, the steps of the reaction have not been clearly elucidated. At least two possible routes exist; proline could be nitrosated to form NPRO which is subsequently decarboxylated during frying to NPYR, or proline is first decarboxylated to the amine pyrrolidine which is then subsequently nitrosated. Both decarboxylation and nitrosation, regardless of order, must occur during frying because uncooked bacon does not contain NPYR or sufficient preformed NPRO (Hansen et al, 1977). Nakamura et al. (1976) have suggested that the mechanism is temperature dependent; at temperatures above 175°C decarboxylation precedes nitrosation and at lower temperatures nitrosation precedes decarboxylation.

In addition to NDMA and NPYR, Kimoto et al. (1982) and Gray et al. (1982) each have reported that fried bacon also contains NTHZ. This VNA was likely missed by many researchers due to its long retention time or its on-column decomposition. We have also confirmed this VNA in fried bacon and have further identified the compound in the fried-out fat from bacon. NDMA and NPYR are, under most frying conditions, found in higher concentration in the fried-out fat than in the edible portion. However, in our experiments NTHZ consistently occurs in higher concentrations in the edible portion regardless of the frying conditions (Table 1). The mutagenicity of NTHZ has been demonstrated (Sekizawa and Shibamoto, 1980) but the compound has not been tested in whole animals for carcinogenicity. The formation of precursors of NTHZ have also not been studied in fried bacon but thiazolidine has been identified as a browning  product in a glucose-cysteamine model system (Mihara and Shibamoto, 1980).

TABLE 1

Volatile N-nitrosamines in Bacon and Bacon Fried-out Fat

| Sample | Edible | | | Fried-out Fat | | |
|---|---|---|---|---|---|---|
| | NDMA | NPYR | NTHZ | NDMA | NPYR | NTHZ |
| A (3)[a] | 1.4[b] | 13 | 18 | 2.1 | 46 | 6 |
| B (3) | 1.7 | 39 | 68 | 2.9 | 88 | 23 |
| C (4) | 2.5 | 59 | 51 | 7.8 | 114 | 17 |
| D (6) | 4.3 | 67 | 30 | 2.0 | 44 | 8 |
| E (6) | 3.2 | 75 | 69 | 2.6 | 75 | 24 |

[a] minutes frying per side at 170°C

[b] µg/kg

Nitrate may be added to certain cheeses to retard the growth of microorganisms which might cause defects.  Concern has been expressed that the nitrate might be reduced to nitrite by reductase containing microflora and that this nitrite could nitrosate amines endogenous to the product.  The Danish government has published the results of a large survey of cheeses in which no correlation between the use of nitrate and concentration of VNA in the product could be made (Anon. 1980).  Only very small amounts (less than 0.7 µg/kg) were found in any cheese.  Sen et al. (1978), however, found 21 of 31 cheeses imported into Canada contained VNA up to 20 µg/kg .  These apparent discrepancies with regard to the use of nitrate in cheese have not been resolved.

DRYING

In addition to the use of nitrite and nitrate as additives, a second general mechanism by which foods may become contaminated with NA is through the drying of foods in air which has been directly heated in an open flame.  The highest levels of VNA resulting from this common method of food processing have been in the kilning of malted barley.  Concentrations of NDMA in the dried malt of over 100 µg/kg have been reported (Hotchkiss et al. 1980; Preussmann et al. 1981).

A number of workers have shown that the NDMA in the malt survives the brewing process and can be detected in the resulting beer in concentrations expected from the dilution of the malt (Havery et al. 1981). This widespread contamination was shown to be the result of the formation of oxides of nitrogen in the air as it is heated in the flame. Oxides of nitrogen have been demonstrated to be effective nitrosating agents over a wide pH range (Challis and Kyrtopoulos, 1978).

Scanlan and coworkers have extensively investigated the formation of NDMA in malt. They have demonstrated that the plant alkaloids hordenine and gramine are effective precursors of NDMA in model systems and that it is likely that NVNA may also be present in direct fired kiln dried malt (Mangino et al. 1981).

The first reports of NDMA in beer indicated average concentrations in the range of 2 to 6 µg/kg (Spiegelhalder et al. 1981). While these levels seem, at first, low it is possible to consume 1 to 2 kg of beer at a single serving. This represents more NDMA exposure than from any other food source. Spiegelhalder et al. (1980) have estimated that 64% of a West German's dietary NDMA came from beer. There is recent evidence, however, that the VNA content of beer has decreased sharply (Mangino et al. 1981). This decrease is due to the widespread use of sulfur dioxide or indirect heating of the drying gases in the malting industry. The application of sulfur dioxide during the early part of the kilning process may be either by direct injection of gaseous sulfur dioxide or by burning elemental sulfur in the drying air. The inhibition by sulfur dioxide is most likely due to the formation of bisulfite which may react with the nitrosating agent in a redox reaction.

Other foods which are dried in direct flame heated air have also been shown to contain trace amounts of VNA, albeit at lower levels than malt. Most notable is the finding that nonfat dried milk may contain traces of NDMA. Several reports have shown NDMA levels of 0.1 to approximately 5 µg/kg (Libbey et al. 1980; Lakritz and Pensabene, 1981). In a recent nationwide survey of 57 nonfat dried milks conducted by the US Food and Drug Administration, an average NDMA level of 0.6 µg/kg was found with 48 samples being positive (Havery et al. 1981). Apparent NPYR and NPIP were also detected in sub µg/kg concentrations. Because nonfat dry milk is diluted 10x before consumption some have considered it not to be a significant problem while others have been concerned because of the widespread use of this product by the young.

Other dried foods have been shown to sporadically contain detectable VNA. Sen and Seaman (1981) analyzed nonfat dry milk, dried soups, and instant coffee and found VNA in all dried milks, 3 of 20 dried soups and 5 of 10 instant coffees, most at levels of less than 1 µg/kg . Perhaps more importantly, 3 of 8 dried infant formulas contained detectable VNA. Fazio and Havery (1981) have observed VNA in soy isolates and concentrates and dried cheeses.

MIGRATION

In addition to formation from direct additives or from the
direct flame drying process, recent evidence indicates VNA may enter
foods through migration from food contact surfaces. In 1981
Spiegelhalder and Preussmann (1981) reported that a number of rubber
products including nursing nipples contained substantial levels of
VNA and that these compounds could migrate to water and milk. Later
Havery and Fazio (1982) investigated one brand of nipple available
in the US. They confirmed the presence of VNA in this product and
demonstrated that when inverted nipples were sterilized in milk or
formula, migration occurred.

We have investigated the VNA content of 8 types of rubber
nipples available in the US from several domestic and foreign manu-
facturers (Babish et al. 1982). One or more VNA were detected in all
nipples tested and when each nipple was boiled for 3 minutes in 150
ml water or incubated 3 hours at $37^{\circ}$C, 6 to 44% migration occurred.
Total VNA contents ranged from 42 to 617 µg/nipple (nipple weight is
approximately 5 gm) and most nipples contained more than one VNA
(Table 2).

TABLE 2
Volatile N-nitrosamines in Nursing Nipples

| Nipple | Weight[a] | N-nitrosamine[b] | | |
|--------|-----------|------|------|------|
|        |           | NDMA | NDEA | NPIP |
| A      | 4.9       | 79   | 61   | --   |
| B      | 8.2       | 127  | --   | --   |
| C      | 5.5       | --[c] | 56  | 380  |
| D      | 4.6       | --   | 46   | 299  |
| E      | 5.0       | --   | 42   | --   |
| F      | 5.2       | --   | 79   | 401  |
| G      | 5.2       | --   | 75   | 541  |
| H      | 4.9       | 120  | 74   | --   |

[a] gm

[b] ng per nipple

[c] not detected

Direct food contact paper and paperboard packaging may also be
a source of VNA and nitrosatable amines in foods. Analyses of 34
food packages by GC-TEA revealed 9 to be contaminated with NMOR
(Table 3). Perhaps more importantly, all packaging materials
examined had levels of the parent amine morpholine ranging from

98 to 842 μg/kg (Hotchkiss and Vecchio, 1982). Morpholine is easily nitrosated and there is evidence that it may be nitrosated in the stomach to produce the carcinogenic N-nitroso derivative (Mirvish, 1975). Two experiments indicated that both the NMOR and morpholine may migrate to dry foods. First, when a food package was found to contain NMOR and morpholine, the food closest in the package often also contained NMOR and morpholine (Hoffmann et al. 1982). Secondly, when paperboards which contained NMOR and morpholine were incubated at elevated temperatures in closed vessels with dry foods migration could be demonstrated. Further research is needed to determine the extent of the contamination and degree of migration under normal conditions.

TABLE 3

Morpholine and N-nitrosomorpholine in Paper and Paperboard Packaging

| Container | NMOR[a] | Morpholine[a] |
|---|---|---|
| Pasta Box | 15 | 430 |
| Cake Box | 3.6 | 560 |
| Flour Bag | 13 | 812 |
| Flour Bag | 3.3 | --[b] |
| Packaged Dinner | 6.1 | -- |
| Pasta Box | 8.9 | 347 |
| Milk Carton | ND[c] | 98 |
| Rice Box | ND | 132 |
| Cornstarch Box | ND | 329 |
| Packaged Dinner | ND | 445 |
| Frozen Vegetable Box | ND | 113 |
| Flour Sack | ND | 842 |

[a] μg/kg

[b] not analyzed

[c] not detected

DIRECT ADDITION

When agricultural chemicals or food additives contain preformed VNA, it is conceivable that a portion of the VNA contaminate could be added to food. For example, meat curing premixes which contained salt, sugar, spices and nitrite and were designed to facilitate the mixing of curing brines were shown to contain relatively high levels of VNA including NPIP (Sen et al. 1974). This VNA resulted from the nitrosation of piperidine ring containing compounds in the spices.

The NPIP was then added along with the cure solution to the meat and could be detected in the product.

Certain agricultural chemicals were shown at one time, to contain mg/kg quantities of VNA (Ross et al. 1977). Although current levels have been greatly reduced (Oliver, 1981) it has been demonstrated under laboratory conditions that when these mixtures are applied to food crops, absorption of the VNA either directly through the plant or indirectly through the soil is possible (Khan, 1981). For example, Dean-Raymond and Alexander (1976) have shown that radiolabeled NDMA incorporated into soil could be taken up by edible plants. A recent survey of dried waste sludge also indicates most sludges contain small amounts of VNA (Mumma et al. 1982). If sludge is incorporated into soil uptake may be possible. It should be noted that no confirmed report of VNA in foods as a result of the use of pesticides or sludge in actual field use has appeared. On the contrary, Ross et al. (1978) analyzed soil, run off water and edible plant tissue after the application of a commercial herbicide containing NDPA and failed to detect the VNA in any sample. As pointed out by Oliver, (1981) it is difficult to draw conclusions about the VNA contamination of foods based on laboratory experiments.

Another potential source of direct addition of NA to food maybe through the use of processing water which has been deionized by anion exchangers. Kimoto et al. (1980) have shown that NDMA and NDEA at levels of less than 1 µg/kg can be detected in water which has been passed through an anion exchange column. This is a common treatment process in food plants.


MISCELLANEOUS

In addition to the above four mechanisms by which food may become contaminated with small amounts of VNA, other less well defined or uncorroborated contamination processes have been reported. For example, one group of Japanese workers have reported that broiling fish under a gas flame may result in substantial increases in the VNA content of the food (Matsui et al. 1980). The average NDMA content of 20 fish and seafood products increased 3 fold after broiling under a gas flame and one dried squid sample increased in NDMA from 84 to 313 µg/kg . When broiled under an electric element or covered with aluminum foil smaller increases in NDMA content were seen. Presumably, VNA are being formed by a mechanism similar to that occurring in dried foods such as malt and nonfat dried milk. Further work is needed to evaluate this source of dietary NA. Smoking fish has also been reported to result in the nitrosation of the amines associated with fish (Kann et al. 1980).

CONCLUSIONS

Nearly all categories of foods have been analyzed for VNA. As I've attempted to point out, this research has reached a point where generalizations can be drawn and the routes of contamination defined.

There is little doubt that this research has led to a reduction in human exposure to carcinogenic VNA. This is especially true for fried bacon (Havery et al. 1978), beer (Havery et al. 1981) and rubber products. What is needed is similar information concerning NVNA in foods. Inadequate analytical methodology has precluded this information.

The important question is, of course, what implications toward human health can be drawn from the occurrence of trace amounts of carcinogenic NA in foods. Unfortunately, a definitive answer to that question is not available but we must not dismiss that some finite risk may exist just because the concentrations of VNA in foods appears small when compared to our everyday experiences. It is still prudent to investigate the occurrence of these undesirable compounds and to reduce exposures when feasible.

REFERENCES

Anon., 1980, Investigations on formation and occurrence of volatile nitrosamines in Danish cheese, The Government Research Institute for Dairy Industry and The National Food Institute, Denmark.

Babish, J. G., Hotchkiss, J. H., Wachs, T., Vecchio, A. J., Gutenmann, W. H., and Lisk, D. J., 1982, N-nitrosamines in rubber nursing nipples, J. Toxicol. Environ. Health, (in press).

Bharucha, K. R., Cross, C. K., and Rubin, L. J., 1979, Mechanism of N-nitrosopyrrolidine formation in bacon, J. Agric. Food Chem., 27:63.

Challis, B. C., and Kyrtopoulos, S. A., 1978, The chemistry of nitroso compounds. Part 12. The mechanism of nitrosation and nitration of aqueous piperidine by gaseous dinitrogen tetroxide and dinitrogen trioxide, J. Chem. Soc. Perkin II, 1978:1296.

Dean-Raymond, D., and Alexander, M., 1976, Plant uptake and leaching of dimethylnitrosamine, Nature, 262:394.

Digenis, G. A., and Issidorides, C. H., 1979, Some biochemical aspects of N-nitroso compounds, Bioorg. Chem. 8:97.

Ender, F., Harve, G., Helgebostad, A., Koppang, N., Madsen, R., and Ceh, L., 1964, Isolation and identification of a hepatotoxic factor in herring meal produced from sodium nitrite preserved herring, Naturwissenschaften, 51:637.

Fazio, T., and Havery, D. C., 1981, Volatile N-nitrosamines in direct flame dried processed foods, in: "Proceedings of the Seventh International Meeting on N-nitroso Compounds: Occurrence and Biological Effects", (in press) International Agency for Research on Cancer, Lyon.

Fine, D. H., and Rounbehler, D. P., 1981, Occurrence of N-nitrosamines in the workplace: Some recent developments, in: "N-nitroso Compounds," R. A. Scanlan and S. R. Tannenbaum, eds., American Chemical Society, Washington, D.C.

Fishbein, L., 1979, Overview of some aspects of occurrence, formation, and analysis of nitrosamines, Sci. Total Environ., 13:157.

Fridman, A. L., Mukhametshin, F. M., and Novikov, S. S., 1971, Advances in the chemistry of aliphatic N-nitrosamines, Russ. Chem. Rev., 40:34.

Gray, J. I., 1976, N-nitrosamines and their precursors in bacon: A review, J. Milk Food Technol., 39:686.

Gray, J. I., and Randall, C. J., 1979, The nitrite/nitrosamine problem in meats: An update, J. Food Protect., 42:168.

Gray, J. I., 1981, Formation of N-nitroso compounds in foods, in: "N-nitroso Compounds," R. A. Scanlan and S. R. Tannenbaum, eds., American Chemical Society, Washington, D.C.

Gray, J. I., Reddy, S. K., Price, J. F., Mandagere, A., and Wilkens, W. F., 1982, Inhibition of N-nitrosamines in Bacon, Food Technology, 36 (6):39.

Hansen, T., Iwaoka, W., Green, L., and Tannenbaum, S. R., 1977, Analysis of N-nitrosoproline in raw bacon: Further evidence that nitrosoproline is not a major precursor of nitrosopyrrolidine, J. Agric. Food Chem., 25:1423.

Havery, D. C., Fazio, T., and Howard, J. W., 1978, Trends in levels of N-nitrosopyrrolidine in fried bacon, J. Assoc. Off. Anal. Chem., 61:1379.

Havery, D. C., Hotchkiss, J. H., and Fazio, T., 1981, Nitrosamines in malt and malt beverages, J. Food Sci., 46:501.

Havery, D. C., and Fazio, T., 1982, Estimation of volatile N-nitrosamines in baby bottle rubber nipples, Food Chem. Toxicol., (in press).

Hoffmann, D., Brunnemann, K. D., Adams, J. D., Rivenson, A., and Hecht, S. S., 1982, N-nitrosamines in tobacco carcinogenesis, in: "Banbury Report 12: Nitrosamines and Human Cancer," (in press) Cold Spring Harbor Laboratory, Cold Spring Harbor.

Hotchkiss, J. H., Barbour, J. F., Scanlan, R. A., 1980, Analysis of malted barley for N-nitrosodimethylamine, J. Agric. Food Chem., 28:680.

Hotchkiss, J. H., 1981, Analytical methodology for sample preparation, detection, quantitation, and confirmation of N-nitrosamines in foods, J. Assoc. Offic. Anal. Chem., 64:1037.

Hotchkiss, J. H., and Vecchio, A. J., 1982, Analysis of direct
    contact paper and paperboard food packaging for N-nitroso-
    morpholine and morpholine, J. Food Sci., (in press).
Kann, J., Tauts, O., Kalve, R., Bogovski, P., 1980, Potential forma-
    tion of N-nitrosamines in the course of technological processing
    of some foodstuffs, in: "N-nitroso Compounds: Analysis, Forma-
    tion and Occurrence," E. A. Walker, L. Griciute, M. Castegnaro,
    and M. Borzsonyi, eds., International Agency for Research on
    Cancer, Lyon.
Khan, S. U., 1981, N-nitrosamine formation in soil from the herbicide
    glyphosate and its uptake by plants, in:  "N-nitroso Compounds,"
    R. A. Scanlan and S. R. Tannenbaum, eds., American Chemical
    Society, Washington, D.C.
Kimoto, W. I., Dooley, C. J., Carre, J., and Fiddler, W., 1980, Role
    of strong anion exchange resins in nitrosamine formation in
    water, Water Res., 14:869.
Kimoto, W. I., Pensabene, J. W., and Fiddler, W., 1982, Isolation
    and identification of N-nitrosothiazolidine in fried bacon,
    J. Agric. Food Chem., 30:757.
Lakritz, L., and Pensabene, J. W., 1981, Survey of fluid and nonfat
    dry milks for N-nitrosamines, J. Dairy Sci., 64:371.
Libbey, L. M., Scanlan, R. A., and Barbour, J. F., 1980, N-nitroso-
    dimethylamine in dried dairy products, Food Cosmet. Toxicol.,
    18:459.
Mangino, M. M., Scanlan, R. A., and O'Brien, T. J., 1981, N-nitro-
    samines in beer, in: "N-nitroso Compounds," R. A. Scanlan and
    S. R. Tannenbaum, eds., American Chemical Society, Washington,
    D.C.
Matsui, M., Ohshima, H., and Kawabata, T., 1980, Increase in the
    nitrosamine content of several fish products upon broiling,
    B. Japan Soc. Sci. Fisheries, 46:587.
Mihara, S., and Shibamoto, T., 1980, Mutagenicity of products ob-
    tained from cysteamine-glucose browning model systems, J.
    Agric. Food Chem., 28:62.
Mirvish, S.S., 1975, Formation of N-nitroso compounds: Chemistry,
    Kinetics and in vivo occurrence, Toxicol. Appl. Pharm., 31:325.
Mumma, R. O., Raupach, D. R., Waldman, J. P., Tong, S. S. C., Jacobs,
    M. L., Hotchkiss, J. H., Babish, J. G., Van Campen, D. R.,
    Wszolek, P. C., Gutenmann, W. H., Bache, C. A., and Lisk, D. J.,
    1982, National survey of elements and other constituents in
    municipal sewage sludges, Environ. Sci. Technol. (submitted).
Nakamura, M., Baba, N., Nakaoka, T., Wada, Y., Ishibash, T., and
    Kawabata, T., 1976, Pathways of formation of N-nitroso-
    pyrrolidine in fried bacon, J. Food Sci., 41:874.
Oliver, J. E., 1981, Pesticide-derived nitrosamines: Occurrence and
    environmental fate, in: N-nitroso Compounds," R. A. Scanlan
    and S. R. Tannenbaum, eds., American Chemical Society,
    Washington, D.C.

Preussmann, R., Spiegelhalder, B., and Eisenbrand, G., 1981, Reduc-
    tion of human exposure to environmental N-nitroso compounds,
    in: "N-nitroso Compounds," R. A. Scanlan and S. R. Tannenbaum,
    eds., American Chemical Society, Washington, D.C.

Ross, R. D., Morrison, J., Rounbehler, D. P., Fan, T. Y., and Fine,
    D. H., 1977, N-nitroso compound impurities in herbicide formu-
    lations, J. Agric. Food Chem., 25:1416.

Ross, R., Morrison, J., and Fine, D. H., 1978, Assessment of
    dipropylnitrosamine levels in a tomato field following
    application of treflan EC, J. Agric. Food Chem., 26:455.

Scanlan, R. A., 1975, N-nitrosamines in foods, CRC Crit. Rev. Food
    Sci. Technol., 5:357

Sekizawa, J., and Shibamoto, T., 1980, Mutagenicity of 2-alkyl-N-
    nitrosothiazolidines, J. Agric. Food Chem., 28:781.

Sen, N. P., Donaldson, B., Charbonneau, C., Miles, W. F., 1974,
    Effect of additives on the formation of nitrosamines in meat
    curing mixtures containing spices and nitrites, J. Agric.
    Food Chem. 22:1125.

Sen, N. P., Donaldson, B. A., Seaman, S. Iyengar, J. R., Miles, W.
    F., 1978, Recent studies in Canada on the analysis and
    occurrence of volatile and nonvolatile N-nitroso compounds
    in foods, in: "Environmental Aspects of N-nitroso Compounds",
    E. A. Walker, M. Castegnaro, L. Griciute, and R. E. Lyle, eds.,
    International Agency for Research on Cancer, Lyon.

Sen, N. P., and Seaman, S., 1981, Volatile N-nitrosamines in dried
    foods, J. Assoc. Off. Anal. Chem., 64:1238.

Spiegelhalder, B., Eisenbrand, G., and Preussmann, R., 1979, Con-
    tamination of beer with trace quantities of N-nitrosodimethyl-
    amine, Food Cosmet. Toxicol., 17:29.

Spiegelhalder, B., Eisenbrand, G., Preussmann, R., 1980, Volatile
    nitrosamines in food, Oncology, 37:211.

Spiegelhalder, B., and Preussmann, R., 1981, Nitrosamines and rubber,
    in: "Proceedings of the Seventh International Meeting on
    N-nitroso Compounds: Occurrence and Biological Effects," (in
    press) International Agency for Research on Cancer, Lyon.

# 15

NUTRITIONAL AND METABOLIC RESPONSE TO PLANT INHIBITORS OF

DIGESTIVE ENZYMES

Daniel Gallaher and Barbara O. Schneeman

Department of Nutrition

University of California, Davis, CA  95616

## INTRODUCTION

Man's food source derives predominantly from plants, from which only a relatively small proportion can be utilized nutritionally.  Of these plants, many must be treated or processed in some way to improve the digestibility or remove naturally occurring toxicants.  Considerable effort has been put forth in elucidating those factors in food that may constitute a hazard to man or his domesticated animals when consumed.  One group of compounds that has received much attention is the digestive enzyme inhibitors.  The distribution of these inhibitors among plants is widespread, and includes most agronomic crops (Liener and Kakade, 1980).  The inhibitor is most often found within the edible portion.  By far the most prominent of these enzyme inhibitors are the inhibitors of the proteolytic enzymes, particularly trypsin and chymotrypsin.  Examples of amylase inhibitors are few, but potentially important.  Although no inhibitors of pancreatic lipase have been definitively established, a compound to be discussed may fit this role.  Finally, there are certain substances that inhibit all the digestive enzymes in a relatively non-specific manner.  Most notable among these are the tannins (Griffiths and Moseley, 1980).  The discussion to follow will consider only the more specific types of inhibitors.

Numerous studies have been conducted on the nutritional response to consumption of these digestive enzyme inhibitors.  For the most part, changes in growth and in the pancreas, which synthesizes the enzymes have been examined.  More recently, the meta-

bolic responses occurring secondarily to those on growth and on the
pancreas have received attention.  These nutritional and metablic
responses will be discussed.

## GROWTH DEPRESSION DUE TO PROTEOLYTIC INHIBITORS

Since the discovery by Osborne and Mendel in 1917 that raw
soybeans would not support growth in rats, numerous studies using
several different animal species have demonstrated that some uncooked plant products, particularly legumes, depress growth rates.
Because of the economic importance of legumes as an animal feedstuff and the potential importance of soy protein in human
nutrition, this effect of raw soybeans on growth has been extensively studied.  Several early reports found that the slight increment in apparent protein digestibililty caused by heating soybean
meal was too small to account for the pronounced improvement in
growth rate that also occurs (Melnick et al., 1946; Hayward et al.,
1936; Mitchell et al., 1945).  However, these studies of digestibility were based on fecal nitrogen excretion.  Carroll et al.
(1952) examined digestibility more directly by measuring net
nitrogen absorption from a section of the small intestine and
found that rats fed raw soybean meal absorbed less than half the
nitrogen of those fed heated meal.  Taken together, these studies
suggest that feeding raw soybean results in a considerable loss of
protein from the site of absorption in the small intestine, but
within the cecum bacteria metabolize the amino acids to nitro-
genous compounds which are absorbed from the cecum but not utilized.
Consequently, although fecal nitrogen excretion is only slightly
increased in animals fed raw vs. heated soybean, less nitrogen is
effectively available.

The discovery of a heat-labile trypsin inhibitor in soybeans
(Bowman, 1944; Ham and Sandstedt, 1941) suggested a hypothesis for
the growth depression effect.  Impairment of intestinal
proteolysis by the trypsin inhibitors present in soybeans could
cause a ration only marginally sufficient in some essential amino
acids to become deficient (Almquist and Merritt, 1951; Almquist
and Merritt, 1953).  Several studies have demonstrated decreased
proteolysis in the intestinal tract of young chicks fed raw soybean meal (Alumot and Nitsan, 1961; Bielorai and Bondi, 1963),
although older chicks quickly adapt and show normal levels of
intestinal proteolysis (Saxena et al., 1963; Nitsan and Alumot,
1964).  In the rat and mouse, two species in which raw soybean
feeding clearly causes growth depression, early studies had
suggested that no difference in intestinal proteolysis occurred
since intestinal proteolytic activity was generally equal or even
greater in animals fed raw soybeans (Lyman, 1957; Lyman and
Lepkovsky, 1957; Haines and Lyman, 1961; Nitsan and Bondi, 1965).
More recently, Lyman et al. (1974) demonstrated that in the rat

the presence of soybean trypsin inhibitor can interfere with
protein digestion by binding trypsin and preventing the complete
activation of chymotrypsinogen.

Proteolysis of endogenously secreted protein as well as
dietary protein could be reduced by inhibitors.  Lyman and
Lepkovsky (1957) found that feeding raw soybean meal or a purified
soybean trypsin inhibitor caused an exaggerated secretion of pancreatic enzymes into the small intestine.  Considerable amounts of
these enzymes were not degraded and reabsorbed but passed into the
cecum and large intestine.  This loss of essential amino acids from
endogenous sources in addition to the incomplete intestinal
proteolysis of dietary proteins contribute to the soy-induced
growth depression.  The loss of sulfur amino acids would be most
critical as the cystine content of pancreatic enzymes is high
(Neurath, 1961).  Supplementing raw soybean meal with certain
essential amino acids, especially methionine, improves the growth
rate associated with raw soybean meal to that of heated meal
(Borchers, 1959; Booth et al., 1960; Borchers, 1961).  Part of this
improvement is undoubtedly due to the fact that soybean meal is
limiting in methionine (Almquist et al., 1942).  However, the
growth response with methionine supplementation is much greater in
animals fed a raw soy flour diet than the heated flour, which is
consistent with an increased loss of cystine-rich endogenous
protein due to trypsin inhibitors (Gertler et al., 1967; Borchers,
1961; Fisher and Johnson, 1958).  In addition, rats fed purified
amino acid diets supplemented with an active anti-trypsin fraction
have considerably more TCA-precipitable nitrogen in their intestinal contents than those fed the same diet containing a
heat-inactivated anti-trypsin fraction (de Muelanaere, 1964).
Elevation of the TCA-precipitable nitrogen is most likely due to
the presence of excess pancreatic enzymes in the intestine.

Several factors will mitigate the deleterious effect of
trypsin inhibitors.  The degree of growth depression is dependent
upon the protein level of the diet.  An increase in raw soybean
meal from 15% to 25% of the diet was found to produce a large
increase in weight gain in rats (Fisher and Johnson, 1958) and
chicks (Gertler et al., 1967).  In another study, mice fed a 24%
casein diet containing a trypsin inhibitor fraction had a growth
rate equivalent to those fed the same diet without the inhibitor
(Roy and Schneeman, 1981).  These results indicate that increasing
the dietary supply of amino acids in terms of either protein
quantity or quality supports better growth by compensating for
endogenous losses.  Antibiotics incorporated into the diet also
have a beneficial effect on growth rates of animals fed a raw
soybean diet (Carroll et al., 1953; Nitsan and Bondi, 1965; Barnes
et al., 1965a).  In rats, prevention of coprophagy eliminates this
effect (Barnes et al., 1965b).  The antibiotics appear to reduce or
alter the intestinal microflora such that more endogenous protein

is excreted intact in the feces.  Coprophagy then makes this
protein available for reabsorption.  Chicks, which do not
coprophagize, experience less growth rate improvement with antibiotics than rats (Nitsan and Bondi, 1965).  Thus they apparently
absorb some amino acids from the lower intestine that are normally
degraded by bacteria, but the amount is quantitatively less significant than that reabsorbed by the rat via coprophagy.

Most of the research involving trypsin inhibitor effects of
growth have been done in rats, chicks, and mice.  However, growth
depression with raw soybean meal feeding has been noted in other
species, such as guinea-pigs (Patten et al., 1973), goslings
(Nitsan and Nir, 1977), growing swine (Yen et al., 1974), and
calves (Gorrill and Thomas, 1967).  The effect may be
age-dependent, as no difference in weight gain was found when raw
soybean meal was fed to laying hens (Saxena et al., 1963) or adult
dogs (Patten et al., 1971).  One study found no weight difference
in rats after six months of raw soybean meal feeding compared to
rats fed heated soybean meal, suggesting eventual catch-up occurs
(Booth et al., 1964).  The effect of age may be related to the
lower dietary protein requirement for adult animals.  Feeding a
lower protein diet may be necessary to test the effects of trypsin
inhibitors on weight gain in adult animals.  In general, it appears
that dietary trypsin inhibitors pose less of a threat to adult
animals.

Although trypsin inhibitors contribute to the growth
depression experienced by growing animals fed raw soybean meal, it
appears that other factors are involved.  The addition of purified
trypsin inhibitors to diets of heated soybean meal in amounts
equivalent to that present in the raw meal did not produce growth
depression to the same extent as did the raw meal (Gertler et al.,
1967).  Kakade et al. (1973) removed the trypsin inhibitors from
the raw meal by affinity chromatography.  When this inhibitor-
devoid raw meal was fed to rats, the degree of growth depression
was only somewhat less than that of the raw meal containing the
inhibitor.  Although this experiment left no doubt that trypsin
inhibitors in the diet can cause growth depression, it suggests
that much if not most of the effect is due to the raw soybean
protein itself.  Plant proteins in their native state are known to
be refractory to proteolytic attack resulting in poor amino acid
availability (Grau and Carroll, 1958; Bozzini and Silano, 1978).
In addition, the prolonged presence in the intestine of protein of
poor digestibility may act in a similar manner to trypsin
inhibitors by causing hypersecretion or reduced degradation of pancreatic enzymes (Green et al., 1973; Percival and Schneeman, 1979).
Heat treatment of soybean meal then would have two effects,
denaturing the trypsin inhibitors and increasing the digestibility

of the protein, both of which would lead to alleviation of growth depression. The degree to which each of these factors is important may be species dependent.

## PANCREATIC EFFECTS OF PROTEOLYTIC INHIBITORS

Pancreatic enlargement due to raw soybean meal feeding was first reported by Chernick et al. in 1948. Since that time the observation has been repeatedly confirmed in chicks (Nitsan and Alumot, 1964; Gertler and Nitsan, 1970; Nitsan and Nir, 1977) and rats (Booth et al., 1960; Konijn and Guggenheim, 1967; Rackis, 1965). Addition of purified trypsin inhibitor fractions to casein- or heated soybean meal- based diets has demonstrated that trypsin inhibitors are responsible for this phenomenon (Rackis, 1965; Khayambashi and Lyman, 1966; Gertler et al., 1967; Roy and Schneeman, 1981; Chan and de Lumen, 1982).

This enlargement has been characterized as primarily hyper-trophic, with some hyperplasia occurring (Melmed et al., 1976; Crass and Morgan, 1982; Kakade et al., 1967). Concomitant with enlargement is an increase in pancreatic proteolytic activity (Chernick et al., 1948; Applegarth et al., 1964; Kakade et al., 1967; Getler and Nitsan, 1970; Patten et al., 1973; Roy and Schneeman, 1981). This increase occurs very quickly, with only a single meal of raw soybeans needed to detect an acceleration of the trypsin synthetic rate in the rat (Dijkhof and Poort, 1978). Amylase activity in the chick pancreas may be either unchanged (Lepkovsky et al., 1966; Gertler and Nitsan, 1970; Nitsan and Nir, 1977) or slightly reduced (Hewitt et al., 1973; Kakade, 1967) with trypsin inhibitor-containing diets, whereas in the mouse it is increased (Roy and Schneeman, 1981). Lipase activity is unaffected in both species (Nitsan and Nir, 1977; Roy and Schneeman, 1981).

How trypsin inhibitors cause pancreatic hypertrophy and increase proteolytic enzyme concentrations is best explained in terms of the mechanism of regulation of pancreatic secretion. It has been shown that the levels of free intestinal trypsin and chymotrypsin act to monitor the secretion rate by a negative feed-back inhibition (Green and Lyman, 1972; Lyman et al., 1974; Schneeman and Lyman, 1975; Schneeman et al., 1977). These investi-gators postulated that the presence of these enzymes in the upper intestine suppress release of a humoral agent that increases pan-reatic secretion. The presence of protein in the tract leads to the formation of enzyme-substrate complexes allowing the release of this humoral agent, thereby increasing pancreatic juice flow. With more trypsin and chymotrypsin in the intestine, protein digestion can occur efficiently, and secretion rates will sub-sequently decrease. Cholecystokinin-pancreozymin (CCK-PZ) is believed to be the humoral agent released from the intestine, based on its known abilities to stimulate pancreatic secretion and cause both pancreatic hypertryphy and hyperplasia (Rothman and Wells, 1967; Mainz et al., 1973; Yanatori and Fijita, 1976).

Trypsin inhibitors in the intestine act analogous to protein, but are a much more potent stimulus (Green and Lyman, 1972; Green et al., 1973). Whereas the protein-enzyme complexes that form are reversible and fleeting, the inhibitor-enzyme complexes are essentially irreversible. Because of this the entire complex is believed to be lost from the animal. In this way a small amount of inhibitor mimics a large amount of protein and, when present in a meal, causes hypersecretion by the pancreas. The finding that purified trypsin inhibitor is a potent stimulator of the release of CCK-PZ-like activity from the upper intestine of the rat supports this hypothesis (Khayambashi and Lyman, 1969; Brand and Morgan, 1981; Schneeman and Lyman, 1975).

In addition to the rat and chick, the mouse (Schingoethe et al., 1970; Roy and Schneeman, 1981) and the young guinea-pig (Patten et al., 1973) have demonstrated pancreatic enlargement with raw soybean meal or purified trypsin inhibitor feeding. However, pancreatic enlargement failed to occur in the adult guinea-pig (Patten et al., 1973), the dog (Patten et al., 1971), the growing swine (Yen et al., 1977), or the calf (Gorrill and Thomas, 1967; Kakade et al., 1976). From this information, it was noted by Kakade et al. (1976), and reiterated by Liener (1979a; 1979b) and Liener and Kakade (1980), that animals which have pancreata that are 0.3% of body weight or more (such as the mouse, chick, rat, and young guinea-pig) show pancratic enlargement with trypsin inhibitor feeding. Those animals in which the pancreas weight is less than 0.3% (the adult guinea-pig, pig, dog, and calf) do not show pancreatic enlargement. These investigators suggested that, if this relationship was a valid one, then the human pancreas would not be expected to enlarge due to trypsin inhibitors, as it weighs about 0.09 to 0.12% of body weight.

A closer examination of the studies upon which this relationship is based suggests that one needs to view this relationship cautiously. To this end, several considerations should be noted. First of all, there is evidence that the negative feedback inhibition mechanism of pancreatic enzyme secretion found in the rat also exists in the pig (Corring, 1974), calf (Davicco et al., 1979) and possibly humans (Ishe et al., 1977). The existence of the same mechanism in these species implies the potential for the same effect of pancreatic enlargement by trypsin inhibitors as occurs in the rat. Another point to consider is the nutritional status of the animal during the experiment. The pancreas is a tissue that rapidly responds to starvation (Fauconneau and Michel, 1970) and to the protein content of the diet. It has been shown in both rats and chicks that increasing the protein content of the diet acts permissively to produce a greater degree of pancreatic

enlargement with trypsin inhibitor feeding, compared to a heated
soybean meal diet of similar protein content (Chernick et al.,
1948; Fisher and Johnson, 1958; Gertler et al., 1967). The quality
of the protein is also involved in the degree of the pancreatic
response.  An animal in good nutriture responds quickly to a change
in diet by altering the digestive enzyme profile (Snook, 1974;
Corring, 1980).  An increase in dietary protein will increase
trypsinogen and chymotrypsinogen in the pancreas within a few days.
Similarly, pancreatic amylase and lipase respond quickly to changes
in the amount of dietary carbohydrate and fat, respectively.  When
only poor quality proteins are fed, however, the ability of the
enzymes in the pancreas to adapt to changes in dietary composition
is impaired (Ikegami et al., 1975; Johnson et al., 1977).  Raw soybean meal can be considered a poor quality protein for two reasons.
First of all, the protein is refractory to digestion (Kakade et al.,
1973).  Thus, much of it is lost to the animal by excretion in the
feces or due to bacterial metabolism to compounds useless to the
animal.  In addition, soybean meal, and legumes in general, are
relatively deficient in sulfur amino acids, a problem which is compounded by the fact that a significant proportion of the cystine
in legumes is present in the trypsin inhibitors, which appear to
be unavailable to the animal (Kakade et al., 1969).  A final point
that is important concerning the pancreatic enzyme profile in
response to inhibitors is whether the animal was in an unfed or a
fed state prior to removing the pancreatic tissue.  The enzyme
profile in the pancreas of a fed animal will reflect the secretion
and synthesis that occurs in response to the diet and will not
necessarily reflect the longer term adaptive changes that can be
observed in an unfed animal.  In many of the studies cited it is
difficult to estimate whether the animal was in a fed or unfed
state when pancreatic enzyme levels were determined.

Keeping in mind the above, it is possible to re-examine the
studies in which no pancreatic hypertrophy was detected from a
different perspective.  For example, in the case of the adult
guinea-pig (Patten et al., 1973), the level of protein in the diet
was adequate, but the animals fed the raw soybean meal lost considerable weight during the experiment.  In the study on growing
swine (Yen et al., 1977), a slight weight loss occurred in one
trial and in another only a very slight weight gain occurred.
Furthermore, the protein level of 16% may have been inadequate to
have supported a hypertrophic response.  In both studies on calves,
the groups fed inhibitor-containing diets lost weight.  In fact,
in the study by Gorrill and Thomas (1967), 3 of 9 calves on the raw
soy diet died.  The failure to detect pancreatic enlargement is
probably related to the poor growth experienced by these animals.
In the study with adult dogs (Patten et al., 1971), pancreatic
enlargement, when expressed per kg body weight, was found in a
group fed raw soybean meal at 15% of the diet for 12 weeks.  This
enlargement was accompanied by significant increases in the con-

centrations of trypsinogen and chymotrypsinogen.  However, 8 weeks of feeding either a 15% or 30% raw soybean meal diet did not produce a pancreatic enlargement.  It may be that the amount of soybean meal in the diet was too little to produce a clear-cut response, as the concentration of trypsin inhibitor used in their 15% soybean meal diet was less than one-tenth that used in the unheated soy flour diet employed by Kakade et al. (1973).  Two additional points are worth making.  One is that the pancreatic growth response in adult animals may be species specific. Adult rats clearly showed pancreatic enlargement due to raw soybean feeding (Crass and Morgan, 1982) whereas laying hens did not (Saxena et al., 1963).  The other point is that the feedback mechanism reported to control the pancreas in various species does not seem to be important in the dog, and the dog pancreas did not respond to trypsin inhibitor in the intestine with pancreatic enzyme hypersecretion (Sale et al., 1977).  If these findings are correct, then the dog pancreas would not be expected to be very sensitive to the presence of trypsin inhibitors in the diet.

Examining the evidence as a whole, it does not appear justified to hypothesize a relationship between the occurrence of pancreatic enlargement with raw soybean feeding and pancreas size as a percent of body weight.  This relationship may, in fact, exist, but to establish its validity more definitive studies will be needed.  Until that time it is difficult to predict whether the human pancreas will enlarge with raw soybean feeding.

Another factor contributing to differences in the pancreatic response to trypsin inhibitors among species may be variations in the predominant forms of trypsin synthesized by the pancreas. One study using purified human enzymes and soybean inhibitor suggested that the predominant trypsin in human pancreatic juice is bound only weakly by the inhibitor (Figarella, 1974).  In contrast studies with crude pancreatic homogenates and crude soybean preparations suggest that substantial interference with human proteolytic activity can occur in vitro (Krogdahl and Holm, 1979). The extent of human enzyme inhibition in vivo by soybean inhibitor preparations and the physiological importance of any interactions are currently unknown.  Humans are unlikely to consume a diet rich in raw soybeans, thus avoiding the extreme responses observed in experimental animal studies.  Data currently available suggests that if the diet is adequate, especially with regard to protein, mechanisms exist in adult humans to adapt to small amounts of inhibitors which might be consumed in the diet (Ihse, 1977).

AMYLASE INHIBITORS

Naturally occurring inhibitors of pancreatic amylase were first discovered in aqueous extracts of wheat, rye, and kidney beans (Kneen and Sandstedt, 1943, 1946; Bowman, 1945). They were found to have the properties of a protein and to be quite thermostable. Inhibition occurred in a noncompetitive manner. The wheat-derived inhibitor is active against the amylase from a number of mammalian species, including the rat, mouse and dog, whereas the kidney bean-derived inhibitor was found to have substantial activity only against dog amylase (Ho et al., 1981). More recently a new class of amylase inhibitors have been isolated from culture broths of <u>Actinomycetes</u> (Schmidt et al., 1977). These inhibitors are complex oligosaccharides with a molecular weight of 500-5000. Some are fully competitive inhibitors, while others are noncompetitive (Folsch et al., 1981; Suchiro et al., 1981). Although it is unlikely that this class of inhibitors would occur in a normal diet, they may find therapeutical use in the treatment of obesity or diabetes (Keup and Puls, 1974).

Due to their thermostability, proteinaceous amylase inhibitors have been detected in significant amounts in the center of baked wheat-flour goods (Bessho and Kurosawa, 1968) and in some wheat-based breakfast cereals (Marshall, 1975). However, relatively few studies have been conducted on the physiological effects of consuming these inhibitors. Using large doses of the oligosaccharide inhibitor, the growth rates of rats was reduced when fed a diet where the carbohydrate source was starch (Puls and Keup, 1974; Folsch et al., 1981). This reduction occurred in spite of an increased food intake. Evidence is conflicting as to whether the proteinaceous amylase inhibitors reduce growth rates. In one study rats were fed a caloric-deficient, starch-free diet; addition of starch to the diet increased growth. When wheat-derived amylase inhibitor was included in with the starch, the increase in growth was much less than with the starch alone (Lang et al., 1974; Saunders, 1975). Heat-inactivated inhibitor was without effect. Furthermore, if sucrose replaced starch as the carbohydrate source, then inclusion of the inhibitor did not retard growth. Conflicting with this study is one by Savaiano et al. (1977). These investigators, using a similar technique of restricting food intake, fed kidney bean-derived amylase inhibitor to rats. In two separate experiments they found no effect on growth using an amount of inhibitor equal that to that found in 11 or 82 g of kidney bean flour per day. The source of the inhibitor may explain the discrepancy between these two reports. Rat pancreatic amylase is very sensitive to the wheat-derived inhibitor, but is only slightly affected by the inhibitor from kidney in vitro (Ho et al., 1981). This differential sensitivity is probably pH related, as the activity of the kidney bean-derived inhibitor drops off rapidly as the pH shifts from its optimum of 5.5 (Marshall and Lauda, 1975). As the intestinal lumen pH is about 6.8, one would expect the activity of the kidney bean-derived inhibitor to be low in vivo.

In partial agreement with the study of Lang et al. (1974), chicks fed inhibitor-containing wheat albumin, encapsulated to prevent gastric destruction, had a reduction in initial growth rate (Macri et al., 1977). Four weeks of this regime, however, led to an apparent adaptation, with growth rates of the inhibitor-fed chicks equalling that of the control animals. Of interest is the finding that the inhibitor preparation without encapsulation had no effect on growth. It is not clear why encapsulation is necessary to produce this effect in the chick, but not in the rat.

A decreased rate of starch hydrolysis in the small intestine would logically lead to a slower absorption of glucose and possibly cause certain metabolic changes. This has been found to be the case. In rats and dogs, a wheat-derived amylase inhibitor added to a starch load attenuated both the rise in blood glucose and the fall in non-esterified fatty acids that normally occurs with such a loading (Puls and Keup, 1973). In humans, a similar effect of the inhibitor on blood glucose levels was found, as well as a flattening of the serum insulin peak. These same effects on blood glucose and serum insulin are found when amylase inhibitors of microbiol origin are used (Puls and Keup, 1974). Long term feeding of this type of inhibitor led to a decreased rate of incorporation of $^{14}$C-starch into lipids of epididymal adipose tissue and of aortic tissue (Puls and Keup, 1974).

Given the dramatic ability of trypsin inhibitors to increase pancreatic trypsin and chymotrypsin, one might expect amylase inhibitors to produce an analogous response. However, regulation of pancreatic amylase levels appears to be fundamentally different than that of trypsin and chymotrypsin. As discussed in the previous section, the presence of protein or a substrate analogue within the intestinal lumen regulates these two proteolytic enzymes. In contrast, regulation of pancreatic amylase levels is related to the quantity of digestion products absorbed (Lavau et al., 1974). Hence, increases in the quantity of absorbed glucose will increase the pancreatic amylase activity. Insulin is involved in the stimulation of pancratic amylase synthesis, although its role may be only a permissive one (Soling and Unger, 1972). Accordingly, if starch is the only carbohydrate source, inhibition of amylase in the intestine would be expected to decrease pancreatic amylase concentrations, the opposite effect of trypsin inhibitors on trypsin and chymotrypsin. In two feeding studies with rats, using the microbial-derived amylase inhibitor, this was found to occur (Puls and Keup, 1974; Folsch et al., 1981). Pancreatic amylase levels dropped to one-half the control value in one experiment and to one-quarter in the other. However, the study by Macri et al. (1977) conflicts with these reports. These workers found that chicks receiving the encapsulated wheat albumin-inhibitor preparation showed pancreatic hypertrophy, an increased amylase activity, and no difference in pancreatic

protease activity.  It was claimed that the trypsin inhibitor
activity in the preparation was quantitatively insignificant
(Petrucci et al., 1974).  Since both types of inhibitors could
produce this effect their results are difficult to explain.  No
reason for the apparent discrepancy is obvious to these authors.
Further investigation will be required to clarify this matter.

Amylase inhibitors would seem to offer, in theory, an
attractive method of decreasing glucose absorption from starch,
thereby reducing the caloric content of a meal.  This possibility
has created interest in the scientific community for use of these
inhibitors in the control of obesity (Marshall, 1975; Sullivan
et al., 1980).  As of yet, no studies have been reported in which
amylase inhibitors have been evaluated in this capacity.  However,
numerous commercial concerns, impatient at the ponderous pace of
scientific investigation, and hearing the clear knock of
opportunity, have flooded the marketplace with amylase inhibitor
preparations, the so-called "starch blockers".  These preparations,
derived from kidney beans, commonly claim to contain enough
inhibitor activity to block the digestion of 100 g of starch, or
the equivalent of 400 kcal.  The enormous popularity of starch
blockers is attested to by the assertion that over one million
starch blocker tablets were consumed daily in the United States in
the first part of 1982 (Sheff, 1982).  Concern about the safety
and efficacy of starch blockers has prompted the Food and Drug
Administration to prohibit their sale until these issues are
resolved.

Only recently has the first study on the effectiveness of the
starch blockers in humans been published (Bo-Linn et al., 1982).
These investigators developed a one-day calorie-balance technique
to allow determination of the number of calories absorbed from a
high-starch test meal with and without starch blockers.  The subjects were first prepared by flushing the entire gastrointestinal
tract free of residual food and fecal matter.  A meal containing
97 g of starch was then consumed by each subject.  Two starch
blocker tablets (Carbo-Lite ) were taken at the beginning of the
meal and one at mid-meal.  Fourteen hours later, the GI tract was
again flushed and the rectal effluent collected.  The caloric
content of the effluent, as well as any stool passed during the
period, was determined by bomb calorimetry.  No difference was
found between subjects consuming the starch blockers or a placebo.
It was concluded that the starch blocker preparation used did not
inhibit the digestion and absorption of starch in vivo.

The failure of the starch blockers to decrease starch hydrolysis may be due to several factors.  It is possible that much of
the inhibitor, a glycoprotein (Marshall and Lauda, 1975), is
destroyed in the stomach by pepsin or acid.  Once in the small
intestine, the pH would be sufficiently far from the optimum for

enzyme-inhibitor complex formation that its activity could be reduced by two-thirds, based on in vitro studies (Marshall and Lauda, 1975). Furthermore, these same in vitro studies showed that the inhibitor must be preincubated with amylase in the absence of starch to achieve inhibition of the enzyme. Finally, one must consider that the pancreas normally secretes ten times the amount of amylase required to affect complete digestion of the starch in a meal (Gray, 1971). Thus, the amount of amylase inhibitor present in the starch blocker preparations, when consumed at the prescribed dose, would at best inhibit only a fraction of the amylase present.

LIPASE INHIBITORS

In contrast to the large number of known proteolytic enzyme inhibitors of plant origin, virtually no lipase inhibitors have been identified. Initial reports of a pancreatic lipase inhibitor from soybean cotyledons described a protein that decreased lipolytic activity (Mori et al., 1973; Satouchi et al., 1974). Pancreatic lipase, as all lipases, acts at the oil-water interface of an emulsified oil droplet. Therefore, the substrate concentration is not equal to the amount of triacylglycerol present in the system, but to the total interfacial area. This area, for a given amount of triacylglycerol, is dependent upon the size of the emulsified oil droplets and the presence of substances that compete for the interfacial area. Inhibitors of pancreatic lipase may act as true inhibitors by forming enzyme-inhibitor complexes, or they may reduce activity by adhering to the oil-water interface and thereby effectively reducing the substrate concentration. Further study of the soybean cotyledon lipase inhibitor revealed its mode of action to be adherence to the interface (Satouchi et al., 1976) and not an enzyme inhibitor in the classical sense. Whether this protein would have any physiological effects if fed is unknown.

Research conducted in our laboratory has uncovered another inhibitor of pancreatic lipase. While examining interactions among various types of dietary fibers and digestive enzyme activities, it was found that cellulose drastically reduced lipase activity in vitro (Schneeman, 1978). In these studies, cellulose was incubated with purified porcine lipase, then filtered and assayed for lipase activity. This method assures that inhibition is due to an affect on the enzyme itself and not to an interference with access of the enzyme to the substrate. This effect of cellulose on lipase occurs when the source of lipase is rat pancreatic juice (Schneeman and Gallaher, unpublished observations) or human pancreatic juice (Dunaif and Schneeman, 1981). Kinetic studies have not been done because of the difficulty of determining substrate concentrations. However, cellulose will reduce lipase activity in the presence of an emulsified oil substrate (Schneeman and Gallaher, unpublished observations). This suggested the possibility that cellulose may inhibit lipase activity in vivo as well.  An experiment was conducted to test this hypothesis (Schneeman and Gallaher, 1980).  Rats were fed either a no fiber diet or a diet containing 20% cellulose.  One group of animals was killed after 12 hours without food and is designated the unfed group.  Another group was killed 3 hours after a 3 g meal and is designated the fed group.  Some results of this study are shown in Table 1.  In the unfed group, both pancreatic weight and the lipase concentration  were the same for both dietary treatments.  Lipase activity in the fed group intestinal contents was considerably lower

TABLE 1

Pancreas weight and lipase activity in the intestines of rats adapted to a diet with cellulose (C) or fiber free (FF).

|  | Diet |  |
| --- | --- | --- |
| Pancreas (unfed) | FF | $245.6 \pm 20.2$ |
| (mg) | C | $241.0 \pm 19.3$ |
| Lipase in intestinal contents (fed) | FF | $43.4 \pm 5.1$[a] |
| (units[1]/contents) | C | $5.7 \pm 1.3$[a] |
| Lipase in intestinal mucosa (fed) | FF | $92.7 \pm 9.9$[b] |
| (units/mucosa) | C | $116.4 \pm 8.8$[b] |

[1]Lipase unit = 1 $\mu$eq OH[-] titrated/min at 30°, [a]P < 0.001;
[b] P < 0.10.

in the cellulose-fed animals.  However, lipase activity of the washed mucosa from the cellulose-fed group was actually slightly higher than the control group.  When the intestinal and mucosal values are combined, there is no statistically significant difference between diets.  Interpretation of these last results is difficult because it is not known to what extent mucosal bound pancreatic enzymes are active in the digestive process.  However, the pancreatic data provides some insight.  Pancretic lipase appears to be regulated in a manner similar to amylase.  That is, it is the amount of triacylglycerol digestion products absorbed, not the presence of substrate in the intestinal lumen, that influences pancreatic lipase activity (Lavau et al., 1974).  The lack of an effect of the dietary cellulose on pancreatic lipase concentrations strongly suggests that cellulose does not signifi-cantly reduce the total amount of triacylglycerol absorbed.  This does not preclude the possibility that cellulose does inhibit lipase in vivo, slowing digestion rate.  In humans, lipase activity must be reduced 90% to produce significant fat malabsorp-tion (DiMago et al., 1973).  Because of this normal hypersecretion, it may be that cellulose does inhibit lipase in vivo, but sufficient enzyme is still present to hydrolyze all the triacylglycerol present. If this is the case, absorption may simply be delayed, with greater amounts of fatty acids absorbed lower in the intestinal tract. As it has been shown that chylomicrons secreted from the distal intestine differ in composition and size from those secreted more proximally (Wu et al., 1980), such a shift in the site of absorption could effect lipid metabolism. Preliminary data indicates that cellulose may slow the disappearance of triacylglycerol from the small intestine (Schneeman et al., 1983).

CONCLUSIONS

Although enzyme inhibitors have been investigated since the early 1900's, we still have little information on their importance in the human diet. One can speculate that since humans have evolved with inhibitors of plant or animal origin in their diet, they have adapted to their presence either through physiological mechanisms or cultural food preparation practices. We are capable of estimating inhibitor levels in commercially prepared products to estimate potential intake levels. Both animal and human studies are needed to determine physiological and adaptive responses to these levels of inhibitor.

Modern science has begun to view enzyme inhibitors from perspectives other than their impact on nutrient availability. For example, naturally occurring levels of amylase inhibitors may not be nutritionally important, but researchers are examining their use as a drug in the treatment of diabetes or obesity. Likewise similar shifts are seen in the study of proteolytic enzyme inhibitors. Experiments with diabetic rats have indicated some enhancement of endocrine function as well as exocrine frunction with trypsin inhibitors in the diet, suggesting a use in diabetic treatment (Ihse et al., 1976). On the other hand, concern has been raised that the pancreatic hypertrophy associated with trypsin inhibitors will enhance the incidence of pancreatic tumors (Morgan et al., 1977). These associations at present time are unconfirmed and probably represent pharmacological responses to enzyme inhibitors, however, they do lead to new areas of investigation and emphasize the importance of designing experiments with humans to determine pancreatic response to the consumption of inhibitors.

REFERENCES

Almquist, H. J., Mecchi, E., Kratzer, F. H., and Grau, C. R., 1942, Soybean protein as a source of amino acids for the chick, J. Nutr., 24:385-392.

Almquist, H. J., and Merritt, J. B., 1953, Accentuation of dietary amino acid deficiency by raw soybean growth inhibitor, Proc. Soc. Exp. Biol. Med., 84:333-334.

Almquist, H. J., and Merritt, J. B., 1951, Effect of soybean anti-
    trypsin on experimental amino acid deficiency in the chick,
    Arch. Biochem. Biophys., 31:450-453.
Alumot, E., and Nitsan, Z., 1961, The influence of soybean anti-
    trypsin on the intestinal proteolysis of the chick,
    J. Nutr., 73:71-77.
Applegarth, A., Furuta, F., and Lepkovsky, S., 1964, Response of
    the chicken pancreas to raw soybeans, Poult. Sci., 43:733-738.
Barnes, R. H., Kwong, E., and Fiala, G., 1965a, Effect of
    penicillin added to an unheated soybean diet on cystine
    excretion in feces of the rat, J. Nutr., 85:123-126.
Barnes, R. H., Fiala, G., and Kwong, E., 1965b, Prevention of
    coprophagy in the rat and the growth stimulating effects of
    methionine, cystine, and penicillin when added to diets con-
    taining unheated soybeans, J. Nutr., 85:127-131.
Bielorai, R., and Bondi, A., 1963, Relationship between
    'antitrypic factors' of some plant protein feeds and products
    of proteolysis precipitable by trichloroacetic acid, J. Sci.
    Food Agric., 14:124-132.
Bessho, H., and Kurosawa, S., 1967, Enzyme inhibitor in foods. III.
    Effect of cooking on the amylase inhibitor in flour.  Eiyo To
    Shokuryo 20:317-319; Chem. Abstr., 68:113474e.
Bo-Linn, G., Santa Ana, C., Morawski, S., and Fordstran, J., 1982,
    Starch blockers - their effect on calorie absorption from a
    high-starch meal, New Engl. J. Med., 307:1413-1416.
Booth, A. N., Robbins, D. J., Ribelin, W. E., and DeEds, F., 1960,
    Effect of raw soybean meal and amino acids on pancreatic
    hypertrophy in rats, Proc. Soc. Exp. Med. Biol., 104:681-
    683.
Booth, A. N. Robbins, D. J., Ribelin, W. E., DeEds, F., Smith,
    A. K., and Rackis, J. J., 1964, Prolonged pancreatic hyper-
    trophy and reversibility in rats fed raw soybean meal, Proc.
    Soc. Exp. Biol. Med., 116:1067-1068.
Borchers, R., 1961, Counteraction of the growth depression of raw
    soybean oil meal by amino acid supplements in weanling rats,
    J. Nutr., 75, 330-334.
Borchers, R., 1959, Tyrosine stimulates growth on raw soybean
    rations, Fed. Proc., 18:517 (abstract).
Bowman, D., 1945, Amylase inhibitors of navy beans, Science, 102:
    358-359.
Bowman, D. E., 1941, Fractions derived from soy beans and navy
    beans which retard tryptic digestion of casein, Proc. Soc.
    Exp. Biol. Med., 57:139-140.
Bozzini, A., and Silano, V., 1978, Control through breeding
    methods of factors affecting nutritional quality of cereals
    and grain legumes, in: "Nutritional Improvement of Food and
    Feed Proteins," M. Friedman, ed., Plenum Press, New York.
Brand, S. J., and Morgan, R. G. H., 1981, The release of rat
    intestinal cholecystokinin after oral trypsin inhibitor
    measured by bio-assay, J. Physiol., 319:325-343.

Carroll, R. W., Hensley, G. W., Graham, Jr., W. R., 1952, The site
    of nitrogen absorption in rats fed raw and heat-treated
    soybean meals, Science, 115:36-39.
Carroll, R. W., Hensley, G. W., Sittler, C. L, Wilcox, E. L., and
    Graham, Jr., W. R., 1953, Absorption of nitrogen and amino
    acids from soybean meal as affected by heat treatment or
    supplementation with aureomycin and methionine, Arch.
    Biochem. Biophys., 45:260-274.
Chan, J., and de Lumen, B. O., 1982, Biological effects of
    isolated trypsin inhibitor from winged bean (Psophocarpus
    tetragonolobus) on rats, J. Agric. Food Chem., 30:46-50.
Chernick, S. S., Lepkovsky, S., and Chaikoff, I. L., 1948, A
    dietary factor regulating the enzyme content of the
    pancreas:  changes induced in size and proteolytic activity
    of the chick pancreas by the ingestion of raw soybean meal,
    Am. J. Physiol., 155:33-41.
Corring, T., 1980, The adaptation of digestive enzymes to the
    diet:  its physiological significance, Reprod. Nutr. Develop,
    20:1217-1235.
Corring, T., 1974, Regulation de la secretion pancreatique par
    retro-action negative chez le porc, Ann. Biol. Anim. Bioch.
    Biophys., 14:487-498.
Crass, R. A., and Morgan R. G. H., 1982, The effect of long-term
    feeding of soya-bean flour diets on pancreatic growth in the
    rat, Br. J. Nutr., 47:119-129.
Davicco, M. J., Lefaivre, J., Thiuend, P., and Barlet, J. P.,
    1979, Feedback regulation of pancreatic secretion in the
    young milk-fed calf, Ann. Biol. Anim. Bioch. Biophys., 19:
    1147-1152.
de Muelenaere, H. J. H., 1964, Studies on the digestion of soy-
    beans, J. Nutr., 82:197-205.
Dijkhof, J., and Poort, C., 1978, Changes in rat pancreatic protein
    synthesis after a single feeding with diets containing raw or
    heated soybeans, J. Nutr., 108:1222-1228.
DiMagno, E., Go, V., and Summerskill, W., 1973, Relations between
    pancreatic enzyme outputs and malabsorption in severe
    pancreatic insufficiency, N. Engl. J. Med., 288:813-815.
Dunaif, G., and Schneeman, B. O., 1981, The effect of dietary
    fiber on human pancreatic enzyme activity in vitro, Am. J.
    Clin. Nutr., 34:1034-1035.
Fauconneau, G., and Michel, M. C., 1970, The role of the gastro-
    intestinal trace in the regulation of protein metabolism,
    in: "Mammalian Protein Metabolism," H. Munro, ed., Academic
    Press, New York.
Figarella, C., Negri, G. A., and Guy, O., 1974, Studies on
    inhibition of the two human trypsins, in: "Proteinase
    Inhibitors," Fritz et al., ed., Springer-Verlag, New York.
Fisher, H., and Johnson, Sr., D., 1958, The effectiveness of
    essential amino acid supplementation in overcoming the growth

depression of unheated soybean meal, Arch. Biochem. Biophys., 77:124-128.

Folsch, U., Grieb, N., Caspary, W., and Creutzfeldt, W., 1981, Influence of short- and long-term feeding of an α-amylase inhibitor (BAY e 4609) on the exocrine pancreas of the rat, Digestion, 21:74-82.

Gertler, A., Birk, Y., and Bondi, A., 1967, A comparative study of the nutritional and physiological significance of pure soybean trypsin inhibitors and of ethanol-extracted soybean meals in chicks and rats, J. Nutr., 91:358-370.

Gertler, A., and Nitsan, Z., 1970, The effect of trypsin inhibitors on pancreotopeptidase E, trypsin, chymotrypsin, and amylase in the pancreas and intestinal tract of chicks receiving raw and heated soya-bean diets, Br. J. Nutr., 24: 803-904.

Gorrill, A. D. L., and Thomas, J. W., 1967, Body weight changes, pancreas size and enzyme activity, and proteolytic enzyme activity and protein digestion in intestinal contents from calves fed soybean and milk protein diets, J. Nutr., 92: 215-223.

Grau, C. R., and Carroll, R. W., 1958, Evaluation of protein quality of processed plant protein foodstuffs, in: "Processed Plant Protein Foodstuffs," A. M. Altschul, ed., Academic Press, New York.

Gray, G., 1971, Intestinal digestion and maldigestion of dietary carbohydrates, Ann. Rev. Med., 22:391-404.

Green, G. H., Olds, B. O., Matthews, G., and Lyman, R. L., 1973, Protein as a regulator of pancreatic enzyme secretion in the rat, Proc. Soc. Exp. Biol. Med., 142:1162-1167.

Green, G. M., and Lyman, R. L., 1972, Feedback regulation of pancreatic enzyme secretion as a mechanism for trypsin inhibitor induced hypersecretion in rats, Proc. Soc. Exp. Biol. Med., 140:6-12.

Griffiths, D. W., and Moseley, G., 1980, The effect of diets containing field beans of high or low polyphenolic content on the activity of digesive enzymes in the intestines of rats, J. Sci. Food Agric., 31:255-259.

Haines, P. C., and Lyman, R. L., 1961, Relationship of pancreatic enzyme secretion to growth inhibition in rats fed soybean trypsin inhibitor, J. Nutr., 74:445-452.

Ham, W. E., and Sandstedt, R. M., 1944, A proteolytic inhibiting substance in the extract from unheated soy bean meal, J. Biol. Chem., 154:505-506.

Hayward, J. W., Steenbock, H., and Bohstedt, G., 1936, The effect of heat as used in the extraction of soy bean oil upon the nutritive value of the protein of soy bean oil meal, J. Nutr., 11:219-234.

Hewitt, D., Coates, M. E., Kakade, M. L., and Liener, I. E., 1973, A comparison of fractions prepared from navy (haricot) beans

(Phaseolus valgaris L.) in diets for germ-free and conventional chicks, Br. J. Nutr., 29:423-435.

Ho, R., Aranda, C., and Venico, J., 1981, Species differences in
response to two naturally occurring α-amylase inhibitors,
J. Pharm. Pharmacol., 33:551.

Ihse, I., Lundquist, I., and Arnesijo, B., 1976, Oral trypsin-
inhibitor-induced improvement of the exocrine and endocrine
pancreatic functions in alloxan diabetic rats, Scand. J.
Gastroent., 11:363-368.

Ikegami, S., Takai, Y., and Iwoa, H., 1975, Effect of dietary
protein on proteolytic activities in the pancreatic tissue
and contents of the small intestine in rats, J. Nutr. Sci.
Vitaminol., 21:287-295.

Ihse, I., Lilja, P., and Lundquist, I., 1977, Feedback regulation
of pancreatic enzyme secretion by intestinal trypsin in man.
Digestion, 15:303-308.

Johnson, A., Hurwitz, R., and Kretchmer, N., 1977, Adaptation of
rat pancreatic amylase and chymotrypsinogen to changes in
diet, J. Nutr., 107:87-96.

Kakade, M. L., Arnold, R. L., Liener, I. E., and Waibel, P. E.,
1969, Unavailability of cystine from trypsin inhibitors as
a factor contributing to the poor nutritive value of navy
beans, J. Nutr., 99:34-41.

Kakade, M. L., Barton, T. L., Schaible, P. J., and Evan, R. J.,
1967, Biochemical changes in the pancreases of chicks fed
raw soybeans and soybean meal., Poult. Sci., 46:1578-1585.

Kakade, M. L., Hoffa, D. E., and Liener, I. E., 1973, Contribution of trypsin inhibitors to the deleterious effects of
unheated soybeans fed to rats, J. Nutr., 103:1772-1778.

Kakade, M. L., Thompson, R. D., Engelstad, W. E., Behrens, G. C.,
Yoder, R. D., and Crave, F. M., 1976, Failure of soybean
trypsin inhibitor to exert deleterious effects in calves,
J. Dairy Sci., 59:1484-1489.

Keup, U., and Puls, W., 1974, Influence of an amylase inhibitor,
BAY e4609, on the conversion of orally applicated starch
into total lipids of rat adipose tissue, in: "Recent
Advances in Obesity Research," A. Howard, ed., Newman
Publishing Co., Ltd., London.

Khayambashi, H., and Lyman, R. L., 1966, Growth depression and
pancreatic and intestinal changes in rats force-fed amino
acid diets containing soybean trypsin inhibitor, J. Nutr.,
89:455-464.

Khayambashi, H., and Lyman, R. L., 1969, Secretion of rat pancreas
perfused with plasma from rats fed soybean trypsin inhibitor,
Am. J. Physiol., 217:646-651.

Kneen, E., and Sandstedt, R., 1946, Distribution and general
properties of an amylase inhibitor in cereals, Arch. Biochem.,
9:235-249.

Kneen, E., and Sandstedt, R., 1943, An amylase inhibitor from
certain cereals, J. Am. Chem. Soc., 65:1247-1248.

Konijn, A. M., and Guggenheim, K., 1967, Effect of raw soybean flour on the composition of rat pancreas, Proc. Soc. Expl. Biol. Med., 126:65-67.

Krogdahl, A., and Holm, H., 1979, Inhibition of human and rat pancreatic proteinases by crude and purified soybean proteinase inhibitors, J. Nutr., 109:551-558.

Lang, J., Chang-Hum, L., Reyes, P., and Briggs, G., 1974, Interference of starch metabolism by α-amylase inhibitors, Fed. Am. Soc. Exp. Biol., 33:718 (abstract).

Lavau, M., Bazin, R., and Herzog, J., 1974, Comparative effects of oral and parenteral feeding on pancreatic enzymes in the rat, J. Nutr., 104:1432-1437.

Lepkovsky, S., Koike, T., Sugiura, M., Dimick, M. K., and Furuta, F., 1966, Pancreatic amylase in chickens fed on soya-bean diets, Br. J. Nutr., 20:421-437.

Liener, I., 1979a, Significance for humans of biologically active factors in soybeans and other food legumes, J. Am. Oil Chemists Soc., 56:121-129.

Liener, I. E., 1979b, Protease inhibitors and lectins, in: "International Review of Biochemistry," A. Veuberger and T. H. Jukes, eds., University Park Press, Baltimore.

Liener, I. E., and Kakade, M. L., 1980, Protease inhibitors, in: "Toxic Constituents of Plant Foodstuffs," I. E. Liener, ed., Academic Press, New York.

Lyman, R. L., 1957, The effect of raw soybean meal and trypsin inhibitor diets on the intestinal and pancreatic nitrogen in the rat, J. Nutr., 62:285-294.

Lyman, R. L., and Lepkovsky, S., 1957, The effect of raw soybean meal and trypsin inhibitor diets on pancreatic enzyme secretion in the rat, J. Nutr., 62:269-284.

Lyman, R. L., Olds, B. O., Green, G. M., 1974, Chymotrypsinogen in the intestine of rats fed soybean trypsin inhibitor and its inability to suppress pancreatic enzyme secretions, J. Nutr., 104:105-110.

Macri, A., Parlamenti, R., Silano, V., and Valfre, F., 1977, Adaptation of the domestic chicken, Gallus Domesticus, to continuous feeding of albumin amylase inhibitors from wheat flour as gastro-intestinal microgranules, Poult. Sci., 56: 434-441.

Mainz, D. L., Black, O., and Webster, P. D., 1973, Hormonal control of pancreatic growth, J. Clin. Invest., 52:2300-2304.

Marshall, J., 1975, α-amylase inhibitors in foods, in: "Physiological Effects of Food Carbohydrates", Am. Chem. Soc. Symp. Ser. 115:244-266.

Marshall, J., and Lauda, C., 1975, Purification and properties of phaseolamin, an inhibitor of α-amylase, from the kidney bean, Phaseolus vulgaris, J. Biol. Chem., 250:8030-8037.

Melmed, R. N., El-Aaser, A. A. A., and Holt, S. J., 1976, Hyper- trophy and hyperplasia of the neonatal rat exocrine pancreas induced by orally administered soybean trypsin inhibitor, Biochim. Biophys. Acta, 421:280-288.

Melnick, D., Oser, B. L., and Weiss, S., 1946, Rate of enzymatic digestion of proteins as a factor in nutrition, Science, 103: 326-329.

Mitchell, H. H., Hamilton, T. S., and Beadles, J. R., 1945, The importance of commercial processing for the protein value of food products, J. Nutr., 29:13-25.

Morgan, R. G. H., Levinson, D. A., Hopwood, D., Saunders, J. H. B., and Wormsley, K. G., 1977, Potentiation of the action of azaserine on the rat pancreas by raw soybean flour, Cancer Let., 3:87-90.

Mori, T., Satouchi, K., and Matsushita, S., 1973, A protein inhibiting pancreatic lipase activity in soybean seeds, Agric. Biol. Chem., 37:1225-1226.

Neurath, H., 1961, Considerations of the occurrence, structure, and function of the proteolytic enzymes of the pancreas, in: "The Exocrine Pancreas," A. V. S. de Reuck and M. P. Cameron, eds., Little, Brown, Boston.

Nitsan, Z., and Alumot, E., 1964, Overcoming the inhibition of intestinal proteolytic activity caused by raw soybean in chicks of different ages, J. Nutr., 84:179-184.

Nitsan, Z., and Bondi, A., 1965, Comparison of nutritional effects induced in chicks, rats and mice by raw soya-bean meal, Br. J. Nutr., 19:177-187.

Nitsan, Z., and Nir, I., 1977, A comparative study of the nutritional and physiological significance of raw and heated soya beans in chicks and goslings, Br. J. Nutr., 37:81-91.

Osborne, T. B., and Mendel, L. S., 1917, The use of soy bean as food, J. Biol. Chem., 32:369-387.

Patten, J. R., Patten, J. A., and Pope II, H., 1973, Sensitivity of the guinea-pig to raw soya bean in the diet, Fd. Cosmet. Toxicol., 11:577-583.

Patten, J. R., Richards, E. A., and Pope II, H., 1971, The effect of raw soybeans on the pancreas of adult dogs, Proc. Soc. Exp. Biol. Med., 137:59-63.

Percival, S. S., and Schneeman, B. O., 1979, Long term pancreatic response to feeding heat damaged casein in rats, J. Nutr., 109:1609-1614.

Petrucci, T., Tomasi, M., Cantagalli, P., and Silano, V., 1974, Comparison of wheat albumin inhibitors of $\alpha$-amylase and trypsin, Phytochemistry, 13:2487-2495.

Puls, W., and Keup, U., 1973, Influence on an $\alpha$-amylase inhibitor (BAY d 7791) on blood glucose, serum insulin, and NEFA in starch loading tests in rats, dogs, and man. Diabetologia 9:97-101.

Puls, W., and Keup, U., 1974, Metabolic studies with an amylase inhibitor in acute starch loading tests in rats and men and its influence on the amylase content of the pancreas, in:

"Recent Advances in Obesity Research," A. Howard, ed.,
Newman Publishing Co., Ltd., London.

Rackis, J. J., 1965, Physiological properties of soybean trypsin
inhibitors and their relationship to pancreatic hypertrophy
and growth inhibition of rats, Proc. Fed. Am. Soc. Exp.
Biol., 24:1488-1493.

Rothman, S. S., and Wells, H., 1967, Enhancement of pancreatic
enzyme synthesis by pancreozymin, Am. J. Physiol., 213:
215-218.

Roy, D. M., and Schneeman, B. O., 1981, Effect of soy protein,
casein and trypsin inhibitor on cholesterol, bile acids and
pancreatic enzymes in mice, J. Nutr., 111:878-885.

Sale, J. K., Goldberg, D. M., Fawcett, A. N., and Wormsley, K. G.,
1977, Chronic and acute studies indicating absence of exocrine
pancreatic feedback inhibition in dogs, Digestion, 15:540-555.

Satouchi, K., and Matsushita, S., 1976, Purification and properties
of a lipase inhibiting protein from soybean cotyledons,
Agric. Biol. Chem., 40:889-897.

Satouchi, K., Tomohiko, M., and Matsushita, S., 1974, Characterization of inhibitor protein for lipase in soybean seeds.
Agric. Biol. Chem., 38:97-101.

Saunders, R. M., 1975, α-amylase inhibitors in wheat and other
cereals, Cereal Foods Wrld., 20:282-285.

Savaiano, D., Powers, J., Costello, M., Whitaker, J., and
Clifford, A., 1977, The effect of an α-amylase inhibitor
on the growth rate of weanling rats, Nutr. Rep. Int., 15:
443-449.

Saxena, H. C., Jensen, L. S., and McGinnis, J., 1963, Influence
of age on utilization of raw soybean meal by chickens, J.
Nutr. 80:391-396.

Schingoethe, D. J., Aust, S. D., and Thomas, J. W., 1970,
Separation of a mouse growth inhibitor in soybeans from
trypsin inhibitors, J. Nutr., 100:739-748.

Schmidt, D. D., Frommer, W., Junge, B., Miller, I., Wingender, W.,
and Truscheit, E., 1977, α-Glucosidase inhibitors,
Naturwissenschaften 64:535-536.

Schneeman, B. O., 1978, Effect of plant fiber on lipase, trypsin
and chymotrypsin activity, J. Food Sci., 43:634-635.

Schneeman, B. O., Forman, L. P., and Gallaher, D., 1983,
Pancreatic and intestinal enzyme activity in rats fed various
fibre sources, in: "Fibre in Human and Animal Nutrition,"
G. Wallace and L. Bell, eds, Royal Society of New Zealand
Bulletin 20, in press.

Schneeman, B. O., and Gallaher, D., 1980, Changes in small
intestinal digestive enzyme activity and bile acids with
dietary cellulose in rats, J. Nutr., 110:584-590.

Schneeman, B. O., Chang, I., Smith, L., and Lyman, R. L., 1977,
Effect of dietary amino acids, casein, and soybean trypsin
inhibitor on pancreatic protein secretion in rats, J. Nutr.,
107:281-288.

Schneeman, B. O., and Lyman, R. L., 1975, Factors involved in
    the intestinal feedback regulation of pancreatic enzyme
    secretion in the rat, Proc. Soc. Exp. Biol. Med., 148:
    897-903.
Sheff, D., 1982, Want to have your pasta and eat it, too?, People
    Weekly, June 28:30-32.
Snook, J. T., 1974, Adaptive and nonadaptive changes in digestive
    enzyme capacity influencing digestive function, Fed. Proc.,
    33:88-89.
Soling, H., and Unger, K., 1972, The role of insulin in the
    regulation of α-amylase synthesis in the rat pancreas, Eur.
    J. Clin. Invest. 2:199-212.
Suchino, I., Otsuki, M., Yamasaki, T., Ohki, A., Sakamoto, C.,
    Yuu, H., Maeda, M., and Baba, S., 1981, Effect of α-
    glucosidase inhibitor on human pancreatic and salivary
    α-amylase, Clin. Chim, Acta, 117:145-152.
Sullivan, A., Comai, K., and Triscari, J., 1980, Novel anti-
    obesity agents whose primary site of action is the gastro-
    intestinal tract, in: "Recent Advances in Obesity Research,"
    P. Bjorntorp, M. Cairella, and A. Howad, eds., John Libby,
    London.
Wu, A-L., Clark, S. B., and Holt, P., 1980, Composition of lymph
    chylomicrons from proximal or distal rat small intestine,
    Am. J. Clin. Nutr., 33:582-589.
Yanatori, Y., and Fujita, T., 1976, Hypertrophy and hyperplasia
    in the endocrine and exocrine pancreas of rats fed soybean
    trypsin inhibitor or repeatedly injected with pancreozymin,
    Arch. Histol. Jap., 39:67-78.
Yen, J. T., Hymowitz, T., and Jensen, A. H., 1974, Effects of
    soybeans of different trypsin-inhibitor activities on
    performance of growing swine, J. Anim. Sci., 38:304-309.
Yen, J. T., Jensen, A. H., and Simon, J., 1977, Effect of dietary
    raw soybean and soybean trypsin inhibitor on trypsin and
    chymotrypsin activities in the pancreas and in small
    intestinal juice of growing swine, J. Nutr., 107:156-165.

## 16

NUTRITIONAL STUDIES OF A CARBOXYPEPTIDASE INHIBITOR FROM

POTATO TUBERS

Gregory Pearce, James McGinnis and Clarence A. Ryan

Institute of Biological Chemistry,
Biochemistry/Biophysics Program and
Program in Nutrition
Washington State University
Pullman, Washington 99164

ABSTRACT

Carboxypeptidase inhibitor from potato tubers was fed to newly hatched chicks at a level equal to that present in diet containing 50% raw potato, which caused severe growth depression and 20% mortality. At this level the effects of the inhibitor on growth was small but the following effects were noted: (a) increased fecal protein (the increase mainly consisting of low molecular weight proteins); (b) poorer feed efficiency; and (c) a significant decrease in pancreatic digestive proenzyme levels, although no hypertrophy was noted. In addition, the inhibitor was not digested readily in the intestinal tract and it increased in concentration in intestinal contents as it progressed down the tract. Potato Inhibitor II, a potent trypsin inhibitor, when fed to chicks, also at the level found in a 50% raw potato diet, was severely growth depressing. It significantly increased fecal protein and caused pancreatic hypertrophy. The trypsin inhibitor may be a major growth depressing agent in raw potatoes whereas the carboxy-peptidase inhibitor probably contributes little to the growth depression.

INTRODUCTION

Proteinase inhibitor proteins are thought to be the most abundant antinutrient proteins found in nature. They are stored in tubers and seeds of plants and have been found in storage organs of nearly every species of plant where they have been

sought.   In major agricultural crops they can comprise 5-15% of
the storage proteins, depending upon the species or variety.

The antinutrient properties of the proteinase inhibitor
fractions have been ascribed primarily to the inhibitors of
trypsin.  However only limited data is available concerning the
effects of inhibitors of chymotrypsin and elastase, and virtually
nothing is known of the effects in animal diets of inhibitors
with specificities toward the carboxypeptidase A and B.  The
major impediment to such research is the availability, for
testing in animal diets, of inhibitors having these latter
specificities.

Potatoes contain a variety of proteinase inhibitors that
inhibit all five of the major pancreatic digestive endo- and
exopeptidases that are responsible for protein degradation in the
intestinal tracts of higher animals (1-4).  The growth depressing
effects on chick growth of a partially purified inhibitor fraction
from Russet Burbank potato tubers, containing several proteinase
inhibitors, has been reported.  Upon autoclaving, this fraction
became growth promoting, presumably because the antinutrient
inhibitor proteins, when denatured, are excellent food proteins
(5).

Most of the inhibitors in potato tubers are destroyed
rapidly by cooking.  Huang et al. (6) demonstrated that two of
the major endopeptidase inhibitors present in potato tubers were
rapidly inactivated during three commonly used methods of cooking,
i.e., boiling, baking and microwaves.  However, the carboxypeptidase A and B inhibitor (CPI) survived all three treatments.
The unique stability of this inhibitor may be partially due to
its small size and its high content of disulfide linkages (3).
The three disulfide bonds involve six of the 39 amino acids
present in the molecule (15% of the total).

Inhibitors of the metallo-carboxypeptidases A and B are
rare.  The first such inhibitor was isolated from potatoes, and
later an inhibitor with similar specificity was isolated from the
intestinal worm, Ascaris lumbricoides.  The latter inhibitor is
difficult to purify, which precludes any nutritional studies in
animals.  Potato CPI, on the other hand, can represent a significant
contribution to the potato proteins, depending upon variety (7).  In
a survey of 106 commercial and experimental varieties, CPI varied
from zero to 846 µg/ml tuber juice (over 4% of the soluble
proteins).  In Russet Burbank potatoes the levels are about 30%
of this value.  The isolation of enough CPI from Russet Burbank
potatoes for a nutritional study was therefore possible, if a
simple efficient method could be developed to isolate it.

In this communication we describe a novel method to isolate large amounts of CPI and evaluate it, for the first time, in an animal diet. The effects of CPI in diets of newly hatched chicks were compared with control diets, diets containing raw and cooked potatoes, and with a diet containing a potent trypsin-chymotrypsin inhibitor from potato tubers, called Inhibitor II.

## PREPARATION OF CARBOXYPEPTIDASE INHIBITOR

CPI is an unusually stable naturally occurring polypeptide that can survive severe heating conditions (80°C) both in the pure state (in solution) and in situ in potato tubers during cooking (3). The inhibitor is also known to be soluble in 80% ethanol (C. A. Ryan personal observation). Thus, the heat stability and ethanol solubility were utilized as major steps to isolate the inhibitor in high yields.

Russet Burbank potatoes were quartered and placed in a water bath at 80°C for 10 min. The tubers were rinsed with cold water and blended to a paste. Absolute ethanol was blended in slowly with the paste until it was 80% ethanol by volume. The mixture was filtered through Whatman No. 1 filter paper with the aid of a vacuum. The ethanol in the filtrate was removed by vacuum evaporation and the resulting solution was dialyzed in a 2,000 MW cut off dialysis bag against several changes of distilled water. The remaining extract in the dialysis bag was lyophilized. The dry material contained one major polypeptide, CPI, and some minor components. On a protein basis the material was over 95% CPI. Purity of this crude CPI fraction was also demonstrated by electrophoresis by the method of Panyim and Chalkley (8). No Inhibitor I nor Inhibitor II could be detected by immunological techniques, however a small quantity of trypsin inhibitor activity (less than 0.5%) was present, presumably from contamination by the polypeptide trypsin inhibitor that is present in potato tubers (4). This CPI preparation was utilized for the chick feeding experiments.

## EFFECTS OF CPI IN CHICK DIETS

Supplementing chick diets with a protein fraction, isolated from raw Russet Burbank potato tubers, enriched in proteinase inhibitors, had previously been shown to severely depress their growth (5). The inhibitor-rich fraction contained at least six well characterized inhibitors of mammalian pancreatic digestive proteinases trypsin, chymotrypsin, elastase and carboxypeptidases A and B (5). Since the fraction contained an array of proteinase inhibitors it was not known if CPI contributed to the growth depressing activities.

Before embarking on a feeding study with CPI, it was important to establish if the inhibitor was indeed an inhibitor of chick carboxypeptidases A and B. Extracts of chick pancreata, after activation with trypsin, contained considerable carboxypeptidase A and B activities. CPI potently inhibited both chick carboxypeptidase A and carboxypeptidase B activities. The inhibition of the chick carboxypeptidases by CPI was not surprising since the inhibitor has a broad specificity toward metallo-carboxypeptidases and has been shown to inhibit all animal carboxypeptidases A and B tested to date including those of insects (9).

The CPI-enriched protein fraction was then fed to newly hatched chicks. The CPI preparation was added to a diet that was designed to provide all nutritional requirements, allowing for the addition of the test materials. Preparations of freeze dried raw and cooked (autoclaved to destroy all inhibitors including CPI) potato tubers were added to the diet in order to establish basal data for comparisons with the test diets containing the individual inhibitors. In Figs. 1A and 1B the effects of the raw and cooked potato tubers are shown with respect to their percentage of the total dietary solids. This data shows that the addition of cooked potatoes to the diets of newborn chicks did not significantly effect growth rates with respect to basal diets. On the other hand, raw potatoes significantly decreased weight gain, even when added as 10% of the total diet. The effects of raw potatoes became more pronounced as their percent of the diet increased (Fig. 1A). Raw potato solids at 50% of the diets had caused the feed efficiency to increase from 1.5 to 3.6 (Fig. 1B). The decrease in weight gain was accompanied by a lower consumption of diet. Severe diarrhea was observed with chicks fed the raw potato diets. While not quantified, it appeared that a correlation existed between the dietary levels of raw potato solids and the severity of the diarrhea. No diarrhea was observed with chicks fed any other diets including those with cooked potatoes.

Having established that the cooked potato solids were not growth depressing, a CPI enriched preparation was supplied to chick diets under identical conditions, i.e., the level of CPI the chicks would consume if fed the 50% raw potato diet. Also fed were two pure trypsin inhibitors, potato Inhibitor II, which potently inhibits both trypsin and chymotrypsin, and soybean trypsin inhibitor (SBTI), also a potent trypsin-chymotrypsin inhibitor. The diets are presented in Table I. The effects of these diets on the weight gain, feed efficiency, fecal protein and mortality of newborn chicks are presented in Table II. The feed efficiency of the CPI diet was only slightly elevated from the control diets and the fecal protein was moderately elevated. Potato Inhibitor II and SBTI both caused a significant increase in feed efficiencies, significant decreases in weight gain and

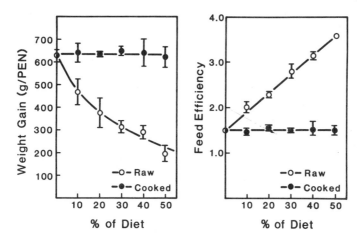

Fig. 1A.   Effects of freeze dried raw and cooked potatoes on the
           growth of newly hatched chicks.  Chicks were fed diets
           with increasing percentages of potato solids as described
           in the text.  Each value represents an average of 3 pens,
           5 chicks per pen, fed each diet for ten days.

Fig. 1B.   The relationships of feed efficiency to the percentage
           of raw or cooked potato solids in diets.  The data
           represents the feed efficiency for the experimental
           points in Fig. 1A.  Feed efficiency is the food consumed
           (g) per weight gain (g).

increases in fecal protein similar to those resulting from
feeding raw potato solids.  The results of the experiments
provide evidence that CPI only moderately affects weight gain,
feed efficiency and fecal protein with respect to control diets.
It indicates that the inhibition of carboxypeptidases A and B may
be occurring but  that this inhibition is probably not severely
affecting protein digestion.  These results also suggest that the
trypsin inhibitors from both potatoes and soybeans significantly
impair chick growth.

TABLE I. Composition of Diets Containing Cooked and Raw Potatoes, CPI, Inhibitor II and SBTI.

| Diet | Basal Diet | Glucose | Potato Solids | Percent of Total Diet | | |
|------|------------|---------|---------------|-----------------------|--------------|------|
| | | | | Inhibitors | | |
| | | | | CPI | Inhibitor II | SBTI |
| 1. Control | 50 | 50 | 0 | 0 | 0 | 0 |
| 2. Autoclaved Potato | 50 | 0 | 50 cooked | 0 | 0 | 0 |
| 3. Raw Potato | 50 | 0 | 50 raw | 0 | 0 | 0 |
| 4. CPI | 50 | 49.7 | 0 | 0.033 | 0 | 0 |
| 5. Inhibitor II | 50 | 49.7 | 0 | 0 | 0.098 | 0 |
| 6. SBTI | 50 | 49.8 | 0 | 0 | 0 | 0.20 |

TABLE II.   Summary of Effects of Various Diets Containing Proteinase
Inhibitors on the Weight Gain, Feed Efficiency, Fecal
Protein and Mortality of Newly Hatched Chicks.[1]

| Diet | Weight Gain[2] (g/pen) | Feed Efficiency[2,3] | Fecal Protein[4] (mg/g) | Mortality[5] (%) |
|---|---|---|---|---|
| 1. Control | 675 | 1.46 | 65 | 0 |
| 2. Autoclaved Potato | 572 | 1.46 | 64 | 0 |
| 3. Raw Potato | 197[6] | 4.82[6] | 153[6] | 20 |
| 4. CPI | 595 | 1.55 | 80[6] | 0 |
| 5. Inhibitor II | 494[6] | 1.82[6] | 135[6] | 0 |
| 6. SBTI | 514 | 1.80 | 135 | 0 |

[1] Chick fed diets described in Table I, for ten days.
[2] Average of three pens, five chicks per pen, except SBTI, which was only one pen.
[3] Feed consumption per weight gain (g/g).
[4] Average of three samples per pen; three pens.  Except SBTI diet, three samples, one pen.
[5] Total from all chicks in the experiments.
[6] Significant difference ($P < 0.05$) from control values.  No statistical analysis done on SBTI diet.

Trypsin inhibitors in animal diets are known to cause pancreatic hypertrophy and hyperplasia. We therefore monitored pancreas weight and enzymic contents in all of the above experiments. The pancreata were removed from the chicks at the termination of the experiments, weighed, and assayed for activities of trypsin, chymotrypsin and carboxypeptidases A and B, after activating their zymogens with either trypsin or enterokinase.

Diets containing raw potato solids caused a significant increase in the weights of pancreata (Table III). The pancreata from chicks fed trypsin inhibitors were also enlarged, but not to the extent of those fed raw potato solids. Chicks fed CPI exhibited no measurable pancreas enlargements.

TABLE III. Activities of Chymotrypsin, Trypsin and Carboxypeptidases A and B in Extracts of Trypsin-Activated Chick Pancreata After Ten Days on Diets Containing Potato Solids or Pure Proteinase Inhibitors CPI, Inhibitor II or SBTI.

| Diet[1] | Pancreas g/100 g Body Wt. | Enzyme Activities (Units/g Pancreas)[2] | | Carboxypeptidases | |
| | | Trypsin | Chymotrypsin | A | B |
| --- | --- | --- | --- | --- | --- |
| 1. Control | 0.501 | 158.0 | 188.2 | 83.9 | 29.5 |
| 2. Autoclaved Potato | 0.529 | 165.2 | 139.0 | 95.2 | 22.1 |
| 3. Raw Potato | 0.859[3] | 51.0[3] | 38.7[3] | 31.2[3] | 11.8[3] |
| 4. CPI | 0.479 | 104.3[3] | 70.0[3] | 69.5 | 19.7 |
| 5. Inhibitor II | 0.604[3] | 146.2 | 60.6[3] | 46.2[3] | 17.2 |
| 6. SBTI | 0.653[3] | 151.0 | 88.4[3] | 70.8 | 19.6 |

[1] See Table I.

[2] One unit is defined as that quantity of enzyme that hydrolyzes one μmole substrate per min at 25°C.

[3] Significant difference (P < 0.05) from control value. All values are an average of 6 chicks except SBTI diet where only 5 chicks were used.

The levels of activatable proteinases in the pancreata were decreased (Table III).  Again, raw potato solids in the diets caused severe decreases in the activities of all four enzymes assayed, suggesting that the secretion of enzyme may have been accelerated, depleting the enzyme content of the pancreas, or that their synthesis had been decreased.  Activatable chymotrypsin, and CPAases A and B activities were significantly lowered in pancreata from chicks fed diets containing CPI, potato Inhibitor II and SBTI.  In contrast, activatable trypsin activity levels were essentially unaffected by Inhibitor II and SBTI, but definitely reduced by CPI.  It appears that the effects of raw potatoes and/or proteinase inhibitors are complex and that the depletion of enzymes from the pancreas by the potato inhibitors is not necessarily related to hyperplasia or to hypertrophy.

Effects of the diets on fecal protein were assessed.  The protein in feces appeared to be significantly increased due to all diets containing proteinase inhibitors.  When protein profiles in fecal matter from chicks fed the experimental diets for ten days were analyzed by electrophoresis (Fig. 2), their profiles in feces from control diet and the diet containing cooked potato solids (lanes 1 and 2) appeared to contain little protein or poly-peptide materials.  Diets containing raw potato solids (lane 3), potato Inhibitor II (lane 5), or SBTI (lane 6), contained large amounts of both high and low molecular weight proteins.  The only apparent effects of CPI (lane 4) on the patterns was an increase in the small molecular weight bands over those present in the control fecal contents.  This data confirms (1) that trypsin-chymotrypsin inhibitors in the diet significantly alter the patterns of intestinal and fecal proteins and (2) CPI in diets appears to alter only the polypeptide content of the feces.

The data indicate that much of the proteinase inhibitors fed probably survive digestion in the stomach and become effective inhibitors in the intestinal tract.  An experiment was performed to monitor the presence of potato Inhibitor II and CPI in the digestive tract by immunological radial diffusion assays, which quantitatively recognize native structural features of the proteins.  Both inhibitors, Inhibitor II and CPI, are heat stable and can be separated from the enzymes they are complexed with by heat denaturation of the enzymes, and both can be subsequently assayed immunologically in the resulting solutions.  Inhibitor II and CPI were analyzed in contents of intestinal segments as shown in Table IV.  It can be seen that the inhibitors were present throughout the entire length of the intestinal tract, increasing in concentration all the way from the stomach to the cecum.  The inhibitors decrease in concentration in the feces, perhaps degraded by enzymes secreted by the bacterial flora.

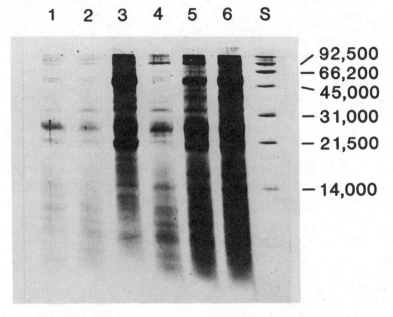

Fig. 2.   Electrophoresis of feces obtained from chicks after ten
          days feeding on various diets containing (1) basal diet
          plus glucose (control) diet; and basal control diets
          plus (2) cooked potato solids, (3) raw potato solids,
          (4) CPI, (5) Inhibitor II, (6) SBTI, and (7) standards.
          Equal quantities of dry matter (400 µg/track) were applied.

     In summary, the availability of a carboxypeptidase inhibitor
in large amounts using our new procedure presented the opportunity
to assess its effects when added to the diets of animals, the
effects on the digestion and growth of animals and its fate in the
intestinal tract were assessed.  From this study we can conclude
that the carboxypeptidase inhibitor from potatoes has measurable,
but minimal, effects on protein digestion and feed efficiency of
newly hatched chicks when supplied at levels up to 0.03% of the
total diet.  However, this level of CPI is much higher than levels
that would be consumed by humans under any reasonable conditions.
Our data suggest that CPI can be considered insignificant as an
antinutrient in animal diets.

TABLE IV. Levels of Inhibitor II and CPI in Contents of Intestinal Tract Segments of Newly Hatched Chicks, Fed Diets Containing Inhibitor II and CPI.[1]

| Intestinal Segment | Dry weight (mg) | | Protein (µg/mg dry wt.) | | Inhibitors (% of protein) | |
|---|---|---|---|---|---|---|
| | Inh. II | CPI | Inh. II | CPI | Inh. II | CPI |
| Gizzard | 230 | 218 | 126 | 131 | .32 | .30 |
| Duodenum | 167 | 225 | 443 | 362 | .14 | 0 |
| Lower Intestine Segment | | | | | | |
| #1 | 103 | 127 | 369 | 277 | .32 | .18 |
| #2 | 112 | 137 | 275 | 222 | .62 | .24 |
| #3 | 113 | 132 | 185 | 159 | 1.24 | .45 |
| #4 | 113 | 120 | 152 | 143 | 1.84 | .71 |
| #5 | 195 | 125 | 142 | 99 | 2.60 | 1.37 |
| #6 | 108 | 107 | 198 | 93 | 2.32 | .99 |
| Cecum | 125 | 88 | 303 | 260 | 3.07 | .96 |
| Colon | 137 | 192 | 158 | 118 | 1.52 | 1.36 |
| Feces | n.d. | n.d. | 135 | 80 | 2.37 | 1.20 |

[1] The tracts were removed from the chicks and the contents of the gizzard, duodanum, cecum and colon were recovered. The remainder of the intestine was cut into six equal segments (approximately 10 cm each segment) and the contents recovered by washing with 10 ml $H_2O$. Contents were freeze dried, weighed and ground in a Wiley mill through a 60 mesh screen. An appropriate sample (25 mg) was dissolved in 0.5 ml of 0.5 M NaOH, incubated 2 h and assayed for protein. A similar sample (50 mg/ml) was dissolved in $H_2O$, heated to 60°C for 10 min and assayed immunologically for Inhibitor II and CPI concentrations. Feces were freeze dried and treated in a similar manner.

ACKNOWLEDGMENTS

The authors acknowledge the support of the Rockefeller
Foundation and the College of Agriculture Research Center,
Washington State University, Pullman, Washington. We thank the
Department of Horticulture and Landscape Architecture for supplying
potatoes.

## REFERENCES

1. Melville JC, Ryan CA. Chymotrypsin Inhibitor I from potatoes.
   J Biol Chem 247:3445-3453, 1972.
2. Bryant J, Green TB, Gurusaddaiah T, Ryan CA. Proteinase
   Inhibitor II from potatoes: isolation and characterization
   of its protomer compounds. Biochemistry 15:3418-3424, 1976.
3. Ryan CA, Hass GM, Kuhn RW. Purification and characterization
   of a carboxypeptidase inhibitor from potatoes. J Biol Chem
   249:5495-5499, 1974.
4. Pearce G, Sy L, Russel C, Ryan CA, Hass GM. Isolation and
   characterization from potato tubers of two polypeptide
   inhibitors of serine proteinases. Arch Biochem Biophys
   213:456-462, 1982.
5. Pearce G, McGinnis J, Ryan CA. Utilization by chicks of
   half-cystine from native and denatured proteinase inhibitor
   protein from potatoes. Proc Soc Exp Biol Med 160:180-184,
   1979.
6. Huang DY, Swanson BG, Ryan CA. Stability of proteinase
   inhibitors in potato tubers during cooking. J Food Sci
   46:287-290, 1981.
7. Ryan CA, Kuo T, Pearce G, Kunkel R. Variability in the
   concentration of three heat stable proteinase inhibitor
   proteins in potato tubers. Am Potato J 53:443-455, 1976.
8. Panyim S, Chalkley R. High resolution acrylamide gel electrophoresis of histones. Arch Biochem Biophys 130:337-346, 1969.
9. Hass GM, Ryan CA. Carboxypeptidase inhibitor from potato.
   In: Lorand L, ed. Methods in Enzymology. Academic Press,
   New York, Vol 80:p778, 1981.

# 17

STABILITY OF ENZYME INHIBITORS AND LECTINS IN FOODS AND THE

INFLUENCE OF SPECIFIC BINDING INTERACTIONS

James C. Zahnley

Western Regional Research Center, Agricultural Research
Service, United States Department of Agriculture,
Berkeley, CA   94710

ABSTRACT

Proteins with actual or potential antinutrient or toxicant activity found in foodstuffs include (1) enzyme inhibitors, especially
those specific for serine proteinases and $\alpha$-amylases, and (2) lectins
(hemagglutinins).  These inhibitors and lectins must be inactivated
during processing or food preparation, usually by heat, to avoid
possible undesirable effects.  Knowledge of their heat stabilities
thus helps determine conditions required for their inactivation or
denaturation. Many are heat-stable proteins, and their conformations
can be stabilized or destabilized by interactions with other constituents present in the food or the digestive tract.  Differential
scanning calorimetric (DSC) results show that specific binding interactions can lead to substantial increases in kinetic thermal stability of proteins.  Examples of such stabilization include serine
proteinase-proteinase inhibitor, $\alpha$-amylase-amylase inhibitor, and
metal ion-lectin complexes.  The extent of thermal stabilization of
proteinases in complexes with inhibitors is correlated with the
equilibrium association constant.  Presence of more than one denaturing unit revealed by DSC in complexes involving multiheaded inhibitors can be interpreted in relation to domain structures of the
inhibitors.  Basic information on stability of the enzyme inhibitors
and lectins is relevant to food processing, quality, and safety.

INTRODUCTION

Protein enzyme inhibitors and lectins (hemagglutinins), and their
interactions with other molecules, have a broad range of ramifications
(Table I).  Both classes of proteins are subject to the same set of

333

questions.  Answering such questions has engaged researchers in
several disciplines, including biochemistry, cellular and molecular
biology, food science and technology, immunology, nutrition, pharmacology, and plant and animal physiology.  This paper will be concerned
primarily with the first and last items listed in Table I.  The others
are considered in the reviews cited below.

Protein inhibitors of a variety of enzymes have been identified
(Whitaker and Feeney, 1973; Richardson, 1981; Whitaker, 1981), but
those that inhibit proteinases and amylases are the best known.
Since 1976, proteinase inhibitors have been reviewed by, among others,
Birk (1976), Kassell and Williams (1977), Richardson (1977,1981);
Laskowski and Kato (1980), Liener and Kakade (1980), Ryan (1981),
Tschesche (1981), Weder (1981), and Whitaker (1981), and in Ramshaw
(1982), Sgarbieri and Whitaker (1982), and Gallaher and Schneeman
(this volume).  Useful earlier reviews include those of Vogel et al.
(1968), Liener and Kakade (1969), Kassell (1970), Laskowski and
Sealock (1971), Tschesche (1974), Means et al. (1974), and the symposium proceedings edited by Fritz and Tschesche (1971) and Fritz et
al. (1974).  This list is not inclusive, but is intended to direct
interested readers to more comprehensive discussions of various
aspects of the subject.

The $\alpha$-amylase inhibitors have been reviewed by Marshall (1975),
Buonocore et al. (1977), and Silano (1978).  More recent references
are included in Richardson (1981), Whitaker (1981), Sgarbieri and
Whitaker (1982), and Gallaher and Schneeman (this volume).

TABLE I.  Questions Asked about Enzyme Inhibitors and Lectins

| Question[a] | Relevant Aspects |
|---|---|
| What? | Structure and activity |
| Where? | Distribution, both inter- and intra-species |
| When? | Evolutionary and developmental relations |
| Why? | Physiological role(s) |
| How? | Mechanism of action |
| Impact? | Possible nutritional effects<br>Factor influencing food processing conditions<br>Applications in research and therapy |

[a]Resemblance of these questions to the points covered in
 a good journalistic lead is not entirely coincidental.

Lectins have been reviewed recently by Liener (1976), Lis and Sharon (1977,1981), Goldstein and Hayes (1978), Jaffé (1980), Toms (1981), and in Ramshaw (1982) and Sgarbieri and Whitaker (1982). Lectins were also a major topic at a recent phytochemical symposium (Loewus and Ryan, 1981).

The actual or potential antinutrient activity of protein enzyme inhibitors and lectins (see reviews cited above) makes their inactivation in foods an important objective of processing.  Since excessive denaturation promotes protein deterioration (Feeney, 1980), minimizing heat damage to other proteins is also desirable.  Information on protein stabilities can suggest ways to improve strategies for meeting these objectives.  Such information may be largely empirical, since protein conformational stability is the net result of many interactions (Schulz and Schirmer, 1979; Jaenicke, 1980; Pfeil, 1981), and seemingly subtle differences in structure may produce significant differences in stability.  Thus, isomorphs of the same inhibitor (Sanderson et al., 1982) as well as distinct inhibitor proteins may differ in stability (e.g., Kassell, 1970; Liener and Kakade, 1980; Huang et al., 1981; Sgarbieri and Whitaker, 1982).  Despite considerable effort, much remains to be learned about the mechanisms of protein folding or unfolding and the detailed basis of protein stability (for references, see Privalov, 1979,1982; Schulz and Schirmer, 1979; Jaenicke, 1980; Pfeil, 1981; Wetlaufer, 1981; Kim and Baldwin, 1982; Creighton, 1983; Pain, 1983).

As indicated, this paper will concentrate mainly on questions related to stability of enzyme inhibitor proteins and lectins.  In particular, differential scanning calorimetric findings on effects of specific interactions on thermal stability will be reviewed. Stabilities of these proteins to other denaturing agents are not considered here.  Older work showed that proteinase inhibitors are generally highly stable; some resist denaturation at 100°C when dry or in neutral or weakly acidic solutions.  Some even survive the action of pepsin, and can thus reach the small intestine intact. Such inhibitors might bind intestinal proteinases and thus be of concern (Richardson, 1981).  Less is known about stability of various lectins (Jaffé, 1980).  Some lectins appear to be less heat-stable, but some reports suggest that others may survive food processing (Nachbar and Oppenheim, 1980; Sgarbieri and Whitaker, 1982; Thompson et al., 1983).  Other topics related to stability of proteinase inhibitors are discussed elsewhere in this volume by Gallaher and Schneeman and by Pearce et al.

EXPERIMENTAL APPROACH -- DIFFERENTIAL SCANNING CALORIMETRY

Differential scanning calorimetry (DSC) measures the excess heat absorbed when proteins or other molecules denature (melt) as they are heated at a constant rate.  Denaturing transitions are recorded as endothermic peaks in the thermograms (plots of heat flow as a function of temperature). DSC thermograms can yield denaturation temperatures ($T_d$) and thermodynamic information, namely, enthalpies of denaturation ($\Delta H_d$) and, in some cases, activation energies ($E_a$) of individual transitions. The denaturation temperature ($T_d$) as used here is the temperature of maximum rate of heat flow (peak). For irreversible denaturation, $T_d$ is a function of heating rate (Donovan and Beardslee, 1975). Denaturation enthalpies ($\Delta H_d$) are calculated from areas of endotherms, and are determined on the basis of active protein for the proteinases and inhibitors when possible. Peaks in complex thermograms can be resolved by a curve-fitting procedure (Zahnley, 1979). Because the amplitude of the recorded signal in DSC is a derivative function, (heat flow, dH/dT), this technique can discriminate between individual denaturing transitions in many cases. This resolution makes the DSC technique well-suited to study of denaturation of multicomponent systems. For more general discussions of DSC and its applications to biopolymers, see reviews by Privalov (1979,1980,1982) and references therein.

A DuPont model 990 thermal analyzer, modified to improve sensitivity and signal-to-noise ratio (Donovan, 1979), was used. Operating procedures and methods for determining denaturation characteristics were as described (Donovan and Beardslee, 1975; Donovan and Ross, 1973,1975). In a typical DSC experiment, components of the reaction mixture were combined and allowed to react in a small tube. An aliquot was weighed into a tared sample pan, and the pan was hermetically sealed. Weights of solvent were matched in the sample and reference pans. Sample and reference were placed in the DSC cell and heated. The heating rate was 10°C/min unless noted.

Descriptions of or references to other methods used, including enzyme and inhibitor assays, and sources of materials are given in the original papers cited.

CONFORMATIONAL (THERMAL) STABILITIES:  EXPERIMENTAL FINDINGS

Free Proteinase Inhibitors

Earlier work on stability of proteinase inhibitors was frequently carried out by heating the inhibitor under specified conditions, usually at constant temperature, then measuring the loss of inhibitory activity or a change in some physicochemical property, solubility, or proteolytic susceptibility. Selected data on inactivation taken from the literature (Table II), including some results for inhibitors found in common foodstuffs, illustrate the high stability of most inhibitors. This list comprises examples from most of the inhibitor families in the Laskowski and Kato (1980) classification. In general, denaturation measured by physicochemical or other techniques (not shown) shows the same pattern as the changes in inhibitory activity (see Kassell, 1970; Laskowski and Sealock, 1971, for summaries of earlier data). Because samples were usually

cooled before assay of remaining inhibitory activity, rapid recovery
could make reversible inactivation indistinguishable from stability
(cf. Pain, 1983). Therefore, the distinction between these results
in Table II may be artificial. Techniques that can measure a property under the perturbing (denaturing) conditions such as high
temperature or extremes of pH can distinguish these possibilities.

Most of the inhibitors listed in Table II or their homologs
have been analyzed by DSC, generally in the neutral pH range, where
disulfide bond cleavage should not complicate denaturation. Changes
in $pK_a$ of buffers with temperature were minimized by adjusting the pH
of the protein solution (Donovan and Beardslee, 1975; Zahnley, 1979,
1980) or by using a buffer (phosphate) that changes little in $pK_a$
with temperature (Takahashi and Sturtevant, 1981). DSC thermograms
of several inhibitors are shown in Figure 1.

Several points regarding stability of the inhibitors listed in
Table II should be noted. They differ considerably in $T_d$ (see also
Fig. 1). The irreversibly denatured fraction of soybean trypsin
inhibitor (STI) increases with time of heating at 90°C and pH 3
(Kunitz, 1947). This progressive loss of reversibility may well

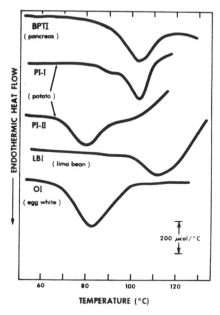

Figure 1. Thermal denaturation of selected proteinase inhibitors
at pH 6.7 in 0.05 M KCl, 0.02 M CaCl$_2$. Inhibitors: BPTI, bovine
pancreatic trypsin inhibitor (Kunitz); PI-I and PI-II, proteinase
inhibitors I and II from potato; LBI, lima bean protease inhibitor;
OI, chicken ovoinhibitor. Heating rate, 10°C/min.

TABLE II. Thermal Stabilities of Selected Protein Inhibitors of Serine Proteinases[a]

| Inhibitor | Family | Inhibitory Activity[b] | | | | | Heat Denaturation (DSC)[c] | | | | |
| | | Temp. (°C) | pH | Time (min) | Result | Refs.[d] | Sample | Heating Rate (°C/min) | $T_d$ (°C) | $\Delta H_d$ (cal/g) | Ref.[d] |
|---|---|---|---|---|---|---|---|---|---|---|---|
| A. From Foodstuffs | | | | | | | | | | | |
| Soybean (Kunitz, STI) | IV | 80 | 3 | 5 | L-R | 1 | A | 1 | 66 | | 3 |
| | | 90 | 3 | 2-60 | L-I | 1 | | 10 | 76 | 5.1 | 3 |
| | | 90 | 7 | 15 | L-?M | 2 | | | | | |
| Soybean (Bowman-Birk) | V | 100 | 7 | 10 | S | 4 | - | -- | -- | --- | - |
| | | 105 | dry | -- | S | 4 | | | | | |
| Lima bean (LBI) | V | 90 | 5,7 | 15 | S | 2 | A | 10 | 114 | 2 | 5 |
| Chicken ovomucoid | II | 80 | 3-7 | 30 | S | 6 | A | 10 | 79 | 6.9 | 3 |
| | | 80 | 9 | 30 | L-I? | 7 | | | | | |
| | | 90 | 5,7 | 15 | S-P | 2,7 | | | | | |
| Chicken ovoinhibitor (OI) | II | 85 | 7 | 10 | S-P | 9 | A | 10 | 82 | 6.6 | 10 |
| | | 90 | 3,5 | 15 | S | 8 | | | | | |
| | | 90 | 7,9 | 15 | L-I? | 8 | | | | | |
| Inhibitor I (potato) | VI | 80 | 2-10 | 10 | S | 11,12 | A | 10 | 104[e] | 8 | 14 |
| | | 100 | 3,8 | 5 | S | 13 | | | 92[e] | | |
| Inhibitor II (potato) | VII | 70 | 2,8 | 30 | S | 15 | A | 10 | 80 | 6 | 14 |

B. From Other Sources

| Inhibitor | | | | | | | | | | |
|---|---|---|---|---|---|---|---|---|---|---|
| Bovine pancreas (BPTI, I Kunitz inhibitor) | 80 | 8.0 | 30 | S | 16 | B | 1? | 100[f] | 9[f] | 18 |
| | 100 | acid | -- | S | 17 | A | 10 | 103 | 10 | 10 |
| Pancreatic secretory inhibitor (Kazal) II | 85 | 7.8 | 360 | S | 19 | - | -- | --- | -- | -- |
| Subtilisin inhibitor III (Streptomyces, SSI) | 100 | 4-6 | 10 | S | 20 | C | 1 | 83 | 5.3 | 21 |

[a] For reviews of older data on stability of these inhibitors, see Kassell (1970) and Laskowski and Sealock (1971). Families are classified according to Laskowski and Kato (1980).

[b] Key to effects on inhibitory activity: L-I, irreversible loss; L-R, reversible loss; L-?M, most of activity lost, reversibility not given; S, stable; S-P, partly stable (substantial activity remained). See original references for details of heating and assay procedures.

[c] Conditions (sample): A, about 20 μl of 1-10% protein in 0.05 M KCl, 0.02 M CaCl$_2$, adjusted to pH 6.7; B, not available, but probably similar to those used in C; C, about 1 ml of 0.2-0.5% protein in 0.06 M phosphate buffer, pH 7.00. Denaturation temperatures (T$_d$) are rounded to the nearest degree. Where possible, ΔH$_d$ was calculated on the basis of active inhibitor.

[d] References: 1, Kunitz (1947); 2, Fraenkel-Conrat et al. (1952); 3, Donovan and Beardslee (1975); 4, Birk (1961); 5, Zahnley (1980); 6, Lineweaver and Murray (1947); 7, Stevens and Feeney (1963); 8, Matsushima (1958); 9, Chu and Hsu (1965); 10, Zahnley (1979); 11, Ryan and Kassell (1970); 12, Melville and Ryan (1972); 13, Plunkett and Ryan (1980); 14, Zahnley (1977); 15, Bryant et al. (1976); 16, Dietl and Tschesche (1976); 17, Green and Work (1953); 18, Tishchenko and Gorodkov (1978?), quoted in Privalov (1979); 19, Menegatti et al. (1981); 20, Sato and Murao (1973); 21, Takahashi and Sturtevant (1981).

[e] Samples were kindly provided by Dr. C. A. Ryan. Inhibitor I: major peak at 104°C.

[f] At pH of maximum stability. From Figs. 10 and 24b of Privalov (1979).

result from slower, secondary steps after the initial unfolding.  In the DSC, Kunitz STI shows a substantial effect of heating rate on $T_d$ (Donovan and Beardslee, 1975).  This emphasizes the kinetic nature of the measurements at higher heating rates used to obtain most of the DSC data shown in Table II.  Lima bean protease inhibitor (LBI) and its homologs of the Bowman-Birk inhibitor family from various beans show highly thermostable inhibitory activities (Fraenkel-Conrat et al., 1952; Birk, 1961; Belew et al., 1975; Sgarbieri and Whitaker, 1982).  The DSC results for LBI (Zahnley, 1980) are consistent with this other evidence.  The high content of disulfide linkages, including some between homologous binding regions, is thought to help stabilize these inhibitors (Liener and Kakade, 1980).  The disulfide bridges may compensate for the lack of some other stabilizing interactions, accounting for the low $\Delta H_d$ of LBI.

The inhibitors from chicken egg white, ovomucoid and ovoinhibitor, show similar stabilities by inactivation (see Refs. to Table II) or DSC criteria (Donovan and Beardslee, 1975; Zahnley, 1979).  This is in accord with chemical and immunological evidence for their structural similarity (Laskowski and Kato, 1980; Laskowski et al., 1980; Matsuda et al., 1983).  Each consists of homologous, largely independent structural domains (independent folding units) differing in reactive site sequences (Kato et al., 1976,1980).  These domains could denature as one cooperative unit or as independent units when heated in the DSC.  The number of cooperative denaturing units can be determined from the ratio of calorimetric to van't Hoff (effective) enthalpies (Privalov, 1979,1980).  A ratio of 1.0 is obtained for a single two-state transition.  Our value for ovoinhibitor was above 4.5, approaching the number of domains, 7 (Kato et al., 1980).  Ovomucoid shows three transitions (Griko and Privalov, 1982), equal to the number of its domains (Kato et al., 1976).  These results suggest that the domains of free ovoinhibitor and ovomucoid denature independently when the protein in heated.

Proteinase inhibitors I and II from potato differ substantially in heat stability (Table II).  The high $T_d$ for inhibitor I agrees with findings of Ryan and coworkers (Plunkett and Ryan, 1980, and references therein).  Presence of the minor peak in thermograms for inhibitor I (Fig. 1) could be due to a small fraction of a less stable type of subunit, since the subunits of inhibitor I are microheterogeneous (Melville and Ryan, 1972; see also Richardson, 1981).  Iwasaki et al. (1971) reported that inhibitors II-a and II-b, which are similar to, but not identical to, inhibitor II of Bryant et al. (1976) are also stable at 80°C below pH 7.

Three inhibitors from non-food sources are included in Table II.  The basic (Kunitz) trypsin inhibitor from bovine pancreas (BPTI) is a small protein that has been used extensively in protein structure studies.  At neutral or acid pH, it denatures near 100°C.  From the ratio of calorimetric to van't Hoff (effective) enthalpies, Privalov

(1979) concluded that its denaturing unit is a dimer.

Although pancreatic secretory trypsin inhibitor (Kazal inhibitor) has not been studied by DSC to our knowledge, it is included because it is homologous to the domains of ovomucoids and ovoinhibitors (Laskowski and Kato, 1980). Both the porcine and bovine Kazal inhibitors are relatively stable (Menegatti et al., 1981; De Marco et al., 1982), but less so than the Kunitz inhibitor (BPTI).

The Streptomyces subtilisin inhibitor is closely related to the Kazal inhibitor family (Laskowski and Kato, 1980). The subunits of this dimeric inhibitor dissociate during thermal unfolding (Takahashi and Sturtevant, 1981), when dilute solutions (2--5 mg/ml) are heated slowly (1°C/min) in a Privalov (1980) type of DSC instrument. The $T_d$ for Streptomyces subtilisin inhibitor is 17°C above that for STI at this heating rate (Table II), suggesting that it is intermediate in stability among the inhibitors tested.

Denaturation of the proteinases themselves will be discussed in connection with the specific enzyme--inhibitor complexes. For additional data on the free enzymes, readers are referred to the original papers.

## Proteinase--Inhibitor Complexes (Monomeric Inhibitors)

Many enzymes or other proteins are stabilized by interaction with small ligands (substrates, cofactors, inhibitors, products). Nonspecific protection by macromolecules and stabilization by specific protein--protein interaction are generally less readily quantified. As a model system in which to determine effects of protein--protein interaction on conformational (thermal) stability, Donovan and Beardslee (1975) selected the proteinase--inhibitor association between bovine β-trypsin and STI or chicken ovomucoid. They observed that both protein components were stabilized in the 1:1 complexes and, furthermore, that each trypsin--inhibitor complex showed one endotherm in the DSC. No stabilization was observed for a weaker electrostatic complex between ovalbumin and lysozyme. They interpreted the stabilization as an increase in kinetic thermal stability, i.e., a decrease in the rate of denaturation at a given temperature. Results obtained by other methods (Edelhoch and Steiner, 1965; Laskowski and Sealock, 1971; Jibson et al., 1981) show that proteinase--inhibitor association generally involves no major conformational changes in the proteins.

These results with singleheaded inhibitors suggested a way to answer questions raised by earlier kinetic findings on complexes formed by multiheaded inhibitors. Using 2:1 complexes between bovine or porcine trypsins and chicken ovoinhibitor, we showed that the two trypsin molecules bound to one ovoinhibitor molecule dissociated at markedly different rates from nonequivalent and apparently noninter-

acting sites (Zahnley and Davis, 1970; Zahnley, 1974,1975). Would
kinetic nonequivalence of binding sites for trypsin on ovoinhibitor
be reflected in thermal stabilities of the complexes?

   Structural similarities between ovomucoid and ovoinhibitor (see
above), suggested that ovoinhibitor also consisted of several homologous domains. If the trypsin-binding sites on chicken ovoinhibitor
were located on essentially independent structural domains, would an
entire 2:1 complex constitute one denaturing unit, or would it denature in two steps, as depicted schematically in Figure 2? Would the
binding of a second trypsin molecule add to stability of the system,
making the 2:1 complex more stable than the 1:1 complex? Because
DSC can discriminate between denaturing transitions and can detect
small changes in $T_d$ (Donovan and Beardslee, 1975), we should be able
to gain clearer insight into the nature of the complex and effects
on stability of binding more than one enzyme molecule.

   Although the 2:1 complex between bovine β-trypsin and chicken
ovoinhibitor did not show two distinct peaks in the DSC thermogram
(Zahnley, 1979), asymmetry of the endotherm and nonlinearity of the
Arrhenius plots for denaturation (log $k_{denat}$ vs. 1/T) suggested that
two closely overlapping endotherms were present. Since the ratio of
dissociation rates for the two porcine trypsin molecules from a 2:1
complex with ovoinhibitor was about 10 times as large as the ratio
for bovine trypsin molecules (Zahnley and Davis, 1970; Zahnley,
1974), porcine trypsin--ovoinhibitor complexes seemed more likely to
to answer the above question about the number of denaturing units.
Porcine β-trypsin itself denatured in two transitions, complicating
the thermograms, unless the less stable form was eliminated by pre-
heating (66°C, 10 min, pH 6.7). Thermograms produced by complexes
formed with the preheated trypsin showed loss of the low-temperature
peaks; therefore, these peaks appear to represent denaturation of
complexes formed by the less thermostable form of trypsin. DSC
thermograms produced by complexes containing stock or preheated

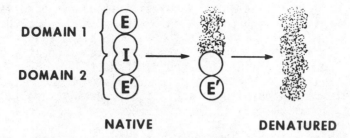

**NATIVE**                          **DENATURED**

Figure 2. Postulated denaturation pathway for a complex between
two enzyme molecules (E, E') and one multiheaded inhibitor molecule
(I). This hypothetical inhibitor consists of two structural
domains, each with one reactive site (binding site for enzyme).
E and E' may be the same or different.

trypsin were similar in the high-temperature region (Zahnley, 1979). Denaturation of complexes at several trypsin:OI ratios (Fig. 3) shows greater stabilization in the 1:1 complex than in the 2:1 complex.

Heating a mixed trypsin complex, which consisted of one mole each of bovine and porcine trypsin per mole of OI, in the DSC produced denaturation peaks at 79° and 85°C, close to those produced by the two 1:1 complexes (Zahnley, 1979). This result is consistent with the slow dissociation previously observed for such complexes (Zahnley, 1975).

Ovoinhibitor can also bind two molecules of chymotrypsin or subtilisin at sites distinct from those for trypsin (Tomimatsu et al., 1966). The affinity for subtilisin is much stronger than that for chymotrypsin (Osuga et al., 1974); therefore, a greater range of stabilization and clearer resolution of the denaturation peaks might be obtainable than was possible with trypsins, if stabilization is related to affinity. The apparent endotherm for the 2:1 complex between bovine $\alpha$-chymotrypsin and chicken ovoinhibitor was fairly sharp, but asymmetric (Fig. 4), suggesting that more than one peak was present. The thermogram could be resolved into two endothermic peaks, with $T_d$ values of 72° and 77°C (Zahnley, 1980). Denaturation of subtilisin--ovoinhibitor complexes showed clearer separation of endotherms (Fig. 5). As the mole ratio of subtilisin to ovoinhibitor was decreased from 2:1 to 1:1, one major peak (near 90°C) disappeared, leaving an asymmetric peak near 100°C. At a 1:1 ratio, a small peak with the same $T_d$ as free ovoinhibitor was also visible.

The relative affinities (Osuga et al., 1974) allow subtilisin to displace chymotrypsin from the complex with ovoinhibitor. Excess chymotrypsin and inhibitor (mole ratio 3:1) were allowed to react for 6 min, then one mole of subtilisin per mole of inhibitor was added and the mixture was heated in the DSC (Zahnley, 1980). The thermogram (Fig. 6) lacks the chymotrypsin--ovoinhibitor (2:1) peak at 72°C and the free subtilisin peak at 84°C, but it shows peaks coinciding with those for the 1:1 complexes (dashed lines) and the mixed chymotrypsin--ovoinhibitor--subtilisin complex (top thermogram). Thus, subtilisin displaced chymotrypsin, in agreement with the enzymatic results of Osuga et al. (1974).

Areas and positions of distinct endothermic peaks in DSC thermograms for trypsin--ovoinhibitor--chymotrypsin complexes (Zahnley, 1980) also did not show interference of binding of trypsin with that of chymotrypsin, or vice versa, in agreement with enzymatic findings of Tomimatsu et al. (1966).

To determine whether this independent stabilization is limited to larger multiheaded inhibitors having proteinase-binding sites on independent structural domains, complexes of trypsin and chymotrypsin with LBI were heated in the DSC. LBI and other members of the Bowman-

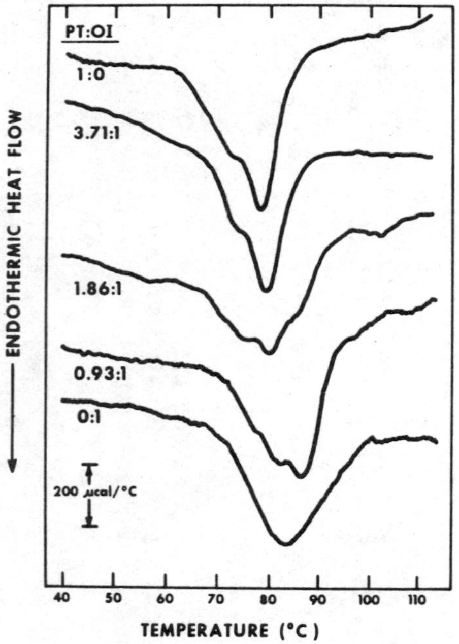

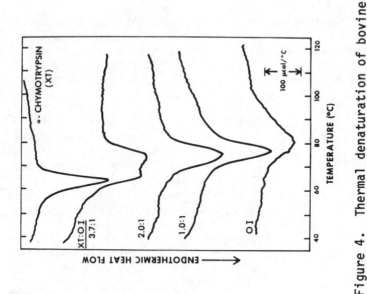

Figure 4.  Thermal denaturation of bovine α-chymotrypsin (XT, top curve), chicken ovoinhibitor (OI, bottom curve), and their complexes.  Mole ratios shown in the Figure are based on active chymotrypsin and inhibitor.  Conditions: pH 6.7; 0.05 M KCl, 0.02 M CaCl2; heating rate, 10°C/min.

Figure 3.  Thermal denaturation of porcine β-trypsin (PT, top curve), chicken ovoinhibitor (OI, bottom curve), and their complexes.  Mole ratios shown in the Figure are based on active trypsin and inhibitor.  Conditions: pH 6.7; 0.05 M KCl, 0.02 M CaCl2; heating rate, 10°C/min.

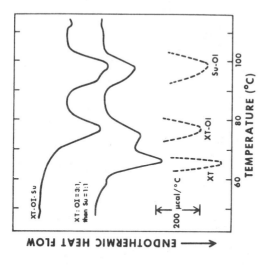

Figure 6. Thermal denaturation of complexes of chicken ovoinhibitor (OI) with both α-chymo-trypsin (XT) and subtilisin BPN' (Su). XT-OI-Su (top curve), complex formed by direct mixing of components (mole ratio XT:OI:Su = 1.1:1.0:1.4). Lower solid curve shows displacement of XT from a 3:1 mixture with OI by Su at 1 mol/mol OI. Dashed lines at the bottom show positions of peaks for free XT and the 1:1 complexes (XT-OI, Su-OI).

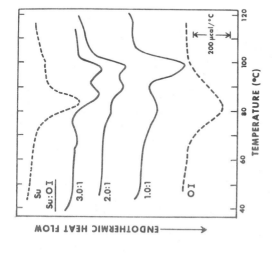

Figure 5. Thermal denaturation of subtilisin BPN' (Su, top, dashed curve), chicken ovoinhibitor (OI, bottom, dashed curve), and their complexes. Mole ratios shown in the Figure are based on active subtilisin and inhibitor. Conditions: pH 6.7; 0.05 M KCl, 0.02 M CaCl2; heating rate, 10°C/min.

Birk family (Laskowski and Kato, 1980) are small proteins with disulfide crosslinks between the proteinase-binding regions.  Bound trypsin
or chymotrypsin was stabilized in its 1:1 complex with LBI; both
enzymes were stabilized in the ternary (trypsin--LBI--chymotrypsin)
complex (Fig. 7).  With highly stable inhibitors, such as LBI or
BPTI, association produced no observable stabilization of the inhibitor (Zahnley, 1979,1980).

## Complexes with Multimeric Inhibitors

Each of the multiheaded inhibitors described above is a monomer,
with one enzyme-binding region per covalently linked domain.  Effects
of association of proteinases with multimeric inhibitors, which have
the enzyme-binding regions on separate subunits (noncovalently associated), on stability have also been measured by DSC.  Considerable
stabilization of chymotrypsin and subtilisin was observed in complexes
with proteinase inhibitors I and II from potato (Table III).  Chymotrypsin showed increases in $T_d$ of 33°C (major peak) on binding to

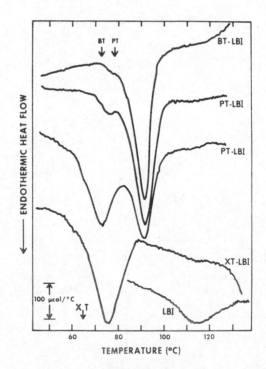

Figure 7.  Thermal denaturation of LBI (lower right) and its
complexes with trypsin (bovine, BT; porcine, PT) or bovine
α-chymotrypsin (XT).  Arrows at top and bottom of Figure show
positions of peaks for the free enzymes.  Conditions: pH 6.7;
0.05 M KCl, 0.02 M CaCl$_2$; heating rate, 10°C/min.

TABLE III. Thermal Stabilization of Serine Proteinases Resulting from Association with Multi-subunit Inhibitors.

| Inhibitor | Proteinase | $\log K_a{}^a$ | Ref. | $\Delta T_d(°C)^b$ |
|---|---|---|---|---|
| Proteinase inhibitor I (potato) | Chymotrypsin | 11 | (1) | 33 (major) 26 |
| | Subtilisin | 10 | | 17 |
| Proteinase inhibitor II (potato) | Chymotrypsin | 8 | (2) | 27 |
| | Subtilisin | | | 9 |
| Subtilisin inhibitor (Streptomyces) | Subtilisin | 10 | (3) | $20^c$ |

[a]Calculated from $K_i$ values. References: (1) Ryan and Kassell (1970); (2) Bryant et al. (1976); (3) Uehara et al. (1978).

[b]Rounded to nearest degree. $\Delta T_d = T_d^{bound} -- T_d^{free}$. The major peak is indicated, if two are present. Heating rate, 10°C/min (except as noted).

[c]Determined at a heating rate of 1°C/min. Data of Takahashi and Sturtevant (1981).

inhibitor I and 27°C on binding to inhibitor II. Subtilisin showed smaller increases when bound to inhibitor I (17°C) or inhibitor II (9°C). Although trypsin is inhibited (Melville and Ryan, 1972; Bryant et al., 1976), results on its stabilization were inconclusive. Takahashi and Sturtevant (1981), using different experimental conditions, found an increase of 20°C in $T_d$ of subtilisin on binding to Streptomyces subtilisin inhibitor, a dimer, in an $E_2I_2$ complex. In these cases, the inhibitor subunits did not appear to dissociate when the enzyme--inhibitor complexes formed.

## Possible Correlation between Stabilization and Affinity

To look for possible relations of thermal stabilization, measured as increase in $T_d$, to other parameters of the proteinase--inhibitor interaction, we plotted increases in $T_d$ of the proteinases resulting from their association with the inhibitors as a function of the logarithms of the equilibrium association constants ($\log K_{assoc}$). Published values of $K_{assoc}$ vary somewhat (see Zahnley, 1980, for references to values used), and these values were not determined under the same conditions as used for the DSC experiments. Accordingly, ranges of $\log K_{assoc}$ were used in Figure 8. Values fell within a narrow

corridor, except for chymotrypsin bound to a trypsin inhibitor, BPTI.
These results for monomeric inhibitors, both singleheaded and multiheaded, show a strong positive correlation (r > 0.89) between stabilization ($\Delta T_d$) of the bound proteinase and the enzyme-inhibitor
affinity (log $K_{assoc}$).

## Non-inhibitory Proteinase--Inhibitor Combinations

Two non-inhibitory proteinase--proteinase inhibitor combinations
exhibit destabilization of inhibitor proteins in the DSC (Zahnley,
unpublished). Ovoinhibitor was rapidly degraded by thermolysin, as
indicated by decreases in both $T_d$ and area ($\Delta H_d$) of its endotherm.
Chicken ovomucoid was slowly degraded by proteinase K (Fig. 9), a
fungal serine proteinase that attacks many native proteins. Proteinase K, in accord with its subtilisin-like specificity (Kraus and
Femfert, 1976), was inhibited by chicken ovoinhibitor, but not by
chicken ovomucoid. In these non-inhibited systems, proteolysis can
destabilize the inhibitor (substrate) protein, provided the susceptible bonds are accessible to the proteinase. Relevance of these examples to food systems derives from the role of proteolysis in protein
deterioration (Feeney, 1980) and digestion.

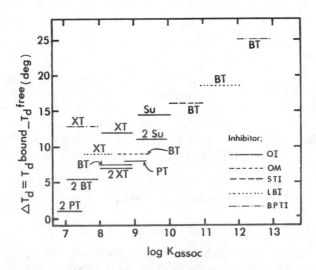

Figure 8. Stabilization of proteinases ($\Delta T_d$) on association
with their protein inhibitors as a function of the logarithms of
the equilibrium association constants (log $K_{assoc}$). Horizontal
bar lines are used to indicate the range of log $K_{assoc}$ values.
Key to inhibitors (inside Figure, at lower right): OM, chicken
ovomucoid; for other abbreviations, see Table II. Key to
proteinases: BT, bovine β-trypsin; PT, porcine β-trypsin;
XT, bovine α-chymotrypsin; Su, subtilisin BPN'.

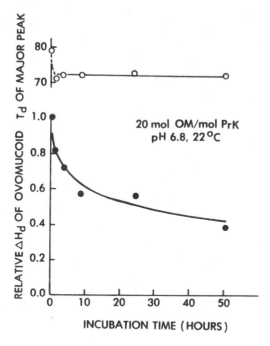

Figure 9.   Time course of thermal destabilization of chicken ovomucoid (OM) by treatment (proteolysis) with proteinase K (PrK).

## Amylase--Inhibitor Complexes

The proteinase--inhibitor complexes described above can be viewed as formal enzyme-substrate complexes (Laskowski and Sealock, 1971). To see whether other types of enzyme--protein inhibitor association also stabilize the enzyme, an amylase--amylase inhibitor system was studied by DSC.  Complexes between either of two α-amylase inhibitors from wheat flour, designated 0.19 and 0.28, and the α-amylase from yellow mealworm (Tenebrio molitor L.), a pest that feeds on wheat flour, were used.  Both complexes denatured at higher temperatures than the free enzyme, but the extent of stabilization differed (Silano and Zahnley, 1978).  Both free inhibitors were heat-stable ($T_d$ = 93°C).

## Lectins: Concanavalin A

Lectins will be represented here by concanavalin A (Con A), obtained from jack bean (Canavalia ensiformis).  Con A is the best-known lectin (hemagglutinin).  Its structure and mechanism of action have been studied intensively (for summaries, see Lis and Sharon, 1977, 1981; Goldstein and Hayes, 1978; Koenig et al., 1982; Hardman et al., 1982), but its stability has received less attention. Doyle et al. (1976) found that metal ions protected Con A against thermal aggregation.  Blumberg and Tal (1976) observed that, under

certain conditions, metals reduced its proteolytic susceptibility.
Metal ion binding and its effects on activity of Con A are strongly
pH-dependent.  For example, EDTA is much more inhibitory to saccha-
ride binding by Con A at pH 5.3 (Doyle and Stroupe, 1978) than at pH
7 (Doyle et al., 1968).

    Our interest lay in the effects of limiting amounts of metal
ions, a condition that might prevail in nature, on conformational
stability of Con A.  Most of these DSC experiments (Zahnley, 1981)
were carried out at pH 5 and ionic strength about 0.5, where Con A
is a dimer (the more soluble form).  Removal of metal ions markedly
lowers the thermal stability of Con A (Fig. 10).  Stability equal to
that of native Con A was regained only after addition of excess $Mn^{2+}$
and $Ca^{2+}$ (Table IV).  Restoration of $Ca^{2+}$ alone, except in excess,
did not restore the stability.  Addition of $Mn^{2+}$ at limiting $Ca^{2+}$
yielded a protein species having intermediate stability ($T_d$ near
85°C). Stoichiometric amounts of both metals restored almost full
stability.  Published $K_a$ values for metal ion binding to Con A
(Stroupe and Doyle, 1980, and references therein) are $<10^5$ $M^{-1}$,
much lower than expected for the extent of stabilization observed,
and far below values for transferrins, which show comparable stabi-

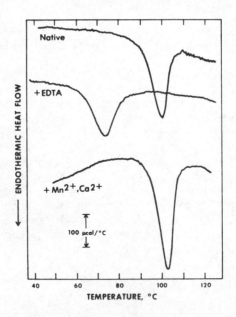

Figure 10.  Effect of metal ion removal and restoration on thermal
stability of concanavalin A (Con A) at pH 5.  Protein in samples:
native Con A, 267 µg; demetallized (+EDTA), 464 µg; remetallized
(plus $Mn^{2+}$, $Ca^{2+}$), 384 µg.  Metal ions in remetallized sample
(mol/mol Con A monomer): $Mn^{2+}$, 10.0; $Ca^{2+}$, 12.4.  Heating rate,
10°C/min.

TABLE IV. Effect of $Mn^{2+}$ and $Ca^{2+}$ on Denaturation Temperatures and Enthalpies of Concanavalin A.

| Metal ion present[a] | | $T_d$[b] | $\Delta H_d$ |
|---|---|---|---|
| $Mn^{2+}$ | $Ca^{2+}$ | | |
| (mol/mol monomer) | | (°C) | (cal/g) |
| 0.18 | 0.55 (native) | 101.0 | 7.4 |
| 0.06 | 0.23-2.0 | 73.5-74.5 | 3.2 ± 0.2 |
| 10.0 | 12.4 | 103.0 | 6.8 |
| 0.01 | 14. | 93.0 | 5.1 |
| 13. | 0.2-0.3 | 94.5 | 5.3 |
| 0.56 | 0.23 | 75.5, 93.5 | 2.9 |
| 0.56 | 0.50 | 76.0, 95.0 | 3.3 |
| 0.56 | 1.00 | 76.0, 96.0 | 3.6 |
| 0.56 | 1.82 | 77.0, 97.0 | --- |
| 1.07 | 0.23 | 77.5, 85.0, 94.5 | 2.9 |
| 1.07 | 0.50 | 77.0, 83.5, 97.5 | 3.5 |
| 1.07 | 1.00 | 97.5 | 5.3 |
| 2.14 | 0.23 | 88.0, 96.0 | 2.9 |

[a]Except for native concanavalin A, samples were treated to remove metal ions and metal ions were then added as indicated.

[b]The $T_d$ values are given to the nearest 0.5°C. Where two or more peaks are listed, the major peak is underlined.

lization by metals (Donovan and Ross, 1975). This apparent anomaly may be explained by the location of the transition-metal and $Ca^{2+}$ binding sites near the amino terminus of each polypeptide chain of Con A (Hardman et al., 1982), a possible key region in determining onset of denaturation.

A DSC study of tetrameric Con A (pH 7) by Borrebaeck and Mattiasson (1980) showed that at 2.5 metal ions per monomer, Con A was stabilized by $Mn^{2+}$, $Ca^{2+}$, or both. Additional stabilization by methyl α-D-mannoside was observed primarily when both metals were present, but some was observed with $Ca^{2+}$. Stabilization of Con A by small saccharides was a function of the association constant. Association with horseradish peroxidase, a glycoprotein, stabilized Con A at pH 7. In contrast, we found no stabilization on association with the egg-white glycoproteins, ovalbumin or ovomucoid, at pH 5.

DISCUSSION

The nutritional significance of protein enzyme inhibitors and
lectins is related to both the physiology of the ingesting species
(biological specificity) and the state of the potential antinutrient
protein (stability).  Knowledge of the heat denaturation behavior of
these proteins has practical importance (Jaffé, 1980; Liener and
Kakade, 1980), since heat is the primary agent used to inactivate
these proteins during food or feed processing.  A large body of
literature on antinutritive effects has developed, particularly for
legumes.  Nutritional and stability aspects of enzyme inhibitors in
foodstuffs have been included in recent reviews (Anderson et al.,
1979; Liener, 1980; Liener and Kakade, 1980; Richardson, 1981; Weder,
1981; Whitaker, 1981; Gallaher and Schneeman, this volume).  These
aspects of lectins are covered by Liener (1976,1979), Jaffé (1980),
and Sgarbieri and Whitaker (1982).

The following examples illustrate effects of differences in
diet, physiology, or processing conditions on activity and possible
antinutrient effects.  Some individuals have abnormal digestive function or atypical dietary preferences or requirements.  Thus, inhibition of human chymotrypsin by native egg white, presumably due to
ovoinhibitor, might be significant for use of uncooked egg white in
some special liquid diets (Imondi et al., 1973).  The assumption
that lectins are heat-labile may not be justified in some cases
(Jaffé, 1980; Sgarbieri and Whitaker, 1982), and some lectins can
reportedly survive processing and conditions in the digestive tract.
Bean lectins may not be destroyed under some cooking conditions
(Grant et al., 1982; Sgarbieri and Whitaker, 1982; Thompson et al.,
1983).  Nachbar and Oppenheim (1980) found significant agglutinating
activity, possibly due to residual active lectins, in 29 of 88
common foodstuffs and processed foods tested.  Since metal ions are
constituents of a number of lectins in foodstuffs, destabilization
by removal of metals as done with Con A (Doyle et al., 1976; Blumberg
and Tal, 1976; Zahnley, 1981) might facilitate inactivation of lectins in some foods.

The data summarized in Table II show that, by both enzymatic and
physicochemical (DSC) criteria, representative protein proteinase
inhibitors are highly stable.  The cases shown include members of
most families of serine proteinase inhibitors as classified by
Laskowski and Kato (1980).  High thermal stability has been observed
for inhibitors from various other food plants, such as cereal grains
(Boisen and Djurtoft, 1982, and references cited therein), millets
(Veerabhadrappa et al., 1978; Chandrasekher et al., 1982; Shivaraj et
al., 1982), peanut (Tur-Sinai et al., 1972), winged bean (Tan et al.,
1982), and adzuki bean (Yoshikawa et al., 1977), to name a few examples.  A novel bifunctional protein purified from finger millet
(ragi) that inhibits both trypsin and $\alpha$-amylase is also highly stable
(Shivaraj and Pattabiraman, 1981).  Differences in stability among

distinct inhibitors or isomorphs of the same inhibitor (see Introduction) can extend even to different inhibitory activities on the same inhibitor protein, as seen with ovoinhibitor (Chu and Hsu, 1965) and the chymotrypsin-trypsin inhibitor from finger millet (Shivaraj et al., 1982). Internal gene duplication during evolution may have led to increased stability in the Bowman-Birk inhibitor family (Ikenaka and Odani, 1978).

## Stability of Crude vs. Purified Inhibitor

Stabilities of crude and purified proteinase inhibitors may or may not match. Similar stabilities of crude and purified inhibitors were found for BPTI (Green and Work, 1953; Dietl and Tschesche, 1976), LBI (Fraenkel-Conrat et al., 1952), proteinase inhibitor II from potato (Bryant et al., 1976; Huang et al., 1981) and chicken ovomucoid (see below). Substantial differences in stability have been reported for STI (Ellenrieder et al., 1980), trypsin-chymotrypsin inhibitor from chick pea (Belew et al., 1975), inhibitors from some other beans (Sgarbieri and Whitaker, 1982), peanut trypsin inhibitor (Tur-Sinai et al., 1972; Perkins and Toledo, 1982), proteinase inhibitor I from potato (Ryan, 1966; Huang et al., 1981), and chicken ovoinhibitor (see below). Although stabilities of the crude and purified inhibitors may differ, characterization of intrinsic stability requires isolation of the inhibitor protein. Effects of other factors on stability of the crude inhibitor may be difficult to define.

Chicken ovomucoid and ovoinhibitor illustrate the possible problems in predicting stabilities from the crude or purified state to the other. Early work on ovomucoid (Lineweaver and Murray, 1947; Fraenkel-Conrat et al., 1952; Stevens and Feeney, 1963) and ovoinhibitor (Matsushima, 1958; Chu and Hsu, 1965) pointed to similar, high thermal stabilities for both inhibitors, as did later DSC results (Donovan and Beardslee, 1975; Zahnley, 1979). Reversible denaturation of the purified proteins in the DSC was observed up to 80°C for ovomucoid (Donovan and Beardslee, 1975) and 75°C or above for ovoinhibitor (Zahnley, 1979), and their $T_d$ values were similar (Table I). Ovomucoid was denatured irreversibly when held above 90°C (Matsuda et al., 1981a). The domains of both proteins appeared to denature independently (Matsuda et al., 1981b; Zahnley, 1979; Griko and Privalov, 1982).

In egg white, however, stabilities of ovomucoid and ovoinhibitor may differ. When Matsuda et al. (1981c) heated egg white and determined the extent of aggregation by gel electrophoresis, they found that ovomucoid bands did not change after heating at 76°C and pH 7, but that ovoinhibitor bands diminished after heating at 62°C. This apparent discrepancy might be explained if ovoinhibitor interacts more readily than ovomucoid with other, readily denatured constituents of egg white to form aggregates. Hence the difference may not reflect irreversible denaturation of ovoinhibitor as such.

Modulation of the intrinsic structural stability of the inhibitors or lectins by the protein environment occurs both in cells and in isolated systems. In the cell, conditions may change with stage of development or metabolic activity, or with changes in integrity of cell structure. Post-harvest storage or processing conditions may also alter accessibility of cell constituents (Schwimmer, 1981), and protein stabilities in cells or homogenates may differ (Alexandrov, 1977). Factors altering the protein milieu can affect biological activity (Liener and Kakade, 1969; Richardson, 1977) or functional properties in foods (Kinsella, 1976,1979). Understanding the basis for such changes in stability depends on knowledge of the stability properties of the isolated protein of interest. In practice, however, changes in stability are often measured empirically.

Chemical modification by disulfide bond cleavage can markedly alter thermal stabilities or other properties of the inhibitors, since disulfide crosslinks help maintain the active conformations of many proteinase inhibitors (Laskowski and Sealock, 1971; Tschesche, 1974; Weder, 1981). Loss of thermal stability resulting from reductive alkylation of disulfide bonds has been observed for BPTI by $^1$H-NMR (Wüthrich et al., 1980), for proteinase inhibitor I from potato by immunological assay (Plunkett and Ryan, 1980), and for STI, LBI, and chicken ovomucoid by DSC (Friedman et al., 1982c). This modification may also decrease the amount of protein structure available for heat denaturation (Friedman et al., 1982a), though it may or may not inactivate the inhibitor (see Plunkett and Ryan, 1980; Friedman et al., 1982a, for references). Use of chromophoric or fluorophoric alkylating agents facilitates determination of the extent of partial modification (Friedman et al., 1980; Zahnley and Friedman, 1982). Proteinase inhibitors from soybean or lima bean are more readily inactivated by heat if thiols, such as cysteine or N-acetylcysteine, are present (Lei et al., 1981; Friedman et al., 1982a--c), suggesting that the combined action of heat and thiols may be of practical value for inactivating enzyme inhibitors in foods.

Most of the findings summarized under Results deal with effects of specific interactions on thermal stabilities of protein enzyme inhibitors and lectins. Specific protein-protein association or metal ion binding to proteins of the types described can result in appreciable stabilization. This stabilization is an increase in kinetic thermal stability (Donovan and Beardslee, 1975; Zahnley, 1979,1980), that is, a decrease in denaturation rate constant at a given temperature. Changes in the nature of the cooperative denaturing unit may also occur as a result of protein-protein association, as suggested by changes in the shape (sharpness) of the denaturing transition (DSC endotherm). In DSC, large-scale changes in proteins are measured, unlike the localized changes detected by some other methods used to study protein conformational stability. In addition to questions of changes used as criteria for denaturation (or inactivation), questions of equilibrium or kinetic nature of the measure-

ments and reversibility of the process, including possible complica-
tions introduced by aggregation or other secondary reaction, need to
be considered.

Possible stabilization by nonspecific means would be relevant
to food systems. No stabilization was observed by DSC for the
ovalbumin-lysozyme interaction (Donovan and Beardslee, 1975). Non-
specific interactions with heat-stable proteins (including ovomucoid)
reportedly stabilize heat-susceptible proteins against inactivation,
possibly by reducing diffusion-limited interactions between unfolded
chains and resulting aggregation, in several model systems (Minton
et al., 1982). In this case, the mechanism is different.

Proteins in solutions are dynamic systems, and their conformations are not fixed (Pfeil, 1981, Huber and Bennett, 1982; Creighton,
1983; Pain, 1983; McCammon and Karplus, 1983). This flexibility can
involve structural elements of various levels, including domains
(Huber and Bennett, 1982). Imparting greater energy to the system
increases the fluctuations. When the stabilizing interactions can
no longer maintain the native structure, cooperative unfolding (denaturation) ensues. Increases in internal mobility of BPTI derivatives
measured from amide proton exchange are correlated with decreases in
$T_d$ (Wüthrich et al., 1980).

Denaturation of a number of multidomain proteins involves more
than a simple two-state ($N \rightleftharpoons D$) transition (see Privalov, 1982, for
references). In such cases, breaking of interdomain contacts may
occur first, followed by unfolding of the domains (breaking of intra-
domain contacts), as outlined by Kim and Baldwin (1982). The domains
may denature independently, as has been observed in several cases,
including ovomucoid and ovoinhibitor. Some of these proteins show
overlapping or simultaneous denaturation of domains, but the ratio
of calorimetric ($\Delta H_{cal}$) to van't Hoff enthalpies ($\Delta H_{vH}$) exceeds one.
Determination of exact numbers of cooperative denaturing units from
this ratio is complicated because calculation of $\Delta H_{vH}$ assumes a
two-state transition (Sturtevant, 1974) and may not be valid in some
cases, where denaturation is irreversible (Takahashi and Sturtevant,
1981). Denaturation of individual domains may be a two-state process.
Reversibility is sometimes difficult to determine. Ovomucoid and
ovoinhibitor denature reversibly when heated below $T_d$ (Donovan and
Beardslee, 1975; Zahnley, 1979), and apparent irreversibility after
heating to higher temperatures may result from secondary reactions
such as aggregation of the unfolded proteins.

Protein-protein association or other stabilizing interactions
may (1) decrease flexibility or mobility of the protein molecules,
as seen in a loss of entropy (Janin and Chothia, 1976; Schulz and
Schirmer, 1979), (2) alter relative stabilities of the folded (native)
and unfolded (denatured) proteins, or (3) lead to changes in sizes of

the cooperative denaturing units, as suggested by the DSC results (Donovan and Beardslee, 1975; Zahnley, 1980). According to Privalov (1982), thermodynamic stability and ease of folding favor protein structural units of molecular weight between 3000 and 30,000. Most domains in globular proteins fall within this range (Schulz and Schirmer, 1979). Strong noncovalent interactions between the pro-teinases and inhibitors, given sizes of structural domains in these proteins, could produce larger denaturing units in the complexes.

Protein conformational stability is not only of fundamental interest for protein chemistry, but also relevant to questions of food safety and quality. High stability of proteinase inhibitors may necessitate use of conditions for their inactivation that denature other proteins first and lead to other reactions that have undesir-able consequences for protein digestibility (Richardson, 1977). In discussing inactivation of enzymes during food processing, Balls (1942) stated, "Until we know more about it, we shall probably over-heat our products rather than underheat them, just to be safe. When we know more about it we may be able to moderate our enthusiasm and our technique". Basic studies on stabilities of protein enzyme inhibitors and lectins, as well as enzymes, can contribute to devel-opment of methods that avoid overheating or underheating food or feed products.

## ACKNOWLEDGMENTS

I thank Dr. John W. Donovan for helpful advice and discussions, Dr. Sigmund Schwimmer for a critical reading of the manuscript, and Dr. Clarence A. Ryan for samples of purified potato proteinase inhibitors I and II.

Reference to a company or product name does not imply approval or recommendation of the product by the U. S. Department of Agriculture to the exclusion of others that may also be suitable.

## REFERENCES

Alexandrov, V. Y. (1977). "Cells, Molecules, and Temperature". Springer-Verlag, Berlin, Heidelberg, and New York.
Anderson, R. L., Rackis, J. J., and Tallent, W. H. (1979). Biolog-ically active substances in soy products. In: "Soy Protein and Human Nutrition", H. L. Wilcke, D. T. Hopkins, and D. H. Waggle, eds., Academic Press, New York, pages 209-233.
Balls, A. K. (1942). The fate of enzymes in processed foods. Fruit Prod. J. 22, 36-39.
Belew, M., Porath, J., and Sundberg, L. (1975). The trypsin and chymotrypsin inhibitors in chick peas (Cicer arietinum L.). Purification and properties of the inhibitors. Eur. J. Biochem. 60, 247-258.

Birk, Y. (1961). Purification and some properties of a highly
    active inhibitor of trypsin and α-chymotrypsin from soy
    beans. Biochim. Biophys. Acta 54, 378-381.
Birk, Y. (1976). Proteinase inhibitors from plant sources.
    Methods Enzymol. 45, 695-739.
Blumberg, S., and Tal, N. (1976). Effect of divalent metal ions
    on the digestibility of concanavalin A by endopeptidases.
    Biochim Biophys. Acta 453, 357-364.
Boisen, S., and Djurtoft, R. (1982). Protease inhibitor from barley
    embryo inhibiting trypsin and trypsin-like microbial proteases.
    Purification and characterisation of two isoforms. J. Sci.
    Food Agric. 33, 431-440.
Borrebaeck, C., and Mattiasson, B. (1980). A study of structurally
    related binding properties of concanavalin A using differential
    scanning calorimetry. Eur. J. Biochem. 107, 67-71.
Bryant, J., Green, T. R., Gurusaddaiah, T., and Ryan, C. A. (1976).
    Proteinase inhibitor II from potatoes: Isolation and charac-
    terization of its protomer components. Biochemistry 15,
    3418-3424.
Buonocore, V., Petrucci, T., and Silano, V. (1977). Wheat protein
    inhibitors of α-amylase. Phytochemistry 16, 811-820.
Chandrasekher, G., Raju, D. S., and Pattabiraman, T. N. (1982).
    Natural plant enzyme inhibitors. Protease inhibitors in
    millets. J. Sci. Food Agric. 33, 447-450.
Chu, T. H., and Hsu, C.-L. (1965). Inhibition of trypsin and chymo-
    trypsin by ovoinhibitor. Acta Biochim. Sin. 5, 434-441.
Creighton, T. E. (1983). An empirical approach to protein conforma-
    tional stability and flexibility. Biopolymers 22, 49-58.
De Marco, A., Menegatti, E., and Guarneri, M. (1982). $^1$H-NMR studies
    of the structure and stability of the bovine pancreatic secre-
    tory trypsin inhibitor. J. Biol. Chem. 257, 8337-8342.
Dietl, T., and Tschesche, H. (1976). Die Disulfidbrucken des
    Trypsin-Kallikrein-Inhibitors K aus Weinbergschnecken (Helix
    pomatia). Thermische Denaturierung und thermolysinolytische
    Inaktivierung. Hoppe-Seyler's Z. Physiol. Chem. 357, 139-145.
Donovan, J. W. (1979). Phase transitions of the starch-water
    system. Biopolymers 18, 263-275.
Donovan, J. W., and Beardslee, R. A. (1975). Heat stabilization
    produced by protein-protein association. A differential
    scanning calorimetric study of the trypsin-soybean trypsin
    inhibitor and trypsin-ovomucoid complexes. J. Biol. Chem.
    250, 1966-1971.
Donovan, J. W., and Ross, K. D. (1973). Increase in the stability
    of avidin produced by binding of biotin. A differential
    scanning calorimetric study of denaturation by heat.
    Biochemistry 12, 512-517.
Donovan, J. W., and Ross, K. D. (1975). Iron binding to conalbumin.
    Calorimetric evidence for two distinct species with one bound
    iron atom. J. Biol. Chem. 250, 6026-6031.

Doyle, R. J., and Stroupe, S. D. (1978). Inhibition of concanavalin A by (ethylenedinitrilo)tetraacetic acid. Carbohydr. Res. 64, 327-333.

Doyle, R. J., Pittz, E. P., and Woodside, E. E. (1968). Carbohydrate--protein complex-formation. Some factors affecting the interaction of D-glucose and polysaccharides with concanavalin A. Carbohydr. Res. 8, 89-100.

Doyle, R. J., Thomasson, D. L., and Nicholson, S. K. (1976). Stabilization of concanavalin A by metal ligands. Carbohydr. Res. 46, 111-118.

Edelhoch, H., and Steiner, R. F. (1965). Changes in physical properties accompanying the interaction of trypsin with protein inhibitors. J. Biol. Chem. 240, 2877-2882.

Ellenrieder, G., Geronazzo, H., and De Bojarski, A. B. (1980). Thermal inactivation of trypsin inhibitors in aqueous extracts of soybeans, peanuts, and kidney beans: Presence of substances that accelerate inactivation. Cereal Chem. 57, 25-27.

Feeney, R. E. (1980). Overview on the chemical deteriorative changes of proteins and their consequences. In: "Chemical Deterioration of Proteins", J. R. Whitaker and M. Fujimaki, eds., Am. Chem. Soc. Symp. Ser. 123, 1-47.

Fraenkel-Conrat, H., Bean, R. C., Ducay, E. D., and Olcott, H. S. (1952). Isolation and characterization of a trypsin inhibitor from lima beans. Arch. Biochem. Biophys. 37, 393-407.

Friedman, M., Grosjean, O. K., and Zahnley, J. C. (1982a). Effect of disulfide bond modification on the structure and activities of enzyme inhibitors. In: "Food Protein Deterioration. Mechanisms and Functionality", J. P. Cherry, ed., Am. Chem. Soc. Symp. Ser. 206, 359-407.

Friedman, M., Grosjean, O. K., and Zahnley, J. C. (1982b). Inactivation of soya bean trypsin inhibitors by thiols. J. Sci. Food Agric. 33, 165-172.

Friedman, M., Grosjean, O. K., and Zahnley, J. C. (1982c). Cooperative effects of heat and thiols in activating trypsin inhibitors from legumes in solution and in the solid state. Nutr. Rep. Int. 25, 743-751.

Friedman, M., Zahnley, J. C., and Wagner, J. R. (1980). Estimation of the disulfide content of trypsin inhibitors as S-$\beta$-(2-pyridyl-ethyl)-L-cysteine. Anal. Biochem. 106, 27-34.

Fritz, H., and Tschesche, H., eds. (1971). "Proceedings of the First International Research Conference on Proteinase Inhibitors, Munich, 1970". Walter de Gruyter, Berlin.

Fritz, H., Tschesche, H., Greene, L. J., and Truscheit, E., eds. (1974). "Proteinase Inhibitors. Bayer Symposium V." Springer-Verlag, Berlin, Heidelberg, and New York. (Proc. 2nd Int. Res. Conf. on Proteinase Inhibitors, 1973).

Gallaher, D., and Schneeman, B. O. (this volume). Nutritional and
    metabolic response to plant inhibitors of digestive enzymes.
    In: "Nutritional and Metabolic Aspects of Food Safety", M.
    Friedman, ed., Advances in Experimental Medicine and Biology,
    Plenum Press, New York.
Goldstein, I. J., and Hayes, C. E. (1978). The Lectins:
    Carbohydrate-binding proteins of plants and animals. Adv.
    Carbohydr. Chem. Biochem. 35, 127-340.
Grant, G., More, L. J., McKenzie, N. H., and Pusztai, A. (1982).
    The effect of heating on the hemagglutinating activity and
    nutritional properties of bean (Phaseolus vulgaris) seeds.
    J. Sci. Food Agric. 33, 1324-1326.
Green, N. M., and Work, E. (1953). Pancreatic trypsin inhibitor. 1.
    Preparation and properties. Biochem. J. 54, 257-266.
Griko, Y. V., and Privalov, P. L. (1982), quoted in Privalov (1982).
Hardman, K. D., Agarwal, R. C., and Freiser, M. J. (1982).
    Manganese and calcium binding sites of concanavalin A.
    J. Mol. Biol. 157, 69-86.
Huang, D. Y., Swanson, B. G., and Ryan, C. A. (1981). Stability
    of proteinase inhibitors in potato tubers during cooking.
    J. Food Sci. 46, 287-290.
Huber, R., and Bennett, W. S., Jr. (1982). Functional significance
    of flexibility in proteins. Pure Appl. Chem. 54, 2489-2500.
Ikenaka, T., and Odani, S. (1978). Structure-function relation-
    ships of soybean double-headed proteinase inhibitors. In:
    "Regulatory Proteolytic Enzymes and Their Inhibitors." (Proc.
    11th FEBS Meeting), S. Magnusson, M. Ottesen, B. Foltmann,
    K. Dano, and H. Neurath, eds., Pergamon, Oxford, pages 207-216.
Imondi, A. R., Ferguson, J. L., and Wolgemuth, R. L. (1973).
    Inhibition of human chymotrypsin by dried egg white. Am.
    J. Dig. Dis. 18, 1051-1054.
Iwasaki, T., Kiyohara, T., and Yoshikawa, M. (1971). Purification
    and partial characterization of two different types of
    proteinase inhibitors (inhibitors II-a and II-b) from
    potatoes. J. Biochem. (Tokyo) 70, 817-826.
Jaenicke, R., ed. (1980). "Protein Folding." Elsevier/North-
    Holland Biomedical Press, Amsterdam and New York.
Jaffe, W. G. (1980). Hemagglutinins (lectins). In: "Toxic
    Constituents of Plant Foodstuffs", 2nd Edn., I. E. Liener,
    ed., Academic Press, New York, pages 73-102.
Janin, J., and Chothia, C. (1976). Stability and specificity of
    protein-protein interactions: the case of the trypsin-trypsin
    inhibitor complexes. J. Mol. Biol. 100, 197-211.
Jibson, M. D., Birk, Y., and Bewley, T. A. (1981). Circular
    dichroism of trypsin and chymotrypsin complexes with Bowman-
    Birk or chickpea trypsin inhibitor. Int. J. Pept. Protein
    Res. 18, 26-32.
Kassell, B. (1970). Inhibitors of proteolytic enzymes. Methods
    Enzymol. 19, 839-906.

Kassell, B., and Williams, M. J. (1976). Proteinase inhibitors.
    In: "CRC Handbook of Biochemistry and Molecular Biology, 3rd
    Edn., Proteins--Vol. II", G. D. Fasman, ed., CRC Press,
    Cleveland, pages 583-668.

Kato, I., Kohr, W. J., and Laskowski, M., Jr. (1980). Preparation
    of intact domains from chicken ovoinhibitor by limited
    cleavage. Fed. Proc. 39, 1688.

Kato, I., Schrode, J., Wilson, K. A., and Laskowski, M., Jr. (1976).
    Evolution of proteinase inhibitors. Protides Biol. Fluids,
    Proc. Colloq. 23, 235-243.

Kim, P. S., and Baldwin, R. L. (1982). Specific intermediates in
    the folding reactions of small proteins and the mechanism
    of protein folding. Annu. Rev. Biochem. 51, 459-489.

Kinsella, J. E. (1976). Functional properties of proteins in
    foods: a survey. CRC Crit. Rev. Food Sci. Nutr. 7, 219-280.

Kinsella, J. E. (1979). Functional properties of soy proteins.
    J. Am. Oil Chem. Soc. 56, 242-258.

Koenig, S. H., Brown, R. D., III, Brewer, C. F., and Sherry, A. D.
    (1982). Conformational equilibrium of demetalized concana-
    valin A: a reexamination of the kinetics of its interactions
    with $Ca^{2+}$-ions and fluorescent saccharide. Biochem.
    Biophys. Res. Commun. 109, 1047-1053.

Kraus, E., and Femfert, U. (1976). Proteinase K from the mold
    Tritirachium album Limber. Specificity and mode of action.
    Hoppe-Seyler's Z. Physiol. Chem. 357, 937-947.

Kunitz, M. (1947). Crystalline soybean trypsin inhibitor. II.
    General properties. J. Gen. Physiol. 30, 291-310.

Laskowski, M., Jr., and Kato, I. (1980). Protein inhibitors of
    proteinases. Annu. Rev. Biochem. 49, 593-626.

Laskowski, M., Jr., and Sealock, R. W. (1971). Protein proteinase
    inhibitors--molecular aspects. In: "The Enzymes", 3rd Edn.,
    Vol. III, P. D. Boyer, ed., Academic Press, New York,
    pages 375-473.

Laskowski, M., Jr., Kato, I., Kohr, W. J., March, C. J., and
    Bogard, W. C. (1980). Evolution of the family of serine
    proteinase inhibitors homologous to pancreatic secretory
    trypsin inhibitor (Kazal). Protides Biol. Fluids, Proc.
    Colloq. 28, 123-128.

Lei, M-G., Bassette, R., and Reeck, G. R. (1981). Effect of
    cysteine on heat inactivation of soybean trypsin inhibitors.
    J. Agric. Food Chem. 29, 1196-1199.

Liener, I. E. (1976). Phytohemagglutinins (phytolectins). Annu.
    Rev. Plant Physiol. 27, 291-319.

Liener, I. E. (1979). Significance for humans of biologically
    active factors in soybeans and other food legumes. J. Am.
    Oil Chem. Soc. 56, 121-129.

Liener, I. E. (1980). Miscellaneous toxic factors. In: "Toxic
    Constituents of Plant Foodstuffs", 2nd Edn., I. E. Liener,
    ed., Academic Press, New York, pages 429-467.

Liener, I. E., and Kakade, M. L. (1969). Protease inhibitors. In: "Toxic Constituents of Plant Foodstuffs", 1st Edn., I. E. Liener, ed., Academic Press, New York, pages 7-68.

Liener, I. E., and Kakade, M. L. (1980). Protease inhibitors. In: "Toxic Constituents of Plant Foodstuffs", 2nd Edn., I. E. Liener, ed., Academic Press, New York, pages 7-71.

Lineweaver, H., and Murray, C. W. (1947). Identification of the trypsin inhibitor of egg white with ovomucoid. J. Biol. Chem. 171, 565-581.

Lis, H., and Sharon, N. (1977). Lectins: their chemistry and application to immunology. In: "The Antigens", Vol. 4, M. Sela, ed., Academic Press, New York, pages 429-529.

Lis. H., and Sharon, N. (1981). Lectins in higher plants. In: "The Biochemistry of Plants. A Comprehensive Treatise. Vol. 6: Proteins and Nucleic Acids", A. Marcus, ed., Academic Press, New York, pages 371-447.

Loewus, F. A., and Ryan, C. A., eds. (1981). The Phytochemistry of Cell Recognition and Cell Surface Interactions. Recent Adv. Phytochem., Vol. 15.

Marshall, J. J. (1975). α-Amylase inhibitors from plants. In: "Physiological Effects of Food Carbohydrates", A. Jeanes and J. Hodges, eds., Am. Chem. Soc. Symp. Ser. 16, 244-266.

Matsuda, T., Watanabe, K., and Nakamura, R. (1983). Ovomucoid and ovoinhibitor isolated from chicken egg white are immunologically cross-reactive Biochem. Biophys. Res. Commun. 110, 75-81.

Matsuda, T., Watanabe, K., and Sato, Y. (1981a). Independent thermal unfolding of ovomucoid domains. Biochim. Biophys. Acta 669, 109-112.

Matsuda, T., Watanabe, K., and Sato, Y. (1981b). Temperature-induced structural changes in egg white ovomucoid. Agric. Biol. Chem. 45, 1609-1614.

Matsuda, T., Watanabe, K., and Sato, Y. (1981c). Heat-induced aggregation of egg white proteins as studied by vertical flat-sheet polyacrylamide gel electrophoresis. J. Food Sci. 46, 1829-1834.

Matsushima, K. (1958). On the naturally occurring inhibitors for Aspergillus protease. III. Ovo-inhibitor. Nippon Nogei Kagaku Kaishi 32, 211-215.

McCammon, A. J., and Karplus, M. (1983). The dynamic picture of protein structure. Acc. Chem. Res. 16, 187-193.

Means, G. E., Ryan, D. S., and Feeney, R. E. (1974). Protein inhibitors of proteolytic enzymes. Acc. Chem. Res. 7, 315-320.

Melville, J. C., and Ryan, C. A. (1972). Chymotrypsin inhibitor I from potatoes. Large scale preparation and characterization of its subunit components. J. Biol. Chem. 247, 3445-3453.

Menegatti, E., Tomatis, R., Guarneri, M., and Scatturin, A. (1981). Conformational stability of porcine-pancreatic secretory trypsin inhibitors (Kazal inhibitors). Int. J. Pept. Protein Res. 17, 454-459.

Minton, K. W., Karmin, P., Hahn, G. M., and Minton, A. P. (1982). Nonspecific stabilization of stress-susceptible proteins by stress-resistant proteins: a model for the biological role of heat shock proteins. Proc. Nat. Acad. Sci. U.S.A. 79, 7107-7111.

Nachbar, M. S., and Oppenheim, J. D. (1980). Lectins in the United States diet: a survey of lectins in commonly consumed foods an a review of the literature. Am. J. Clin. Nutr. 33, 2338-2345.

Osuga, D. T., Bigler, J. C., Uy, R. L., Sjoberg, L., and Feeney, R. E. (1974). Comparative biochemistry of penguin egg white proteins--I. Ovomucoids: composition and inhibitory activities for trypsin, α-chymotrypsin and subtilisin. Comp. Biochem. Physiol. B 48B, 519-533.

Pain, R. H. (1983). Protein folding, denaturation and stability. Biochem. Soc. Trans. 11, 15-16.

Pearce, G., McGinnis, J., and Ryan, C. A. (this volume). Nutritional studies of a carboxypeptidase inhibitor from potato tubers. In: "Nutritional and Metabolic Aspects of Food Safety", M. Friedman, ed., Advances in Experimental Medicine and Biology, Plenum Press, New York.

Perkins, D., and Toledo, R. T. (1982). Effect of heat treatment for trypsin inhibitor inactivation on physical and functional properties of peanut protein. J. Food Sci. 47, 917-923.

Pfeil, W. (1981). The problem of the stability of globular proteins. Mol. Cell. Biochem. 40, 3-28.

Plunkett, G., and Ryan, C. A. (1980). Reduction and carboxamidomethylation of the single disulfide bond of proteinase inhibitor I from potato tubers. Effects on stability, immunological properties, and inhibitory activities. J. Biol. Chem. 255, 2752-2755.

Privalov, P. L. (1979). Stability of proteins. Small globular proteins. Adv. Protein Chem. 33, 167-241.

Privalov, P. L. (1980). Scanning microcalorimeters for studying macromolecules. Pure Appl. Chem. 52, 479-497.

Privalov, P. L. (1982). Stability of proteins. Proteins which do not present a single cooperative system. Adv. Protein Chem. 35, 1-104.

Ramshaw, J. A. M. (1982). Structures of plant proteins. In: "Nucleic Acids and Proteins in Plants I: Structure, Biochemistry, and Physiology of Proteins". (Encyclopedia of Plant Physiology, New Series, Vol. 14, Pt. A), D. Boulter and B. Parthier, eds., Springer-Verlag, Berlin, Heidelberg, and New York, pages 229-290.

Richardson, M. (1977). The proteinase inhibitors of plants and microorganisms. Phytochemistry 16, 159-169.

Richardson, M. (1981). Protein inhibitors of enzymes. Food Chem. 6, 235-253.

Ryan, C. A. (1966). Chymotrypsin inhibitor I from potatoes: reactivity with mammalian, plant, bacterial, and fungal proteinases. Biochemistry 5, 1592-1596.

Ryan, C. A. (1981). Proteinase inhibitors. In: "The Biochemistry of Plants. A Comprehensive Treatise. Vol. 6: Proteins and Nucleic Acids", A. Marcus, ed., Academic Press, New York, pages 351-370.

Ryan, C. A., and Kassell, B. (1970). Chymotrypsin inhibitor I from potatoes. Methods Enzymol. 19, 883-889.

Sanderson, J. E., Freed, R. C., and Ryan, D. S. (1982). Thermal denaturation of genetic variants of the Kunitz soybean inhibitor. Biochim. Biophys. Acta 701, 233-241.

Sato, S., and Murao, S. (1973). Isolation and crystallization of microbial alkaline protease inhibitor, S-SI. Agric. Biol. Chem. 37, 1067-1074.

Schulz, G. E., and Schirmer, R. H. (1979). "Principles of Protein Structure." Springer-Verlag, Berlin, Heidelberg, and New York.

Schwimmer, S. (1981). "Source Book of Food Enzymology." AVI, Westport, CT.

Sgarbieri, V. C., and Whitaker, J. R. (1982). Physical, chemical, and nutritional properties of common bean (Phaseolus) proteins. Adv. Food Res. 28, 93-166.

Shivaraj, B., and Pattabiraman, T. N. (1981). Natural plant enzyme inhibitors. Characterization of an unusual α-amylase/trypsin inhibitor from ragi (Eleusine coracana Gaertn.). Biochem. J. 193, 29-36.

Shivaraj, B., Rao, H. N., and Pattabiraman, T. N. (1982). Natural plant enzyme inhibitors. Isolation of a trypsin/α-amylase inhibitor and a chymotrypsin/trypsin inhibitor from ragi (Eleusine coracana) grains by affinity chromatography and study of their properties. J. Sci. Food Agric. 33, 1080-1091.

Silano, V. (1978). Biochemical and nutritional significance of wheat albumin inhibitors of α-amylase. Cereal Chem. 55, 722-731.

Silano, V., and Zahnley, J. C. (1978). Association of Tenebrio molitor L. α-amylase with two protein inhibitors--one monomeric, one dimeric--from wheat flour. Differential scanning calorimetric comparison of heat stabilities. Biochim. Biophys. Acta 533, 181-185.

Stevens, F. C., and Feeney, R. E. (1963). Chemical modification of avian ovomucoids. Biochemistry 2, 1346-1352.

Stroupe, S. D., and Doyle, R. J. (1980). Kinetic analysis of calcium binding to concanavalin A. J. Inorg. Biochem. 12, 173-178.

Sturtevant, J. M. (1974). Some applications of calorimetry in biochemistry and biology. Annu. Rev. Biophys. Bioeng. 3, 35-51.

Takahashi, K., and Sturtevant, J. M. (1981). Thermal denaturation of Streptomyces subtilisin inhibitor, subtilisin BPN', and the subtilisin--inhibitor complex. Biochemistry 20, 6185-6190.

Tan, N.-H., Lowe, E. S. H., and Iskandar, M. (1982). The extractability of winged bean (Psophocarpus tetragonolobus) seed trypsin inhibitors. J. Sci. Food Agric. 33, 1327-1330.

Thompson, L. U., Rea, R. L., and Jenkins, D. J. A. (1983). Effect of heat processing on hemagglutinin activity in red kidney beans. J. Food Sci. 48, 235-236.

Tishchenko, V. M., and Gorodkov, B. G. (1978?), quoted in Privalov (1979).

Tomimatsu, Y., Clary, J. J., and Bartulovich, J. J. (1966). Physical characterization of ovoinhibitor, a trypsin and chymotrypsin inhibitor from chicken egg white. Arch. Biochem. Biophys. 115, 536-544.

Toms, G. C. (1981). Lectins in Leguminosae. In: "Advances in Legume Systematics", R. M. Polhill and P. H. Raven, eds., Royal Botanic Gardens, Kew, pages 561-577.

Tschesche, H. (1981). Proteolytic enzyme inhibitors. Med. Chem. Adv., Proc. Int. Symp., 7th, 1980, F. G. De Las Heras and S. Vega, eds., Pergamon Press, Oxford, pages 411-422.

Tschesche, H. (1974). Biochemistry of natural proteinase inhibitors. Angew. Chem. Int. Ed. Engl. 13, 10-28.

Tur-Sinai, A., Birk, Y., Gertler, A., and Rigbi, M. (1972). A basic trypsin- and chymotrypsin-inhibitor from groundnuts (Arachis hypogaea). Biochim. Biophys. Acta 263, 666-672.

Uehara, Y., Tonomura, B., and Hiromi, K. (1978). Direct fluorometric determination of a dissociation constant as low as $10^{-10}$ M for the subtilisin BPN'--protein proteinase inhibitor (Streptomyces subtilisin inhibitor) complex by a single photon counting technique. J. Biochem. (Tokyo) 84, 1195-1202.

Veerabhadrappa, P. S., Manjunath, N. H., and Virupaksha, T. K. (1978). Proteinase inhibitors of finger millet (Eleusine coracana Gaertn.). J. Sci. Food Agric. 29, 353-358.

Vogel, R., Trautschold, I., and Werle, E. (1968). "Natural Proteinase Inhibitors." Academic Press, New York.

Weder, J. K. P. (1981). Protease inhibitors in the Leguminosae. In: "Advances in Legume Systematics", R. M. Polhill and P. H. Raven, eds., Royal Botanic Gardens, Kew, pages 533-560.

Wetlaufer, D. B. (1981). Folding of protein fragments. Adv. Protein Chem. 34, 61-92.

Whitaker, J. R. (1981). Naturally occurring peptide and protein inhibitors of enzymes. In: "Impact of Toxicology on Food Processing", J. C. Ayres and J. C. Kirschman, eds., AVI, Westport, CT, pages 57-104.

Whitaker, J. R., and Feeney, R. E. (1973). Enzyme inhibitors in foods. In: "Toxicants Occurring Naturally in Foods", 2nd Edn., F. M. Strong, ed., National Academy of Sciences, Washington, D.C., pages 276-298.

Wüthrich, K., Wagner, G., Richarz, R., and Braun, W. (1980). Correlations between internal mobility and stability of globular proteins. Biophys. J. 32, 549-560.

Yoshikawa, M., Ogura, S., and Tatsumi, M. (1977). Some properties of proteinase inhibitors from adzuki beans (Phaseolus angularis). Agric. Biol. Chem. 41, 2235-2239.

Zahnley, J. C. (1974). Evidence that the two binding sites for trypsin on chicken ovoinhibitor are not equivalent. Dissociation of the complexes with porcine trypsin. J. Biol. Chem. 249, 4282-4285.

Zahnley, J. C. (1975). Preferred binding of bovine and porcine trypsins at two different sites on chicken ovoinhibitor. Reduced dissociation of mixed trypsin complexes. J. Biol. Chem. 250, 7879-7884.

Zahnley, J. C. (1977). Independence of domains in protease--inhibitor complexes formed by singleheaded and multiheaded inhibitors. Fed. Proc. 36, 765.

Zahnley, J. C. (1979). Independent heat stabilization of proteases associated with multiheaded inhibitors. Trypsins bound at nonequivalent sites on chicken ovoinhibitor. J. Biol. Chem. 254, 9721-9727.

Zahnley, J. C. (1980). Independent heat stabilization of proteases associated with multiheaded inhibitors. Complexes of chymotrypsin, subtilisin and trypsin with chicken ovoinhibitor and with lima bean protease inhibitor. Biochim. Biophys. Acta 613, 178-190.

Zahnley, J. C. (1981). Effects of manganese and calcium on conformational stability of concanavalin A: a differential scanning calorimetric study. J. Inorg. Biochem. 15, 67-78.

Zahnley, J. C., and Davis, J. G. (1970). Determination of trypsin--inhibitor complex dissociation by use of the active site titrant, p-nitrophenyl p'-guanidinobenzoate. Biochemistry 9, 1428-1433.

Zahnley, J. C., and Friedman, M. (1982). Absorption and fluorescence spectra of S-quinolylethylated Kunitz soybean trypsin inhibitor. J. Protein Chem. 1, 225-240.

## 18

PROTEIN-ALKALI REACTIONS: CHEMISTRY, TOXICOLOGY, AND NUTRITIONAL CONSEQUENCES

Mendel Friedman*, Michael R.Gumbmann*
and Patricia M. Masters**

*Western Regional Research Center, Agricultural Research
Service, U. S. Department of Agriculture, Berkeley,
California 94710

**Scripps Institution of Oceanography, University of
California, La Jolla, California 92093

ABSTRACT

Heat and alkali treatment of proteins catalyzes formation of crosslinked amino-acid side chains such as lysinoalanine, ornithino-alanine and lanthionine, and concurrent racemization of L-isomers of all amino acid residues to D-analogues. Factors that favor these transformations include high pH and temperature, long exposure, and certain inductive or steric properties of the various amino acid side chains. Factors that minimize crosslink formation include the presence of certain additives, such as cysteine or sulfite ions, and acylation of $\varepsilon$-NH$_2$ groups of lysine side chains. Free and protein-bound lysinoalanine and D-serine induce nephrocytomegaly in rat kidney tissues. The presence of lysinoalanine and D-amino acid residues along a protein chain decreases its digestibility and nutritional quality. Understanding the factors that govern the formation of potentially harmful unnatural amino acid residues in food proteins and the toxic and nutritionally antagonistic action of these compounds in animals should lead to better and safer foods.

INTRODUCTION

Food proteins have been treated with alkali or heat for preparing meat analogue vegetable (soy) protein (Hewitt et al. 1979; Sternberg and Kim, 1977), preparing protein concentrates and tortillas (Tannenbaum et al., 1970; Saunders et al., 1975; Wu et al.,

1977; Sanderson et al., 1978), destroying aflatoxin (Goldblatt, 1969), peeling fruits and vegetables (Gee et al., 1974), sterilizing milk (Fritsch et al., 1983; de Koning and van Rooijen, 1982) and in home and commercial processing of foods. Such treatments, however, cause undesirable changes. These include formation of crosslinked peptide chains through formation of amino acids such as lysinoalanine (LAL), ornithinoalanine, and lanthionine, and concurrent racemization of optically active L-amino acids to D-isomers (Friedman, 1977; 1982a; Masters and Friedman 1979; 1980; Friedman et al., 1981). Foods containing the crosslinked and D-amino acids are widely consumed (Sternberg and Kim, 1977; Sternberg et al., 1975; Friedman et al., 1981; Bunjapamai et al., 1982; Steinig and Montag, 1982).

Feeding alkali-treated soy proteins to rats induces changes in kidney cells. These changes are characterized by enlargement of the nucleus and cytoplasm, increased nucleoprotein content, and disturbances of DNA synthesis and mitosis. The lesions affect epithelial cells of the straight portion (pars recta) of the proximal renal tubules. (Feron et al., 1977; Gould and MacGregor, 1977; Karayiannis et al., 1979a, b; O'Donovan, 1976; Reyniers et al., 1974; Woodard et al., 1975; 1979). See Figure 1.

These observations cause concern about the nutritional quality and safety of alkali-treated foods. Chemical changes that govern formation of unnatural amino acids during alkali treatment of proteins need to be studied and explained. The nutritional and toxicological significance of these changes need to be defined. Finally, appropriate strategies to minimize or prevent these reactions need to be developed.

In this paper we describe some of the factors that influence lysinoalanine formation and racemization of amino acid residues in food proteins. We also discuss toxicological and nutritional consequences of feeding alkali-treated food proteins, lysinoalanine, and D-amino acids.

The findings have practical implications for nutrition and food safety. They may be useful to food processors and consumers who wish to minimize the formation and consumption of unnatural, deleterious compounds during commercial and home food processing.

## FACTORS GOVERNING THE FORMATION OF LYSINOALANINE

pH effects. The effect of pH on the lysinoalanine content of soy protein is given in Table 1. The results show that lysinoalanine began to appear at pH 9.0 and increased continuously with pH up to

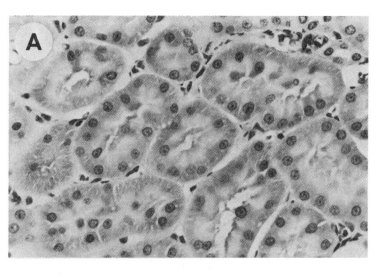

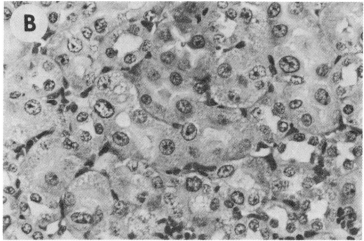

Figure 1.  Photomicrograph of outer medullary stripe of kidneys from rats fed 20% soyprotein for eight weeks.  Hematoxylin and eosin.  (A) control, pair-fed, untreated soyprotein diet.  Note uniformity of the pars recta cells.  (B) alkali-treated soy-protein (2630 ppm total dietary LAL).  Note the nuclear and cytoplasmic enlargement of many of the pars recta cells. Magnification both 440X.

Reproduced from Gould and MacGregor (1977).

pH 12.5, and then decreased at pH 13.9. The following amino acids were also degraded, destroyed, or modified during the alkaline treatment of soy protein: threonine, serine, (half) cystine, lysine, and arginine (results not shown).

Time effects. Data in Table 2 show that heating a 1% (w/v) solution of soy protein in 0.1 N NaOH at 75°C for various time periods progressively increased lysinoalanine formation for about 3 hr; LAL formation then leveled off and started to decrease.

These results and those mentioned above on the pH-dependence of lysinoalanine formation indicate that for each protein, conditions may exist in which lysinoalanine is formed as fast as it is destroyed. Additional studies are needed to verify this hypothesis.

Temperature effects. Exposure of a 1% (w/v) solution of soy protein in 0.1 N NaOH (pH 12.5) for 3 hr at temperatures ranging from 25°C to 85°C caused progressive degradation of the following labile amino acids: threonine, serine, lysine, and arginine (Table 3). Serine and threonine destruction started at about 45°C and lysine and arginine at about 35°C. Thus, the value of threonine decreased from 4.23 mole % for the untreated soy protein to 2.56 mole % for the sample treated at 85°C. The corresponding values for serine are 6.41 and 2.73 mole %; for lysine, 5.55 and 3.24 mole %; and for arginine, 5.84 and 3.15 mole %; respectively. The treatment also induced the progressive disappearance of (half) cystine (not shown) and the appearance of lysinoalanine residues. These residues started to appear at 25°C (0.29 mole %) and continuously increased up to 2.24 mole % at 85°C.

Table 1. Effect of pH on lysinoalanine content of soy protein. [1]

| pH | | | | | | | | |
|---|---|---|---|---|---|---|---|---|
| 8.0 | 9.0 | 9.5 | 10.0 | 10.5 | 10.9 | 12.0 | 12.5 | 13.9 |
| 0.00 | 0.28 | 0.40 | 0.61 | 0.68 | 0.95 | 1.14 | 1.86 | 1.47 |

[1]Conditions: 1% protein; 75°C; 3 hours. Numbers in mole % are averages from two separate experiments.

Table 2.   Effect of time of treatment on LAL content of alkali-treated soy protein.[1]

| Time (min) | Lysinoalanine (Mole %) |
|------------|------------------------|
| 10         | 0.46                   |
| 20         | 0.63                   |
| 40         | 1.01                   |
| 60         | 1.10                   |
| 120        | 1.75                   |
| 180        | 2.39                   |
| 300        | 2.98                   |
| 480        | 2.05                   |

[1]Conditions:  average of two experiments; 0.5 g protein in 50 ml 0.1 $\underline{N}$ NaOH; 75°.

Effect of carbohydrates.  The ε-amino group of lysine reacts under the influence of heat with the carbonyl groups of sugars such as glucose to form Schiff's bases that are further transformed to incompletely characterized Maillard browning products (Friedman, 1982b).  If such reactions were to occur in strongly alkaline media, they would be expected to reduce the formation of lysinoalanine because the blocked lysine amino groups would be unable to combine with dehydroalanine, the lysinoalanine precursor (see below).

Exposure of a 1% (w/v) solution of soy protein to 0.1 N NaOH for 1 hr at 45 to 75°C in the absence and presence of 5 to 20% (w/v) glucose did not affect the amount of lysinoalanine produced (Figures 2,3; Tables 4).  Glucose, however, did produce two noticeable effects:  1) a sharp decrease in the arginine content, which varied directly with the amount of glucose present; and 2) the formation of a new compound eluting just before lysinoalanine.

Table 3. Effect of temperature on amino acid composition of alkali-treated soy protein.[1]

| Sample Number | THR Wt.% | THR Mole% | SER Wt.% | SER Mole% | HIS Wt.% | HIS Mole% | LYS Wt.% | LYS Mole% | LAL Wt.% | LAL Mole% | ARG Wt.% | ARG Mole% |
|---|---|---|---|---|---|---|---|---|---|---|---|---|
| 1. Untreated control | 3.40 | 4.23 | 4.55 | 6.41 | 2.25 | 2.15 | 5.47 | 5.55 | 0.00 | 0.00 | 6.87 | 5.84 |
| 2. H$_2$O-dialyzed control | 3.40 | 4.12 | 4.78 | 6.57 | 2.27 | 2.11 | 5.54 | 5.48 | 0.00 | 0.00 | 6.96 | 5.77 |
| 3. 25°C | 3.39 | 4.49 | 4.47 | 6.33 | 2.28 | 2.27 | 4.91 | 5.32 | 0.69 | 0.29 | 6.49 | 5.86 |
| 4. 35°C | 3.35 | 4.32 | 4.48 | 6.24 | 2.16 | 2.14 | 4.88 | 5.08 | 0.81 | 0.41 | 6.52 | 5.81 |
| 5. 45°C | 3.25 | 4.29 | 4.07 | 6.08 | 2.17 | 2.19 | 4.63 | 4.97 | 1.04 | 0.79 | 6.29 | 5.67 |
| 6. 55°C | 3.00 | 4.16 | 4.08 | 6.13 | 2.12 | 2.16 | 4.29 | 4.64 | 1.64 | 0.85 | 6.11 | 5.54 |
| 7. 65°C | 2.83 | 3.87 | 3.32 | 5.14 | 2.06 | 2.15 | 3.72 | 4.13 | 2.64 | 1.40 | 5.48 | 5.11 |
| 8. 75°C | 2.24 | 2.92 | 2.69 | 4.97 | 2.17 | 2.18 | 3.46 | 3.68 | 3.66 | 1.86 | 4.86 | 4.33 |
| 9. 85°C | 1.84 | 2.56 | 1.73 | 2.73 | 2.29 | 2.45 | 2.86 | 3.24 | 4.13 | 2.24 | 3.30 | 3.15 |

[1]Conditions: 0.5 g protein; 50 cc 0.1 N NaOH; 3 hours. Numbers are averages from two experiments.

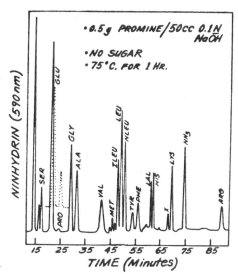

Figure 2. Amino acid analysis of alkali-treated soy protein. Note peaks of lysinoalanine (LAL) and unknown (x) appear before lysine (LYS). Conditions: 0.5 g Promine/50 ml 0.1 N NaOH; no sugar; and 75°C for 1 h.

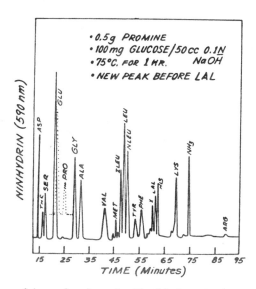

Figure 3. Amino acid analysis of alkali-treated soy protein in the presence of glucose. Note the new peak before lysinoalanine (LAL). Conditions: 0.5 g Promine; 100 mg glucose/50 ml 0.1 N NaOH; and 75°C for 1 h.

Table 4. Effect of glucose on amino acid composition of alkali-treated soy protein (Promine-D).[1]

| Sample Number | THR | SER | MET | HIS | LYS | ARG | X | LAL | N % |
|---|---|---|---|---|---|---|---|---|---|
| 1. Untreated control | 4.23 | 6.41 | 1.08 | 2.15 | 5.55 | 5.84 | 0.00 | 0.00 | 13.47 |
| 2. H$_2$0-dialyzed control | 4.12 | 6.57 | 1.00 | 2.11 | 5.48 | 5.77 | 0.00 | 0.00 | 14.33 |
| 3. NaOH only | 4.02 | 6.34 | 0.555 | 2.16 | 5.05 | 5.48 | 0.00 | 0.41 | 13.44 |
| 4. NaOH plus 25 mg glucose | 4.03 | 6.27 | 0.815 | 2.24 | 4.90 | 4.83 | 0.51 | 0.50 | 13.37 |
| 5. NaOH plus 50 mg glucose | 4.03 | 6.28 | 0.755 | 2.22 | 5.05 | 4.50 | 0.45 | 0.44 | 13.17 |
| 6. NaOH plus 100 mg glucose | 4.12 | 6.31 | 0.655 | 2.16 | 5.07 | 3.63 | 0.33 | 0.40 | 13.15 |

[1]Conditions: 0.5 g protein; 50 cc 0.1 N NaOH; 1 hour; 45°C. Numbers are averages from two experimental treatments in mole per cent.

Additional studies indicate that the concentration of the unknown compound X increased sharply when the temperature was raised to 75°C. This concentration ranged for 0.645 mole % in the presence of 5% glucose to 1.87 mole % with 20% glucose. The arginine content dropped from 5.84 mole % for untreated soy protein to 0.805% for soy protein treated with base in the presence of 20% glucose.

Although the conditions in this experiment were more severe than generally used in food processing, the results of this temperature study and those cited earlier strikingly demonstrate that significant amounts of lysinoalanine appear to form in soy protein under relatively mild conditions of pH, time, and temperature. Consequently, preventive measures against lysinoalanine formation may be justified even if proteins are subjected even to mild alkaline conditions during food processing.

## MECHANISM OF LYSINOALANINE FORMATION

A postulated mechanism of lysinoalanine formation (Figures 4-6) is at least a two-step process (Asquith and Otterburn, 1977; Friedman, 1977; Whitaker and Feeney, 1977). First, hydroxide ion-catalyzed elimination reactions of serine, threonine, and cystine (and to a lesser extent probably also cysteine) give rise to a dehydroalanine intermediate. Since such elimination reactions are second-order reactions that depend directly on the concentration of both hydroxide ion and susceptible amino acids, the extent of lysinoalanine formation should vary directly with hydroxide ion concentration. This is indeed the case within certain pH ranges. Second, a dehydroalanine residue, which contains a carbon-carbon double bond, reacts with the $\varepsilon$-amino group of lysine to form a lysinoalanine crosslink. This combination is governed not only by the number of available amino groups but also by the location of the amino group and dehydroalanine potential partners in the protein chain.

When convenient sites have reacted, additional lysinoalanine (or other) crosslinks form less readily. Each protein, therefore, may have a limited fraction of sites for forming crosslinked residues. The number of such sites is presumably dictated by the protein's size, composition, conformation, chain mobility, steric factors, and extent of ionization of reactive amino (or other) nucleophilic centers.

Stereochemistry. Figure 6 illustrates nucleophilic elimination-addition reactions of amino acid residues, which form crosslinked proteins that yield, on hydrolysis, complex amino acids. Note that the addition reaction may restore or reverse the configuration of the original asymmetric center in the alanine part of 3-N$^6$-lysinoalanine. Note, too, that the derivatives of 3-methyldehydroalanine

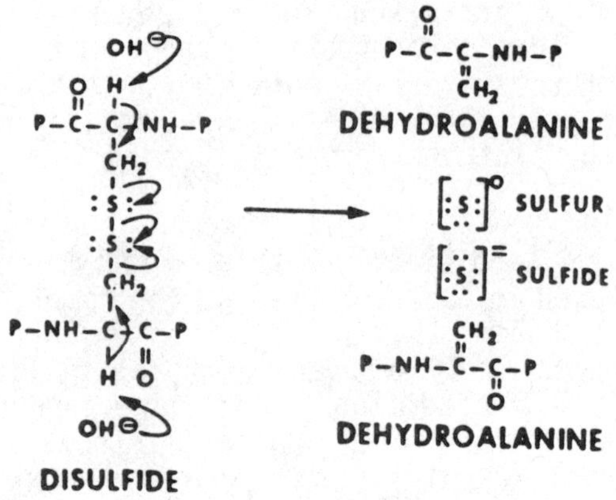

Figure 4.  Postulated mechanism for base-catalyzed transformation of a protein-disulfide bond to two dehydroalanine side chains, a sulfide ion, and sulfur.

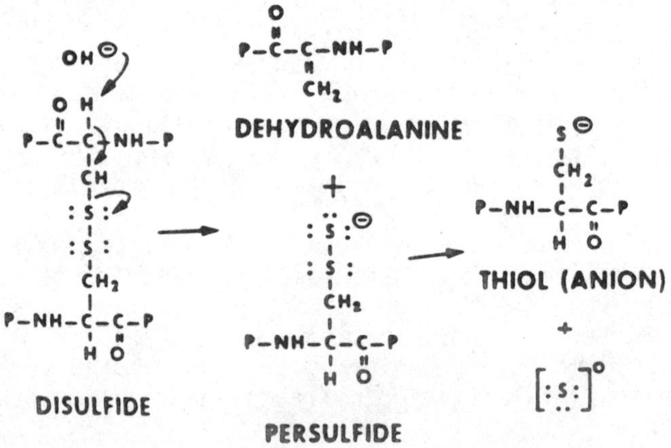

Figure 5.  Postulated mechanism for base-catalyzed formation of one dehydroalanine side chain and one persulfide anion from a protein-disulfide bond.  The persulfide can decompose to a thiol anion (cysteine) and elemental sulfur.

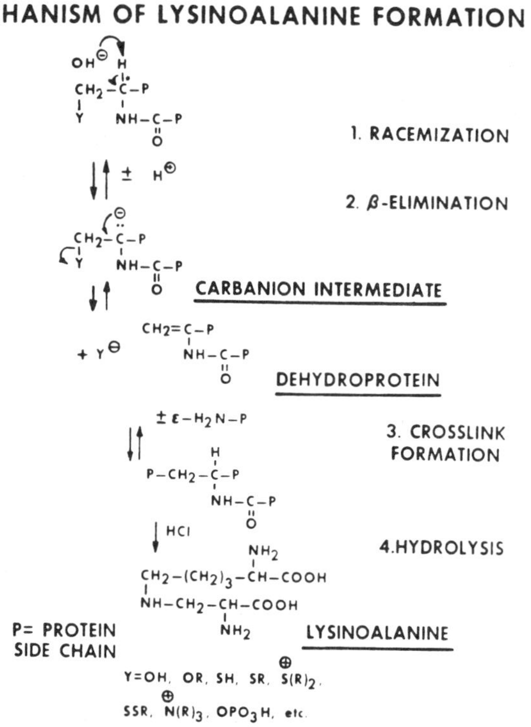

## MECHANISM OF LYSINOALANINE FORMATION

Figure 6. Transformation of a reactive protein side chain to a lysinoalanine side chain via elimination and crosslink formation. Note that the intermediate carbanion has lost the original asymmetry. The carbanion can combine with a proton to regenerate the original amino acid which is then racemic, or undergo an elimination reaction to form dehydroalanine.

would have at least one more asymmetric carbon atom than those of dehydroalanine (Alworth, 1976; Friedman, 1977).

Since lysinoalanine has two asymmetric carbon atoms, Bohak (1964) originally designated it as:

$N^{\varepsilon}$-(DL-2-amino-2-carboxyethyl)-L-lysine.

A more up-to-date name with R and S designations for the systematically named substituent would be:

$N^6$-[(2RS-2-amino-2-carboxyethyl]-L-lysine.

D-Lysine would give rise to the following diastereoisomers:

$N^6$-[(2RS-2-amino-2-carboxyethyl]-D-lysine.

Interaction between the double bond of threonine-derived methyldehydroalanine with the $\varepsilon$-NH$_2$ group of L-lysine can, in principle, produce the following methyl-lysinoalanine isomers:

$N^6$-[(1R,2R)-2-amino-2-carboxy-1-methylethyl]-L-lysine

$N^6$-[(1R,2S)-2-amino-2-carboxy-1-methylethyl]-L-lysine

$N^6$-[(1S,2S)-2-amino-2-carboxy-1-methylethyl]-L-lysine

$N^6$-[(1S,2R)-2-amino-2-carboxy-1-methylethyl]-L-lysine

An analogous D-lysine series is derived from D-lysine

Additional crosslinked amino acids can, in theory, arise from interactions of the NH group of histidine, the aliphatic OH groups of serine and threonine, and the phenolic group of tyrosine with the double bond of dehydroalanine and methyl-dehydroalanine (Figure 7).

INHIBITING LYSINOALANINE FORMATION

Effect of sulfur amino acids. Lysinoalanine and related cross-linked amino acids may be derived from reaction of lysine with dehydroalanine residues formed by elimination reactions from serine, cystine, and possibly cysteine residues in proteins. Threonine residues can, in principle, react similarly to form methylated homologues (Friedman, 1977). The double bond of dehydroalanine, which

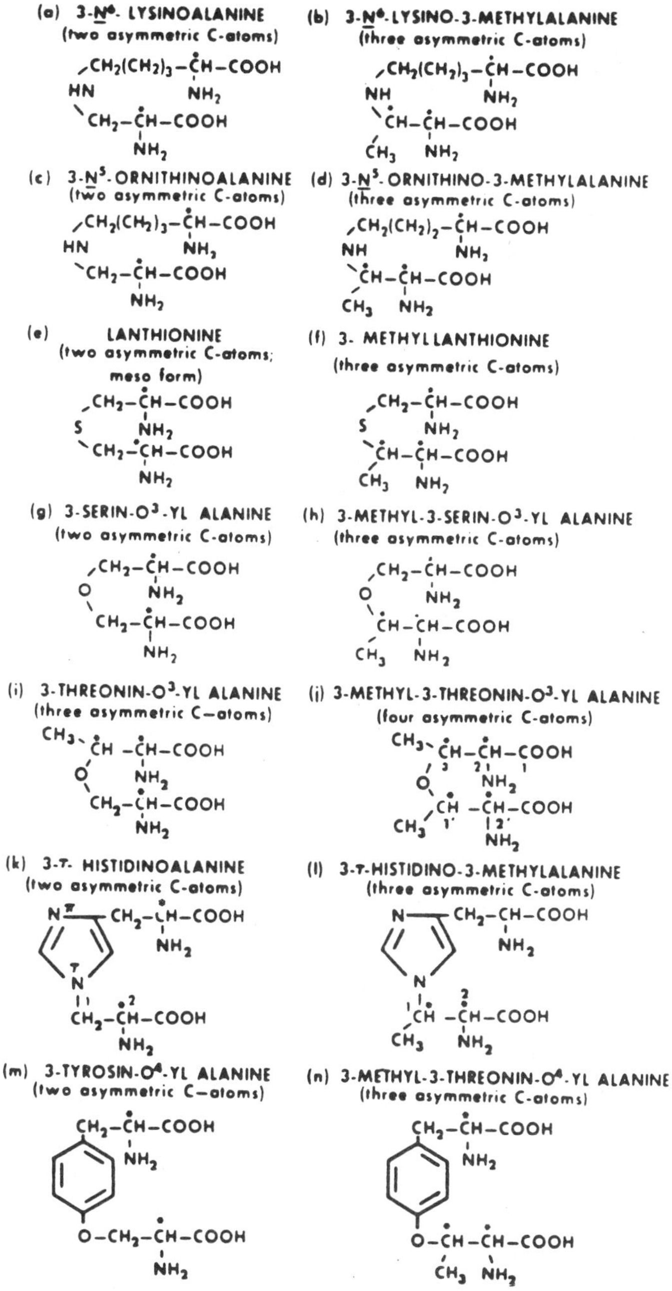

Figure 7. Postulated structures of crosslinked amino acids derived from interaction of protein functional groups (NH, NH2, OH, SH) with the double bond of dehydroalanine and methyldehydroalanine side chains. Asymmetric centers are designated by asterisks.

is part of an activated conjugated double-bond system, reacted more rapidly with SH groups of cysteine than $\varepsilon$-NH$_2$ groups of lysine. (Table 5).

Thus, depending on the pH of the reaction, sulfhydryl groups of cysteine react about 34 to 5000 times faster than the $\varepsilon$-amino groups of lysine with vinyl compounds such as N-acetyl dehydroalanine methyl ester (Friedman and Wall, 1964; Friedman et al., 1965; Cavins and Friedman, 1967; Snow et al., 1976). Therefore, by adding thiols such as cysteine it may be possible to prevent the formation of dehydroalanine residues during alkali-treatment of proteins.

Table 5.  Second-order rate constants ($k_2$ in $M^{-1}$ min $^{-1}$) for reaction of N-acetyldehydroalanine methyl ester with the $\varepsilon$-NH$_2$ group of N-$\alpha$-acetyl-L-lysine and with the SH groups of cysteine at various pH's.

| pH | $\alpha$-N-acetyl-L-lysine (A) | L-cysteine (B) | Rate ratio (B/A) |
|------|------|------|------|
| 7.0 | 5.14 x 10$^{-3}$ | 25.62 | 5000 |
| 8.0 | 6.48 x 10$^{-2}$ | 148.8 | 2300 |
| 9.0 | 0.763 | 314.4 | 410 |
| 10.0 | 2.58 | 344.6 | 133 |
| 11.0 | 8.66 | 357.2 | 43 |
| 12.0 | 11.22 | 358.6 | 34 |

(Snow et al., 1976)

Mechanistic considerations (Asquith and Otterburn, 1977; Whitaker and Feeney, 1977; Friedman, 1977) suggest that added thiol or sulfite ions can inhibit lysinoalanine formation by at least three distinct mechanisms. The first is by direct competition. The added nucleophile (mercaptide, sulfite, bisulfite, thiocyanate, thiourea, etc.) can trap dehydroalanine residues derived from protein amino acid side chains, forming their respective adducts. In particular, lanthionine side chains (Figure 8) are formed from added cysteine and N-acetyl-cysteine. The second possible mechanism is best described as indirect competition. The added nucleophile can cleave protein disulfide bonds and thus generate free protein SH groups. These may, in turn, combine with dehydroalanine residues, (Figure 9). The third possible mechanism can be described as suppression of dehydroalanine formation. The added nucleophile, by cleaving disulfide bonds, can diminish a potential source of dehydroalanine, inasmuch as the resulting ionized and thus negatively charged cysteine residues undergo elimination reactions to form dehydroalanine much less readily than the original cystine (disulfide) precursor residues (Figure 10).

Tables 6 and 7 compare the effects of several organic and inorganic compounds on the lysinoalanine and lysine contents of alkali-treated wheat gluten and soybean protein. These results show that all these compounds partly inhibit lysinoalanine formation. The extent of inhibition may vary from protein to protein and should be related to both the content and reducibility of the disulfide bonds (Friedman, 1978a; Finley et al., 1978 a,b; Masri and Friedman, 1982).

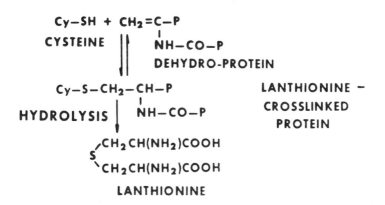

Figure 8. Inhibition of lysinoalanine (crosslink) formation. Added cysteine combines at a faster rate with the double bond of a dehydro-protein (to form a lanthionine crosslink) than does the amino group of a lysine residue (to form a lysinoalanine crosslink).

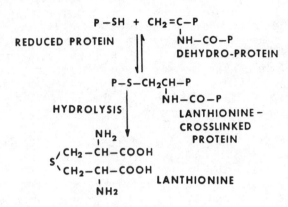

Figure 9.   Inhibition of lysinoalanine formation by indirect competition. The SH group of a reduced protein combines with the
double bond of a dehydro-protein, thus preventing it from reacting
with the amino group of lysine to form lysinoalanine.

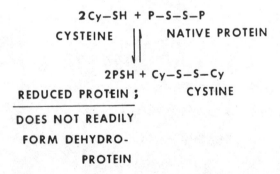

Figure 10. Inhibition of lysinoalanine by suppression of dehydro-
protein formation. The added cysteine can perform two functions.
It can combine with a dehydro-protein and/or reduce a protein disulfide.

Table 6.  Effect of additives on lysinoalanine content of alkali-treated wheat gluten.[1]

| Compound added | Lysinoalanine (Mole %) |
|---|---|
| None | 0.65 |
| Sodium sulfite | 0.55 |
| Sodium bisulfite | 0.25 |
| L-cysteine | 0.26 |
| N-acetyl-L-cysteine | 0.30 |
| Thiourea | 0.36 |

[1]Conditions:  1% wheat gluten in 1N NaOH;  1 hour; 65°C.


Table 7.  Effect of thiols on lysinoalanine content of alkali-treated soy protein.[1]

| Additive (1 mM) | Lysinoalanine (Mole %) |
|---|---|
| None | 1.09 |
| 2-Aminoethanethiol | 0.58 |
| L-Cysteine | 0.36 |
| N-Acetyl-L-cysteine | 0.38 |
| Reduced glutathione | 0.38 |
| DL-Penicillamine | 0.60 |

[1]Conditions: 1% protein in 0.1 N NaOH;  65°C;  3 hours.

Table 8.   Effect of acetylation on lysinoalanine content of alkali-
           treated gluten.   Numbers are in mole percent.

| Amino Acid | Gluten Control (untreated) | Gluten + NaOH | Acetylated gluten + NaOH |
|---|---|---|---|
| LAL | 0.00 | 0.634 | 0.00 |

Conditions: 1% protein; 1 N NaOH; 3 hours; 65°C.


Effect of acylation.   Formation of lysinoalanine from lysine requires the participation of an ε-amino group of a lysine side
chain.   Therefore, it was expected that protection of amino groups
by acylation (acetylation, succinylation, etc.), would reduce lysinoalanine formation  under alkaline conditions, if the protective group(s) survived the treatment.   Results in Table 8 strikingly demonstrate that acetylation of wheat gluten prevents lysinoalanine formation.

        Analogous results were noted when acetylated and succinylated
soy protein was subjected to alkaline treatment (Friedman, 1978b).


Table 9.   Protein efficiency ratios (PER) of rats fed soy protein
           and acylated soy protein.

| Dietary Source of Protein | PER[1] | % Digestibility[2] | |
|---|---|---|---|
| | | Diet | Nitrogen |
| Casein standard | 3.54 ± 0.04 | 96 | 95 |
| Soy protein Untreated | 1.58 ± 0.06 | 94 | 88 |
| Soy protein, acetylated | 1.60 ± 0.09 | 94 | 91 |
| Soy protein, succinylated | -0.82 ± 0.09 | 94 | 9 |

[1]PER = weight gain/protein intake
[2]Diet Digestibility = (feed intake - fecal weight)/feed intake X
100.
   Nitrogen Digestibility = (N intake - fecal N)/N intake X 100.

Preliminary studies shown in Table 9 suggest that the nutritive quality of soy protein was not affected by acetylation. In contrast, succinylated soy protein did not support the growth of rats.

## DEHYDROALANINE

As already mentioned, dehydroalanine is the postulated reactive precursor for lysinoalanine. Direct evidence for dehydroalanine reactivity was obtained by Friedman et al. (1977). They showed that dehydroalanine derivatives convert lysine side chains in casein, bovine serum albumin, lysozyme, wool, or polylysine to lysinoalanine residues at pH 9 to 10. Related studies showed that protein SH groups generated by reduction of disulfide bonds are completely alkylated at pH 7.6 to lanthionine side chains. These studies demonstrate that lysinoalanine and lanthionine residues can be introduced into a protein under relatively mild conditions, without strong alkaline treatment. They also imply that it should be possible to explore nutritional and toxicological consequences of lysinoalanine and lanthionine consumption in the absence of racemization (see below).

In a related study, Masri and Friedman (1982) developed a procedure for detecting dehydroalanine in alkali-treated proteins based based on addition of the SH group of 2-mercaptoethylpyridine to the double bond of dehydroalanine to form S-β-(2-pyridylethyl) cysteine. The cysteine derivative can be assayed, after acid protein hydrolysis, by standard amino acid analysis techniques (Friedman et al., 1979).

This method revealed significant amounts of dehydroalanine in alkali-treated casein and acetylated casein (Table 10).

Because dehydroalanine residues:

(a) occur naturally in peptide antibiotics and hormones (Gross, 1977; Gaveret et al.,).

(b) are present in alkali-treated proteins (Masri and Friedman, 1982); and,

(c) act as alkylating agents in vitro (Friedman, et al., 1977), a need exists to:

(a) establish whether they act as biological alkylating agents in vivo; and

(b) assess their nutritional and toxicological effects, both in pure forms and in foods.

The nutritional and toxicological significance of dehydroalanine residues in proteins has apparently not yet been investigated.

Table 10.    Dehydroalanine content of alkali-treated casein and acetylated casein.[a,b]

| PROTEIN | DEHYDROALANINE |
|---|---|
| 1.  Casein, untreated | 0.00 |
| 2.  Casein + alkali | 0.33 |
| 3.  Acetylated casein | 0.00 |
| 4.  Acetylated casein + alkali | 1.39 |

[a]Conditions of alkali-treatment:   1% protein; 0.1 N NaOH;   70°C; 3 hrs.

[b]Listed values for dehydroalanine (determined as S-β-(2-pyridylethyl)-cysteine) are in g/16g N.   (Masri and Friedman, 1982).

FACTORS AFFECTING RACEMIZATION

Mechanism of racemization.   Since the early part of this century alkali and heat treatment have been reported to racemize amino acid residues and degrade proteins (Kossel and Weiss, 1909; Dakin, 1912; Dakin and Dudley, 1913; Levene and Bass, 1927, 1928; 1929, Horn et al., 1941; Nicolet, 1931; Nicolet et al., 1942).   Because all amino acid residues in a protein undergo racemization simultaneously to various degrees, assessment of the extent of racemization in a protein requires quantitation of at least 36 optical isomers (L and D).   This is a difficult problem not yet solved.

An amino acid is racemized by removal of a proton from the $\alpha$-carbon atom to form a carbanion intermediate.   The trigonal carbon atom of the carbanion, having lost the original asymmetry of the $\alpha$-carbon, recombines with a proton from the environment to regenerate a tetrahedral structure.   The reaction is written as:

$$\text{L-amino acid} \quad \underset{k'_{rac}}{\overset{k_{rac}}{\rightleftharpoons}} \quad \text{D-amino acid} \tag{1}$$

where $k_{rac}$ and $k'_{rac}$ are the first-order rate constants for the forward and reversible racemization of the stereoisomers.

The product is racemic if recombination can take place equally well on either side of the carbanion, giving an equimolar mixture of L- and D-isomers.  If the molecule has more than one asymmetric center, recombination may be biased, resulting in an equilibrium mixture slightly different from a 1:1 enantiomeric ratio.

Effects of the following parameters on racemization in casein have been investigated:  pH, time, temperature of treatment, concentration of casein and structural effects (Masters and Friedman, 1979; 1980; Friedman and Masters, 1982; Friedman et al., 1981).

Effect of time.  The course of racemization of seven amino acid residues in casein treated as a 1% solution in 0.1 N NaOH (pH 12.5) at 65°C from 10 min to 24 hr is plotted as ln $[(1 \mp D/L)/ (1-D/L)]$ vs. time in Figure 11.  This plot is based on the first-order kinetic equation for reversible amino acid racemization (see Masters and Friedman, 1980, for derivation):

$$\ln \frac{|1+D/L|}{|1-D/L|}t - \ln \frac{|1+D/L|}{|1-D/L|}t=0 = 2 k_{obs} \cdot t \qquad (2)$$

The results show two regions of different apparent racemization rates.  Rapid initial rates to about 3 hr were followed by slower rates, up to 24 hours.  The fastest reaction was in the first hour; about one-third to one-half of the D-amino acids produced during a 24 hr exposure to alkali accumulate within the first hour.  None of the amino acids reached equilibrium ($D/L = 1$).

Although racemization appeared to obey first-order kinetics during the early stages of the reaction, the apparent rate changed with time.  The rate constants we used for calculation of thermodynamic and linear free-energy parameters were derived from the initial, linear rates.

Structure-reactivity correlations.  Because racemization of any amino acid proceeds through a carbanion intermediate, the differing reaction rates of the various amino acids depend on structural and electronic factors that are specific to each amino acid, and that facilitate the formation of the carbanion or act to stabilize this negatively-charged $\alpha$-carbon.  The inductive strengths of the R-substituents have been invoked to explain differing racemization rates in the various amino acids (Bada, 1972; Masters and Friedman, 1979; Bada and Shou, 1980).

The relationship between the electron-withdrawing ability of the R-group as measured by inductive substituent constant $\sigma^*$ (Charton, 1964) and the $k_{rac}$ has been questioned recently (Smith et al., 1978). Plotting racemization rates vs. the substituent $\sigma^*$ values for six amino acids (Figure 12), however, clearly demonstrates a strong correlation.

A correlation coefficient of 0.995 is calculated from Figure 13 when $\sigma^*$ values for the amides of aspartic and glutamic acids are used. With our analytical methods, asparagine and glutamine are deamidated prior to enantiomeric determinations. Hence, they are included in the overall D-aspartic acid and D-glutamic acid measurements. Previous calculations, however, (Masters and Friedman, 1979) have indicated that most of the asparaginyl and glutaminyl residues would remain intact during the 3 hr treatment with heated alkali used in the present study. Thus, it seems likely that the amides are the racemizing forms of aspartic and glutamic acids in casein under our treatment conditions.

The linear plot shown in Figure 12 is a free energy relationship that can be described by

$$\log \ (k_{aa}/k_{ala}) = 1.738 \ \sigma^* + 0.045 \tag{3}$$

This equation relates the logarithm of the rate constant for the racemization of any amino acid ($k_{aa}$) normalized by that of alanine ($k_{ala}$) to an inductive constant ($\sigma^*$) of the R-substituent. As defined by Charton (1964) and Exner (1978), $\sigma^*$ is zero for the $CH_3$ R-group of alanine, the simplest amino acid with an asymmetric center. Eq. 3 can be used to calculate a predicted rate of racemization for any casein-bound amino acid whose R-group $\sigma^*$ is known. For example, a $\sigma^*$ value of 0.555 for $HOCH_2$ (Taft, 1956) would imply a racemization rate for serine more than twice that described for asparagine in casein.

Recently, Liardon and Hurrell (1983) reported D-serine measurements which indicate that serine is indeeed racemizing about twice as fast as aspartic acid, thus verifying our earlier predictions. This rate relationship should be true only in base-catalyzed racemization, however, because the $COO^-$ (delta-carboxylate) ionic species of the carboxylic acid side chain is the predominant racemizing form under alkaline conditions.

In a preliminary study we found that alkali-induced D-serine formation in soy protein parallels lysinoalanine production (Figure 13).

Effects of pH. Figure 14 shows the the pH dependence of the racemization for the three most rapidly racemizing amino acids: aspartic acid, glutamic acid, and phenylalanine. Racemization rates in the

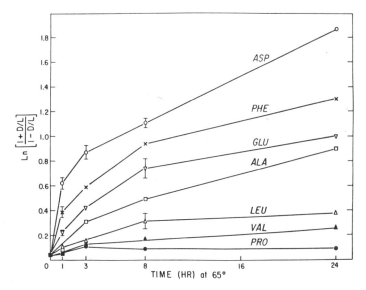

Figure 11.   Time course of amino acid racemization of casein in 0.1N NaOH at 65°C.   Vertical bars denote range of values from several determinations.

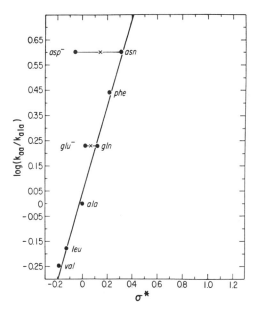

Figure 12.   Relationship between the inductive strength $\sigma^*$ of the amino acid side chain and the racemization rate constant ($k_{aa}$) relative to alanine ($k_{ala}$) for leucine, valine, phenylalanine, glutamic acid, and aspartic acid in 0.1 N NaOH at 65°C for 3 hr.   x is the weighted average for Asx (aspartic plus asparagine) and Glx (glutamic and glutamine) residues, respectively.

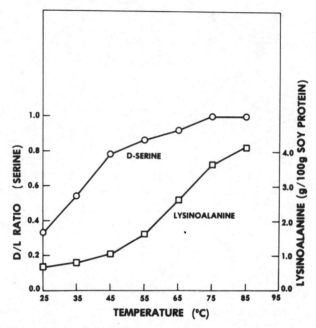

Figure 13.   Preliminary results on the effect of temperature of treatment on D-serine and lysinoalanine formation in soybean protein. Conditions of treatment:   1% (w/v) protein in 0.1 N NaOH; 3 hours.   Note rapid initial rate of production of D-serine which then levels off at about 75°C, where the D/L ratio approaches the maximum theoretical value of 1.0.   We thank Dr. D. L. Schwass for his help with the D-serine measurements.

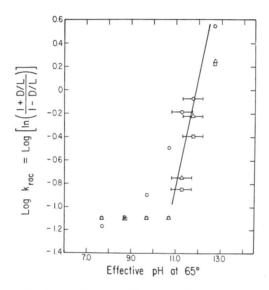

Figure 14. The pH dependence of racemization for aspartic acid (0), phenylalanine (Δ) and glutamic acid ( □ ) in casein in the pH range 8 to 13. The expression for log $k_{rac}$ shown in the ordinate is derived from Eq. 2. Under our experimental conditions, Eq. 2 can be reduced to this one term when solving for k (Masters and Friedman, 1979). The line is a first-order kinetic plot superimposed on the data points. Horizontal bars denote range of values from several determinations.

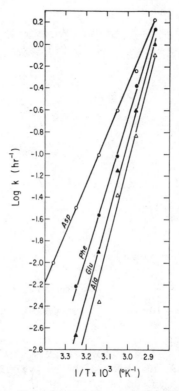

Figure 15.   Arrhenius plot for the racemization of four amino acid residues in casein determined at 10°C intervals in the temperature range 25°C–75°C.

buffers at 65°C were calculated using the temperature data of Masters and Friedman (1980). The solid line represents findings from rates that are first-order with respect to hydroxide ion concentration. This line is a reasonable fit to the data above pH 10. The results show that significant racemization takes place above pH 9, with the rates increasing exponentially above pH 11.

Temperature effects. Arrhenius plots of log k vs. 1/T are shown in Figure 15. Plots of log k vs. 1/T in the temperature range 25 to 75°C are linear within experimental error.

The calculated activation energy for aspartic acid racemization (20.8) is similar to the value reported by Darge and Thiemann (1971) for racemization of protein-bound aspartic acid. However, this value is about 10 kcal/mole less than the corresponding value reported by Bada (1971) for free aspartic acid determined at pH 7.6. In contrast, activation energies for alanine and phenylalanine residues in casein, determined at pH 12.5 were similar to those found for the same amino acids at pH 7.6.

Asparagine, comprising 50% of the Asx (aspartic acid plus asparagine) residues in casein acid (Brunner, 1977) was apparently the main component of the D-aspartic acid measured. The $\sigma^*$ value for the R-group ($CONH_2$) of asparagine (0.31) reflects a considerably larger inductive effect than that exerted by the $COO^-$ ionic form of aspartic acid (-0.06), which would be the predominant ionic species at pH 12.5. Consequently, racemization rates for aspartic acid may be inflated by the high asparagine content of casein. On the other hand, glutamine at 30% of the Glx (glutamic acid plus glutamine) residues, did not differ as greatly in inductive strength from glutamic acid. A smaller difference in relative contribution of glutamine versus glutamic acid to the Arrhenius plot is, therefore, expected compared to asparagine versus aspartic acid.

## Discriminating between racemization and lysinoalanine formation

As already mentioned, racemization occurs by abstraction of a proton by hydroxide ion (or any general base) from an optically-active carbon atom to form a carbanion, which thus loses the original asymmetry. The pair of unshared electrons on the carbanion can undergo two reactions:

(a) it can recombine with a proton from the solvent to regenerate either the original amino acid side chain or its optical antipode, so that it is racemized;

(b) it can undergo the indicated β-elimination reaction to form a dehydroalanine derivative, which can then combine with an ε-amino group of a lysine side chain to form a lysinoalanine crosslink.

Since lysinoalanine and at least one D-amino acid (D-serine) are toxic to some animals (see below), we wished to distinguish their effects in alkali-treated proteins. We have found that acylating the ε-amino group of lysine proteins seems to prevent lysinoalanine formation (Friedman, 1978; Friedman et al., 1982). Because lysinoalanine formation from lysine requires participation of the ε-amino group of lysine side chains, acylation of the amino group with acetic anhydride is expected to prevent lysino-alanine formation under alkaline conditions, if the protective effect survives the treatment.

Although acetylation appears to minimize or prevent lysino-alanine formation, our findings (Friedman and Masters, 1982) indicate that acylation does not significantly change the extent of racemization after 3 hr at 65°C (Table 11). These results show that it is possible, in principle, to discriminate between the alkali-induced effects of racemization and lysinoalanine formation.

Recently, Bunjapamai et al., (1982) have shown that citracony-lated casein treated with alkali is as highly racemized as treated non-citraconylated casein, although the former contains no LAL. In vitro digestion of the racemized but non-crosslinked casein was just as restricted as was the digestion of the racemized crosslinked casein. Thus, racemization alone can inhibit in vitro digestibility of alkali-treated proteins.

In summary, alkali treatment of casein catalyzes racemization of optically active amino acids. Factors that influence racemiza-tion include pH, temperature, and time of treatment. Further stud-ies are desirable to assess (a) how these factors operate in struc-turally different proteins and (b) the presence of D-amino acids in foods and feeds. (See Table 12 for some preliminary findings).

Table 11. Aspartic acid racemization and lysinoalanine (LAL) con-tent of alkali-treated casein and acetylated casein.

| Sample | D/L ASP Ratio | LAL (Mole %) |
|--------|:-------------:|:------------:|
| Casein control | 0.023 | 0.0 |
| Casein + alkali[1] | 0.387 | 2.35 |
| Acetylated casein + alkali[1] | 0.336 | 0.0 |

[1] 1% protein; 0.1N NaOH; 65°C; 3 hrs.

Table 12.   D-Aspartic  acid  content  in  commercial  food  products.

| Commercial Product | D/L ASP | $\dfrac{\text{D-ASP}}{\text{D-ASP+L-ASP}}$ |
|---|---|---|
| Texturized soy protein | 0.095 | 0.09 |
| Baby formula (soy protein) | 0.108 | 0.10 |
| Simulated bacon (soy protein) | 0.143 | 0.13 |
| Corn chips | 0.164 | 0.14 |
| Dairy creamer (sodium caseinate) | 0.208 | 0.17 |

## CYTOTOXIC CONSEQUENCES

Alkali-treated soy proteins, when fed to rats, induce changes in kidney cells. These changes are characterized by enlargement of the nucleus and cytoplasm, increase in nucleoprotein, and disturbances in DNA synthesis and mitosis. These lesions, first described by Newberne and Young (1966), attributed to lysinoalanine by Woodard (1969), and now designated nephrocytomegaly, affects the epithelial cells of the straight portion (pars recta) of the proximal renal tubules. Renal cytomegaly of the pars recta is also induced by feeding rats synthetic lysinoalanine (Woodard and Short, 1973; Woodard et al., 1975; 1977). Since this unusual amino acid is formed in proteins during alkaline treatment, the nephrotoxic action of the treated proteins was ascribed to the presence of lysinoalanine (LAL).

Enlarged nuclei tend to have more than the diploid complement of DNA, unusual chromatin patterns, and proteinaceous inclusions (Gould and MacGregor, 1977; Woodard et al., 1975). Increases in total nonchromosomal protein parallel the increases in nuclear volume. These observations indicate disruption of normal regulatory functions in the pars recta cells.

Findings by several laboratories suggest that factor(s) other than LAL content may affect the biological response to alkali-treated proteins. With rats, DeGroot and coworkers (DeGroot and Slump, 1969; DeGroot et al., 1976) initially did not observe nephrocytomegaly when feeding alkali-treated soy protein (although in 1976 they did report the renal changes when feeding free LAL). This contrasts with the findings of Woodard and others (Woodard, 1969; Woodard and Short, 1973; Reyniers et al., 1974; Leegwater, 1978; Struthers et al., 1979, 1980; Karayiannis, 1976; Karayiannis et al., 1979 a,b; Newberne and Young, 1966; Newberne et al., 1968).

These authors have all seen the renal lesions in rats fed alkali-treated soy proteins.

Another difficulty in formulating a simple relationship between LAL and nephrocytomegaly is that proteins of equivalent LAL content produce different biological responses. O'Donovan (1976) reported that feeding rats alkali-treated soy protein led to severe nephrocytomegaly, while a different protein with the same LAL content did not produce lesions. Karayiannis (1976) found that alkali-treated soy protein (supplying 1400-2600 ppm LAL) resulted in nephrocytomegaly, whereas 2500 ppm LAL derived from alkali-treated lactalbumin did not.

The divergent observations about relative potencies of various alkali-treated proteins in inducing kidney lesions could arise from dietary factors. Adding high-quality, untreated proteins such as casein and lactalbumin to diets containing alkali-treated proteins appears to prevent the lesions (Feron et al., 1977). One possible explanation for this effect is that amino acids (e.g. lysine, methionine) derived from added proteins may prevent the binding of lysinoalanine to metalloproteins in the kidney. The observed long residence time of lysinoalanine in the kidney (Finot et al., 1977) may be related to a possible chelating action of lysinoalanine, which has three amino and two carboxyl groups that participate in binding metal ions of metallo-enzymes (Friedman, 1977; Hayashi, 1982).

The conflicting reports from various laboratories might also be explained by amino acid racemization during alkaline processing of the test proteins. Our studies show that four different proteins subjected to the same alkaline treatment exhibited varying degrees of racemization (Masters and Friedman, 1979). One result of the presence of D-amino acids would be to decrease enzymatic digestion of the proteins, thus restricting the amount of free LAL released. In order of cytotoxic effect, the most pronounced response is with free LAL, then low molecular weight LAL-containing peptides, then LAL-containing proteins. Thus, different proteins having the same quantities of bound LAL may release different amounts of free LAL or low molecular weight LAL-peptides depending upon their extent of racemization. Some of the discrepancies thus may be attributed to the use of different fractions of soy protein in the experimental diets. In addition, the alanine part of lysinoalanine is a potential precursor for D,L-serine, and D-serine and one of the lysinoalanine diasteroisomers offer similar configurations to potential receptor sites (Wachstein, 1947; Ganote et al., 1974; Kaltenbach et al., 1979).

We have also shown that small differences in the conditions of alkaline treatment can produce relatively large differences in the extent of racemization in casein. Temperature and length of treatment are critical. Protein concentration, however, does not appear to influence significantly the extent of racemization (Friedman and Masters, 1982). Therefore, treatment conditions may generate comparable contents of LAL but varying D-amino acid contents.

D-Amino acids may have other effects in nephrocytomegaly, beyond influencing release of LAL. Soy protein, the most easily racemized of the four proteins studied, is more cytotoxic than lactalbumin (Karayannis et al., 1979), the least racemized. If, as postulated, D-amino acids inhibit release of LAL, then lactalbumin would be expected to yield more free LAL than soy protein. These considerations suggest that proteolytic release of LAL may be influenced by both the presence and location of D-amino acid residues along a polypeptide chain.

It is possible that D-amino acid (s) may act synergistically with LAL in the expression of nepherocytomegaly. D-serine can induce renal lesions when fed to rats (Table 13). See also Figure 12.

Table 13.   Rat nephrotoxicity of compounds structurally related to serine injected intraperitoneally.

| Compound | Dose range[1] | PST necrosis[2] |
|---|---|---|
| D-serine | 0.30-0.76 | + |
| D-cysteine | 0.76 | 0 |
| DL-serine ethyl ester | 0.76 | + |
| DL-2-amino-1-propanol (alaninol) | 0.34-2.45 | +/0 |
| DL-2,3-diaminopropionic acid | 0.50-1.00 | + |
| DL-glyceric acid | 0.58-0.76 | 0 |
| DL-1,2-diaminopropane | 0.30-0.62 | +/0 |
| DL-1-amino-2-propanol | 0.38-1.52 | 0 |

[1]Dose values are nmoles/100 g body weight of the D-isomer.

[2]Proximal straight tubular (PST) necrosis= +; no necrosis = 0; possible necrosis = +/0.

(Kaltenbach et al., 1979)

If the protein-induced renal cytomegaly is precancerous, it
is a cause for concern, since alkali-processed protein is used extensively in commercial food preparations (O'Donovan, 1976; Sternberg and Kim, 1977; Sternberg et al., 1975). Although lysinoalanine concentrations in foods are usually lower than amounts needed
to induce nephrocytomegaly in rats, a health hazard may exist since
human tolerance to chronic exposure of the 'unnatural' amino acids
generated during commercial processing are not known. Whether or
not the protein-induced lesion proves to be precancerous, a greater
understanding of its etiology is essential to help evaluate implications to human health.

Maillard products such as lysine-sugar complexes are generated
by heating proteins in the presence of carbohydrates. When casein
heated with glucose under mild conditions (to maximize formation of
$\varepsilon$-fructosyllysine without inducing LAL crosslinks) was fed to rats,
histopathological renal changes were observed (Erbersdobler, et al.,
1978). According to preliminary evidence, heating free aspartic
acid in an aqueous solution containing glucose increases the racemization rate two- to threefold (Zumberge, 1978). Roasting casein
in the presence of glucose (Hayase et al., 1975) also enhanced the
extent of racemization of 6 amino acids, including aspartic acid,
approximately 1.3 fold. Therefore, these conditions may induce significant amounts of racemization. The cytotoxic response observed
by Erbersdobler et al. (1978) thus may be due D-enantiomers as well
as to Maillard products. In general, our knowledge about the
occurrence and conditions of formation of 'unnatural' amino acid
derivatives is directly related to our understanding of how food
quality is related to commercial and home processing conditions.

The mechanism of the observed cellular action of lysinoalanine is not well understood (Finot et al., 1977, Finot, 1983;
Engelsma et al., 1979; Reyniers, 1979; Leegwater and Tas, 1980).
The possible interaction of LAL with metal ions needs to be explored, however, in view of the recent observations both here and
by Hayashi (1982) that LAL inhibits the enzymatic activity of
metallo-enzymes such as carboxypeptidase, which contains zinc as
part of its active site. The inhibition appears reversible since
carboxypeptidase activity was regenerated following the addition
of zinc sulfate to the LAL-inactivated enzyme. Inhibition is not
surprising since LAL contains three amino and two carboxyl groups
and structurally resembles ethylenediaminetetraacetic acid (EDTA),
a well-known metal chelator.

In fact, as already mentioned, the long residence time of LAL
in the kidney of rats (Finot, 1977) may be related to chelating
action of LAL on metal ions of kidney structural proteins or metal-
loenzymes. This hypothesis raises several important questions,
including:

1. Can LAL toxicity be prevented by giving equivalent amounts of zinc or other metal salt, i.e. does co-administration of LAL with metal ions affect toxicity?

2. Are organisms deficient in zinc or other metals more susceptible to LAL toxicity?

3. Does LAL compete with physiologically active proteins for essential metal ions?

4. Does LAL bind to specific metallo-containing biomolecules such as hemoglobin, etc. in vitro and in vivo?

5. Does administration of LAL have any observable (transient) effect on excretion of metal ions?

## NUTRITIONAL IMPLICATIONS

The nutritive quality of any protein depends on three factors: amino acid composition, digestibility, and utilization of the released amino acids. Racemization brought about by processing can impair the nutritive value of proteins by (1) generating non-metabolizable forms of amino acids (some D-enantiomers), (2) creating peptide bonds inaccessible to proteolytic enzymes, possibly restricting release of essential amino acids (3) toxic action (or interaction) of specific D-enantiomers. Little is known concerning the nutritional consequences of human consumption of racemized proteins.

Alkali-treated proteins have been shown to have reduced digestibility and bioavailability. (Friedman et al., 1981; Cheftel, 1979; Gauther et al., 1982; Freimuth et al., 1978; Provansal et al., 1975; Robbins and Ballew, 1982; de Groot and Slump, 1969; L. Savoie, this volume) and in vivo (de Groot and Slump, 1969; Finot et al., 1977), and to produce symptoms of dietary inadequacy such as poor growth, hair loss, and diarrhea (Woodard and Alvarez, 1967; Van Beek et al., 1974; Karayiannis, 1976). These studies focused attention on the crosslinked amino acid derivative, lysinoalanine (LAL), and its possible toxicity. Although the treatments used almost certainly caused significant racemization, the possible effect of D-amino acids on the results was not evaluated.

Possible effects of D-amino acid and lysinoalanine contents on food casein digestibility was examined (Friedman et al., 1981). The following variables, which were expected to influence amino-acid composition and utilization, were evaluated: (a) pH (8-14); (b) temperature (25-75°C); and (c) time of treatment (10 min to 24 hr). An approximately inverse relationship was observed between D-amino acid and lysinoalanine content and the extent of proteolysis by trypsin. These studies helped to define the conditions which lower the nutritional quality and safety of foods.

The biological utilization of lysinoalanine as a source of lysine was determined in a growth assay in weanling male mice in which all L-lysine in a synthetic amino acid diet was replaced by a molar equivalent of lysinoalanine-HCl (Friedman et al., 1982). Replacement of all of the 1.35% of L-lysine in the synthetic amino acid diet (Table 14) by lysinoalanine produced an amount of weight gain equivalent to that expected from a diet containing 0.05% of L-lysine.

Table 14.  Weight gain of mice fed lysinoalanine

| Amino Acid Tested | Percent in Diet | Number in Mice | Mean Weight Gain |
|---|---|---|---|
| L-Lysine-HCl | 0 | 4 | -2.00 ± .00 |
| | 1.35 | 8 | 12.50 ± .53 |
| Lysinoalanine – 2 HCl | 2.25[1] | 6 | -.80 ± .20 |

[1]Equimolar with 1.35% L-lysine.

On a molecular equivalent basis, lysinoalanine when fed at the full replacement level for L-lysine was 3.8% as potent as lysine in supporting weight gain.

Our results contrast with previously reported studies on the utilization of lysinoalanine as a source of lysine. Thus, Karayiannis et al. (1979 b) found that lysinoalanine could satisfy the lysine requirements of the lysine-requiring microorganisms Erwinia chrysanthemi and Escherichia coli. Sternberg and Kim (1979) report that the protein efficiency ratio (PER) determined with the lysine-requiring microorganism Tetrahymena pyriformis increased from 0.65 for wheat gluten to 1.19 for wheat gluten supplemented with lysinolanine. This compares to a value of 1.62 for wheat gluten supplemented with an equivalent amount of lysine. In contrast, supplementation of wheat gluten with lysinoalanine decreased the PER of mice from 0.52 to 0.31. The authors ascribe this adverse effect to the toxic action of lysinoalanine in the rodents' kidneys. Finally, Robbins et al. (1980) report that lysinoalanine is completely unavailable as a source of lysine to the rat although it is 37% available to the chick.

The 3.8% replacement value for lysinoalanine in mice compares with the previously observed 3.8% value of ε-N-trimethyl-L-lysine, 4.6% for ε-N-dimethyl-L-lysine, and 7.5% for ε-N-methyl-L-lysine also in mice (Friedman and Gumbmann, 1979; 1981). Although the metabolism of lysinoalanine appears to be complex and not well understood (Finot, 1977; Reyniers, 1979; Leegwater and Tas, 1980), a fraction of the compound must undergo in vivo dealkylation at the nitrogen-carbon bond joining lysine to the alanine part if it is to serve as a source of lysine. Structural similarities between methyl-lysine and lysinoalanine may lead to similar rates of dealkylation.

Observations, then, suggest that synthetic lysinoalanine derived from lysine can be partly utilized in vivo as a source of lysine. However, the extent of biological utilization is species-specific, and appears to be quite low in mice. Additional studies are desirable to elucidate this dependence reliably for different animals, especially primates.

Additional studies are also needed to define possible differences in biological and nutritional responses to four possible individual lysinoalanine isomers, and to assess possible synergistic effects of lysinoalanine, D-serine, and dehydroalanine.

With respect to racemization, a major consideration is whether humans can utilize the D-enantiomers of essential amino acids (Stegink, 1977; Masters and Friedman, 1980). Berg (1959) reviewed human and animal utilization of free D-amino acids. Uptake of L-amino acids is invariably faster than that of the D-enantiomers in the intestine (Gibson and Wiseman, 1951; Finch and Hird, 1960) and kidney (Rosenhagen and Segal, 1974). Once absorbed, D-amino acids can be utilized by two pathways: (1) racemases (or epimerases) may convert D-isomers to DL-forms; or (2) D-amino acid oxidases may catalyze oxidative deamination to α-keto acids, which can be specifically reaminated to the natural L-forms (Meister, 1965). Only the latter activity has been demonstrated in mammals. However, when mixtures of D-amino acids are fed to rats, the oxidase system can be overloaded so that the D-enantiomers of essential amino acids cannot be transaminated in sufficient quantity to support growth (Wretlind, 1952).

This enzyme system should permit human use of D-amino acids for growth and maintenance. Of the D-isomers of the eight essential amino acids, however, only D-phenylalanine and D-methionine were found to maintain human nitrogen equilibrium in early studies (Rose, 1949). More recently, Kies et al. (1975) and Zezulka and Calloway (1976) presented evidence that D-methionine is, in fact, poorly utilized by humans.

Finally, in a preliminary study, Friedman and Gumbmann (1982) found that the bioavailabilities in mice of D-forms of nine essential amino acids ranged from -4.4% for D-lysine to partial utilization of the other D-amino acids. Additional studies, however, revealed, that in a number of cases, bioavailabilities of D-amino acids relative to the corresponding L-forms are dose-dependent.

In summary, understanding chemical events governing protein-alkali reactions provides insight into events occurring during food processing. Such understanding can be decisive for producing better and safer foods. This area of food science, with many unsolved problems, affords a fascinating interplay among chemistry, nutrition, and toxicology of importance to human nutrition and health.

## ACKNOWLEDGEMENTS

We are grateful to Carol J. Levin for excellent technical assistance.

## REFERENCES

Alworth, W. L. (1972). "Stereochemistry and Its Application in Biochemisty," Wiley, Interscience, New York.

Asquith, R. S. and Otterburn, M. S. (1977). Cystine-alkali reaction in relation to protein crosslinking. In "Protein Crosslinking: Nutritional and Medical Consequences," M. Friedman, Ed., Plenum Press, New York, pp. 93-121.

Bada, J. L. (1972). Kinetics of racemization of amino acids as a function of pH. J. Amer. Chem. Soc., 94, 1371-1372.

Bada, J. L. and Shou, M. Y. (1980). Kinetics and mechanism of amino acid racemization in aqueous solution and in bones. In "Biologeochemistry of Amino Acids," P. E. Hare, Ed., John Wiley & Sons, Inc., p. 235.

Berg, C. P. (1959). Utilization of D-amino acids. In "Protein and Amino Acid Nutrition," A. A. Albanese, Ed., Academic Press, N. Y., pp. 57-96.

Bohak, Z. (1964). N-Epsilon-(DL-2-amino-2-carboxyethyl)-L-lysine (lysinoalanine), a new amino acid formed on alkaline treatment of proteins. J. Biol. Chem., 239, 2878-2887.

Brunner, J. R. (1977). Milk proteins. In "Food Proteins," J. R. Whitaker and S. R. Tannenbaum, Eds., Avi, Westport, CT., p. 175.

Bunjapamai, S., Mahoney; R. R. and Fagerson, I. S. (1982). Determination of D-amino acids in some processed foods and effect of racemization on in vitro digestibility of casein. J. Food Sci., 47, 1229-1234.

Cavins, J. F. and Friedman, M. (1967). New amino acids derived from reactions of ε-amino groups in proteins with α,β-unsaturated compounds. Biochemistry, 6, 3766-3770.

Charton, M. (1964). Definition of "inductive substituent constants." J. Org. Chem., 29, 1222.

Cheftel, J. C. (1979). Proteins and amino acids. In "Nutritional and Safety Aspects of Food Processing", S. R. Tannenbaum, Ed., Marcel Dekker, New York, pp. 154-215.

Dakin, H. D. (1912-1913). The racemization of proteins and their derivatives resulting from tautomeric change. J. Biol. Chem., 13, 357-362.

Dakin, H. D. and Dudley, H. W. (1913). The action of enzymes on racemized proteins and their fate in the animal body. J. Biol. Chem., 15, 271-276.

Darge, W. and Thiemann, W. (1971). Hydrolysis and racemization of oligopeptides studied by optical rotation measurement and ion exchange chromatography. In "First European Biophysics Congress Proceedings", E. Broda, A. Locker, and H. Springer-Lederer, Eds., Baden, Vol. 1, 133-137.

de Groot, A. P. and Slump, P. (1969). Effects of severe alkali treatmeent of proteins on amino acid composition and nutritive value. J. Nutr., 98, 45-56.

de Groot, A. P., Slump, P., Feron, V. J. and Van Beek, L. (1976). Effects of alkali-treated proteins. Feeding studies with free and protein-bound lysinoalanine in rats and other animals. J. Nutr., 106, 1527-1538.

de Groot, A. P., Slump, P., Van Beek, L. and Feron, V. J. (1976). Severe alkali treatment of proteins, "Evaluation of Proteins for Humans, C. Bodwell, Ed., Avi Publishing Co., Westport, CT.

de Koning, P. J. and van Rooijen, P. J. (1982). Aspects of the formation of lysionalanine in milk and milk products. J. Dairy Res. 49, 725-736.

Erbersdobler, H. F., von Wangenheim, B., Hänichen, T. (1978). Adverse effects of Maillard products - especially of fructoselysine in the organism. Paper presented at XI Internat. Congress of Nutrition, Rio de Janeiro, Brazil, and Private Communication.

Engelsma, J. W., Muelen, J. D. v.d., Slump, P. and Haagsma, N. (1979). The oxidation of lysinoalanine by L-amino acid oxidase. Identification of Products. Food Sci. & Technol. 12, 203-207.

Exner, O. (1978). A critical compilation of substiuent constants. In "Correlation Analysis in Chemistry." N. B. Chapman and J. Shorter, Eds., Plenum Press, New York, pp. 439-540.

Feron, V. J., van Beek, L., Slump, P. and Beems, R. B. (1977). Toxicological aspects of alkali treatment of food proteins. In "Biological Aspects of New Protein Food", Pergamon Press, Oxford, England, J. Adler Nissen, Ed., pp. 139-147.

Finch, L. R. and Hird, F. J. R. (1960). The uptake of amino acids by isolated segments of rat intestine. II. A survey of affinity for uptake from rates of uptake and competition of uptake., Biochim. Biophys. Acta, 43, 278-287.

Finley, J. W., Snow, J. T., Johnston, P. H. and Friedman, M. (1978 a). Inhibitory effect of mercaptoamino acid on lysinoalanine formation during alkali treatment of proteins. In "Protein Crosslinking: Nutritional and Medical Consequences:, M. Friedman, Ed., Plenum Press, New York, pp. 85-92.

Finley, J. W., Snow, J. T., Johnston, P. H., and Friedman, M. (1978 b). Inhibition of lysinoalanine formation in food proteins. J. Food Sci., 43:619-621.

Finot, P.-A. (1983). Lysinoalanine in food proteins. Nutrition Abstracts and Reviews, Clin. Nutr., 53, 67-80.

Finot, P.-A., Bujard, E., and Arnaud, M. (1977). Metabolic transit of lysinoalanine (LAL) bound to protein and of free radioactive [$^{14}$C]-lysinoalanine. In: "Protein Crosslinking -- Nutritional and Medical Consequences," M. Friedman, Ed., Plenum Press, N.Y., pp. 51-71.

Freimuth, U., Krause, W. and Doss, A. (1978). On the alkali treatment of proteins. II. Enzymatic hydrolysis of alkali-treated β-casein and acid casein. Nahrung, 22, 557-568. (German).

Friedman, M. (1982a). Lysinoalanine formation in soybean proteins: kinetics and mechanisms. In "Food Protein Deterioration Mechanisms and Functionality", J. P. Cherry (Ed.), ACS Symposium Series, Washington, D. C. 206; 231-273.

Friedman, M. (1982b). Chemically reactive and unreactive lysine as an index of browning. Diabetes, 31, 5-14.

Friedman, M. (1980). Inhibition of lanthionine formation during alkaline treatment of keratinous fibers. U. S. Patent 4,212,800.

Friedman, M. (1978a). Wheat gluten-alkali reactions. (1978 a). In "Proceedings of the 10th National Conference on Wheat Utilization Research", U. S. Department of Agriculture, Science and Education Administration, Western Regional Research Center, Berkeley, California 94710, ARM-W-4, pp. 81-100.

Friedman, M. (1978b). Inhibition of lysinoalanine synthesis by protein acylation. In "Nutritional Improvement of Food and Feed Proteins", M. Friedman, Ed., Plenum Press, New York, pp. 613-648.

Friedman, M. (1977). Crosslinking amino acids-stereochemistry and namenclature. In "Protein Crosslinking: Nutritional and Medical Consequences", M. Friedman, Ed., Plenum Press, New York, pp. 1-27.

Friedman, M. and Gumbmann, M. R. (1982). Bioavailability of D-amino acids in mice. Fed. Proc. 41, 392.

Friedman, M. and Gumbmann, M. R. (1981). Bioavailability of some lysine derivatives in mice. J. Nutrition, 111, 1362-1369.

Friedman, M., and Gumbmann, M. R. (1979). Biological availability of epsilon-N-methyl-L-lysine, 1-N-methyl-L-histidine, and 3-N-methyl-L-histidine in mice. Nutrition Reports International, 19, 437-443.

Friedman, M. and Masters, P. M. (1982). Kinetics of racemization of amino acid residues in casein. J. Food Sci., 47, 760-764.

Friedman, M., and Wall, J. S. (1964), Application of a Hammett-Taft relation to kinetics of alkylation of amino J. Amer. Chem. Soc., 86, 37735-3741.

Friedman, M., and Wall, J. S. (1966). Additive linear free energy relationships in reaction kinetics of amino groups with $\alpha$-, $\beta$-unsaturated compounds. J. Org. Chem., 31, 2888-2874.

Friedman, M., Gumbmann, M. R. and Savoie, L. (1982a). The nutritional value of lysinoalanine as a source of lysine for mice. Nutrition Reports International, 26, 939-947.

Friedman, M., Wehr, C. M., Schade, J. E. and MacGregor, J. T. (1982b). Inactivation of aflatoxin $B_1$ mutagenicity by thiols. Food and Chemical Toxicology, 20: 887-892.

Friedman, M., Zahnley, J. C. and Masters, P. M. (1981). Relationship between in vitro digestibility of casein and its content of lysinoalanine and D-amino acids. J. Food Sci. 46, 127-131 and 134.

Friedman, M., Finley, J. W. and Yeh, Lai-Sue. (1977). Reactions of proteins with dehydrolanines. In "Protein Crosslinking: Nutritional and Medical Consequences", M. Friedman, Ed., Plenum Press, New York, pp. 213-224.

Friedman, M. Cavins, J. F., and Wall, J. S. (1965). Relative nucleophilic reactivities of amino groups and mercaptide ions in addition reactions with α, β-unsaturated compounds. J. Amer. Chem. Soc., 87, 3572-3682.

Fritsch, R. J., Hoffman, H. and Klostermeyer, H. (1983). Formation of lysinoalanine during heat treatment of milk. Z. Lemensm. Unters. Forsch. 176, 341-345.

Ganote, C. E., Peterson, D. R. and Carone, F. A. (1974). The nature of D-serine induced nephrotoxicity. Am. J. Pathol., 77, 269-176.

Gauthier, S. F., Vachon, C., Jones, J. D. and Savoie, L. (1982). Assessment of protein digestibility by in vitro enzymatic hydrolysis with simultaneous dialysis. J. Nutr., 112, 1717-1725.

Gee, M., Huxsoll, c. C. and Graham, R. P. (1974). Acidification of dry caustic peeling by lactic acid fermentation. Am. Potato J. 51, 126-131.

Gibson, O. H. and Wiseman, G. (1951). Selective absorption of stereoisomers of amino acids from loops of the small intestine of the rat. Biochem. J., 48, 426-429.

Goldblatt, L. A. (1969). "Aflatoxin: Scientific Background, Control, and Implications", Academic Press, New York.

Gross, E. (1977). α, β-Unsaturated and related amino acids in peptides and proteins. Adv. Exp. Med. Biol., 86A, 131-153.

Gould, D. H. and MacGregor, J. T. (1977). Biological effects of alkali-treated protein and lysinoalanine: an overview. In "Protein Crosslinking: Nutritional and Medical Consequences," M. Friedman, Ed., Part B, Plenum Press, N. Y., pp. 29-48.

Hayase, F., Kato, H., and Fujimaki, M. (1975). Racemization of amino acid residues in proteins and poly(L-amino acids) during roasting. J. Agric. Food Chem., 23, 491-494.

Hayashi, R. (1982). Lysinoalanine as a metal chelator. An implication for toxicity. J. Biol. Chem., 257, 13,896-13,898.

Hayashi, R. and Kameda, I. (1980). Decreased proteolysis of alkali-treated protein: consequences of racemization in food processing. J. Food Sci., 45, 1430-1431.

Hewitt, D., Ford, J. E. and Porter, J. W. G. (1979). Nutritional quality of spun soya-bean protein: comparison of biological and microbiological tests. Qualitas Plantarum, 29, 253-260.

Horn, M. J., Jones, D. B., Ringel, S. J. (1941). Isolation of a new sulfur-containing amino acid (lanthionine) from sodium carbonate treated wool. J. Biol. Chem., 138, 141-149.

Kaltenbach, J. P., Ganote, G. E. and Carone, F. A. (1979). Renal tubular necrosis induced by compounds structurally related to D-serine. Exp. and Mol. Pathol., 30, 209-214.

Karayiannis, N. (1976). Lysinoalanine Formation in Alkali Treated Proteins and their Biological Effects, Ph.D. Thesis. University of California, Berkeley.

Karayiannis, N. I., MacGregor, J. T. and Bjeldanes, L. F. (1979a). Lysinoalanine formation in alkali-treated proteins and model compounds. Food Cosmet. Toxicol., 17, 585-590.

Karayiannis, N. I., MacGregor, J. T. and Bjeldanes, L. F. (1979b). Biological effects of alkali-treated soy protein and lactalbumin in the rat and mouse. Food and Cosm. Toxicol., 17, 509-604.

Karayiannis, N. I., Panopoulos, N. J., Bjeldanes, L. F. and MacGregor, J. T. (1979c). Lysinoalanine utilization by Erwinia Chrysanthemi and Escherichia Coli. Food Cosmet. Toxicol., 17, 319.

Kies, C, Fox, H. and Aprahamian, S. (1975). Comparative value of L-, DL-, and D-methionine supplementation of an oat-based diet for humans. J. Nutr., 105, 809-814.

Kossel, A., and Weiss, F. (1909). Über Einwirkung von Alkalien auf Proteinstoffe. Z. Physiol. Chem., 59, 492-498.

Leegwater, D. C. and Tas, A. C. (1980). Identification of an enzymic oxidation product of lysinoalanine. Lebensm.-Wiss. Technol., 13, 87-91.

Leegwater, D. C. (1978). The nephrotoxicity of lysinoalanine in the rat. Food Cosmet. Toxicol., 16, 405.

Levene, P. A. and Bass, L. W. (1927). Studies on racemization. V. The action of alkali on gelatin. J. Biol. Chem., 74, 715-725.

Levene, P. A. and Bass, L. W. (1928). Studies on racemization. VII. The action of alkali on casein. J. Biol. Chem., 78, 145-157.

Levene, P. A. and Bass, L. W. (1929). Studies on racemization. VIII. The action of alkali on proteins: racemization and hydrolysis. J. Biol. Chem., 82, 171-190.

Liardon, R. and Hurrell, R. F. (1983). Amino acid racemization in heated and alkali-treated proteins. J. Agric. Food Chem., 31, 432-437.

Masri, M. S. and Friedman, M. (1982). Transformation of dehydroalanine to S-β-(2-pyridylethyl)-L-cysteine side chains. Biochem. Biophys. Res. Commun., 104, 321-325.

Masters, P. M. and Friedman, M. (1980). Amino acid racemization in alkali-treated food proteins--chemistry, toxicology, and nutritional consequences. In "Chemical Deterioration of Proteins", J. R. Whitaker and M. Fujimaki, Eds. ACS Symposium Series, Washington, D. C. 123, 165-194.

Masters, P. M. and Friedman, M. (1979). Racemization of amino acids in alkali-treated food proteins. J. Agric. Food Chem., 27, 507-511.

Meister, A. (1965). "Biochemistry of the Amino Acids," Vol. I, Academic Press, N.Y., pp. 338-369.

Munro, H. N. (1978). Nutritional consequences of excess amino acid intake. In "Nutritional Improvement of Food and Feed Proteins", M. Friedman, Ed., Plenum Press, New York, pp. 119-129.

Newberne, P. M. and Young, V. R. (1966). Effects of diets marginal in methionine and choline with and without vitamin $B_{12}$ on rat liver and kidney. J. Nutr., 89, 69-79.

Newberne, P. M., Rogers, A. E. and Wogan, G. N. (1968). Hepatorenal lesions in rats fed a low lipotrope diet and exposed to aflatoxin. J. Nutr., 94, 331-343.

Nicolet, B. H. (1931). The mechanism or sulfur lability in cysteine and its derivatives. I. Some thio ethers readily split by alkali. J. Am. Chem. Soc., 53, 3066-3072.

Nicolet, B. H. (1932). The mechanism of sulfur lability in cystein and its derivatives. II. The addition of mercaptan to benzoylaminocinnamic acid derivatives. J. Biol. Chem., 95, 389-392.

Nicolet, B. H., Shinn, L. A., and Saidel, L. J. (1942). The lability toward alkali of serine and threonine in proteins, and some of its consequences. J. Biol. Chem., 142, 609-613.

O'Donovan, C. J. (1976). Recent studies of lysinoalanine in alkali-treated proteins. Fd. Cosmet. Toxicol., 14, 483-489.

Patchornik, A. and Sokolovsky, M. (1964a). Chemical interactions between lysine and dehydroalanine in modified bovine pancreatic ribonuclease. J. Am. Chem. Soc., 86, 1860-1861.

Patchornik, A. aand Sokolovsky, M. (1964b). Nonenzymatic cleavages of peptide chains at the cysteine and serine residues through their conversion into dehydroalanine residues. J. Am. Chem. Soc., 86, 1206-1212.

Provansal, M. M. P., Cuq, J. L. A. and Cheftel, J. C. (1975). Chemical and nutritional modifications of sunflower proteins due to alkaline processing. Formation of amino acid crosslinks and isomerization of lysine residues. J. Ag. Food Chem., 23, 938-943.

Reyniers, J. P., (1979). Renal toxicity of lysinoalanine and its potential catabolites in germfree and conventional animals. Diss. Abstr. Int. B. 39(10), 1820.

Reyniers, J. P., Woodard, J. C. and Alvarez, M. R. (1974). Nuclear cytochemical alterations in $\alpha$-protein induced nephrocytomegalia. Lab. Invest., 30, 582-588.

Robbins, K. R. and Ballew, J. E. (1982). Effect of alkaline treatment of soy protein on sulfur amino acid bioavailability. J. Food Sci. 47, 2070-2071.

Robbins, K. R., Baker, D. H. and Finley, J. W. (1980). Studies on the utilization of lysinoalanine and lanthionine. J. Nutr., 110, 907-914.

Rose, W. C. (1949). Amino acid requirements of man. Fed. Proc. 8, 546-552.

Rosenhagen, M. and Segal, S. (1974). Stereospecificity of amino acid uptake by rat and human kidney cortex slices. Am. J. Physiol., 227, 843-847.

Sanderson, J., Wall, J. S. Donaldson, G. L. and Cavins, J. F. (1978) Effect of alkaline processing of corn on its amino acids. Cereal Chem., 55, 204-213.

Saunders, R. M., Connor, M. A., Edwards, R. H. and Kohler, G. O. (1975). Preparation of protein concentrates from wheat shorts and wheat millrun by a wet alkaline process. Cereal Chem., 52, 93-101.

Smith, G. G., Williams, K. M. and Wonnacot, D. M. (1978). Factors affecting the rate of racemization of amino acids and their significance of geochronology. J. Org. Chem., 43, 1-5.

Snow, J. T., Finley, J. W. and Friedman, M. (1976). Relative reactivities of sulfhydryl groups with N-acetyl dehydroalanine and N-acetyldehydroalanine methyl ester. Int. J. Petide Protein Res., 7, 461-466.

Stegink, L. D. (1977). D-Amino acids. In "Clinical Nutrition Update: Amino Acids," American Medical Association, Chicago, Illinois, p. 198.

Steining J. and Montag, A. Studies on the alteration of lysine of food proteins. II. Formation of lysinoalanine. Z. Lebensm. Unter. Forsch. 175, 8-12. (German).

Sternberg, M. and Kim, C. Y. (1979). Growth response of mice and Tetrahymena pyriformis to lysinoalanine-supplement wheat gluten. J. Agric. Food Chem., 27, 1130.

Sternberg, M. and Kim, C. Y. (1977). Lysinoalanine formation in protein food ingredients. In "Protein Crosslinking: Nutritional and Medical Consequences", M. Friedman, Ed., Plenum Press, New York, pp. 73-84.

Sternberg, M., Kim, C. Y., and Schwende, F. J. (1975). Lysinoalanine: presence in foods and food ingredients. Science, 190, 992-994.

Struthers, B. J., Brielmaier, J. R., Raymond, M. L., Dahlgren, R. R. and Hopkins, D. T. (1980). Excretion and tissue distribution of radioactive lysinoalanine. J. Nutr., 110, 2065-2077.

Struthers, B. J., Dahlgren, R. R., Hopkins, D. T. and Raymond, M. L. (1979). Lysinoalanine: biological effects and significance. In In "Soy Protein and Human Nutrition: H. L. Wilcke, D. T. Hopkins, and D. H. Waggle, Eds., Aademic Press, New York, pp. 235-260.

Taft, R. Jr. (1956). In "Steric Effects in Organic Chemistry," M. S. Newman, Ed., John Wiley and Sons, Inc. New York, N.Y., Chapter 13.

Tannenbaum, S. R., Ahern, M., and Bates, R. P. (1970). Solubilization of fish protein concentrate. 1. An alkaline process. Food Tech., 24, 96-99.

Tas, A. C. and Kleipool, R. J. C. (1979). The stereoisomers of lysinoalanine. Lebensm. Wiss. und Technol., 9, 360-362.

Van Beek, L., Feron, V. J. and de Groot, A. P. (1974). Nutritional effects of alkali-treated soy protein in rats. J. Nutr., 104, 1630-1636.

Wachstein, M. (1974). Nephrotoxic action of D,L-serine in the rat. Arch. Pathol., 43, 503-514.

Whitaker, J. R. and Feeney, R. E. (1977). Behaviour of O-glycosyl and O-phosphoryl proteins in alkaline solution. In "Protein Crosslinking: Nutritional and Medical Consequences", M. Friedman, Ed., Plenum Press, New York, pp. 155-175.

Woodard, J. C. (1969). On the pathogenesis of alpha protein-induced nephrocytomegalia. Lab. Invest., 20, 9-16.

Woodard, J. C. and Alvarez, M. R. (1967). Renal lesions in rats fed diets containing alpha protein. Arch. Path., 84, 153-162.

Woodard, J. C., and Short, D. D. (1973). Toxicity of alkali-treated soy protein in rats. J. Nutr., 103, 569-574.

Woodard, J. C. and Short, D. D. (1977). Renal toxicity of Lysinoalanine in rats. Food Cosmet. Toxicol., 15, 117-119.

Woodard, J. C., Short, D. D., Alvarez, M. R. and Reyniers, J. (1975). Biologic effects of N-ε-(DL-2-amino-2-carboxyethyl)-L-lysine, lysinoalanine. In "Protein Nutritional Quality of Foods and Feeds" M. Friedman, Ed., Marcel Dekker, New York, pp. 595-616.

Wretlind, J. A. J. (1952). The effect of D-amino acids on the stereonaturalization of D-methionine. Acta Physiol. Scand., 25, 267-275.

Wu, Y. V., Sexson, K. R., Cluskey, J. E. and Inglett, G. E. (1977). Protein isolates from high protein oats preparation. Composition and properties. J. Food Sci., 42, 1383-1386.

Zezulka, A. Y. and Calloway, D. H. (1976). Nitrogen retention in men fed isolated soybean protein supplemented with L-methionine, D-methionine, N-acetyl-L-methionine, or inorganic sulfate. J. Nutr., 106, 1286-1291.

Zumberge, J. E. The effect of D-glucose on aspartic acid racemization. Paper presented at the Carnegie Institution of Washington Conference: Advances in the Biogeochemistry of Amino Acids, Warrenton, Virginia, Oct. 29-Nov. 1, 1978.

## 19

EFFECT OF PROTEIN TREATMENT ON THE ENZYMATIC HYDROLYSIS

OF LYSINOALANINE AND OTHER AMINO ACIDS[1]

Laurent Savoie

Centre de recherche en nutrition and
Département de nutrition humaine
Faculté des sciences de l'agriculture et de
l'alimentation
Université Laval
Québec G1K 7P4, Canada

ABSTRACT

Lysinoalanine (LAL) release from alkali-treated proteins by proteolytic enzymes could be measured on a large scale using an in vitro digestion method based on a two-step hydrolysis with pepsin and pancreatin. The hydrolysis was carried out in a dialysis bag, and the digestion products collected and analyzed. Applying this procedure to alkali-treated soybean and rapeseed protein showed that the release of LAL was not related to its concentration in the protein. Instead, LAL release depended on the nature of the protein, the length of the treatment, and the presence of untreated protein. The enzymatic procedure was also used to measure the release of other amino acids in treated and untreated proteins. This in vitro method could directly measure beneficial and adverse effects of processing on amino acid digestibility.

---

[1] This research was supported in part by grants from NSERC (Canada strategic grant no G0619, FCAC program (EQ-1188) Québec and contract No 09279-00150 (Agriculture Canada).

# INTRODUCTION

Food processors treat proteins in various ways - for example, heating, extraction and irradiation - in order to modify the proteins' functional properties or their nutritive values, or to obtain high yields of protein from unconventional sources. These treatments also can be used also to improve digestibility and bioavalaibility of proteins for human beings. One of these processes, alkaline treatment, has been used, for example, in the preparation of tortillas from maize and of tofu from soybeans. Modern food processors used this process for dissolving protein in the preparation of concentrates and isolates, or for obtaining proteins suitable for specific uses, such as foaming, emulsifying stabilizing, or spinning fibers.

Such treatments, however, can induce chemical changes in the secondary as well as primary structure of the protein. These includes racemization of amino acids and crosslinking that leads to the formation of new amino acids (Masters and Friedman, 1980; Friedman et al., 1981). Lysinoalanine (LAL) is one of these amino acids (Friedman, 1982). These chemical changes are of nutritional concern because alkali-treated protein containing protein-bound LAL can produce renal lesions when fed to rats (Woodard et al., 1975). These cytomegalic alterations are characterized by an enlargment of cells of the pars recta epithelium, with disturbances in mitosis and DNA synthesis.

The importance of LAL as a potentially toxic compound in foods is controversial among nutritionists and food chemists. This is due in part to incomplete knowledge of two aspects of the problem: first, the conditions that promote LAL formation and the susceptibility of a protein to form LAL; and second, the capacity of organisms to react to the intake of a protein containing LAL.

As reported by Friedman (1982), the major factor controlling the production of lysinoalanine, once the dehydroalanine precursors are formed, could be the location and availability of "partners" for crosslink formation. When they are adjacent or close by, lysinoalanine formation could be facilitated. When treatment allows it, more lysinoalanine could be formed, even involving cross-chain links. Thus, the capacity for forming LAL would vary not only with the treatment applied but with the nature of the protein, involving its secondary or tertiary structure as well as its primary one.

The problem of interpreting different reports on the toxic potential of protein-bound LAL could also result from the different rates of LAL release by gut proteolytic enzymes and the subsequent availability of this amino acid for absorption. Free

LAL is, in fact, 15 to 25 times more nephrotoxic than protein-bound LAL (De Groot et al., 1976). Thus, any decrease in the digestibility of the protein due to alkali treatment could then influence the degree of availability of LAL. The release of biologically active LAL from a treated protein is determined by the proteins' accessibility to proteolytic enzymes. As demonstrated by Masters and Friedman (1980), alkaline treatment induces amino acid racemization, with each protein and even each protein fraction reacting differently to the same treatment. Since racemization decreases the digestibility of protein, these differences could explain discrepancies among results reported by different authors.

The key to the problem seems then to be the measurement of the enzymatic releasability of protein-bound LAL, as mentioned by Slump (1978). Little on this topic can be found in the literature. Using a closed system (pepsin-pancreatin and pepsin-pronase-prolidase-aminopeptidase), Slump (1978) found an LAL release with treated soybean of 3% (1.3 g LAL/16 g N), and of 2% with treated casein (5.5 g LAL/16 g N content) or 0.5% (1.0 g LAL/16 g N content). Using a different in vitro digestion system, Finot et al. (1978) reported a 27% release of LAL from alkali-treated lactalbumine.

Determination of the potential toxicity of alkali-treated protein as regards LAL action, therefore, may require measurement of the amount of lysinoalanine hydrolyzed by digestive proteolytic enzymes. This measurement requires a method that:
1) reproduces in vivo conditions as closely as possible;
2) permits variation of the conditions of hydrolysis without interfering with digestion;
3) permits the collection of hydrolysis products as soon as they are generated;
4) allows rapid and reproducible tests.

The present experiments used a method based on the work of Mauron et al. (1955) and Steinhart and Kirchgessner (1973) and consisting of a two-step hydrolysis, using first pepsin in a closed system, and then pancreatin in a dialysis bag (Savoie et al., 1980; Gauthier et al., 1982).

MATERIALS AND METHODS

The protein sources, a soybean protein isolate (Promine F[R], Central Soya, Chic.) and a rapeseed protein concentrate (prepared by FRI-71 process at the Food Research Institute, Agriculture Canada, Ottawa) were suspended (10% W/V) in 0.1 N NaOH and heated at 60°C for 4 or 8 h with continuous agitation. Treatment was stopped by the rapid addition of 6 N HCl followed by cooling, adjustment of pH to 4.5, and centrifugation for 15 min at

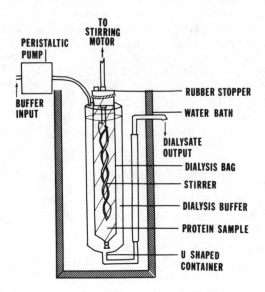

Figure I.   Schematic drawing of digestive unit with simultaneous
            dialysis; from Gauthier et al., J. Nutr., 112, 1718
            (1982).

2 600 *g*.   The residue was washed once with tap water and
freeze-dried until used.

In vitro digestion standard procedures used samples of
proteins (300 mg N x 6.25) untreated, treated as described above
or mixed with the corresponding untreated protein (2:1).   These
samples were suspended in 75 ml of 0.1 N HCl in a beaker at 37°C,
and stirred for 30 min.   After adjustment of pH at 1.9 with N
NaOH, 15 ml of pepsin solution (5 mg/ml) of hog stomach mucosa
(1:20 000) was added and digestion was carried out for 30 min.
Digestion was stopped by adding 5 ml N NaOH and adjusting
pH to 8.15.

The mixture was then introduced in the digestion unit, schematically reproduced in Fig. I. The products of peptic digestion were poured into a cellulose dialysis bag (27 cm long x 45 mm flat width) with a molecular weight cut off of 1000 (Spectra/Por[R] 6, Spectrum Med. Ind. Inc. L.A.). The dialysis bag was suspended in a U-shaped glass container filled with 5 mM sodium phosphate buffer, pH 8.15, and immersed in a water bath kept at 37°C. The bag was closed at its top by a rubber stopper with two holes, one for the stirring rod (a flat helicoidal glass rod), and the other for access to the digestion medium, allowing addition or removal of samples.

The buffer in the glass container was continuously replaced at a rate of 212 ml/h by a peristaltic pump. Dialysates were collected in beakers or in fraction collectors. Six such digestion units were placed in a row in the same water bath, with agitation by the stirring rods controlled mechanically by a six-paddle stirrer. Up to six simultaneous and controlled assays were thus possible.

To start the second step of the hydrolysis, 15 ml of pancreatin solution (5 mg/ml, hog pancreas 5X) in phosphate buffer was introduced in the bag and mixed with the products of pectic digestion. Pancreatic hydrolysis products then were continuously collected as they were dialyzed for various periods of up to 24 hours. The amount of nitrogen released was determined by an Auto Analyzer (using the method no 329-74 W/B of Technicon, Tarrytown, N.Y.). Samples of dialysate and undigested protein fractions were hydrolyzed for 24 h in 6 N HCl at 100°C. Amino acid content was evaluated by high performance liquid chromatography (Vachon et al., 1982).

The digestibility of total protein and of individual amino acids were then calculated from the following equation:

$$\% \text{ digestibility} = \frac{\text{N (or a.a.) in dialysate}}{\text{N (or a.a.) in dialysate} + \text{N (or a.a.) in dialysis bag}} \times 100$$

RESULTS AND DISCUSSION

Table I shows that a 4 h alkaline treatment elicited different amounts of lysinoalanine, depending on the nature of the protein. The 8 h treatment doubled lysinoalanine production from rapeseed while the increase was only 50% with soybean protein.

Total nitrogen liberated as small peptides or free amino acids (dialysate) was lower in rapeseed protein, but was not

Table I.  Effects of alkaline treatment on protein and lysinoalanine digestibility[1].

| Alkaline treatment | LAL content g/16 g N | % of N released | % of LAL released | % LAL released / % N released |
|---|---|---|---|---|
| **Rapeseed protein** | | | | |
| none | 0 | 72.7 | - | - |
| 4 hours | 1.34 | 74.3 | 32.9 | 0.44 |
| 4 hours (2:1)[2] | 0.9 | 68.3 | 15.7 | 0.23 |
| 8 hours (2:1) | 1.8 | 66.1 | 18.2 | 0.27 |
| 8 hours | 2.8 | 70.8 | 43.1 | 0.61 |
| **Soybean protein** | | | | |
| none | 0 | 78.7 | - | - |
| 4 hours | 1.9 | 75.4 | 33.9 | 0.45 |
| 4 hours (2:1) | 1.4 | 69.2 | 11.6 | 0.17 |
| 8 hours (2:1) | 2.0 | 68.6 | 25.4 | 0.38 |
| 8 hours | 3.0 | 67.4 | 24.7 | 0.37 |

1 Values are means of 3 digestions (24 hours)
2 Mixture of treated protein and untreated protein (2:1, w:w) adapted from Savoie et al. (1981) and Vachon et al. (1983).

significantly affected by alkaline treatment. The 8-h treatment markedly diminished the digestibility of soybean protein.

In comparison with the mean release of other amino acids (total N), LAL was less readily hydrolyzable. Its capacity of enzymatic release varied from 25 to 43% of that of other amino acids. LAL was more available (i.e. digestible) from rapeseed than from soybean; this availability increased with LAL content in rapeseed, and decreased with LAL content in soybean.

Unexpectedly, adding 33% untreated protein reduced the total digestibility of both proteins, especially rapeseed, where LAL availability also was reduced by 50%, regardless of LAL protein content and length of treatment. In soybean protein, the availability of LAL produced by a 4-hour treatment was drastically reduced by the addition of untreated protein. However, the availability of LAL produced by an 8-hour treatment was not affected.

Finally, the initial rate of release of LAL was slower in rapeseed protein, although the total amount hydrolyzed within 24 h was equal or higher than in soybean protein.

Table II shows the effect of alkali treatment on some other amino acid digestibility. The hydrolytic release rate for individual amino acids was different from the mean rate of release for protein nitrogen and varied according to the protein. For example, lysine from rapeseed protein was liberated slowly. Lysine was more available from soybean protein and its release rate was equivalent to that of serine. The latter amino acid is equally available from both protein sources.

Although these amino acids act as precursors of lysinoalanine, their enzymatic release was not uniformly decreased after treatment. In fact, 4- or 8-hour alkali treatment markedly increased the release of lysine in both proteins, especially rapeseed. Serine liberation was not affected in rapeseed, but significantly decreased in soybean. These variations in amino acid availability were more marked in the first stages of digestion. They tended to disappear with time in treated and untreated proteins.

In the two proteins studied, the quantity of LAL formed by alkaline treatment varied with the nature of the protein, but the differences diminished as treatment was lengthened.

The rate of _in vitro_ enzymatic release of protein nitrogen varied with the nature of the protein, and was more affected by alkaline treatment in rapidly digestible protein.

Table II. Effect of alkaline treatment on the relative rate of amino acids release after 3, 6 and 24 h of protein digestion[1] (protein (N) digestibility at a given time = 1).

| Protein | Duration of digestion (h) | Lysine Released | | | Serine released | | |
|---|---|---|---|---|---|---|---|
| | | Untreated | 4 h treatment | 8 h treatment | Untreated | 4 h treatment | 8 h treatment |
| Rapeseed | 3 | 0.63 | 1.32 | 1.29 | 1.02 | 1.01 | 0.94 |
| | 6 | 0.74 | 1.24 | 1.23 | 0.93 | 0.97 | 0.90 |
| | 24 | 0.97 | 1.08 | 1.08 | 1.00 | 1.03 | 1.01 |
| Soybean | 3 | 0.95 | 1.34 | 1.25 | 1.03 | 0.88 | 0.89 |
| | 6 | 1.05 | 1.29 | 1.26 | 1.00 | 0.91 | 0.86 |
| | 24 | 1.06 | 1.12 | 1.10 | 1.05 | 1.01 | 1.01 |

[1] Adapted from Vachon et al. (1983).

The potency of LAL to be released by enzymatic hydrolysis is lower than that of other amino acids and not proportional to the amount formed, but depends on the protein in which it is produced and the length of the treatment applied.

When mixed with corresponding untreated protein, protein-bound LAL was generally less releasable. It follows the reduction of total digestibility when present.

These preliminary experiments give new insights and open new avenues of research into protein modification by alkaline treatment.

REFERENCES

De Groot, A.P., Slump, P., Feron, V.J., and Van Beck, C. (1976) Effects of alkali-treated proteins: feeding studies with free and protein bound lysinoalaline. J. Nutr. 106, 1527.

Finot, P.A., Magnenot, E., Mottu, F. and Bujard, E. (1978) Biological availability and metabolic fate of amino acids modified by technological processing. Ann. Nutr. Alim. 32, 325.

Friedman, M. (1982) Lysinoalanine formation in soybean protein: kinetics and mechanism. In: "Food Protein Deterioration, Mechanism and Functionality". J.P. Cherry, ed. American Chemical Soc., Wash. p. 231.

Friedman, M., Zahnley, J.C. and Masters, P.M. (1981) Relationship between in vitro digestibility of casein and its content of lysinoalanine and D-amino acids. J. Food Sci., 46:127.

Gauthier, S.F., Vachon, C., Jones, J.D., and Savoie, L. (1982) Assessment of protein digestibility by in vitro enzymatic hydrolysis with simultaneous dialysis. J. Nutr. 112, 1718.

Master, P., and Friedman, M. (1980) Amino acid racemization in alkali treated food proteins - Chemistry, toxicology, and nutritional consequence. In: "Chemical Deterioration of Proteins". Whitaker, J.R. and Fujimaki, M. Eds. ACS Symposium Series 123, Washington, D.C. p. 165.

Mauron, J., Mottu, F., Bujard, E., and Egli, R.H. (1955) The availability of lysine, methionine and tryptophane in condensed milk and milk powder in vitro digestion studies. Arch. Biochem. Biophys. 59, 433.

Savoie, L., Brassard, A., Gauthier, S., Vachon, C. et Parent, G. (1980, 1981) Etude toxicologique et métabolique des substances produites durant le traitement de transformation de protéines végétales: toxicité rénale de la lysinoalanine. Research reports, Agriculture-Canada.

Slump, P. (1978) Lysinoalanine in alkali treated protein and factors influencing its biological activity. Ann. Nutr. Alim. 32, 271.

Steinhart, H., and Kirchgessner, R.   (1973)   In vitro verdau
    sapparatur zur enzymatischen Hydrolyse von Protein. Arch.
    Tierernaehr. 23, 449.
Vachon, C., Gauthier, S.F., Jones, J.D., and Savoie, L.   (1982)
    Enzymatic digestion method with dialysis to assess protein
    damage application to alkali-treated protein containing
    lysinoalanine. Nutr. Res. 2, 675.
Vachon, C., Gauthier, S.F., Jones, J.D., and Savoie, L.   (1983)
    In vitro enzymatic release of amino acids from
    alkali-treated proteins containing lysinoalanine. Nutr. Rep.
    Intern. 27:119.
Woodard, C.J., Short, D.D., Alvarez, M.R., and Reynion, J.
    (1979) Biologic effects of lysinoalanine.   In:   "Protein
    Nutritional Quality of Foods and Feeds. Part 2. Friedman,
    M., Ed., Marcel Dekker, New York, p. 595.

## 20

NUTRITIONAL CONSEQUENCES OF THE REACTIONS BETWEEN PROTEINS AND

OXIDIZED POLYPHENOLIC ACIDS

Richard F. Hurrell and Paul-André Finot

Nestlé Products Technical Assistance Co. Ltd.
Reseach Department, P.O. Box 88
CH-1814 La Tour de Peilz (Switzerland)

ABSTRACT

The chemical and enzymatic browning reactions of plant polyphenols and their effects on amino acids and proteins are reviewed. A model system of casein and oxidizing caffeic acid has been studied in more detail. The effects of pH, time, caffeic acid level and the presence or not of tyrosinase on the decrease of FDNB-reactive lysine are described. The chemical loss of lysine, methionine and tryptophan and the change in the bioavailability of these amino acids to rats has been evaluated in two systems : pH 7.0 with tyrosinase and pH 10.0 without tyrosinase. At pH 10.0, reactive lysine was more reduced. At pH 7.0 plus tyrosinase methionine was more extensively oxidized to its sulphoxide. Tryptophan was not chemically reduced under either condition. At pH 10.0 there was a decrease in the protein digestibility which was responsible for a corresponding reduction in tryptophan availability and partly responsible for lower methionine availability. Metabolic transit of casein labelled with tritiated lysine treated under the same conditions indicated that the lower lysine availability in rats was due to a lower digestibility of the lysine-caffeoquinone complexes.

INTRODUCTION

Enzymic browning reactions commonly occur when fruits and vegetables are bruised or peeled. They are due to the polymerization of phenolic compounds on exposure to air. The enzyme responsible for this browning is the polyphenol oxidase or phenolase enzyme which converts the phenol to its corresponding qui-

none. The quinones, which can also be formed non-enzymatically
under alkaline conditions are highly reactive substances and can
polymerize with other quinones or combine with certain amino
acids in food proteins. Such reactions are oxygen dependent, may
produce a variety of different coloured compounds and may be of
importance both organoleptically and nutritionally during the
production of vegetable protein concentrates. Although the
nutritional consequences of the Maillard browning reactions have
been widely studied (Hurrell and Carpenter, 1977; Hurrell, 1980;
Hurrell and Finot, 1983), few nutritional studies have been made
on protein-polyphenol reactions.

Phenolic compounds are widely distributed in plant parts
from the roots to the seeds and include phenolic acids, flavonoids and tannins. The tannins may reduce protein digestibility
(Ford and Hewitt, 1979) and perhaps the bioavailability of other
nutrients. The flavonoids have been reported to have a number of
nutritional and pharmacological activities (Kühnau, 1976). Phenolic acids include benzoic and cinnamic acid derivatives. The
benzoic acid derivatives include p-hydroxy-benzoic, protochate-
chuic, vanillic, gallic and syringic acids. The cinnamic acids,
p-coumaric, caffeic, ferulic and sinapic are found in most oilseeds used to prepare protein concentrates and frequently occur
in the form of esters with quinic acid or sugars. Chlorogenic
acid for example is an ester of caffeic acid and quinic acid and
is found in several isomeric and derivatized forms.

One of the major chemical properties of the phenolic acids
is the ease with which they oxidize. In the presence of oxygen,
chlorogenic acid, caffeic acid and other related o-diphenols can
oxidize by the action of polyphenol oxidase or in alkaline solution. These polyphenol oxidases are copper-containing proteins
and are widespread in nature. Some of them oxidize only selected
o-dihydroxyphenols while others, such as tyrosinase, not only
oxidize dihydroxyphenols but convert monohydroxyphenols, such as
tyrosine, to oxidizable dihydroxyphenols. The quinones formed
are highly reactive substances which normally react further with
other quinones to produce coloured compounds of high molecular
weight. They may also react however with lysine, methionine,
cysteine and tryptophan residues in the protein chain (Fig. 1).

The reaction of quinones with amino acids and proteins was
reviewed by Mason (1955). A more recent study of the reaction of
enzymatically generated caffeoquinone and chlorogenoquinone with
amino acids and proteins was made by Pierpoint (1969ab, 1971).
The sulphydryl groups of cysteine and the ε-amino groups of
lysine as well as α-terminal amino acid groups appear to combine most readily with quinones. Methionine (Vithayathil and
Murphy, 1972; Bosshard, 1972) and tryptophan (Synge, 1975) could
also react and, in addition methionine, cysteine and tryptophan

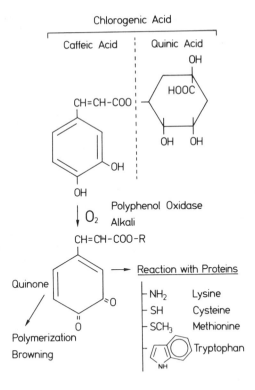

Fig. 1 : Proposed reactions of phenolic acids with amino acids

might be oxidized by the quinones (Synge, 1975). The only direct evidence of the influence of these reactions on protein quality comes from a study performed by Horigome and Kandatsu (1968). They reacted a casein solution with enzymatically-oxidized caffeic acid, isochlorogenic acid and other unidentified polyphenols and reported that the resulting brown protein had lower biological values, protein digestibilities and reactive lysine values than the original casein.

Although phenolic acids are commonly present in soya, cottonseed and peanut meals (Arai et al, 1966; Maga and Lorenz,

1974), it is the high level of chlorogenic acid in sunflower meal which causes most problems during protein extraction. Chlorogenic acid comprises 1.5-2 % of the dry weight of the sunflower kernel (Milic et al., 1968; Pomenta and Burns, 1971) and accounts for about 60 % of the total phenolic compounds (Sabir et al., 1974). Discoloration during protein extraction at alkaline pH restricts the use of sunflower products (Cater et al., 1972) although colourless protein isolates can be produced after making a preliminary extraction of the polyphenols (Sodini and Canella, 1977; Bau and Debry, 1980). Chlorogenic acid has similarly been reported to produce off-colours in leaf protein concentrates (Free and Satterlee, 1975).

We summarize here our studies on the nutritional implications of the reactions between proteins and oxidized phenolic acids. We have used chemical, nutritional and metabolic approaches to study the reactions of oxidized caffeic acid with the lysine (Hurrell et al., 1982), methionine and tryptophan (Hurrell et al., 1981) residues in food proteins and we have investigated the influence of these reactions on protein digestibility.

CHEMICAL STUDIES

For all our studies, we used a model aqueous system containing 5 % casein and 0.2-1.5 % caffeic acid (corresponding to 4-30 % of the casein weight), which was oxygenated and stirred at different pH values, temperatures, times, and in the presence or absence of tyrosinase. The casein-caffeoquinone complexes were precipitated and dried before analysis.

The influence of pH, time and temperature of treatment was investigated only in relation to the loss of fluorodinitrobenzene FDNB-reactive lysine. Fig. 2 shows the influence of pH in solutions containing 5 % casein and 0.2 % caffeic acid stirred for 3 h at room temperature. In the presence of tyrosinase, there was an extremely narrow region for maximum reactive lysine loss at pH 7. At pH 6.8 and pH 7.5, the loss was already half that at pH 7. In the absence of tyrosinase, there was little loss of reactive lysine up until pH 8.8; however, at pH 10.0 only 73 % of the original reactive lysine remained. The loss only occurred in systems which were stirred and oxygenated.

The major part of the loss of reactive lysine occurring at pH 10.0 took place during the first 30 minutes (Fig. 3). The reaction at pH 7.0 with tyrosinase was much slower, probably due to either the slower formation of caffeoquinone or the slower fixation of the lysine residue onto the quinone ring. At pH 10.0, an increase of temperature greatly increased the lysine

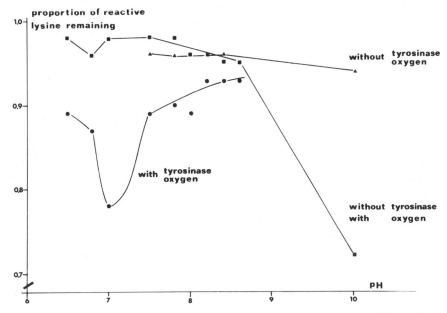

Fig. 2 : Influence of pH on the reaction of lysine residues in
casein with oxidized caffeic acid

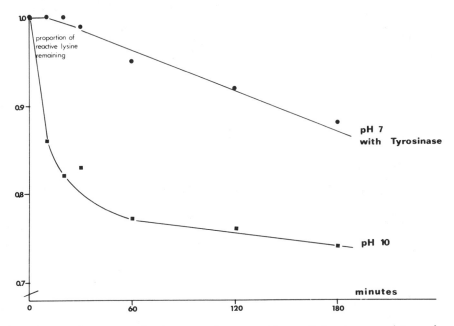

Fig. 3 : Influence of time on the reaction of lysine residues in
casein with oxidized caffeic acid

loss from 27 % at room temperature to around 40 % at 40 °C and
55 % at 60 °C.

The influence of the concentration of caffeic acid on
FDNB-reactive lysine, methionine and tryptophan is shown in
Table 1. Methionine was measured as methionine sulphone by ion-
exchange chromatography after preliminary performic acid oxidation and acid hydrolysis. Methionine sulphoxide was measured by
ion-exchange chromatography after barium hydroxide hydrolysis.
Tryptophan was measured by HPLC after sodium hydroxide hydrolysis. Total methionine and tryptophan values were not reduced
although methionine sulphoxide was progressively and extensively
formed. At pH 7.0 plus tyrosinase and with 1.5 % caffeic acid,
all methionine residues were transformed to the sulphoxide. The
loss of FDNB-reactive lysine and the formation of methionine sulphoxide increased with increasing concentration of caffeic acid
both under enzymic and alkaline conditions. The oxidation of
methionine is catalysed by the presence of caffeic acid probably
through the caffeoquinone formed. Caffeoquinone is involved in
at least three competitive reactions : polymerisation, fixation
on the lysine residues and oxidation of methionine. At both pH
values, the oxidation of methionine appears to be of greater importance than the loss of lysine but more efficient at pH 7.0
than at pH 10.0.

TABLE 1

Lysine, methionine and tryptophan levels (g/16 g N) estimated
chemicaly in casein (5 % in an aqueous solution) reacted with
different concentrations of caffeic acid for 3 h at room
temperature in the presence of oxygen.
(values in brackets represent the value of the test sample
as a percentage of the original unmodified amino acid value).

| Treatment | FDNB-lysine | Methionine | Methionine sulphoxide | Tryptophan |
|---|---|---|---|---|
| Casein control | 8.2 | 2.66 | 0 | 1.38 |
| pH 10.0 | | | | |
| +0.2% caffeic acid | 5.91 (72) | 2.58 (97) | 1.40 (53) | 1.37 (100) |
| +0.5% caffeic acid | 5.00 (61) | 2.64 (99) | 2.12 (80) | 1.38 (100) |
| +1.5% caffeic acid | 3.37 (41) | 2.65 (100) | 2.16 (81) | 1.45 (105) |
| pH 7.0 + tyrosinase | | | | |
| +0.2% caffeic acid | 7.22 (88) | 2.57 (97) | 1.75 (66) | 1.39 (101) |
| +0.5% caffeic acid | 5.85 (71) | 2.56 (96) | 2.06 (77) | 1.32 (96) |
| +1.5% caffeic acid | 4.91 (60) | 2.60 (98) | 2.94 (110) | 1.39 (101) |

NUTRITIONAL STUDIES

Some casein-caffeic acid samples were analysed by rat assays for available lysine, methionine and tryptophan and for true nitrogen digestibility. The results are summarized in Table 2 for samples prepared from aqueous systems of 5 % casein and 0.5 % caffeic acid reacted for 3 h at room temperature. There was a small reduction in protein digestibility at pH 10.0 but none in the sample produced at pH 7.0 with tyrosinase. The losses of available lysine and methionine were far greater than those for available tryptophan which at pH 10.0 could simply be explained by a reduction in protein digestibility. The losses of available methionine are probably explained by a slightly poorer utilization of methionine sulphoxide coupled with reduced protein digestibility. Almost 80 % of the methionine residues in the products tested were as methionine sulphoxide thus confirming the high utilization of this product as a methionine source (Cuq et al., 1978; Marable et al., 1980). In a subsequent test, the sample in which all methionine was converted to its sulphoxide (5 % casein, 1.5 % caffeic acid, pH 7.0 plus tyrosinase) (Table 1) gave a methionine bioavailability 76 % of that of the control casein. Lysine was substantially reduced in bioavailability, especially at pH 10.0, confirming the results of the chemical studies and indicating that lysine residues bound to caffeoquinone are unavailable to the rat.

TABLE 2

Rat assay values for available lysine, methionine, tryptophan (g/16 g N) and for true N digestibility in casein reacted with oxidized caffeic acid (5 % casein, 0.5 % caffeic acid in aqueous solution) for 3 h at room temperature.
(values in brackets represent the value of the test sample as a percentage of the corresponding value for untreated casein)

| Test materials | % N diges-tibility | Lysine | Methionine | Tryptophan |
|---|---|---|---|---|
| Casein | 95.0 | 8.5 | 2.89 | 1.23 |
| + caffeic acid pH 10.0 | 83.7 (88) | 4.8 (56) | 2.14 (74) | 1.07 (87) |
| + caffeic acid pH 7.0 + tyro-sinase | 94.9 (100) | 6.9 (81) | 2.53 (88) | 1.14 (92) |

Caffeic acid itself added at 4 % of the casein had no influence on protein digestibility or NPR in rats. Eklund (1975) has reported similar findings for chlorogenic acid. Caffeic acid which had been polymerized in the absence of protein at pH 7.0 with tyrosinase or at pH 10.0 similarly had no influence on protein digestibility or NPR when added at 4 % of the casein. It had been thought that polymerized caffeic acid might behave like a tannin, also a polymerized polyphenol (flavonoid or gallic acid), and reduce protein digestibility. Eggum and Christensen (1975) have shown that tannic acid added to the diet at a comparable level significantly reduced the N digestibility of soya protein.

METABOLIC STUDIES.

Goat's milk casein containing tritiated lysine was used to study the metabolic transit of the lysine-caffeoquinone complexes in rats. As before, an aqueous system containing 5 % casein and 0.2 % caffeic acid (4 % by weight of the casein) was stirred and oxygenated for 3 hours at room temperature either at pH 10.0 or at pH 7.0 with tyrosinase. After precipitation and drying, the test materials were fed to rats. The results (Table 3) show that rats fed materials prepared at pH 10.0 or at pH 7.0 with tyrosinase excreted far more radioactivity in their feces than

TABLE 3

Distribution of radioactivity in rats 72 h after feeding a test meal containing goat's milk casein labelled with tritiated lysine, or the same material after reaction with oxidized caffeic acid

|                                                          | Control casein | Casein-caffeic acid (pH 7 + tyrosinase) | Casein-caffeic acid (pH 10) |
|----------------------------------------------------------|----------------|-----------------------------------------|------------------------------|
| Percentage of ingested radioactivity excreted in :       |                |                                         |                              |
| Urine                                                    | 12             | 12                                      | 11                           |
| Feces                                                    | 5              | 12                                      | 26                           |
| Activity (dpmx10$^{-4}$ per g) of :                      |                |                                         |                              |
| Liver                                                    | 12.2           | 11.3                                    | 8.7                          |
| Kidney                                                   | 10.8           | 10.9                                    | 7.8                          |
| Muscle                                                   | 5.7            | 5.8                                     | 4.2                          |

rats fed the untreated control casein. Rats fed the casein-caffeic acid complex prepared at pH 10.0 excreted 26 % of the ingested radioactivity in their feces compared to 12 % by the rats
fed the complex prepared at pH 7.0 with tyrosinase and 5 % by
those fed control casein. In addition, correspondingly lower
levels of radioactivity were found in the tissues of rats fed
the pH 10.0 material. There were no differences in the percentage of ingested radioactivity excreted in the urine of the rats
fed the different test diets.

These results indicate that the lysine-caffeoquinone complexes are not absorbed by the rat but that they are excreted
undigested in the feces.


REACTION MECHANISMS

From our results, there is no evidence that methionine or
tryptophan combine covalently with alkali- or enzymically-gene-
rated quinones. Methionine however may be extensively oxidized
by the quinones to its sulphoxide. Lysine, on the other hand,
undoubtedly forms covalent linkages with quinones and two possible reaction pathways are shown in Fig. 4. The proposed schemes
take into consideration our findings (Hurrell et al., 1982) that
lysine-caffeoquinone complexes formed under alkaline and enzymic
conditions are probably different, and that each bound lysine

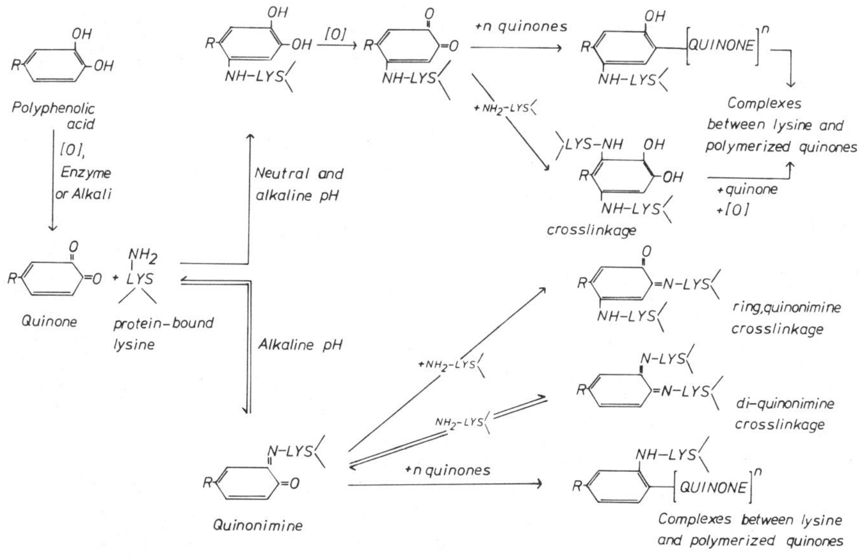

Fig. 4 : Some possible reactions of protein-bound lysine with
         o-diphenolic acids.

unit appears to be associated with several quinone units.

The first step is the conversion of the polyphenolic acid to its corresponding quinone. This can occur enzymically or under alkaline conditions. The subsequent reactions of quinones with lysine are non-enzymic covalent reactions. We have proposed two pathways. Pathway 1 occurs under both neutral and alkaline conditions and by this route lysine is substituted into the quinone ring. Pierpoint (1969a) has suggested that the primary point of substitution is most probably in the 6 position although, under certain conditions, substitutions in the 5 position have also been reported. The most likely route is the one which would lead to complexes between lysine and polymerized caffeoquinone. Crosslinkages could also be formed if further lysine units from an adjacent protein chain substitute into the 2-position of the quinone ring. This is less likely however since we found only small reductions in protein digestibility in rats.

Pathway 2 occurs under alkaline conditions only. In this pathway, lysine groups react with the quinone group itself to form quinonimines (Cheftel, 1979). As in pathway 1, further reactions can then lead to complexes between lysine and polymerized quinones or to the formation of crosslinkages.

CONCLUSIONS

It can be concluded that during alkaline extraction of proteins from vegetable sources, caffeoquinone, chlorogenoquinone and other related o-quinones, formed on the oxidation of the corresponding phenolic acids can react with lysine residues in proteins to make them biologically unavailable. Other phenolic compounds may also react in a similar way particularly the flavonols, such as epicatechin, and their derivatives and even gossypol, the polyphenol of cottonseed, which until now has always been thought to react with lysine via its aldehyde groups.

Is is highly likely that the discoloured sunflower protein concentrates produced at alkaline pH (Pomenta and Burns, 1971; Cater et al., 1972) also have a decreased lysine availability due to the reaction of sunflower protein with chlorogenoquinone and other oxidized phenols. The conversion of phenolic acids to quinones can also occur enzymically under the action of naturally-occurring phenolase enzymes at neutral pH and these quinones appear to react with lysine in a similar manner. The enzymic reactions, however, are much less rapid than those occurring under alkaline conditions and would probably have less influence on the nutritional quality of protein concentrates.

Protein-quinone reactions seem to have little influence on protein digestibility or on the bioavailability of tryptophan. Methionine does not appear to react covalently but may be extensively oxidized to its sulphoxide with some reduction in bioavailability. The reactions of cysteine were not investigated in our study because of the low level of cyst(e)ine in casein. Pierpoint (1969ab) has reported that cysteine reacts like lysine via a substitution into the quinone ring and it is also possible that cyst(e)ine residues may be oxidized (Finley and Lundin, 1980).

REFERENCES

Arai, S., Suzuki, H., Fujimaki, M. and Sakwiai, Y. (1966). Studies on the flavor components in soybean. Part II. Phenolic acids in defatted soybean flour. Agric. Biol. Chem., 30:364

Bau, H.M. and Debry, G. (1980). Colourless sunflower protein isolates : chemical and nutritional evaluation of the presence of phenolic compounds. J. Fd Technol. 15:207

Bosshard, H. (1972). Ueber die Anlagerung von Thioäthern an Chinone und Chinonimine in stark säuren Medien. Helv. Chim. Acta 55:32

Cater, C.M., Gheyasuddin, S. and Mattil, K.F. (1972). The effect of chlorogenic, quinic and caffeic acids on the solubility and color of protein isolates, especially from sunflower seed. Cereal Chem. 49:508

Cheftel, J.C. (1979). Proteins and amino acids, in : Nutritional and safety aspects of food processing, S.R. Tannenbaum, ed., Marcel Dekker, New York, p. 153

Cuq, J.L., Besançon, P., Chartier, L. and Cheftel, J.C. (1978). Oxidation of methionine residues of food proteins and nutritional availability of methionine sulphoxide. Food Chem. 3:85

Eggum, B.O. and Christensen, K.D. (1975). Influence of tannin on protein utilization in feedstuffs with special reference to barley, in Breeding for seed protein improvement using nuclear techniques, p. 135, International Atomic Energy Agency, Vienna

Eklund, A. (1975). Effect of chlorogenic acid in a casein diet for rats. Nutr. Metab. 18:259

Finley, J.W. and Lundin, R.E. (1980). Lipid hydroperoxide induced oxidation of cysteine in peptides, in Autoxidation in food and biological systems, p. 223, M.G. Simic and M. Karel, eds, Plenum Press, New York

Ford, J.E. and Hewitt, D. (1979). Protein quality in cereals and pulses. 3. Bioassays with rats and chickens on sorghum, barley and field beans. Influence of polyethylene glycol on digestibility of the protein in high tannin grain. Br. J. Nutr. 42:325

Free, B.L. and Satterlee, L.D. (1975). Biochemical properties of alpha protein concentrate. J. Fd Sci. 40:85

Horigome, T. and Kandatsu, M. (1968). Biological value of protein allowed to react with phenolic compounds in presence of o-diphenol oxidase. Agric. Biol. Chem. 32:1093

Hurrell, R.F. (1980). Interaction of food components during processing, in Food and Health : Science and Technology, p. 369, G.G. Birch and K.J. Parker, eds., Applied Science, London

Hurrell, R.F. and Carpenter, K.J. (1977). Maillard reactions in foods, in Physical, chemical and biological changes in food caused by thermal processing, p. 168, T. Hoyem and O. Kvale, eds., Applied Science, London

Hurrell, R.F., Finot, P.A., Jaussan, V. and Cuq, J.L. (1981). Nutritional consequences of protein-polyphenol browning reactions. XIIth International Congress of Nutrition, San Diego, California, Abstract 260

Hurrell, R.F., Finot, P.A. and Cuq, J.L. (1982). Protein-polyphenol reactions. 1. Nutritional and metabolic consequences of the reaction between oxidized caffeic acid and the lysine residues of casein. Br. J. Nutr. 47:191

Hurrell, R.F. and Finot, P.A. (1983). Food processing and storage as a determinant of protein and amino acid availability, in Nutritional Adequacy, Nutrient Availability and Needs, p. 135, J. Mauron, ed., Birhäuser Verlag, Basel

Kühnau, J. (1976). The flavonoids. A class of semi-essential food components : their role in human nutrition. Wld Rev. Nutr. Diet 24:117

Maga, J.A. and Lorenz, K. (1974). Gas-liquid chromatography separation of free phenolic acid fractions in various oilseed protein sources. J. Sci. Fd Agric. 25:797

Marable, N.L., Todd, J.M., Korslund, M.K. and Kennedy, B.W. (1980). Human urine and/or plasma levels of methionine, methionine sulphoxide and inorganic sulphur after consumption of a soy isolate : A preliminary study. Qual. Plant Pl Fds Hum. Nutr. 30:155

Mason, H.S. (1955). Comparative biochemistry of the phenolase complex. Adv. Enzym. 16:105

Milic, B., Stojanovic, S., Vucurevic, N. and Turcic, M. (1968). Chlorogenic and quinic acids in sunflower meal. J. Sci. Fd Agric.19:108

Pierpoint, W.S. (1969a). o-Quinones formed in plant extracts. Their reaction with amino acids and peptides. Biochem. J. 112:609

Pierpoint, W.S. (1969b). o-Quinones formed in plant extracts.
     Their reaction with bovine serum albumin. _Biochem_. _J_.
     112:619
Pierpoint, W.S. (1971). Formation and behaviour of o-quinones in
     some processes of agricultural importance. _Rep_. _Rothamsted_
     _exp_. _Stn_, _part 2_, p. 199
Pomenta, J.V. and Burns, E.E. (1971). Factors effecting chloro-
     genic, quinic and caffeic acid levels in sunflower kernels.
     _J_. _Fd_ _Sci_. 36:490
Sabir, M.A., Sosulski, F.W. and Kernan, J.A. (1974). Phenolic
     constituents in sunflower flour. _J_. _Agric_. _Fd_ _Chem_. 22:572
Sodini, G. and Canella, M. (1977). Acid butanol removal of color-
     forming phenols from sunflower meal. _J_. _Agric_. _Fd_ _Chem_.
     25:822
Synge, R.L.M. (1975). Interactions of polyphenols in plants and
     plant products. _Qual_. _Plant_. _Pl_. _Fds_ _hum_. _Nutr_. 24:337
Vithayathil, P.J. and Murphy, G.S. (1972). New reaction of
     o-benzoquinone at the thioether group of methionine. _Nature_
     _New_ _Biol_. 236:101.

## 21

RELATION BETWEEN STRUCTURE OF POLYPHENOL OXIDASE AND PREVENTION OF

BROWNING

Avi Golan-Goldhirsh and John R. Whitaker

Department of Food Science and Technology
University of California
Davis, CA

and Varda Kahn

Department of Food Technology
Volcani Center
Bet-Dagan, Israel

INTRODUCTION

Polyphenol oxidase (monophenol, dihydroxyphenylalanine:oxygen oxidoreductase, E.C. 1.14.18.1), often called tyrosinase, catalyzes two distinctly different chemical reactions.  One reaction is the orthohydroxylation of monophenols; the second reaction is dehydrogenation of orthodihydroxyphenols.  These reactions are shown in Equations 1 and 2.

$$2 \underset{\text{OH}}{\bigcirc} + O_2 \xrightarrow{\text{PPO}} 2 \underset{\text{OH}}{\bigcirc}\text{-OH} \tag{1}$$

$$2 \underset{\text{OH}}{\bigcirc}\text{-OH} + O_2 \xrightarrow{\text{PPO}} 2H_2O + 2 \underset{\text{O}}{\bigcirc}\text{=O} \xrightarrow{O_2} \xrightarrow{O_2} \xrightarrow{O_2} \text{Melanin} \tag{2}$$

The benzoquinone formed undergoes non-enzyme catalyzed condensation and further oxidation to form a brown product called melanin.  This reaction (Eqn. 2), initiated by PPO, is responsible for the browning of some fruits and vegetables upon cutting or bruising.  The browning reaction leads to deterioration of flavor, color and nutritional quality of foods.  It is the most undesirable enzyme-catalyzed reaction in foods, resulting in major economic losses in fresh fruits and vegetables.

Because of undesirability of the browning reaction, many
methods have been proposed for control of PPO. These methods
include heat treatment, lowering of pH and addition of ascorbate,
sodium chloride and sodium bisulfite. While diethyldithiocarbamate
is effective in controlling browning, it is not permitted to be
added to foods. Methods of enzymatic methylation of the hydroxyl
groups of the phenols, while effective in vitro, are not commercially feasible and can't be used on intact foods (Walker, 1977).
With the exception of heat, none of these methods are completely
effective in control of the enzymatic browning. Therefore, workers
continue to look for an effective control of browning in fruits and
vegetables.

In this paper, we describe the effect of $k_{cat}$ type inhibition
(Rando, 1974) as a possible method for controlling enzymatic brown-
ing. $k_{cat}$ inhibition (or suicide type inhibition; Abeles and
Maycock, 1976), is defined by Equation 3 where E is enzyme, S is

$$E + S \underset{}{\overset{K_s}{\rightleftharpoons}} E \cdot S \underset{k_{-2}}{\overset{k_2}{\rightleftharpoons}} E \cdot P \underset{k_{-3}}{\overset{k_3}{\rightleftharpoons}} E + P \qquad (3)$$

$$\downarrow k_4$$

$$E\text{-}P$$
$$(\text{inactive})$$

substrate, E·S is the Michaelis Complex with enzyme and substrate,
E·P is the Michaelis Complex with enzyme and product, P is product
and E-P is a covalent enzyme-product compound which is inactive and
irreversible.

Once formed, E·P can either dissociate to E + P and the active
E can recycle through the pathway again (catalyst) or covalently
react with P, while P is still in the active site, to give an inactive derivative. $k_{cat}$ type inhibition of enzymes is the most
specific type of inhibition known since it requires that the compound be bound in the active site of the enzyme (E·S) and to undergo conversion to form E·P before inhibition occurs. The relative
magnitudes of $k_3$ and $k_4$ will determine the rate at which enzyme
activity is lost.

Early studies showed that polyphenol oxidase rapidly lost its
activity as it acted upon naturally occurring substrates (Nelson
and Dawson, 1944). Addition of substrate under anaerobic conditions (see Equations 1 and 2) or of product to PPO did not cause
inactivation. Inactivation of PPO during the early part of the
reaction was shown not to be due to instability caused by dilute
solution, salts, temperature or pH effects (Asimov and Dawson,

1950). Therefore, PPO loses activity during catalysis because of $k_{cat}$ inhibition, even though the term $k_{cat}$ inhibition was not coined by Rando until 1974.

While $k_{cat}$ inhibition is characteristic of PPO reactions, the mechanism is poorly understood. Wood and Ingraham (1965) found that mushroom PPO was covalently labeled with [14]C when it was incubated with 1-[14]C phenol in the presence of $O_2$. They proposed that a nucleophilic group at or near the active site of PPO reacted with the benzoquinone, via a Michael-type addition, to form an inactive enzyme derivative (Eqn. 4).

$$(4)$$

Such reactions are well-known.

In support of their proposal, Wood and Ingraham (1965) found that 4,5-substituted catechols did not inactivate mushroom PPO although they were converted to the corresponding benzoquinones. This would be expected if inactivation of PPO occurs by nucleophilic interaction between PPO and benzoquinone at the 4 position as shown in Equation 4.

Reaction of Neurospora PPO with uniformly labeled [14]C phenol, under conditions similar to those used by Wood and Ingraham (1965), led to covalent incorporation of < 0.03 mol labeled phenol/mol PPO (Pfiffner et al., 1981). The enzyme lost its activity during the reaction; there was also a loss of one mol of histidine and one mol of copper per mol of Neurospora PPO. Brooks and Dawson (1966) had shown earlier that $k_{cat}$ inactivation of mushroom PPO was accompanied by loss of copper from the enzyme. Lerch (1978) proposed that inactivation of Neurospora PPO may have occurred as a result of the oxidation of histidine due to formation of singlet oxygen or hydroxyl radicals during the reaction. However, Kahn et al. (1982) could not detect superoxide ion ($O_2^-$) or hydroxyl radical (OH·) formation during the reaction of mushroom PPO with substrates.

The active site of Neurospora PPO is thought to contain two mols copper/mol active site (Solomon, 1981). Elucidation of the complete amino acid sequence of Neurospora PPO (Lerch, 1978), coupled with dye-sensitized photooxidation (Pfiffner and Lerch, 1981) and $k_{cat}$ inhibition (Pfiffner et al., 1981) indicated that one of the coppers is liganded to $His_{188}$, $His_{193}$ and $His_{289}$ and the second copper is liganded to $His_{306}$ as shown below.

In the diagram, R is a postulated bridging ligand. $His_{188}$, $His_{193}$ and $His_{289}$ are destroyed by dye-sensitized photooxidation while $His_{306}$ is destroyed by $k_{cat}$ inhibition. The other two ligands attached to the second copper have not been identified.

With these models in mind, it is appropriate to search for ways of maximizing the inactivation of PPO through $k_{cat}$ inhibition as a means of preventing the browning reaction. The same rationale would be important in the use of PPO for therapeutic purposes, especially in the treatment of Parkinson's disease (Letts and Chase, 1974). Direct administration of DL-dopa (dihydroxyphenyl-alanine), produced by chemical synthesis, to patients results in undesirable side effects. The use of PPO to produce pure L-dopa from L-tyrosine may have value (Letts and Chase, 1974) provided the reaction can be controlled by $k_{cat}$ inhibition, at this stage.

The important requirements for an effective $k_{cat}$ inhibitor of PPO include:

(1)  The substrate must be stable in the absence of PPO; that is, it should not be readily oxidized by $O_2$ or by other compounds in the food.
(2)  It should be a good substrate for PPO, with a low $K_m$ and a high $V_{max}$.
(3)  The substrate must be converted into a chemically reactive product that immediately reacts covalently with a part of the active site of PPO to inactivate it, so that there is no product left to diffuse from the active site.
(4)  The PPO of interest must have a reactive residue at the active site to combine with the product.
(5)  One turnover of substrate should lead to stoichiometric inactivation of PPO.

We shall now describe some of the work we have done on the structure of mushroom PPO and toward meeting the requirements for effective $k_{cat}$ inhibition of mushroom PPO.

PURIFICATION AND STRUCTURE OF MUSHROOM POLYPHENOL OXIDASE

Partially purified mushroom PPO obtained from Sigma Chemical Company (lot 31F-9535, 2000 units/mg) was further purified by affinity chromatography using antibodies prepared by injecting a rabbit with mushroom PPO. The antibodies (Igg) and affinity columns were prepared by the general procedures of Livingston (1974). Electrophoretic techniques and other experimental details of the purification were described by Golan-Goldhirsh et al. (1982). PPO activity was determined at pH 6.5 and 25° using 0.33 m$\underline{M}$ L-tyrosine as substrate. One unit of enzyme activity is defined as the amount of enzyme required to cause an increase in absorbance at 280 nm of 0.001 per minute.

Chromatography of mushroom tyrosinase on an Igg-Sepharose-4B column is shown in Fig. 1. Protein peak A, which came through in the void volume, contained a small amount of catecholase activity (peak $A_1$, Fig. 1). Tightly bound proteins (peak B) were eluted by increasing the molarity of the sodium phosphate buffer to 100 m$\underline{M}$. Increasing the pH to 8.0, without change in ionic strength, did not elute any more protein (data not shown). Further elution with 0.5 $\underline{M}$ NaCl in 10 m$\underline{M}$ phosphate buffer gave an additional small peak (C, Fig. 1). Protein recovery varied between 80-98%, while only 13% of the applied activity was recovered (peak B).

There was a slight decrease in specific activity over the starting material. This is hard to explain especially based on the purification achieved, and on the electrophoretic patterns of peak B (Figs. 2 and 3). A possible explanation is that peak B contains protein which is inactive as an enzyme, but still retains its anti-genic activity toward the rabbit antibodies. Chromatography of the partially purified Sigma mushroom PPO (passed through G-50 and G-10 columns to remove small molecular weight inactive material) on a $\underline{p}$-amino benzoate-Sepharose column (data not shown) gave a single peak which was eluted by 100 m$\underline{M}$ phosphate buffer, pH 7.0.

Chromatography of partially purified Sigma mushroom PPO on the Igg-Sepharose affinity column resulted in some purification (Fig. 2). Protein bands $a_1$ and $a_2$ in the original material which did not have activity (lanes 2, 3 and 4) were not present in peak B (lanes 3 and 6).

A major dye binding band (protein?) was present at the track-ing dye position (band c, Fig. 2; at the bottom of the gel) in peaks A and B (lanes 5 and 6) and in unpurified samples (lanes 7 and 8). An electrophoretically similar protein band was found in purified mushroom PPO by other investigators using denaturing gels (Robb, 1981a; Robb and Gutteridge, 1981). Its role in the enzy-matic activity of PPO is unknown. Our preliminary estimate of its molecular weight is in the range of 3,000, much smaller than the

small subunit of 13,000 daltons reported by Strothkamp et al.
(1976). The 3,000 molecular weight component is similar in weight
to the difference between the $H^a$ and $H^b$ forms of the large subunit
of mushroom PPO (Robb, 1981a).

While the data are preliminary (unpublished data), the amino
acid pattern of the electrophoretically eluted band c (Fig. 2),
after hydrolysis in ~6N HCl in evaluated sealed glass tube for 22
hr at 11°C, showed an unusual peak following asparagine. This
chromatographic behavior corresponds precisely to the position of a
2-thiohistidine, resulting from a cysteinyl-histidyl, bound via a
thioether bond (Lerch, 1981)(both Lerch and we used a Durrum D-500
amino acid analyzer) in the mushroom PPO structure.

The zone b proteins (lane 2 vs. lane 3, Fig. 2) were retained
preferentially on the Igg-Sepharose column (peak B, Fig. 1) and
some faint activity (diffused) occurred in the top part of the gel
(between zone a and the top). The specific activity of zone a proteins was much higher in peak B than in peak A (compare activity
and protein for lanes 3 and 6 with lanes 2 and 5 in Fig. 2).

Comparison of peaks A and B, obtained from the Igg-Sepharose
column, on denaturing polyacrylamide gel (Fig. 3), clearly shows
the purification obtained by the Igg-Sepharose column. There are
only three major regions of protein in peak B (a, b and c, in lanes
4, 5, 7 and 8, Fig. 3). Additional bands of protein are seen in
peak A (lanes 2, 3, 9 and 10, Fig. 3). The small molecular weight
band (estimated 3,000) seen on the non-denaturing gel (band c, Fig.
2), was lost during destaining. It was visible during early stages
of destaining.

It appears that the protein(s) in peak B contains one or more
S-S bonds (band a, lanes 4 and 5, Fig. 3), which are reduced by
β-mercaptoethanol (lanes 7 and 8, Fig. 3). This protein band (a,
Fig. 3) is missing in peak A (lanes 2, 3, 9, and 10). Cystine in
mushroom PPO was reported before by Jolley et al. (1969).

The molecular weight estimates for zones a, b and c are
105,000, 53,000 and 28,000, respectively. None of these values fit
the current model of Strothkamp et al. (1976) for mushroom PPO.
Strothkamp et al. (1976) suggested there is a large subunit (H) of
45,000 and a small subunit (L) of 13,000. This model by Strothkamp
et al. (1976) also fails to accommodate the small peptide found by
Robb (1981a), Robb and Gutteridge (1981) and by us (discussed
above). Duckworth and Coleman (1970) reported an apparent molecular weight of 26,300 by sedimentation equilibrium, a value which
they did not include in their calculations because, as they state:
"It has generally been assumed that the molecular weight of the
monomer of mushroom tyrosinase is near 32,000." Our estimate of
the molecular weight of mushroom PPO using gel filtration

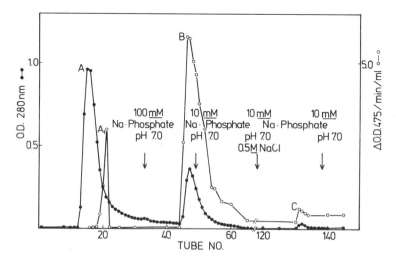

Fig. 1.  Chromatography of mushroom PPO on a preparative Igg-
Sepharose 4B column.  The column consisted of purified rabbit Igg
(34 mg), coupled via cyanogen bromide activation to Sepharose 4B
(50 ml), and equilibrated in 10 mM Na phosphate buffer, pH 7.0.
The enzyme (22.5 mg; Sigma Chemical Co.) in 3.0 ml equilibration
buffer, was applied to the column.  Chromatography was performed as
shown on the figure.  Approximately 4 ml fractions were collected.

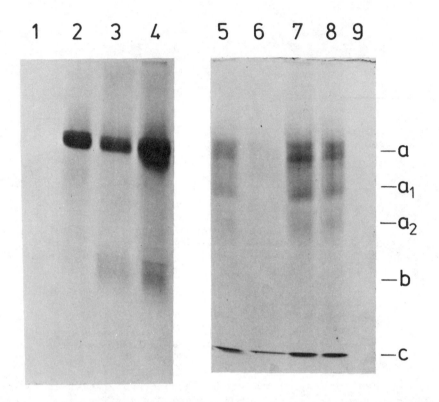

Fig. 2.   Purified mushroom PPO fractions run on a non-denaturing
polyacrylamide gel (10% PAGE).   Lanes 1 and 9 are blanks where
sample buffer only was applied; lanes 2, 5 and 3, 6 correspond to
aliquots of peaks A and B of the Igg-Sepharose column, respectively.   Lanes 4 and 7 are of unpurified enzyme and lane 8 is an
aliquot of redissolved freeze dried unpurified enzyme (Sigma
Chemical Co.).   The left part of the gel (lanes 1, 2, 3 and 4) was
stained by substrate (DL-dopa; 5 mM) reaction in 0.1 M sodium phosphate buffer, pH 7.0, with oxygen replenished by bubbling through
the incubation solution.   The band in zone a stained brown immediately, while there was a delay of 10 min in the start of color
appearance in zone b.   The right hand side of the gel (lanes 5, 6,
7, 8, 9) was stained for protein with Coomassie brilliant blue.

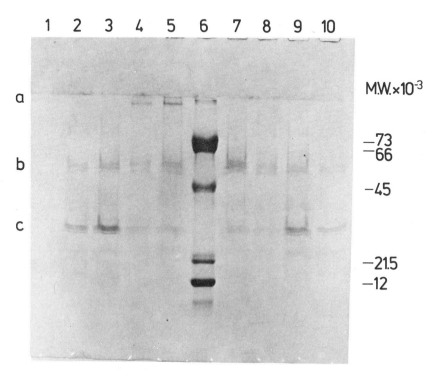

Fig. 3. Peaks A and B from the Igg-Sepharose column on a denaturing polyacrylamide gel (12.5% PAGE-SDS), following staining for protein with Coomassie brilliant blue. Lane 1 represents sample buffer only, lanes 2, 3 and 9, 10 are of an aliquot of peak A, untreated and treated, respectively, with β-mercaptoethanol. Similarly, lanes 4, 5 and 7, 8 are for peak B. Lane 6 is of proteins of known molecular weight, in decreasing order of molecular weight (indicated on the figure). These proteins were: human transferrin, BSA, ovalbumin, soybean trypsin inhibitor, and cytochrome c.

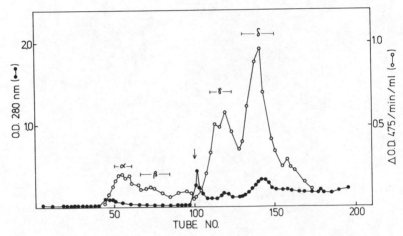

Fig. 4. Chromatography of partially purified mushroom tyrosinase (Sigma Chemical Co.) on a hydroxylapatite column (Robb, 1979). After sample application, column was eluted with a gradient of 5-50 mM Na phosphate buffer, pH 7.0. Arrow indicates start of gradient 50-200 mM, same buffer.

(unpublished data) gave a value of 33,000. These differences in molecular weights, depending on methodology used, may reflect differential interactions of the subunits of the enzyme.

Three and four protein bands are discernable in regions b and c, respectively (Fig. 3), indicating possibly small variations in the molecular weight of each subunit. This multiplicity of bands for mushroom PPO could reflect artifacts produced during purification, such as reaction of benzoquinone with the $\varepsilon$ amino group of lysine of the subunits. If this is so, one must be very careful during purification of the enzyme. The need for a fast and efficient method for purification of PPO, one which prevents any oxidation of phenols during the early stages of purification, is obvious.

The electrophoretic pattern on denaturing polyacrylamide gel of the two major peaks of activity obtained from chromatography on a hydroxylapatite column (Fig. 4) is shown in Fig. 5. Band c, present in peak B obtained from the Igg-Sepharose column (Fig. 2), is not present in either of the major activity peaks obtained from the hydroxylapatite column (lanes 2, 3 and 4 and lanes 5, 6 and 7 correspond to peaks $\gamma$ and $\delta$, respectively). A protein band at approximately 12,000 daltons present in both hydroxylapatite peaks is not present in the Igg-Sepharose purified protein (peak B, Fig. 1). Additional protein bands at higher molecular weights were not present in the Igg-Sepharose purified enzyme, reflecting the higher specificity of binding of the Igg column.

MECHANISM OF PPO INACTIVATION IN THE PRESENCE OF ASCORBIC ACID

In a food system where browning is undesirable, $k_{cat}$ inhibition of the enzyme must lead to rapid inactivation of PPO. Since we have not yet found a substrate where immediate inactivation occurs, we decided to try a combination of treatments in which substrate (pyrocatechol) is added to the enzyme in the presence of ascorbic acid. The ascorbic acid acts as a reducing agent to prevent development of brown color while the enzyme is catalyzing the reaction and inactivating itself.

The mechanism of inactivation of PPO in the presence of ascorbic acid is shown in Fig. 6. The mechanism of the effect of ascorbic acid on PPO is controversial. Some workers reported that ascorbic acid inactivated the enzyme (Baruah and Swain, 1953; Ponting, 1954; Mihályi and Vámos-Vigyázó, 1976) while other workers reported activation (Krueger, 1950). Ingraham (1956) and Scharf and Dawson (1958) reported no effect of ascorbic acid on PPO activity. More recently, Markakis and Embs (1966) and Padrón et al. (1975) reported no direct effect of ascorbic acid on the enzymatic activity. We believe ascorbic acid functions primarily

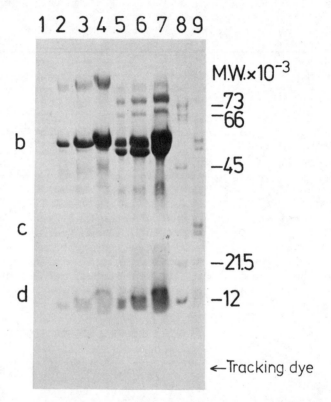

Fig. 5.  Peaks γ and δ from the hydroxylapatite column (Fig. 4),
run by electrophoresis on a denaturing polyacrylamide gel (12.5%
PAGE-SDS) and then stained for protein with Coomassie brilliant
blue.  Lane 1, blank sample buffer only; lanes 2, 3 and 4, peak γ
at increasing protein concentration; similarly, peak δ was run in
lanes 5, 6 and 7 at increasing protein concentrations.  Lane 8 contained proteins of known molecular weight (see legend to Fig. 3).
Lane 9 is crude mushroom PPO (~16 μg).

as a reductant to reduce the quinones back to the dihydroxy
phenolic compound.

Mushroom PPO is rapidly inactivated when pyrocatechol is the
substrate. Within 2 min, there is a substantial reduction in rate
of the reaction and after 7 min, the rate is essentially zero (Fig.
7). Addition of another aliquot of substrate (at point S, Fig. 7)
did not result in further reaction indicating that substrate was
not limiting in the reaction. To further prove that substrate was
not limiting, an aliquot of enzyme was added to the reaction mixture after approximately 7 min of reaction. The reaction rate was
again similar to that at the beginning (Fig. 8).

When the reaction is measured spectrophotometrically, the
progress curve is the net product of the formation by the enzyme of
compounds absorbing at 390 nm and non-enzymatic reduction of the
compounds absorbing at 390 nm. As a result, addition of more
substrate after approximately 1050 sec (Fig. 8) resulted in a
decrease in absorption. We believe this is due to reduction of the
compounds absorbing at 390 nm by pyrocatechol.

The minimum number of turnovers of enzyme before complete
inactivation occurred was calculated to be 10,000. It took 5,000
turnovers of enzyme for complete inactivation of Neurospora PPO
(Dietler and Lerch, quoted by Solomon, 1981).

The effect of different concentrations of ascorbic acid on the
reaction rate of mushroom PPO with pyrocatechol is shown in Fig.
9. No brown color (absorbing at 390 nm) was formed as long as
ascorbic acid was present. Semilogarithmic plots of the data of
Figure 9 gave a linear relationship (Fig. 10), indicating that the
reactions are first order. When extrapolated to zero time, all the
reactions gave the same initial velocity ($0.76 \text{ min}^{-1}$), indicating
there was little or no inhibitory effect of ascorbic acid (0.1-0.8
m$\underline{M}$) on the enzyme per se. The initial velocity in the absence of
ascorbic acid (control) was $0.89 \text{ min}^{-1}$, however, somewhat higher
than the extrapolated value in the presence of ascorbic acid.

The rate of inactivation ($k_{obs}$) of PPO was slightly lower in
the presence of ascorbic acid ($0.72 \text{ min}^{-1}$) than in the absence of
ascorbic acid ($0.84 \text{ min}^{-1}$). The same results were found when the
reaction was monitored polarographically (oxygen consumption).
This small difference may indicate some direct effect of ascorbic
acid on the enzyme. At higher concentrations (>100 m$\underline{M}$) of ascorbic
acid, fast inactivation of mushroom PPO occurred.

The relationship between lag time before color appearance and
concentration of ascorbic acid (Fig. 11) indicates there is an
optimum ascorbic acid concentration at a given enzyme and substrate
concentration at which the lag time is infinite. In other words,

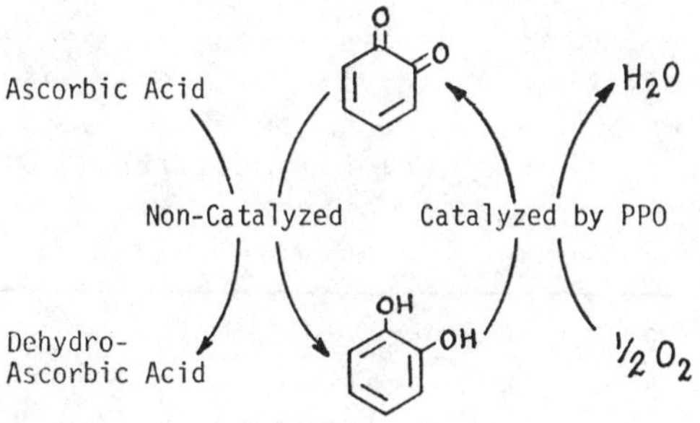

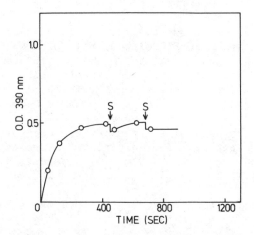

Fig. 6.  Mechanism of prevention of color formation by ascorbic acid.

Fig. 7.  Rate of oxidation of pyrocatechol (1.0 m$\underline{M}$) by mushroom PPO in 0.1 $\underline{M}$ sodium phosphate buffer, pH 7.1.  The arrow and the S indicate the time of addition of a new aliquot of substrate (1 µmol).

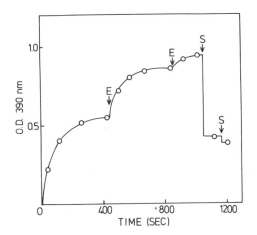

Fig. 8. Rate of oxidation of pyrocatechol (1.0 mM) by mushroom PPO
in 0.1 M sodium phosphate buffer, pH 7.1. The arrows and the
letters E and S indicate the time of addition of new enzyme (18
picomoles by weight) and substrate (1 μmol) aliquots, respectively.

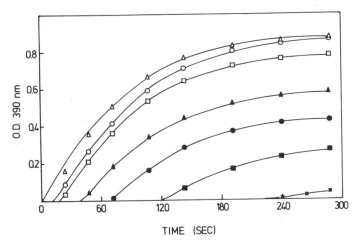

TIME (SEC)

Fig. 9. Observed rate of oxidation of pyrocatechol by mushroom PPO
in the presence of ascorbic acid in 0.1 M sodium phosphate buffer,
pH 7.1. Ascorbic acid concentrations (mM) in the reaction mixtures
were: (△) 0; (○) 0.1; (□) 0.2; (▲) 0.3; (●) 0.5; (■) 0.7; (*) 0.8.

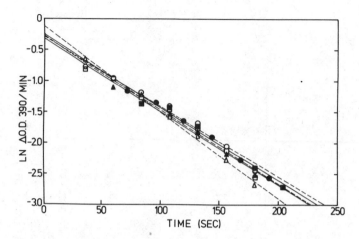

Fig. 10.   Semilogarithmic plot of color formation ($A_{390}$) at different ascorbic acid concentrations (0.1-0.6 m$\underline{M}$) vs time.  The data
are taken from Figure 9.

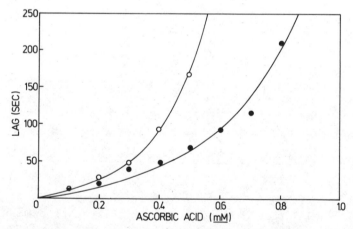

Fig. 11.   Relationship between lag time before color development
and ascorbic acid concentration.  Pyrocatechol concentration, 2.5
m$\underline{M}$ (o); 5.0 m$\underline{M}$ (●).   PPO concentration 18 n$\underline{M}$ (based on weight and
molecular weight of 128,000).

no color will be formed before the enzyme is completely inactivated.


DISCUSSION AND CONCLUSIONS

Polyphenoloxidase (PPO) was the first oxygenase discovered
(Vanneste and Zuberbühler, 1974). However, the two distinctively
different reactions catalyzed by the enzyme and their mechanistic
relationships are complicated and not yet fully understood.

It was shown recently that the Neurospora enzyme contains a
binuclear copper center at the active site. The copper ions acti-
vate molecular oxygen in both the hydroxylation and dehydrogenation
reactions and probably contribute to the three-dimensional conformation of the active site during catalysis (Solomon, 1981).

The complete amino acid sequence of Neurospora PPO (Lerch,
1978) and identification of the amino acid ligands of the binuclear
copper center of Neurospora PPO (Pfiffner and Lerch, 1981) are most
significant in understanding the mechanism of PPO. It is still to
be determined whether the active sites of mushroom PPO and other
PPO's are similar to that of Neurospora PPO.

Although mushroom and Neurospora PPO's catalyze the same reactions, they are quite different in protein structure. Strothkamp
et al. (1976) reported two different size subunits in mushroom PPO,
$H \simeq 45,000$ and $L \simeq 13,000$ daltons. Robb (1979) reported there are
at least two types of heavy subunits, termed $H^a$ and $H^b$, with
slightly different molecular weights.

Our results using Igg-affinity chromatography and partially
purified Sigma mushroom PPO indicated there are at least two additional peptides, one of molecular weight ~28,000 and a second of
molecular weight ~3,000.

The different size subunit structure of mushroom PPO, •and the
multiplicity of bands found for each subunit, as revealed by electrophoresis, indicate a more complicated structure of mushroom PPO
compared to that from Neurospora. There is still reservation about
the interpretation of the meaning of the different subunits in vivo
(Robb, 1981b), as some artifacts may be produced during purification.

An important advantage of $k_{cat}$ inactivation of PPO to prevent
browning, compared to other types of inhibition, is its high
specificity. This permits one to study the inhibition in the presence of other compounds.

In this paper, we showed that $k_{cat}$ inhibition of PPO occurs in the presence of ascorbic acid. Therefore, it should be possible to find an optimum concentration of ascorbic acid at which complete inactivation of PPO by product will occur without color development. The ideal $k_{cat}$ substrate for PPO, one which will inactivate the enzyme in one turnover, has not been found yet. Therefore, use of a combination of ascorbic acid and a substrate which gives a high $k_4/k_3$ ratio (Eqn. 3) may be the only practical way to prevent browning in food products by this method.

ACKNOWLEDGMENTS

This research was supported in part by BARD Project No. I-223-80. We thank Virginia DuBowy and Clara Robison for typing the manuscript and Ruth Golan-Goldhirsh for drawing the figures.

REFERENCES

Abeles, R. H., and Maycock, A. L., 1976, Suicide enzyme inactivators, Acc. Chem. Res. 9:313.

Asimov, I., and Dawson, C. R., 1950, On the reaction inactivation of tyrosinase during the aerobic oxidation of catechol, J. Amer. Chem. Soc. 72:820.

Baruah, P., and Swain, T., 1953, The effect of L-ascorbic acid on the in vitro activity of polyphenoloxidase from potato, Biochem. J. 55:392.

Brooks, D. W., and Dawson, C. R., 1966, Aspects of tyrosinase chemistry, in: "The Biochemistry of Copper," J. Peisach, P. Aisen, and W. Blumberg, eds., Academic Press, New York.

Dietler, C., and Lerch, K., in "Proceedings of the Third International Symposium on Oxidases and Related Redox Systems," (Albany, New York, Aug. 1979), quoted by Solomon (1981). (see below)

Duckworth, H. W., and Coleman, J. E., 1970, Physicochemical and kinetic properties of mushroom tyrosinase, J. Biol. Chem. 245:1613.

Golan-Goldhirsh, A., Kahn, V., and Whitaker, J. R., 1982, Immunoglobulin (Igg) affinity chromatography of mushroom tyrosinase. In preparation.

Ingraham, L. L., 1956, Effect of ascorbic acid on polyphenoloxidase, J. Amer. Chem. Soc. 78:5095.

Jolley, R. L., Robb, D. A., and Mason, H. S., 1969, The multiple forms of mushroom tyrosinase. Association-dissociation phenomena, J. Biol. Chem. 244:1593.

Kahn, V., Lindner, P., Golan-Goldhirsh, A., and Whitaker, J. R., 1982, Evidence against the involvement of superoxide ions in the hydroxylation of tyrosine by mushroom tyrosinase, Phytochem. In Press.

Krueger, R. L., 1950, The effect of ascorbic acid on the enzymatic
    oxidation of monohydric and o-dihydric phenols, J. Amer. Chem.
    Soc. 72:5582.
Lerch, K., 1978, Amino acid sequence of tyrosinase from Neurospora
    crassa, Proc. Natl. Acad. Sci. USA 75:3635.
Lerch, K., 1981, Structure and function of an unusual thioether in
    Neurospora tyrosinase, in: "Invertebrate Oxygen-Binding
    Proteins. Structure, Active Site and Function," J. Lamy and J.
    Lamy, eds., Marcel Dekker, Inc., New York and Basel.
Letts, D., and Chase, T., Jr., 1974. Chemical modification of
    mushroom tyrosinase for stabilization to reaction inactivation,
    Adv. Exp. Med. Biol. 42:317.
Livingston, D. M., 1974, Immunoaffinity chromatography of proteins,
    in: "Methods in Enzymology," Vol. 34, W. E. Jakoby and M.
    Wilchek, eds., Academic Press, New York.
Markakis, P., and Embs, R. J., 1966, Effect of sulfite and ascorbic
    acid on mushroom phenol oxidase, J. Food Sci. 31:807.
Mihályi, K., and Vámos-Vigyázó, L., 1976, A method for determining
    polyphenoloxidase activity in fruits and vegetables applying a
    natural substrate, Acta Alimen. 5:69.
Nelson, J. M., and Dawson, C. R., 1944, Tyrosinase, Adv. Enzymol.
    4:99.
Padrón, M. P., Lozano, J. A., and González, A. G., 1975, Properties
    of o-diphenol:$O_2$ oxidoreductase from Musa cavendishi,
    Phytochem. 14:1959.
Pfiffner, E., Dietler, C., and Lerch, K., 1981, Coordination of the
    active site copper in Neurospora tyrosinase, in: "Invertebrate
    Oxygen-Binding Proteins. Structure, Active Site and Function,"
    J. Lamy and J. Lamy, eds., Marcel Dekker, Inc., New York and
    Basel.
Pfiffner, E., and Lerch, K., 1981, Histidine at the active site of
    Neurospora tyrosinase, Biochemistry 20:6029.
Ponting, J. C., 1954, Reversible inactivation of polyphenoloxidase,
    J. Amer. Chem. Soc. 76:662.
Rando, R. R., 1974, Chemistry and enzymology of $k_{cat}$ inhibitors,
    Science 183:320.
Robb, D. A., 1979, Subunit differences among the multiple forms of
    mushroom tyrosinase, Biochem. Soc. Transact. 7:131.
Robb, D. A., 1981a, The subunits of mushroom tyrosinase, in:
    "Invertebrate Oxygen-Binding Proteins. Structure, Active Site
    and Function," J. Lamy and J. Lamy, eds., Marcel Dekker, Inc.,
    New York and Basel.
Robb, D. A., 1981b, Molecular properties of plant tyrosinases, in:
    "Recent Advances in Biochemistry of Fruits and Vegetables," J.
    Freoid and M. Rhodes, eds., Academic Press, New York.
Robb, D. A., and Gutteridge, S., 1981, Polypeptide composition of
    two fungal tyrosinases, Phytochem. 20:1481 (1981).
Scharf, W., and Dawson, C. R., 1958, The effect of ascorbic acid on
    the inactivation of tyrosinase, J. Amer. Chem. Soc. 80:4627.

Solomon, E. I., 1981, Binuclear copper active site. Hemocyanin, tyrosinase, and type 3 copper oxidases, in: "Copper Proteins," T. S. Spiro, ed., Wiley-Interscience Publication, John Wiley and Sons, New York, Chichester, Brisbane, Toronto, Singapore.

Strothkamp, K. B., Jolley R. L., and Mason, H. S., 1976, Quaternary structure of mushroom tyrosinase, Biochem. Biophys. Res. Commun. 70:519.

Vanneste, W. H., and Züberbühler, A., 1974, Copper-containing oxygenases, in: "Molecular Mechanisms of Oxygen Activation," O. Hayaishi, ed., Academic Press, New York and London.

Walker, J. R. L., 1977, Enzymic browning in foods, Food Technol. New Zealand 12:19.

Wood, B. J. B., and Ingraham, L. L., 1965, Labelled tyrosinase from labelled substrate, Nature 205:291.

## 22

# CHEMISTRY AND SAFETY OF PLANT POLYPHENOLS

S.S. Deshpande, S.K. Sathe[1], and D.K. Salunkhe[2]

Department of Food Science
University of Illinois
Urbana, Illinois  61801

## INTRODUCTION

Phenolic compounds are widely distributed in plant kingdom and are considered to be secondary metabolites.  They do not seem to be essential for plant life, at least at the cellular level.  Plants provide nearly all the phenols found in higher animals, since the latter can not synthesize compounds with benzenoid rings from aliphatic precursors.  The present discussion is mainly confined to polymeric phenols commonly found in cereals and legumes.

To the consumer, the most evident properties of phenolic compounds are the colors and the astringent taste they impart to foods.  With few exceptions such as safrole and coumarin, most low molecular weight plant phenols have been shown to be non-toxic/non-carcinogenic in experimental animals.  Further, plant phenols form a very small portion of total food intakes under normal food consumption patterns and would not be likely to have any serious toxic or antiphysiological effects.  It is therefore reasonable to ask, why consider food safety of plant polyphenols?  Some of the reasons include:

[1]Department of Food Science, University of Arizona, Tucson, AZ 85721.

[2]Vice-Chancellor, Mahatma Phule Agricultural University, Rahuri 413 722, Maharashtra State, India.

1. Both cereals and legumes contain appreciable quantities of phenolic compounds (notably condensed tannins).
2. On a global basis, cereals and legumes make a substantially significant contribution to human diet.
3: In economically disadvantaged countries, plant foods account for the major portion (if not 100%) of human diets, thus proportionally increasing the total intake of phenolic compounds.
4. Circumstances may force relatively high intakes of unusual plant phenols.
5. In reacting with food proteins and minerals, tannins may interfere with protein utilization and mineral absorption.
6. Health consequences of long-term intakes of "above normal" levels of phenolic compounds are not fully understood.

Although vast literature data exist on phenolic compounds of cereals and legumes, attempts to evaluate their safety in human diets have just started. Two excellent reviews (Singleton, 1981; Price and Butler, 1980) on certain aspects of this topic have recently appeared and the present review draws heavily on them.

Structure and Classification of Plant Phenolics

Phenolic compounds embrace a large array of chemical compounds that have an aromatic ring bearing one or more hydroxyl groups, together with a number of other substituents. Even within the confines of the plant kingdom, Harborne and Simmonds (1964) have classified plant phenolics into 15 major groupings. This review considers the structural aspects of only those groups of phenolics that commonly occur in certain cereals, legumes, oilseed protein products, and forage plants.

The major or common phenolic constituents of plants could be divided broadly into two main groups.

1. Phenolic acids and coumarins ($C_6$-$C_1$ and $C_6$-$C_3$ structures) and
2. Flavonoid compounds, including anthocyanidins ($C_6$-$C_3$-$C_6$ structures).

Phenolic acids and coumarins: Two families of phenolic acids are widely distributed in plants - a range of substituted benzoic ($C_6$-$C_1$) acid derivatives and those derived from cinnamic ($C_6$-$C_3$) acid. Both types of phenolic acids usually occur in conjugated or esterified form. The simpler types of benzoic acid derivatives include p-hydroxybenzoic, protocatechuic, vannilic, gallic and syringic acids, and the o-hydroxy salicylic and gentisic acids (Fig. 1). The cinnamic acids: p-coumaric, caffeic, ferulic, and sinapic, are found in most oilseeds and occur frequently in the form of esters with quinic acid or sugars (Fig. 1). Chlorogenic

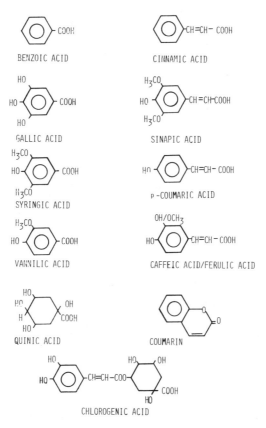

Fig. 1.   Structures of common phenolic compounds found in plant
products.

acid, which commonly occurs in sunflower seeds and coffee beans, is an ester of caffeic and quinic acids and is found in several isomeric and derivatized forms. The coumarins possess the same $C_6$-$C_3$ configuration of the cinnamic acids, but the $C_3$ chain is formed into an oxygen heterocycle (Fig. 1). They are less widely distributed in seeds and occur primarily as glucosides. They exhibit limited chemical reactivity and have little effect on organoleptic or nutritive value of seed products.

The aromatic amino acids, phenylalanine and tyrosine, occur in plants and animals where they are involved in the biosynthesis of a number of other phenolic compounds. Glycosides of dihydroxyphenyl-alanine (DOPA) occur in bean plants. DOPA is also the precursor of the brown melanin pigments of animals.

Flavonoid compounds: This group of compounds shares a basic $C_6$-$C_3$-$C_6$ structure and includes by far the largest and most diverse range of plant phenolics (Walker, 1975). Flavonoids include the red and blue anthocyanin pigments of flowers, the yellow flavones, flavonols, flavanols, and, less commonly, the aurones, chalcones, and isoflavones. Most flavonoids occur as glycosides in which the $C_6$-$C_3$-$C_6$ aglycone part of the molecule is esterified with a number of different sugars. The flavanols are unique in that they do not occur as glycosides but show reactivity through polymerization into "condensed tannins."

Polymeric phenols could be distinguished into two broad groups: tannins and lignin. Tannins comprise a heterogenous group of plant polyphenols, all of which are able to combine with skin proteins in such a way as to render them resistant to putrefaction or, in other words, to 'tan' them into leather. More specifically, tannins are high molecular weight compounds (M.W. 500-5000) containing sufficient phenolic hydroxyl groups to permit the formation of stable cross-links with proteins (Swain, 1965). Although the presence of ortho dihydroxy phenols seems to be essential (Hathway and Seakins, 1958; Bauer-Staob and Niebes, 1976), presumably to form hydrogen bonds with groups on the protein, hydrophobic binding may be an important contribution to the stability of the complex. Tannins usually give rise to a dry, puckery, astringent sensation in the mouth. Because of their protein-binding properties, tannins are of considerable importance in food processing, fruit ripening, and the manufacture of tea, cocoa, and wine.

The tannins have been classified by Freudenberg (1920) into two groups based on structural types, the hydrolyzable tannins and the condensed tannins. That scheme is the most widely accepted classi-fication so far proposed (White, 1957; Haslam, 1966; Swain, 1965; Walker, 1975). Of the two groups, the condensed tannins are the more widely distributed in higher plants (Schanderl, 1970).

Most hydrolyzable tannins contain a central core of glucose or other polyhydric alcohol esterified with gallic acid (gallotannins) or hexahydroxydiphenic acid (ellagitannins) (Fig. 2). The latter is isolated as its stable dilactone, ellagic acid. These types of tannins are readily hydrolyzed by acids, bases or certain enzymes.

The condensed tannins, also referred to as procyanidins (Weinges et al., 1969), and formerly as leucoanthocyanidins (Rosenheim, 1920) because many form cyanidin upon acid hydrolysis, are mostly flavolans or polymers of flavan-3-ols (catechins) and/or flavan 3:4-diols (leucoanthocyanidins) (Fig. 3). Both catechins and leucoanthocyanidins are readily converted by dehydrogenating enzymes or even by very dilute mineral acids at room temperature into flavonoid tannins (Weinges, 1968). Heating in acid solution converts leucoanthocyanidins to the corresponding anthocyanidins and brown "phlobaphene-like" polymers (Swain and Hillis, 1959). The condensed tannins are not readily degraded by acid treatment but polymerize to form amorphous phlobaphenes or "tannin-reds" (Haslam, 1966). A typical structure of a condensed tannin isolated from sorghum is shown in Fig. 3 (Haslam, 1977).

Lignins are complex, three-dimensional polymers of phenyl-propanoid ($C_6$-$C_3$) units which encrust and penetrate the cellulose cell walls of higher plants, thus contributing to their mechanical strength and rigidity. Alkaline hydrolysis of lignin releases a variety of benzoic and cinnamic acid derivatives as well as other unrelated compounds. Lignins from monocotyledons, dicotyledons, and gymnosperms may show slight structural differences, since different lignins are known to yield different aromatic aldehydes when subjected to oxidation by alkaline nitrobenzene (Walker, 1975).

Biogenesis of Plant Phenols

Benzenoid compounds in plants are synthesized by two main pathways: the shikimic acid pathway and the acetate-malonate pathway. In higher plants, a large number of aromatic compounds are derived from phenylalanine, tyrosine, and tryptophan, end-products of the shikimic acid pathway.

Basically, the shikimic acid pathway involves initial conden-sation of phosphoenolpyruvate (PEP) from the glycolysis process with erythrose-4-phosphate derived from the oxidative pentose phosphate cycle. A series of reactions leads to shikimic acid, which is then phosphorylated. The phosphorylated shikimic acid combines with a second molecule of PEP to yield prephenic acid via chorismic acid intermediate. Prephenic acid is then decarboxylated to form phenyl-pyruvate or p-hydroxyphenylpyruvate. On transamination, these two compounds yield phenylalanine and tyrosine, respectively.

The shikimic acid pathway contains several branch points. The

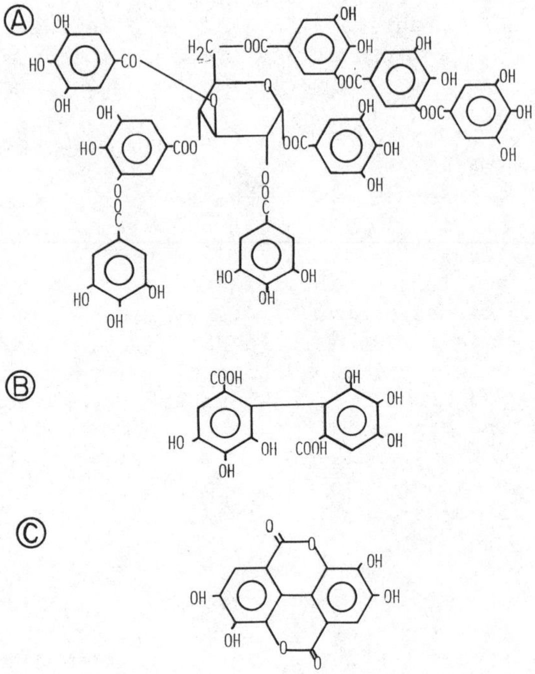

Fig. 2.
  Hydrolyzable tannins found
  in plant products.
  A.  A possible structure of
      a gallotannin.
  B.  Hexahydroxydiphenic acid,
      a constituent of ellagi-
      tannin.
  C.  Ellagic acid, a stable
      dilactone of hexahydroxy-
      diphenic acid.

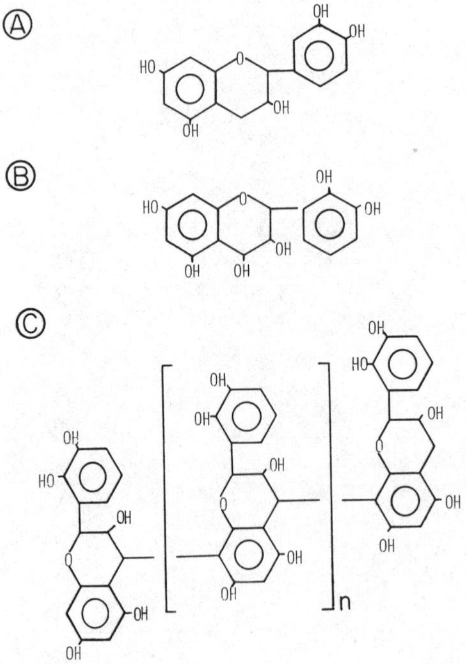

Fig. 3.
  Flavonoid constituents of con-
  densed tannins.
  A.  Flavan-3-ol (catechin).
  B.  Flavan-3:4-diol.
  C.  A possible structure of
      grain sorghum condensed
      tannin.

dehydroquinic acid can be converted either to 3-dehydroshikimate, which continues the pathway or to quinic acid (Fig. 4). Protocatechuic acid and gallic acid are synthesized by direct aromatization of 3-dehydroshikimate, although they can also be formed by hydroxylation of cinnamic and benzoic acid derivatives.

Higher plants accumulate a wide range of phenolic compounds that are derivatives or metabolites of cinnamic acid. The phenylpropanes, generally denoted as $C_6$-$C_3$ compounds, are important intermediates in the biosynthesis of lignin and flavonoids. The common phenylpropane derivatives: cinnamic acid, p-coumaric acid, caffeic acid, ferulic acid, 5-hydroxyferulic acid, and sinapic acid, are interrelated and are synthesized from phenylalanine (Fig. 5).

The widespread chlorogenic acid is an ester of quinic acid, a product of the shikimic acid pathway (Fig. 4), and caffeic acid. Chlorogenic acid is particularly abundant in coffee beans, which contain as much as 13% by weight of chlorogenic acid.

The benzoic acid derivatives such as 4-hydroxybenzoic acid, vanillic acid, and syringic acid are generally synthesized from the corresponding cinnamic acid derivatives by β-oxidation of the propenyl side chain after ester formation with coenzyme A (Fig. 5).

The coumarins are synthesized after an o-hydroxyl group is added to a cinnamic acid derivative. The o-hydroxyl group subsequently undergoes glucosylation, which facilitates the isomerization of *trans*-cinnamic acid derivatives to the *cis* form. The hydrolysis of the sugar then leads to spontaneous cyclization.

The flavonoid and isoflavonoid ring structures are of mixed biosynthetic origin, ring A being derived from three acetate units condensed head-to-tail, while ring B and the three carbon atoms of the central ring are derived from cinnamic acid (Fig. 6). As the acetate units are first converted to malonyl CoA, both the acetate-malonate and the shikimic acid pathways contribute to flavonoid biosynthesis. Several flavone compounds occurring in the plant kingdom differ slightly from one another by the number and orientation of hydroxyl, methoxyl, glycosyl, and other substituents in the aromatic rings A and B.

The first product of the addition reaction between malonyl CoA and p-coumarate, a cinnamic acid derivative, is a chalcone derivative. Chalcones can be considered as the precursors of all other classes of flavonoids (Fig. 7). In general, all flavonoid compounds originate from a $C_6$-$C_3$-$C_6$ chalcone unit, of which the central $C_3$ moiety shows structural variations leading to the formation of various types of flavonoids. The number of hydroxyl groups introduced in ring A at the stage of cyclization depends upon the degree of reduction of the triacetate chain. The first

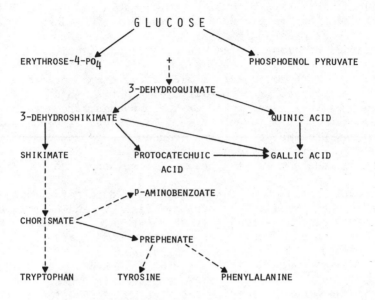

Fig. 4.   Shikimic acid pathway.

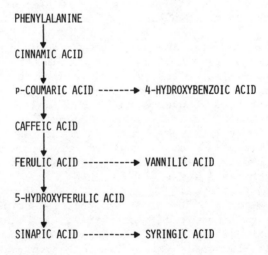

Fig. 5.   Biosynthesis of cinnamic acid and some benzoic acid
          derivatives.

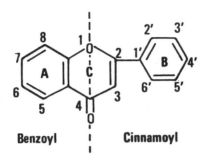

Fig. 6.   Basic flavone skeleton.

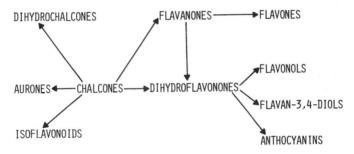

Fig. 7.   The biogenetic relationship of the flavonoids.

OH (4') in ring B is introduced even before the formation of the
$C_6-C_3-C_6$ unit and comes from the shikimic acid pathway. However,
the second hydroxylation occurs at the chalcone level (Grisebach,
1967). Similarly, C-glycosylation also occurs at this level.
Based on these steps, the biosynthetic pathway for C-glycosyl-
flavones, which commonly occur in pearl millet, is outlined in
Fig. 8.

The gallic acid (gallotannins) and ellagic acid (ellagitannins)
units of hydrolyzable tannins are derived from the shikimic acid
pathway. The mechanism by which esterification of sugar with the
various phenolic acids leads to hydrolyzable tannins depends upon
the specificity of the enzyme system involved. The diphenyl linkage
in ellagic acid and in related phenolic moieties of the ellagitannins is more likely formed by the condensation of gallic acid
and is a typical coupling reaction between phenols.

The condensed tannins are synthesized by the condensation of
two or more molecules of flavonoids such as flavan-3-ols or
flavan-3:4-diols, or a mixture of both. The flavan-3-ols exist as
two series of stereoisomers: the compounds in which the 2,3-hydrogen
atoms are *trans*, are known as catechins; when these hydrogen atoms
are *cis*, the epi-catechins result. Flavan-3:4-diols are also known
as leucoanthocyanidins because, when treated with acid, they give
the corresponding anthocyanidin. In plants, however, the latter
compounds are not biosynthesized from their leuco counterparts.

The mechanism of flavan polymerization is not known, although
several theories have been put forward. According to Freudenberg
and Weinges (1962), the flavan-3:4-diol can produce a carbonium ion
at the reactive benzylic $C_4$-OH. The ion then could react with the
nucleophilic centers of another flavan to give a $C_4-C_6$ or $C_4-C_8$
linkage. Flavan-3-ol is less reactive and can form a $C_2-C_6$ or
$C_2-C_8$ linkage. Hathway and coworkers suggested that the condensed
tannins are formed either by autoxidation (Hathway and Seakins,
1957a; Hathway, 1958a) or enzymatically by polyphenoloxidase
(Hathway and Seakins, 1957b). The products formed in each case
seem to be very similar and indistinguishable from each other.
Polyphenoloxidase brings about an oxidation of a flavan to an
o-quinone that can react with the A or B ring of another oxidized
molecule. As a result, catechin condenses to give a "head-to-tail"
polymer, while gallocatechin polymerizes in a "tail-to-tail" manner
(Hathway and Seakins, 1957b; Hathway, 1958b). Malan and Roux (1975)
considered that the condensation patterns of catechin and flavan-
3:4-diols found in several plant sources follow, most likely,
electrophilic substitution pathways. Roux and Paulus (1961)
suggested that the reactive $C_4$-hydroxyl group may be involved in
the biochemical condensation of flavan-3:4-diols. The knowledge
of structure, stereochemistry, distribution, and metabolism of plant
procyanidins or flavan-3-ol dimers and oligomers (Thompson et al.,

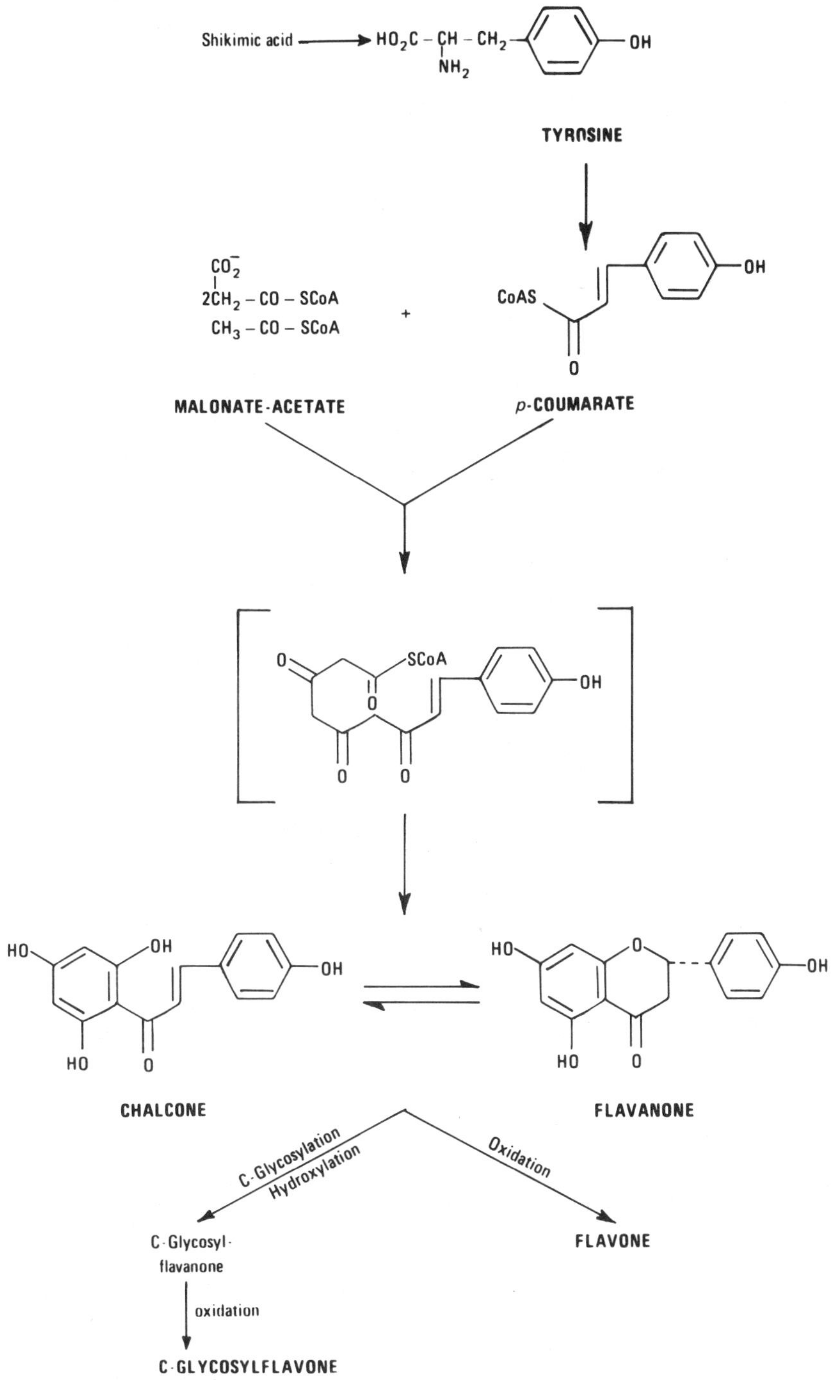

Fig. 8.   Biogenesis of C-Glycosylflavone.

1972; Jacques and Haslam, 1974; Fletcher et al., 1977; Jacques et al., 1977; Haslam et al., 1977) led Haslam (1977) to propose the biosynthetic pathways of the plant procyanidins and associated flavan-3-ols (Fig. 9). The degree of polymerization of the procyanidins appears to be characteristic of specific plants and is controlled by metabolic activities during maturity (Haslam et al., 1977).

Occurrence

The phenolic compounds are widely distributed in the plant kingdom. They vary considerably from plant to plant and may be present in small to very large amounts. Based on chemotaxonomic examination, at least 25 phenolics are of universal occurrence in animal diets derived from plants (Singleton and Kratzer, 1969). The polyphenolic tannins are of little importance in the lower orders of plants such as fungi, algae, and mosses, and in most of the monocotyledons such as the grasses (McLeod, 1974). Most of the tannins appear to be found in a few families of the dicotyledons such as the Leguminosae (White, 1957). Singleton and Kratzer (1969) also suggest that specific toxic phenols, and phenols with closely related structures, may be confined to certain families and not infrequently to certain genera, certain species, or even to certain varieties at certain times. Tannins are believed to occur in the vacuoles of intact plant cells (Kramer, 1955; Forsyth, 1964).

Of all the phenolic compounds, the polymeric phenols or the tannins seem to be the most widely distributed and are also of nutritional concern. Tannins have been reported in a variety of plants utilized for food and feed purposes. These include fruits (Van Buren, 1970); barley (Bate-Smith and Rasper, 1969); millets (Ramachandra et al., 1977; Reichert et al., 1980); sorghum (Bate-Smith and Rasper, 1969; Chavan et al., 1977; Rooney et al., 1980); peas, cowpeas, and pigeonpeas (Lindgren, 1975; Price et al., 1980); winged bean (Sathe and Salunkhe, 1981); common or dry beans (Ronnenkamp, 1977; Ma and Bliss, 1978; Bressani and Elias, 1979; Deshpande et al., 1982); horse, field, or faba beans (Marquardt et al., 1977; Martin-Tanguy et al., 1977); and forage legumes (Jones et al., 1976; McLeod, 1974).

The anthocyanidins and flavonoids seem to be responsible for the intense color of the seed pericarp. Tannins in cereals and legumes are located in the seed coat (Thayer et al., 1957; Maxson et al., 1973; Reichert et al., 1980; Ma and Bliss, 1978; Bressani and Elias, 1980; Deshpande et al., 1982). Maxson et al. (1972) reported that tannin contents were higher in grains with red pericarps than in grains with white pericarps, in both non-related material and in near isogenic lines of sorghum. They further found that the total tannin content of grain sorghum is influenced by pericarp color, presence or absence of a pigmented testa, the extent

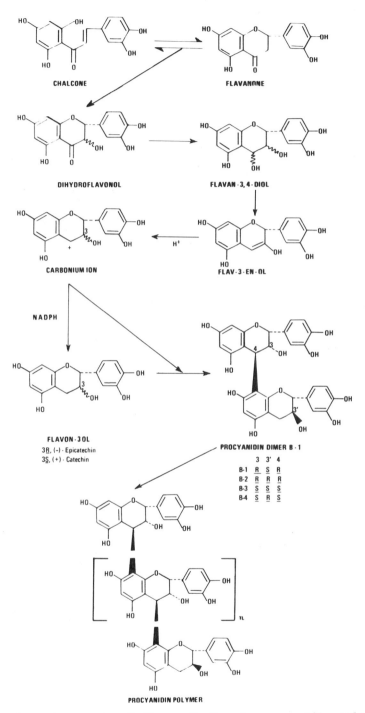

Fig. 9.   Biogenesis of flavan-3-ol and procyanidin polymer in sorghum.

to which the pigmented testa is developed, and plant color.
Similarly, a higher tannin content was reported for pigmented
cultivars of dry beans, with few or no tannins occurring in white
colored beans (Bressani and Elias, 1979; Ma and Bliss, 1978;
Deshpande et al., 1982; Price et al., 1980). It appears from these
reports that white-seeded strains contain no detectable tannins
compared to colored seeds. Ma and Bliss (1978) reported that tannins
were mostly associated with the testa layer of common dry beans, and
that black seeds were usually higher than other seeds in tannin
content. Although black seeds contained more tannin, few plants
with colored, non-black seeds had high tannin contents. These
observations were later supported by Deshpande et al. (1982).
Similarly, several researchers (Thayer et al., 1957; Blessin et al.,
1963; Stephenson et al., 1968; Harris, 1969; Damron et al., 1968;
Mabbayad and Tipton, 1975; Nelson et al., 1975) reported no direct
relationship between pericarp colors and tannin contents of grain
sorghums.

Contrary to plants, higher animals can not synthesize compounds
with benzenoid rings from aliphatic precursors, the very few
exceptions include estrone and related phenolic steroids (Singleton,
1981). Plants are the source of nearly all the phenols found in
animals. Even the phenols that are essential for animals (such as
the catechol amines and phenolic indole amines involved in nerve
action and associated effects), the vitamin E tocopherols, the
vitamin K napthoquinones or menadiones, the ubiquinone benzoquinones,
thyroxine, the tyrosine of proteins, and the tyrosine-DOPA derivatives involved in melanin pigment formation, are all drawn either
directly or indirectly from plants or are modified from an essential
plant precursor, usually phenylalanine (Singleton and Kratzer, 1969).

Functional Aspects of Phenolic Compounds in Plants

In the life cycle of plants, phenols are considered secondary
metabolites because, except for tyrosine, they do not seem to be
essential for plant life, at least at the cellular level (Neish,
1960; Ramwell et al., 1964). Phenols are probably the most important
group of substances useful in chemotaxonomic differentiation among
plant species (Harborne, 1975). This may explain why plants have
been free to develop genetically controlled, chemotaxonomically
significant, but highly variable qualitative differences between
families. Similarly, plants are free to respond to environmental
and physiological changes by qualitatively varying their phenol
synthesis (Singleton and Kratzer, 1969). Thus, phenolic compounds
probably constitute one of the most widespread and diverse groups
of secondary plant metabolites. Nonetheless, many aspects of their
functions in the life of the plants still seem to be a matter of
conjecture, as is their evolutionary significance.

One of the important roles assigned to the anthocyanin and

flavonoid pigments is that of attracting pollinating agents. Walker
(1975) suggests that even this may be a secondary development, since
anthocyanins also occur in leaves, stems, and other organs not
involved in pollination. Anthocyanin pigments are known to be
responsible for nearly all the pink, scarlet, red, mauve, violet,
and blue colors of flowers, leaves, fruits, fruit juices, and wines.
Nonetheless, flavones and flavonoids do make a significant contribution either as yellow pigments or as co-pigments to anthocyanins
(Harborne, 1967). A less obvious but commercially important function
of flavonoids follows from the fact that fruit color gives immense
aesthetic pleasure to man. As a result, cyanic fruit varieties seem
to be more popular than others. The polymers that form with time
in red wine by combination of anthocyanins and other flavonoids are
important in color retention by the wine during aging (Singleton and
Esau, 1969). Anthocyanins responsible for fruit color are thus of
considerable economic importance in influencing consumer appeal, and
the preservation of fruit color can create considerable problems for
food processors.

Phenolic compounds appear to contribute little to flavor.
However, compounds such as the tannins and chlorogenic acids may
markedly affect both the flavor and astringency of the fruit. The
majority of phenolics found in fruits have no particular taste
characteristics when tasted at low concentration in the pure form.
However, the astringency of condensed flavans (leucoanthocyanidins)
and the bitterness due to some flavonoids such as naringin and
neohesperidin of citrus fruits are unique properties (Van Buren,
1970). According to Horowitz (1964), the taste of flavonones has a
definite relation to their chemical structures. Astringency is
closely connected to the tanning reactions of condensed tannins
(Rossi and Singleton, 1966a,b) and their ability to inhibit salivary
enzymes. The relation between astringency and polyphenolics has
been comprehensively reviewed by Joslyn and Goldstein (1964).

Phenolic compounds are also involved in the enzymatic and non-
enzymatic browning of food products. The occurrence of browning
reactions is often disadvantageous in the majority of fruit processing operations. Chlorogenic acid and catechin serve as the
principal substrates for the phenolase complex that is responsible
for the enzymic browning exhibited by many fruits and vegetables.
The o-quinones formed from phenolics further enhance the intensity
of browning by oxidation of other substrates, complexing with amino
acids and protein, and polymerization. Non-enzymatic discoloration
is believed to involve metal-polyphenol complexing as reported in
the processed potato (Bate-Smith et al., 1958), cauliflower (Donath,
1962), and asparagus (DeEds and Couch, 1948), conversion of leucoanthocyanidins to pink anthocyanidins in the processed broad bean
(Dikinson et al., 1957), green bean puree (Roseman et al., 1957),
and canned Bartlett pear (Luh et al., 1960), and protein-polyphenol
complexing in chilled or stored beer (Schuster and Raab, 1961).

Another problem related to the polyphenols is their considerable reactivity towards the animal body as exerted by gossypol (isoprenoid) in cottonseed, tangeretin (methylated flavonoid) in citrus, rhein (anthroquinone) in rhubarb (Singleton and Kratzer, 1973), isoflavone in legumes (Harborne, 1967), and phlorizin (dihydrochalcone glycoside) in apples (Durkee and Poapst, 1965; Williams, 1966). On the contrary, in the processing of tea, coffee, and cocoa; certain baked cereal foods or during the manufacture of cider, the enzyme-catalyzed oxidation of the original phenolics contributes much to the desirable qualities of the finished products such as the color, flavor, and taste.

Bioflavonoids seem to have some therapeutic value. For example, the isoflavones possess weak oestrogenic activity and cause reduced fertility in livestock when fed on certain strains of subterranean clover (Walker, 1975). Dicoumarol is a haemorrhagic factor and is used medicinally to prevent blood clotting. Several aspects of bioflavonoids and flavonoids as drugs were recently reviewed by Singleton (1981).

Toxicity seems to be an important attribute of phenols in plants. Essentially, every phenolic compound has some antimicrobial and antifungal activity (Jenkins et al., 1957; Walker, 1975). Antibiotic, antiviral, anticellular, antineoplastic, and related effects of tannins, mostly in vitro, have been reported by several researchers (Alexander, 1965; Herz and Kaplan, 1968; Kreber and Einhellig, 1972; Konowalchuk and Speirs, 1976; Chan et al., 1978; Loub et al., 1973; DeOliveira et al., 1972).

It has long been suspected that phenolics might be involved in plant defenses against invading phytopathogens. One of the earliest examples of disease resistance associated with the phenolic constituents of plant tissues was observed in onions. The outer neck scales of colored onion bulbs were found to be resistant to infection by Colletotrichum circinans, whereas unpigmented varieties lacked resistance. This resistance was subsequently correlated with the presence of 3:4-dihydroxybenzoic (protocatechuic) acid in the scales. Later on, several examples of varieties of a plant species being more or less resistant to pests or pathogens depending upon their content of natural tannin or other phenols were reported (Harris and Burns, 1973; Cruickshank and Perrin, 1964; Kuc, 1966; McMillan et al., 1972; Tipton et al., 1970; Bullard and Elias, 1980). These reports suggested that the general response to attack by pathogens commonly involves production of high levels of phenols adjacent to the injury and an increase in the activity of phenolase enzymes. It is now commonly believed that the quinones and phytomelanin brown polymer complexes, which are the oxidation products of phenols, are responsible for disease resistance. Such oxidation products are often more toxic to potential invaders than are the original phenols, and their polymerization helps to seal off the injured surface and begins the healing process (Singleton, 1981). In some instances, unusual phenols and other toxic compounds are clearly a part of the normal resistance mechanism of a plant such as gossypol glands in cottonseed. In some cases, phytoalexins are produced in response to microbial or fungal attack (Cruickshank, 1963). These compounds specifically inhibit plant pathogens. According to the Phytoalexin Theory of Plant Disease, first put forward by Muller and Borger (1940), phytoalexins are absent from healthy plant, but they occur whenever the plant is invaded by a fungus and, if produced in high concentration, they are toxic to the invader. Such a resistant state is not inherited but appears as a result of infection. However, the sensitivity and subsequent speed of response of the plant to the invader seem to be genetically controlled.

Phenols appear to be involved in hypersensitive self-toxic reactions within a given plant and in allelopathic competition among plants (Muller and Chou, 1972). These self-toxic reactions include the natural inhibitors of seed germination such as coumarins, which must be leached or destroyed during dormancy period before the seed can sprout (Berrie et al., 1968; Pridham, 1960). High-tannin grain sorghum types resist adverse climatic conditions and preharvest germination (Harris and Burns, 1970). The thick testa layer of their seeds has been thought to retard or delay seed germination by retarding the absorption of water. It is quite likely that tannins present in this layer enter into the endosperm with water and retard the metabolic processes by inactivating several hydrolytic enzymes essential for seed germination.

## Nutritional Effects of Plant Phenols

The nutritional effects of plant phenolics differ greatly with the species under consideration (Arnold and Hill, 1972). Certain types of phenols are known to be quite toxic to some species while having few or no effects on others. Since carnivores have the least exposure to plant phenols, phenols seem to exert a great toxicity in strict carnivores (and insectivores), less in omnivores, and least in strict herbivores (Singleton and Kratzer, 1969). These observations were supported by comparisons of the $LD_{50}$ values collected from several sources. When a new toxin is initially contacted in small amounts, animal metabolism can often adapt in order to detoxify it during any subsequent higher exposures (Feeny, 1976). Thus owing to the evolution of efficient detoxification, the common dietary plant phenols are not considered as toxicants under normal amounts and conditions.

The condensed tannins are one possible exception. They are widespread in fruits and in certain grains. The tannin in plant materials used commercially as sources amounts to about 5-50% of the dry weight (Singleton, 1981). The pigmented varieties of cereals and legumes contain 2-4% condensed tannins, although values up to

7-8% are reported for red, high-tannin sorghum varieties (Mabbayad and Tipton, 1975; Martin-Tanguy et al., 1976). Humans also consume a number of other foods containing considerable amounts of condensed tannins, especially in beverages such as cider, cocoa, tea, and red wine. Strong versions of these beverages may have as much as a gram of tannin per liter. The intake of dimeric flavans has been indicated as about 400 mg/day in human diets (Kuhnau, 1976). The total tannin intake would be somewhat higher. The per capita consumption of red wines in some countries would guarantee an average intake of nearly that amount from this source alone (Singleton, 1981).

Most physiological investigations of tannins have been made with non-ruminants. When fed at levels that commonly occur in cereals and legumes (approximately 1-2%), tannins have depressed the growth rate and caused a poor feed efficiency ratio (PER), i.e., an increase in the grams of food required per gram of weight gain (Price and Butler, 1980). The other deleterious effects of tannins include damage to the mucosal lining of the gastrointestinal tract, alteration in the excretion of certain cations, and increased excretion of proteins and essential amino acids. Higher dietary levels (about 5%) can cause death (Singleton and Kratzer, 1969).

Since grain sorghums, beans, and carob pods are commonly used in poultry feed, most of the investigations into the nutritional aspects of tannins have been done with chickens. Although the minimum threshold levels for harmful effects of condensed tannins in poultry feed have not yet been established, several reports suggest that levels between 0.5-2.0% can cause a depressed growth rate and decreased egg production (Chang and Fuller, 1964; Vohra et al., 1966; Fuller et al., 1966; Ringrose and Morgan, 1970; Potter et al., 1967; Martin-Tanguy et al., 1977). A high mortality rate was reported in chicks when tannins constituted 3-7% of their diet (Vohra et al., 1966; Joslyn and Glick, 1969). Similar effects have followed the addition of tannins to rat and swine diets (Glick and Joslyn, 1970a,b; Schaffert et al., 1974; Thrasher et al., 1975; Wallace et al., 1974). A higher protein level in the diet seems to overcome the deleterious effects of tannins (Schaffert et al., 1974; Glick and Joslyn, 1970b; Combs and Wallace, 1976).

There is no clear evidence that tannins are harmful in ruminant diets. On the contrary, many reports suggest possible beneficial effects. Condensed tannins can effectively protect dietary protein from microbial destruction in the rumen and prevent bloat in cattle (Donnelly and Anthony, 1969; Saba et al., 1972; Driedger and Hatfield, 1972; Clark and Reid, 1974; Kendall, 1966). These beneficial effects may arise due to a lower microbial competition for starch and proteins in the rumen (Hatfield, 1970). Driedger and Hatfield (1972) observed that the treatment of soybean meal with 10% tara tannic acid reduced deamination of the meal by rumen fluid by 90%, but improved the average daily gain by and nitrogen balance in lambs.

In ruminants, the principal cause of bloat (a sometimes fatal condition of high rumen pressure) appears to be the formation of a stable foam in the reticulo-rumen (Reid et al., 1974). The foaming, which causes pressure retention, appears to be due to the soluble plant proteins in the rumen (Clark and Reid, 1974) and is prevented by protein precipitation with tannins (Jones et al., 1973). The production of gas by microbial fermentation is also inhibited. Jones et al. (1973) also reported a complete absence of soluble proteins in the rumen of cattle grazing on species that contained condensed tannins. In an another study, Jones and Mangan (1977) observed that the site of nitrogen metabolism is transferred from the rumen to the intestine of sheep when they are fed sainfoin with condensed tannins, which include the formation of tannin-protein complexes that are stable in the rumen (pH 6.5) but not in the abomasum-duodenum (pH 2.5).

Although tannins appear to protect protein from microbial attack in the rumen, their ultimate effects on ruminant nutrition may depend upon how available this bypass protein is to the animal. According to Price and Butler (1980), the same factors that cause tannins to have a deleterious effect on monogastric nutrition, will presumably be important in the post-rumen digestive tract. If the tannin-protein complex does not dissociate in the abomasum or intestine, there will be no benefit to the animal from the protein having been protected in the rumen. If it does dissociate, the liberated tannin may damage the intestinal tract or form new complexes at some point with endogenous proteins.

## Mechanisms of Tannin Toxicity

The tannins have in common a polyphenolic macromolecular nature, which binds with protein to insolubilize it. Most of the data on the toxicity and physiological effects of tannins have been obtained on mixtures that contain components representing a wide range of molecular weights and isomerism. Since highly purified and well-characterized pure tannins seldom have been available for animal testing, many researchers have resorted to commercial preparations of tannic acid, which again is a mixture of specific hydrolyzable gallotannins. Biological effects of hydrolyzable and condensed tannins vary considerably (Singleton, 1981). Since hydrolyzable tannins are relatively uncommon in human food, only the biological effects of condensed tannins will be discussed below.

Price and Butler (1980) recently summarized the deleterious effects of condensed tannins into six categories.

1.  Depression of food/feed intake,
2.  Formation of tannin-dietary protein complexes,
3.  Inhibition of digestive enzymes,
4.  Increased excretion of endogenous protein,
5.  Malfunctions in digestive tract, and
6.  Toxicity of absorbed tannin or its metabolites.

Depression of food/feed intake:  Since tannins are strongly astringent (due to their protein-binding properties), a depression of feed intake, which lowers animal productivity, is to be expected. Several studies indicate higher feed intakes and superior weight gains by animals on tannin-free diets relative to those fed on endogenous or externally added, tannin-containing diets (Ringrose and Morgan, 1970; Glick and Joslyn, 1970a,b; Vohra et al., 1966; Handler and Baker, 1944; Blakeslee and Wilson, 1979; Zombade et al., 1979).  Compared to the effects of commercial preparations of tannic acid, palatability of feeds containing naturally occurring tannins does not seem to be a major problem.  Animals fed on high-tannin sorghum diets had a higher total feed intakes than those fed low-tannin control diets, although the former showed slower growth rates (Jambunathan and Mertz, 1973; Schaffert et al., 1974; Featherston and Rogler, 1975).  Bornstein et al. (1963) also reported increased feed consumption when carob pods were added to chick diets while weight gain decreased.

Although astringency seems to be the major cause of lower feed intakes, several other factors may also contribute.  The formation of tannin-protein complexes that make the protein unavailable, inhibition of the digestive enzymes, increased synthesis of digestive enzymes due to inadequate enzymic digestion, and increased loss of endogenous proteins such as the mucoproteins of the gastrointestinal tract may be functioning, although they do not involve direct toxic effects of the tannins or the absorption of tannins into the tissues. Singleton (1981) suggests that it is unfair to consider the effects of tannins on animal feed intake as "toxicity," since the result is due to a failure to consume rather than consumption.

Tannin-dietary protein complexes:  Tannins are defined as those phenolic compounds of sufficiently high molecular weight (>500) that form reasonably strong complexes with proteins and other polymers under suitable conditions of concentration and pH (Goldstein and Swain, 1963).  This definition excludes many phenolic compounds that do not form strong associations with proteins such as phenols of low molecular weight.  White (1957) suggested that for a substance to act as a tannin, it must be small enough to penetrate the interfibrillar region of protein fibers but at the same time large enough to cross-link chains at more than one point.  As measured with gelatin, the minimum molecular weight is about 350 for effective protein precipitation (Singleton, 1981).  In the condensed series, this would begin with dimeric flavonoids.  In the

hydrolyzable series, a minimum of two gallic acid precursor units
or one ellagic acid precursor unit would be the theoretical minimum.
Along with the molecular size or molecular weight, the molecular
configuration and the number of reactive groups on the tannin may
determine the reactivity of a tannin.  Russell et al. (1968)
suggested that the removal of a tannin molecule from the tannin-
protein complex requires the simultaneous breaking of all effective
bonds between the phenolic molecule and substrate.  The number of
linkages increase with the number of reactive groups on the tannin.
The increased reactivity also indicates a corresponding increase
in spatial compatability between pairs of active sites on the
adjacent tannin and protein surfaces.  Several reports suggest that
the reactive tannins are usually of larger molecular size than the
less reactive, low molecular weight phenols (Gustavson, 1949;
Schanderl, 1970; Rossi and Singleton, 1966b).

The tannin-protein complexes are believed to be responsible
for the growth depression, low protein digestibility, and increased
fecal nitrogen.  Thus, once consumed, the deleterious effects of
tannins in the diet seem to be related to their binding of the
dietary protein.  The tannin-protein complexes may resist *in vivo*
enzymatic digestion.  Casein (Feeney, 1969), bovine serum albumin
and the Gl protein from beans (*Phaseolus vulgaris*) (Romero and
Ryan, 1978), and carob pod proteins (Tamir and Alumot, 1970) resist
proteolytic digestion when complexed with tannins.  Such complexes
may not be dissociated at the physiological pH and may pass out in
feces.  The nitrogen content of the feces generally rises in
proportion to the amount of tannin fed.

The nutritional problems caused by tannins are thus almost
certainly related in part to a decreased nitrogen retention.
Increased nitrogen excretion and/or reduced protein digestibility
have been reported for rats that have been fed tannin-containing
carob pods (Tamir and Alumot, 1970), sorghum (Featherston and
Rogler, 1975; Martin-Tanguy et al., 1976), or have had tannin added
to their basal diets (Mitjavila et al., 1977; Glick and Joslyn,
1970a,b); for chicks fed tannin-containing sorghum (Nelson et al.,
1975; Stephenson et al., 1971), barley (Eggum and Christensen,
1975), or horse beans (Martin-Tanguy et al., 1977); and for pigs
fed high-tannin sorghum (Thrasher et al., 1975, Glick and Joslyn,
1970b).

The greater affinity of tannins for proteins than for carbohydrates has been attributed to the strong hydrogen bond affinity
of the carboxyl oxygen of the peptide group (Russell et al., 1968).
However, protein is probably not the only dietary component whose
digestion and/or absorption is being adversely affected.  Featherston
and Rogler (1975) observed a decreased protein and carbohydrate
digestibility for rats and chicks fed sorghum with an above average
tannin content.  Tannins also form complexes with starch.  The

formation of such tannin-starch complexes are shown to decrease the
in vitro amylolysis of sorghum starch (Davis and Hoseney, 1979) and
several legume starches (Deshpande and Salunkhe, 1982).

Protein supplementation of the diet reportedly alleviated the
growth-depressing effects of tannic acid and dietary tannins (Glick
and Joslyn, 1970b; Schaffert et al., 1974). Even at optimum levels,
however, a poorer growth rate was observed in animals fed a tannin-
containing diet as compared to those fed on tannin-free diets. The
sulfur amino acids, notably methionine, are considered to have a
special role in detoxicative methylation of tannin breakdown products,
particularly gallic acid and the phenolic acids from condensed
tannin flavonoid fragments (Singleton, 1981). Supplementation of
tannin-containing diets with methyl donors, methionine and choline,
has counteracted a major part of the deleterious effects of dietary
condensed and hydrolyzable tannins, particularly in chicks (Vohra
et al., 1966; Armstrong et al., 1974; Fuller et al., 1967;
Featherston and Rogler, 1975).

Protein supplementation of a diet to overcome the harmful
effects of tannin may not be feasible in practical terms. In the
semi-arid tropics of Asia and Africa, high-tannin grain varieties
are more popular with the farmers because of their resistance to
bird attacks and adverse climatic conditions. On the other hand,
protein malnutrition is also a severe problem in this part of the
world. Thus, growing high-tannin grain varieties because of their
beneficial agronomic effects, may add to the already existing
nutritional problem.

Inhibition of digestive enzymes: The inhibition of digestive
enzymes by dietary tannins may be expected in view of the general
protein binding properties of tannins. This is also indicated by
a decreased digestive yield of metabolizable energy (Featherston
and Rogler, 1975; Gohl and Thomke, 1976; Griffiths and Jones, 1977).
Yapar and Clandinin (1972) observed a 57% increase in the metabolizable energy for 2-week-old chicks fed on a tannin-free rapeseed
meal diet as compared to those fed on a rapeseed meal diet
containing about 3% tannin.

The tannins, particularly of the condensed types, inhibit
several enzymes including pectinases (Hathway and Seakins, 1958),
cellulases (Griffiths and Jones, 1977), amylases (Tamir and Alumot,
1969; Davis and Hoseney, 1979; Griffiths and Moseley, 1980; Davis
and Harbers, 1974; Maxon et al., 1973), several proteolytic enzymes
(Tamir and Alumot, 1969; Milic et al., 1972; Griffiths and Moseley,
1980), β-galactosidase (Goldstein and Swain, 1965), lipases (Tamir
and Alumot, 1969), and those involved in microbial fermentation of
cereal grains (Watson, 1975). The inhibition of several enzymes
by tannins is reported to be of non-competitive type (Goldstein
and Swain, 1965; Tamir and Alumot, 1969) and is believed to be

caused by the non-specific binding of tannins with the enzyme protein.  Decreased enzyme activity may also be due to the binding of tannins with the substrate (Davis and Hoseney, 1979; Deshpande and Salunkhe, 1982).  Feeney (1969) observed 83% hydrolysis of casein when it was incubated *in vitro* with mammalian trypsin, whereas casein complexed with oak-leaf tannins was only 1% hydro- lyzed under similar conditions.

The extent of inhibition of digestive enzymes by tannins may depend upon several factors.  Price and Butler (1980) suggested that the amount of dietary protein, the formation of tannin-dietary protein complexes prior to ingestion and the extent to which they are broken down in the intestine, the relative amounts of the various enzymes and the order in which they are encountered, differing affinities between tannin components and the various enzymes and pH, the type of tannin, and the species and age of the animal may possibly influence the inhibition of digestive enzymes by tannins.

Increased excretion of endogenous protein:  As stated earlier, the presence of dietary tannin causes decreased nitrogen retention in animals.  This may be due to two factors: a lowered digestibility of dietary protein and either an inhibition of digestive enzymes or an increased excretion of endogenous proteins.  If the former were the sole cause of the tannin-induced growth depression, replacement of a significant portion of the protein with free amino acids should result in some improvement in animal growth.  However, this does not seem to be the case.  Elkin et al. (1978) studied the effects of high-tannin and low-tannin sorghums on the growth rate in chicks using a sorghum-soybean meal diet.  They observed, in fact, a significantly lower feed efficiency, when soybean meal was replaced with its equivalent in free amino acids.  Earlier, Glick and Joslyn (1970b) reported good utilization of dietary protein in rats fed rations containing 5% tannic acid and [14]C-casein.  These authors observed an increased excretion of proteins (approximately 25% of intake) by rats fed a tannin diet compared 5% by the controls.  However, only 5% of the [14]C was excreted during a 84-hour collection period.  These observations, together with their finding of a three- to four-fold increase in levels of intestinal proteolytic activity led them to conclude that proteins from enzymatic or other endogenous origins constituted a major portion of the excreted nitrogen compounds.  Mitjavila et al. (1974) also observed a high level of enzyme production after binding with tannin, which contributed to the high nitrogen levels of the feces.  Tannin consumption caused a loss in body weight and the loss of nitrogen in the feces was proportional to the loss of body weight.  The results of the above cited studies, indicating a greater total excretion of nitrogen in the feces than is ingested in the diet, suggest that not just dietary protein is involved in tannin binding.  Mitjavila et al. (1977) suggested mucus hypersecretion as the main source of the

excessive fecal nitrogen.  The loss of endogenous proteins may
sometimes result in a deficiency of one or more essential amino
acids.  This would primarily depend upon the amino acid composition
of the endogenous proteins.  Khayambashi and Lyman (1966) demonstrated the potential for selective removal of endogenous proteins
rich in essential amino acids by rats fed soybean trypsin inhibitor.
These authors reported that proteins in the small intestine
contained seven times the level of essential amino acids and
seventeen times the amount of cysteine as compared to those in
control animals.

The binding of tannins with dietary proteins may in fact
prevent other toxic effect of tannins.  Guillaume and Gomez (1977)
observed a decreased toxic effect of tannin in animals fed cooked
field beans, perhaps because they enable tannin-protein complexing
to take place.  Silano (1976) stated that, "the possible natural
occurrence of tannins in cereals in the form of tannin-protein
complexes could be considered a condition which limits the adverse
nutritional effects of cereal tannins; actually, tannin-protein
complexes may represent partially detoxified tannins."  Singleton
(1981) suggested that intimate mixing of protein and tannin prior
to their reaching more sensitive portions of the gut should
diminish direct membrane effects, absorption, and systemic toxicity
of tannins.

Effects of tannins on the digestive tract:  Few have directly
studied the deleterious effects of tannins on the alimentary canal
of animals.  The increased excretion of mucoproteins, sialic acid,
and glucosamine in the feces of rats fed on tannic acid-containing
diets suggests that tannins may have a destructive action on the
mucosal lining of the digestive tract (Mitjavila et al., 1977).
These authors also observed increased secretion of gastric and
duodenal mucus.  However, it was insufficient to protect the
mucosal lining.  Vohra et al. (1966) reported sloughing off of the
mucosa in the esophagus, subcutaneous edema, and thickening of the
crop when chicks were fed 5% tannic acid in the diet.  Thus, when
present in sufficient amounts, tannins may cause removal of mucus,
epithelial edema, irritation, and tissue breakdown of the alimentary
canal.  This may in turn facilitate greater tannin absorption, thus
increasing its toxicity.  Also, hypersecretion of gastric and
duodenal mucus by the animal may place an additional load on its
nutritional status and health.

Toxicity of absorbed tannin or its metabolites:  Reported data
on whether tannins are absorbed through the normal gastrointestinal
tissues of animals to cause toxic effects are conflicting.  Direct
absorption of protein-precipitating molecules such as tannins in
healthy animals seems unlikely, probably due to membrane barriers.
However, chronic ingestion of large amounts of tannin can damage
the gastrointestinal surface.  Under such conditions, tannins might

be absorbed, producing possible harmful effects.

The acute toxicity of tannins given orally is low, but increases greatly when tannins are administered parenterally. The $LD_{50}$ values for mice, rats, and rabbits after a single large dose of tannic acid given orally ranged from 2.25-6.0 g/kg body weight (Singleton and Kratzer, 1969). Rectal toxicity of tannic acid is about twice its oral toxicity (Boyd et al., 1965). The $LD_{50}$ values for subcutaneous or intraperitoneal administration of condensed tannins from different sources are reported to be 70-300 mg/kg body weight of the animal (Rewerski and Lewak, 1967; Patay et al., 1962). Thus most tannins appear to be quite toxic when injected.

The liver and kidneys are subjected to severe damage from tannin ingestion and injection (Price and Butler, 1980; Vohra et al., 1966). At cellular and biochemical levels, injection of tannic acid can cause liver fibrosis and necrosis, polyribosome disaggregation, inhibition of microsomal enzymes, and synthesis of nucleic acid and protein (Badawy et al., 1969; Reddy et al., 1970; Oler et al., 1976).

The monomeric units of hydrolyzable and condensed tannins may not produce similar toxic effects, since they do not have protein-precipitating properties. However, the low molecular weight breakdown products of tannins may be easily absorbed across the membrane and thus may be more toxic than the monomeric tannin units. Detailed descriptions of the absorption of various hydrolysis products of tannins and of the general toxicity of several phenols are contained in two excellent reviews (Singleton and Kratzer, 1969; Singleton, 1981). Tannins applied to burns or injected subcutaneously are reported to cause tumors in experimental animals (Korpassy, 1961; Bichel and Bach, 1968). Tannins and tannic acid have been listed as tentative carcinogens of category I under the generic carcinogen policy of OSHA (OSHA, 1978). According to Singleton (1981), carcinogenicity of tannins may involve irritation and cellular damage, resulting (over a prolonged period) in a high malignancy rate, instead of producing a true DNA mutagenic carcinogenesis.

Thus, although tannins have been implicated in carcinogenesis, the data appear to be sketchy. Epidemeological studies conducted on a global basis (Morton, 1970, 1972) suggest a possible correlation between unusual consumption of plant materials rich in condensed tannins (particularly high-tannin sorghums and the dark beers prepared from them, tea, red wines, and areca nuts in certain parts of the world) with unusual frequency of cancer of the oesophagus and mouth. Singleton (1981) suggests that such population-diet correlations, even if statistically evaluated as significant, may not indicate a true cause-and-effect relationship. Usually, environmental influence, occupation, heavy smoking, high consumption of alcoholic beverages, poor nutrition, and high use of emetics better correlate with the incidence of cancer than does diet. However, it seems possible that chronic ingestion of high amounts of tannins in the diet may increase the risk of such tumerogenic diseases.

## Attempts to Overcome the Antinutritional Effects of Tannins

Several methods have been attempted to overcome the antinutritional effects of tannins in the diet. The most common are summarized by Price and Butler (1980) along the following lines.

1. Physically remove tannins by extraction or milling,
2. Add agents that complex with dietary tannin,
3. Add agents that aid in metabolic detoxification of tannins,
4. Chemically treat food/feed, and
5. Breeding them away.

Physical removal of tannins: Tannins are characteristically located in the testa or seed coats of cereals and legumes. Thus, physical removal of seed coats, particularly of high-tannin seeds, may be expected to improve their nutritional quality. Significant reductions in tannin contents have been reported following removal of seed coats of sorghum, finger millet, and dry beans (Jambunathan and Mertz, 1973; Ramachandra et al., 1977; Deshpande et al., 1982). Several researchers have reported the possible beneficial effects of dehulled high-tannin grains in the animal diet (Featherston and Rogler, 1975; Armstrong et al., 1974; Jambunathan and Mertz, 1973). Removing seed coats (and thus the tannins therein) generally improves feed efficiency and body weight gain. The degree of improvement, however, is by no means sufficient to raise the nutritive value of high-tannin seeds to that of low-tannin types. Mechanical dehulling of seeds also results in substantial losses of protein and other nutrients. Thus the beneficial effects of tannin removal by dehulling may be partially offset by these nutrient losses.

Agents that complex with dietary tannin: Several chemicals are known to complex or interact with tannins. When used as dietary supplements, such chemicals may prevent the formation of tannin-protein complexes or liberate proteins from such complexes. Iron, caffeine, and non-ionic and cationic detergents such as Tween-80, polyvinylpyrrolidone, and polyethylene glycol, when added to the diet, reportedly overcome the growth depressing effects of tannins (Dollahit and Camp, 1962; Jease and Mitchell, 1940; Armstrong et al., 1973; Marquardt et al., 1977). Protein supplementation of tannin-containing diets also improves their nutritional quality (Schaffert et al., 1974; Glick and Joslyn, 1970a; Combs and Wallace, 1976). Although such improvement may be directly related to the nutritive value of the supplemental protein, it is possible that the added protein may cause effective removal of free tannins in the diet by complexation.

Agents that aid in metabolic detoxification of tannins:  Some
of the byproducts of phenol metabolism are excreted as monomethyl
ethers.  This could potentially deplete two primary methyl group
donors in the body, namely, methionine and choline.  Methionine has
a special role in the detoxificative methylation of tannin breakdown
products, notably gallic acid and the phenolic acids from condensed
tannin flavonoid fragments.  Several studies indicate that supplementing tannin-containing diets with the methyl donors, methionine
and choline, can counteract a major part of the antinutritional
effects of dietary tannins of both condensed and hydrolyzable types
(Armstrong et al., 1974; Fuller et al., 1967; Vohra et al., 1966;
Featherston and Rogler, 1975).  Methionine is also the first
limiting amino acid of several legume proteins.  Consumption of
high-tannin legumes, particularly in areas where they constitute a
substantial part of the diet, may therefore be expected to create
additional nutritional problems in these populations.

Chemical treatment of food/feed:  The treatment of high-tannin
cereals and legumes with chemicals such as sodium hydroxide, ammonia,
hydrogen peroxide, formaldehyde, ferrous sulfate, and ferric chloride
apparently reduces their assayable tannin contents and improves the
nutritional quality of the grains (Wah et al., 1977; Price et al.,
1979; Sathe and Salunkhe, 1981; Price et al., 1978; Bornstein et
al., 1963).  Such treatments may have many advantages over dietary
additives, which must act *in vivo* to be effective.  Price et al.
(1979) reported that moist alkaline conditions generally detoxify
tannins.  The mechanism of tannin detoxification under such
conditions, however, has not been elucidated.  Tannins may become
altered in some manner so as to become nutritionally unreactive,
perhaps by forming insoluble phlobaphenes (Swain, 1965).  Alternatively, they may become permanently bound to some nearby component
in the grain, perhaps protein, during treatment, thus rendering
both substances insoluble and nutritionally inert.  The fact that
available protein in the treated high-tannin grain increases, but
not to levels found in low-tannin grain, even though the effect of
tannin on weight gain is completely overcome, lends some support
to the idea of the formation of an indigestible tannin-protein
complex during the treatment.

Plant breeding methods:  The presence of tannins and other
pigments in the testa of cereal and legume seeds is genetically
controlled (Ma and Bliss, 1978; Quinby and Martin, 1954).  Thus it
is feasible to breed varieties with low tannin contents.  Similarly,
seeds with colored testae and low tannin contents may be obtained
through hybridization and selection.  It remains to be seen,
however, whether reducing the tannin content in the colored testa
of the seed also alters other desirable traits such as pest
resistance and preharvest germination of seeds.  Colored seeds are
greatly preferred to white coated beans by most populations in
Latin America (Bressani and Elias, 1980).  Whether the acceptability

is based only on the color of the seed coat or also on the small
amounts of tannins present is not known.  Thus it may be important,
from the consumer point of view, to determine whether phenolic
compounds in common beans and other grains are related to acceptability, before genetic selection is undertaken by plant breeders to
eliminate them.

## Detoxification of Dietary Phenols

A complete elucidation of the mechanisms involved in the
detoxification of dietary phenols is beyond the scope of this
review.  Detailed descriptions are contained in several excellent
articles (Williams, 1959; Fairburn, 1959; Ramwell et al., 1964).
The salient features are discussed below.

The metabolic fate of many simple, low molecular weight phenols
has been fairly well investigated.  However, little is known about
the detoxification of more complex, polymeric phenols.  The two
major metabolic pathways involved in the detoxification of phenols
include: conjugation of the hydroxyl group with glucuronic acid to
form aryl glucosiduronic acids.  The second one involves reactions
with sulfuric acid to form ethereal sulfates or mono-esters of
sulfuric acid.  Other minor reactions include hydroxylation of
phenols which adds an additional phenolic group to the aromatic
system.  Methylation of the phenolic hydroxyl group has also been
reported.  The reaction, however, is limited to certain phenolic
acids and heterocyclic phenols.

At higher levels of phenols in the diet, glucuronic acid
conjugation of phenols predominates over sulfate conjugation.  This
is attributed to the fact that the rate of glucuronic acid conjugation of phenols is proportional to the body levels of phenols.
Sulfate conjugation, on the other hand, is independent of the phenol
level and primarily depends upon the availability of sulfate in the
body.  Under normal circumstances, the byproducts of phenol metabolism are readily removed from the body by excretion via the
kidneys.

## CONCLUSION

Polyphenols, when present, appear predominantly in the pericarp
and/or testa of certain cereals, millets, and legumes.  Agronomically,
the high-polyphenol seed types are more immune to attack by birds
and resistant to certain diseases and to preharvest germination.
Many high-polyphenol seed types, however, are of lower nutritional
quality when compared to low-polyphenol varieties of otherwise
similar composition.

Much of the evidence of the antinutritional properties of

polyphenols derives from *in vivo* studies with laboratory and farm animals. The deleterious effects of these compounds include, among others, a lower feed efficiency, diminished body weight gain per unit weight of feed intake, reduced overall digestibility and protein utilization, and damage to the mucosal lining of the gastrointestinal tract. Tannins are also implicated in the cancer of mouth and esophagus.

Several physical and chemical methods have been attempted to overcome the harmful effects of tannins in the diet. Dehulling seems to be the most effective and least expensive method. A long-term approach and the most satisfactory one would be to breed for varieties with fewer or no tannins.

Continued research is needed to determine differences in composition, protein-binding, and enzyme inactivation properties of tannins from different sources. The mechanisms involved in the detoxification of polyphenols are not well understood. Simpler processing methods are required to improve the nutritional quality of high-polyphenol seed types. Sorghums, millets, and legumes provide the main subsistence of peoples in the under-developed countries. Any efforts to improve their nutritional quality may therefore be expected to alleviate the severe problem of malnutrition existing in this part of the world.

ACKNOWLEDGMENT

The authors gratefully acknowledge the assistance of Dr. W.C. Kleese in preparing figures, Patty Nagel for typing certain figures, and Dr. D.E. Goll for providing financial assistance for travel to the meeting. All of them are affiliated with the Department of Food Science, University of Arizona, Tucson, AZ 85721. The authors also thank Dr. Munir Cheryan of University of Illinois for providing assistance to SSD.

REFERENCES

Alexander, M., 1965, Biodegradation: Problems of molecular recalcitrance on microbial fallability, Adv. Appl. Microbiol. 7:35.
Armstrong, W.D., Featherston, W.R., and Rogler, J.C., 1973, Influence of methionine and other dietary additives on the performance of chicks fed bird resistant sorghum grain diets, Poultry Sci., 52:k592.
Armstrong, W.D., Rogler, J.C., and Featherston, W.R., 1974, Effect of tannin extraction on the performance of chicks fed bird resistant sorghum diets, Poultry Sci., 53:714.
Arnold, G.W., and Hill, J.L. 1972, Chemical factors affecting selection of food plants by ruminants, in: "Phytochemical

Ecology," J.B. Harborne, ed., Academic Press, New York.

Badawy, A.A.B., White, A.E., and Lathe, G.H., 1969, The effect of tannic acid on the synthesis of protein and nucleic acid by rat liver, Biochem. J., 113:307.

Bate-Smith, E.C., Hughes, J.C., and Swain, T., 1958, After-cooking discoloration in potatoes, Chem. Ind., p. 627.

Bate-Smith, E.C., and Rasper, V., 1969, Tannins of grain sorghum. Luteoforol (Leucoluteolinidin), 3',4,4',5,7 pentahydroxy flavan, J. Food Sci., 34:203.

Bauer-Staob, G., and Niebes, P., 1976, The binding of polyphenols (Rutin and some of its o-β-hydroxyethyl derivatives) to human serum proteins, Experientia, 32:367.

Berrie, A.M.M., Parker, W., Knights, B.A., and Hendrie, M.R., 1968, Lettuce seed germination. I. Coumarin-induced dormancy, Phytochem., 7:567.

Bichel, J., and Bach, A., 1968, Investigation on the toxicity of small chronic doses of tannic acid with special reference to possible carcinogenicity, Acta Pharmacol. Toxicol., 26:41.

Blakeslee, J.A., and Wilson, H.R., 1979, Response of hens to various dietary levels of tannic acid, Poultry Sci., 58:255.

Blessin, C.W., Vanetten, C.H., and Dimler, R.J., 1963, An examination of anthocyanogens in grain sorghums, Cereal Chem., 40:241.

Bornstein, S., Alumot, E., Nakadi, S., Nachtomi, E., and Nahari, O., 1963, Trials for improving the nutritional value of carobs for chicks, Israel J. Agric. Res., 13:25.

Boyd, E.M., Bercezky, K., and Godi, I., 1965, The acute toxicity of tannic acid administered intragastrically, Can. Med. Assoc. J., 92:1292.

Bressani, R., and Elias, L.G., 1979, The nutritional role of poly-phenols in beans, in: "Proc. Symp. Polyphenols in Cereals and Legumes," IFT, St. Louis, MO.

Bressani, R., and Elias, L.G., 1980, The nutritional role of poly-phenols in beans, in: "Polyphenols in Cereals and Legumes," J.H. Hulse, ed., IDRC, Canada.

Bullard, R.W., and Elias, D.J., 1980, Sorghum polyphenols and bird resistance, in: "Polyphenols in Cereals and Legumes," J.H. Hulse, ed., IDRC, Canada.

Chan, B.G., Waiss, A.C., Jr., and Lukefahr, M., 1978, Condensed tannin, an antibiotic chemical from Gossypium hirsutum, J. Insect Physiol., 24:113.

Chang, S.I., and Fuller, H.L., 1964, Effect of tannin content of grain sorghums on their feeding values for growing chicks, Poultry Sci., 43:30.

Chavan, J.K., Ghonsikar, C.P., and Ingle, U.M., 1977, Distribution of proteins and tannin in grain sorghum, Res. Bull. M.A.U., Parbhani, India, 1:88.

Clark, R.T.J., and Reid, C.S.W., 1974, Foamy bloat of cattle: A review, J. Dairy Sci., 57:753.

Combs, G.E., and Wallace, H.D., 1976, Grain sorghums in starter and grower diets, Fla. Agric. Esp. Sta. Res. Rept. No. A1-1976-3.

Cruickshank, I.A.M., 1963, Phytoalexins, Ann. Rev. Phytopathol., 1:351.

Cruickshank, I.A.M., and Perrin, D.R., 1964, Pathological function of phenolic compounds in plants, in: "Biochemistry of Phenolic Compounds," J.B. Harborne, ed., Academic Press, London.

Damron, B.L., Prine, G.M., and Harms, R.H., 1968, Evaluation of various bird-resistant and non-resistant varieties of grain sorghum for use in broiler diets, Poultry Sci., 47:1648.

Davis, A.B., and Harbers, L.H., 1974, Hydrolysis of sorghum grain starch by rumen microorganisms and purified α-amylase as observed by scanning electron microscopy, J. Anim. Sci., 38:900.

Davis, A.B., and Hoseney, R.C., 1979, Grain sorghum condensed tannins. I. Isolation, estimation, and selective adsorption by starch, Cereal Chem., 56:310.

DeEds, F., and Couch, J.F., 1948, Rutin in green asparagus, Food Res., 13:378.

DeOliveira, M.M., Sampaio, M.R.P., Simon, F., Gilbert, B., and Mors, W.B., 1972, Antitumor activity of condensed flavanols, An. Acad. Bras. Cienc., 44:41.

Deshpande, S.S., and Salunkhe, D.K., 1982, Interactions of tannic acid and catechin with legume starches, J. Food Sci., 47:2080.

Deshpande, S.S., Sathe, S.K., Salunkhe, D.K., Cornforth, D.P., 1982, Effects of dehulling on phytic acid, polyphenols, and enzyme inhibitors of dry beans (Phaseolus vulgaris L.). J. Food Sci., 47:1846.

Dikinson, D., Knight, M., and Rees, D.I., 1957, Varieties of broad beans suitable for canning, Chem. Ind., p. 1503.

Dollahit, J.W., and Camp, B.J., 1962, Calcium hydroxide: An antidote for tannic acid poisoning in rabbits, Amer. J. Vet. Res., 23:1271.

Donath, H., 1962, Discoloration phenomena in sterile cauliflower packs, Nahrung, 6:470.

Donnelly, E.D., and Anthony, W.B., 1969, Relationship of tannin dry matter digestibility and crude protein in Sericea lespedeza, Crop Sci., 9:361.

Driedger, A., and Hatfield, E.E., 1972, Influence of tannins on the nutritive value of soybean meal for ruminants, J. Anim. Sci., 34:465.

Durkee, A.B., and Poapst, P.A., 1965, Phenolic constituents in core tissues and ripe seed of McIntosh apples, J. Agric. Food Chem., 13:137.

Eggum, B.O., and Christensen, K.D., 1975, Influence of tannin on protein utilization in feedstuffs with special reference to barley, Breed. Seed Protein Improv. using Nucl. Tech. Proc. Res. Co-ord. Meet., 2nd, p. 135.

Elkin, R.G., Featherston, W.R., and Rogler, J.C., 1978, Investigations of leg abnormalities in chicks consuming high-tannin sorghum grain diets, Poultry Sci., 57:757.

Fairbairn, J.W., 1959, "The Pharmacology of Plant Phenolics,"

Academic Press, London.

Featherston, W.R., and Rogler, J.C., 1975, Influence of tannins on the utilization of sorghum grain by rats and chicks, Nutr. Rep. Intl., 11:491.

Feeney, P.P., 1969, Inhibitory effect of oak leaf tannins on the hydrolysis of proteins by trypsin, Phytochem., 8:2119.

Feeney, P.P., 1976, Plant apparency and chemical defense, Recent Adv. Phytochem., 10:1.

Fletcher, A.C., Porter, L.J., Haslam, E., and Gupta, R.K., 1977, Plant proanthocyanidins. Part III. Conformational and configurational studies of natural procyanidins, J. Chem. Soc. Perkin I., 14:1628.

Forsyth, W.G.C., 1964, Physiological aspects of curing plant products, Ann. Rev. Plant Physiol., 15:443.

Freudenberg, K., 1920, "Die Chemie der Naturlichen Gerbstoffe," Springer, Berlin, Germany.

Freudenberg, K., and Weinges, K., 1962, Acid-catalyzed autocondensation of hydroxyflavans. Condensed proanthocyanidins, Tetrahedron Letters, p. 1073.

Fuller, H.L., Chang, S.I., and Potter, D.K., 1967, Detoxification of dietary tannic acid by chicks, J. Nutr., 91:477.

Fuller, H.L., Potter, D.K., and Brown, A.R., 1966, The feeding value of grain sorghums in relation to their tannin content, Georgia Agric. Exp. Sta. Bull. NS., 176.

Glick, Z., and Joslyn, M.A., 1970a, Food intake depression and other metabolic effects of tannic acid in the rat, J. Nutr., 100:509.

Glick, Z., and Joslyn, M.A., 1970b, Effect of tannic acid and related compounds on the absorption and utilization of proteins in the rat, J. Nutr., 100:516.

Gohl, B., and Thomke, S., 1976, Digestibility coefficients and metabolizable energy of barley diets for layers as influenced by geographical area of production, Poultry Sci., 55:2369.

Goldstein, J.L., and Swain, T., 1963, Changes in tannins in ripening fruits, Phytochem., 2:371.

Goldstein, J.L., and Swain, T., 1965, The inhibition of enzymes by tannins, Phytochem., 4:184.

Griffiths, D.W., and Jones, D.I.H., 1977, Cellulase inhibition by tannins in the testa of field beans (Vicia faba), J. Sci. Food Agric., 28:983.

Griffiths, D.W., and Moseley, G., 1980, The effect of diets containing field beans of high or low polyphenolic content on the activity of digestive enzymes in the intestines of rats, J. Sci. Food Agric., 31:255.

Grisebach, H., 1967, "Biosynthetic Patterns in Microorganisms and Higher Plants," John Wiley and Sons, New York.

Guillaume, J., and Gomez, J., 1977, Effect des proanthocyanidines extractibles au methanesabsolu (PEMA) sur la digestibilite des proteins chez le poussin (essai preliminaire), Ann. Biol. Anim. Biochim. Biophys., 17:549.

Gustavson, K.H., 1949, Some protein-chemical aspects of tanning processes, Adv. Protein Chem., 5:353.

Handler, P., and Baker, R.D., 1944, Toxicity of orally administered tannic acids, Science, 99:393.

Harborne, J.B., 1967, "Comparative Biochemistry of Flavonoids," Academic Press, London and New York.

Harborne, J.B., 1975, Biochemical systematics of flavonoids, in: "The Flavonoids," J.B. Harborne, T.J. Mabry, and H. Mabry, eds., Academic Press, New York.

Harborne, J.B., and Simmonds, N.W., 1964, The natural distribution of the phenolic aglycones, in: Biochemistry of Phenolic Compounds," J.B. Harborne, ed., Academic Press, London.

Harris, H.B., 1969, Bird resistance in grain sorghum, Proc. 24th Ann. Corn and Sorghum Res. Conf., 24:113.

Harris, H.B., and Burns, R.E., 1970, Influence of tannin content on preharvest seed germination in sorghum, Agron. J., 62:835.

Harris, H.B., and Burns, R.E., 1973, Relationship between tannin content of sorghum grain and preharvest seed molding, Agron. J., 65:957.

Haslam, E., 1966, "Chemistry of Vegetable Tannins," Academic Press, London.

Haslam, E., 1977, Review: Symmetry and promiscuity in procyanidin biochemistry, Phytochem. 16:1625.

Haslam, E., Opie, C.T., and Porter, L.J., 1977, Procyanidin metabolism - a hypothesis, Phytochem., 16:99.

Hatfield, E.E., 1970, Achieving the potential of high quality dietary protein for ruminants, Feedstuffs, 42(52):23.

Hathway, D.E., 1958a, Autoxidation of polyphenols. IV. Oxidative degradation of the catechin autoxidative polymer, J. Chem. Soc., p. 1520.

Hathway, D.E., 1958b, Oak-bark tannins, Biochem. J., 70:34.

Hathway, D.E., and Seakins, J.W.T., 1957a, Autoxidation of poly-Phenols. III. Autoxidation in neutral aqueous solution of flavans related to catechin, J. Chem. Soc., p. 1562.

Hathway, D.E., and Seakins, J.W.T., 1957b, Enzymic oxidation of catechin to a polymer structurally related to some phloba-tannins, Biochem. J., 67:239.

Hathway, D.E., and Seakins, J.W.T., 1958, The influence of tannins on the degradation of pectin by pectinase enzymes, Biochem. J., 70:158.

Herz, F., and Kaplan, E., 1968, Effects of tannic acid on erythro-cyte enzymes, Nature, 217:1258.

Horowitz, R.M., 1964, Relation between the taste and structure of some phenolic glycosides, in: "Biochemistry of Phenolic Compounds," J.B. Harborne, ed., Academic Press, New York.

Jacques, D., and Haslam, E., 1974, Biosynthesis of plant proantho-cyanidins, J. Chem. Soc. Chem. Commun., 6:231.

Jacques, D., Opie, C.T., Porter, L.J., and Haslam, E., 1977, Plant proanthocyanidin. IV. Biosynthesis of procyanidins and observations on the metabolism of cyanidin in plants, J. Chem. Soc. Perkin I., 14:1637.

Jambunathan, R., and Mertz, E.T., 1973, Relationship between tannin
    levels, rat growth and diets of proteins in sorghum, J. Agric.
    Food Chem., 21:692.
Jease, E.J., and Mitchell, J.H., 1940, A study of tannins of
    Lespedeza sericea, 53rd Ann. Rep. S. Carol. Exp. Sta., Clemson
    Agric. College, p. 71.
Jenkins, G.L., Hartung, W.H., Hamlin, K.E., Jr., and Data, J.B.,
    1957, "The Chemistry of Organic Medicinal Products," John
    Wiley and Sons, New York.
Jones, W.T., Anderson, L.B., and Ross, M.D., 1973, Bloat in cattle,
    detection of protein precipitants (flavolans) in legumes,
    New Zealand J. Agric. Res., 16:441.
Jones, W.T., Broadhurst, R.B., and Lyttleton, J.W., 1976, The
    condensed tannins of pasture legume species, Phytochem.,
    15:1407.
Jones, W.T., and Mangan, J.L., 1977, Complexes of the condensed
    tannins of sainfoin (Onobrychis viciifolia Scop.) with fraction
    1 leaf protein and with submaxillary mucoprotein and their
    reversal by polyethylene glycol and pH, J. Sci. Food Agric.,
    28:126.
Joslyn, M.A., and Glick, Z. 1969, Comparative effects of gallo-
    tannic acid and related phenolics on the growth of rats,
    J. Nutr., 98:119.
Joslyn, M.A., and Goldstein, J.L., 1964, Astringency of fruits and
    fruit products in relation to phenolic content, Adv. Food
    Res., 13:179.
Kendall, W.A., 1966, Factors affecting foams with forage legumes,
    Crop Sci., 6:487.
Khayambashi, H., and Lyman, R.L., 1966, Growth depression and
    pancreatic and intestinal changes in rats force-fed amino
    acid diets containing soybean trypsin inhibitor, J. Nutr.,
    89:455.
Konowalchuk, J., and Speirs, J.I., 1976, Virus inactivation by
    grapes and wines, Appl. Environ. Microbiol., 32:757.
Korpassy, B., 1961, Tannins as hepatic carcinogens, Prog. Exp.
    Tumor Res., 2:245.
Kramer, P.J., 1955, "Encyclopedia of Plant Physiology," Vol. 1,
    W. Rhuland, ed., Springer, Berlin, Germany.
Kreber, R.A., and Einhellig, F.A., 1972, Effects of tannic acid
    on Drosophila larval salivary gland cells, J. Insect Physiol.,
    18:1089.
Kuc, J., 1966, Resistance of plants to infectious agents, Ann.
    Rev. Microbiol., 20:337.
Kuhnau, J., 1976, The flavonoids. A class of semi-essential food
    components: Their role in human nutrition, World Rev. Nutr.
    Dietet., 24:117.
Lindgren, E., 1975, The nutritive value of peas and field beans
    for hens, Swed. J. Agric. Res., 5:159.
Loub, W.D., Fong, H.H.S., Theiner, M., and Farnsworth, N.R., 1973,
    Partial characterization of antitumor tannin isolated from

*Calycogonium squamulosum* (Melastomatoceae), J. Pharm. Sci., 62:149.

Luh, B.S., Leonard, S.J., and Patel, D.S., 1960, Pink discoloration in canned Bartlett pears, Food Technol., 14:53.

Ma, Y., and Bliss, F.A., 1978, Tannin content and inheritance in common bean, Crop Sci., 18:201.

Mabbayad, B.B., and Tipton, K.W., 1975, Tannin concentration and *in vitro* dry matter disappearance of seeds of bird-resistant sorghum hybrids, Philipp. Agric., 59:1.

Malan, E., and Roux, D.G., 1975, Flavonoids and tannins of *Acacia* species, Phytochem., 14:1835.

Marquardt, R.R., Ward, T., Campbell, L.D., and Cansfield, P.E., 1977, Purification, identification, and characterization of growth inhibitor in faba beans (*Vicia faba* L. var. Minor), J. Nutr., 107:1313.

Martin-Tanguy, J., Vermovel, M., Lenoble, M., Martin, C., and Gallet, M., 1976, Sorghum seed tannins. Importance in nutritional nitrogen use in growing rats, Ann. Biol. Anim. Biochim. Biophys., 16:879.

Martin-Tanguy, J., Guillaume, J., and Kossa, A., 1977, Condensed tannins in horse bean seeds: Chemical structure and apparent effects on poultry, J. Sci. Food Agric., 28:757.

Maxson, E.D., Clark, L.E., Rooney, L.W., and Johnson, J.W., 1972, Factors affecting the tannin content of sorghum grain as determined by two methods of tannin analysis, Crop Sci., 12:233.

Maxson, E.D., Rooney, L.W., Lewis, R.W., Clark, L.E., and Johnson, J.W., 1973, The relationship between tannin content, enzyme inhibition, rat performance, and characteristics of sorghum grains, Nutr. Rep. Intl., 8:145.

McLeod, M.N., 1974, Plant tannins: Their role in forage quality, Nutr. Abs. Rev., 44:803.

McMillan, W.W., Wiseman, R.R., Burns, R.E., Harris, H.B., and Greene, G.L., 1972, Bird resistance in diverse germplasm of sorghum, Agron. J., 64:821.

Milic, B.L., Stojanovic, S., and Vucurevic, N., 1972, Lucerne tannins. II. Isolation of tannins from lucerne, their nature and influence on the digestive enzymes *in vitro*, J. Sci. Food Agric., 23:1157.

Mitjavila, S., Mitjavila, M.T., and Derache, R., 1974, Rise in fecal metabolic nitrogen in rats fed diets containing tannic acid, Ann. Nutr. Aliment., 28:189.

Mitjavila, S., Lacombe, C., Carrera, G., and Derache, R., 1977, Tannic acid and oxidized tannic acid on the functional state of rat intestinal epithelium, J. Nutr., 107:2113.

Morton, J.F., 1970, Tentative correlations of plant usage and esophageal cancer zones, Econ. Bot., 24:217.

Morton, J.F., 1972, Further associations of plant tannins and human cancer, Q.J. Crude Drug Res., 12:1829.

Muller, C.H., and Chou, C.H., 1972, Phytotoxins: An ecological phase of phytochemistry, in: "Phytochemical Ecology," J.B. Harborne, ed., Academic Press, London and New York.

Muller, K.O., and Borger, H., 1940, Arb. Biol. Abt., (Ansl.-Reichsanst), Berlin, Germany, 23:189.

Neish, A.C., 1960, Biosynthetic pathways of aromatic compounds, Ann. Rev. Plant Physiol., 11:55.

Nelson, T.S., Stephenson, E.L., Burgos, A., Floyd, J., and York, J.O., 1975, Effect of tannin content and dry matter digestion on energy utilization and average amino acid availability on hybrid sorghum grains, Poultry Sci., 54:1620.

Oler, A., Neal, M.W., and Mitchell, E.K., 1976, Tannic acid: Acute hepatotoxicity following administration by feeding tube, Food Cosmet. Toxicol., 14:565.

OSHA, 1978, OSHA issues tentative carcinogen list, C & E.N., 56(31);20.

Patay, R., Masquelier, J., Danon, G., and Bernard, J.P., 1962, Sur la toxicite d'un tannoide extrait des parties ligneuses de la grappe de raisin, C.R. Cong. Natl. Soc. Savantes, Sect. Sci., 86:69.

Potter, D.K., Fuller, H.L., and Blockshear, C.D., 1967, Effect of tannic acid on egg production and egg yolk mottling, Poultry Sci., 46:1307.

Price, M.L., Butler, L.G., Featherston, W.R., and Rogler, J.C., 1978, Detoxification of high tannin sorghum grain, Nutr. Rep. Intl., 17:229.

Price, M.L., Butler, L.G., Rogler, J.C., and Featherston, W.R., 1979, Overcoming the nutritionally harmful effects of tannin in sorghum grain by treatment with inexpensive chemicals, J. Agric. Food Chem., 27:441.

Price, M.L., and Butler, L.G., 1980, "Tannins and Nutrition," Purdue Univ. Agric. Exp. Stn. Bull. No. 272, W. Lafayette.

Price, M.L., Hagerman, A.E., and Butler, L.G., 1980, Tannin content of cowpeas, chick peas, pigeon peas, and mung beans, J. Agric. Food Chem., 28:459.

Pridham, J.B., 1960, "Phenolics in Plants in Health and Disease," Pergamon Press, New York.

Quinby, J.R., and Martin, J.H., 1954, Sorghum improvement, Adv. Agron., 6:305.

Ramachandra, G., Virupaksha, T.K., and Shadaksharaswamy, M., 1977, Relationship between tannin levels and in vitro protein digestibility in finger millet (Eleucine coracana Gaertn.), J. Agric. Food Chem., 25:1101.

Ramwell, P.W., Sherratt, H.S.A., and Leonard, B.E., 1964, The physiology of phenolic compounds in animals, in: "Biochemistry of Phenolic Compounds," J.B. Harborne, ed., Academic Press, London and New York.

Reddy, K.J., Chiga, M., Harris, C.C., and Svoboda, D.J., 1970, Polyribosome disaggregation in rat liver following administration of tannic acid, Cancer Res., 30:58.

Reichert, R.D., Youngs, C.G., and Christensen, D.A., 1980,

Polyphenols in *Pennisetum* millet, in: "Polyphenols in Cereals and Legumes," J.H. Hulse, ed., IDRC, Canada.

Reid, C.S.W., Ulyatt, M.J., and Wilson, J.M., 1974, Plant tannins, bloat, and nutritive value, Proc. New Zealand Soc. Anim. Prod., 34:82.

Rewerski, W., and Lewak, S., 1967, Einige pharmakologische eigenschaften der aus weissdorn (*Crataegus oxyacantha*) isolierten flavan-polymeren, Arzneim. Forsch, 17:490.

Ringrose, R.C., and Morgan, C.L., 1970, The nutritive value of *Lespedeza* for poultry feeding, S. Carol. Agric. Exp. Sta. Ann. Rept., 53:91.

Romero, J., and Ryan, D.S., 1978, Susceptibility of the major storage proteins of the bean, *Phaseolus vulgaris* L., to *in vitro* enzymatic hydrolysis, J. Agric. Food Chem., 26:784.

Ronnenkamp, R.R., 1977, The effect of tannins on nutritional quality of dry beans. Ph.D. Thesis, Purdue Univ., Lafayette.

Rooney, L.W., Blakely, M.E., Miller, F.R., and Rosenow, D.T., 1980, Factors affecting the polyphenols of sorghum and their development and location in the sorghum kernel, in: "Polyphenols in Cereals and Legumes," J.H. Hulse, ed., IDRC, Canada.

Roseman, A.S., Livingston, G.E., and Esselen, W.B., 1957, Non-enzymatic discoloration of green bean puree. II. Studies on the thermal discoloration of acetone-extracted tissues, Food Res., 22:542.

Rosenheim, O., 1920, Observations on anthocyanins, I. The anthocyanins of the young leaves of the grape vine, Biochem. J., 14:178.

Rossi, J.A., Jr., and Singleton, V.L., 1966a, Contribution of grape phenols to oxygen absorption and crowning of wines, Amer. J. Enol. Viticul., 17:231.

Rossi, J.A., Jr., and Singleton, V.L., 1966b, Flavor effects and adsorptive properties of purified fractions of grape-seed phenols, Amer. J. Enol. Viticul., 17:240.

Roux, D.G., and Paulus, E., 1961, Condensed tannins. II. Isolation of a condensed tannin from black-wattle heartwood and synthesis of (±)-7:3':4'-trihydroxyflavan-4-ol, Biochem. J., 80:476.

Russell, A.E., Shuttleworth, S.G., and Williams-Wynn, D.A., 1968. Mechanism of vegetable tannage. IV. Residual affinity phenomena on solvent extraction of collagen tanned with vegetable extracts and syntans, J. Soc. Leather Trades Chem., 52:220.

Saba, W.J., Hale, W.H., and Theurer, B., 1972, *In vitro* rumen fermentation studies with a bird resistant sorghum grain, J. Anim. Sci., 35:1076.

Sathe, S.K., and Salunkhe, D.K., 1982, Investigations on winged bean [*Psophocarpus tetragonolobus* L. DC] proteins and anti-nutritional factors, J. Food Sci., 46:1389.

Schaffert, R.S., Oswalt, D.L., and Axtell, J.A., 1974, Effect of supplemental protein on the nutritive value of high and low tannin *Sorghum bicolor* (L) Moench grain for the growing rat, J. Anim. Sci., 39:500.

Schanderl, S.H., 1970, in: "Methods in Food Analysis," 2nd ed., M.A. Joslyn, ed., Academic Press, London.

Schuster, K., and Raab, H., 1961, Polyphenol derivatives of barley and malt and their importance for the preparation of malt and beer, Brauwissenschaft, 14:306.

Silano, V., 1976, Factors affecting digestibility and availability of proteins in cereals in nutritional evaluation of cereal mutants, Proc. Advisory Group Meeting in Vienna, July 26-30, IAWA, Vienna.

Singleton, V.L., 1981. Naturally occurring food toxicants: Phenolic substances of plant origin common in foods, Adv. Food Res., 27:149.

Singleton, V.L., and Esau, P., 1969, Phenolic substances in grapes and wine, and their significance, Adv. Food Res. Supp., 1:1.

Singleton, V.L., and Kratzer, F.H., 1969, Toxicity and related physiological activity of phenolic substances of plant origin, J. Agric. Food Chem., 17:497.

Singleton, V.L., and Kratzer, F.H., 1973, Plant phenolics, in: "Toxicants Occurring Naturally in Foods," National Academy of Sciences, Washington, D.C.

Stephenson, E.L., York, J.O., and Bragg, D.B., 1968, Comparative feeding values of brown and yellow grain sorghum, Feedstuffs, 40:112.

Stephenson, E.L., York, J.O., Bragg, D.S., and Ivy, C.A., 1971, The amino acid content and availability of different strains of grain sorghum to the chick, Poultry Sci., 50:581.

Swain, T., 1965, The tannins, in: "Plant Biochemistry," J. Bonner and E.V. Varner, eds., Academic Press, New York.

Swain, T., and Hillis, W.E., 1959, Phenolic constituents of Prunus domestica: I. Quantitative analysis of phenolic constituents. J. Sci. Food Agric., 10:63.

Tamir, M., and Alumot, E., 1969, Inhibition of digestive enzymes by condensed tannins from green and ripe carobs, J. Sci. Food Agric., 20:199.

Tamir, M., and Alumot, E., 1970, Carob tannins: growth depression and levels of insoluble nitrogen in the digestive tract of rats, J. Nutr., 100:573.

Thayer, R.H., Sieglinger, J.B., and Heller, U.G., 1957, Oklahoma grain sorghum for growing chicks, Oklahoma Agric. Exp. Sta. Bull., B-487.

Thompson, R.S., Jacques, D., Haslam, E., and Tanner, R.J.M., 1972, Plant proanthocyanidins, I. Introduction. The isolation, structure, and distribution in nature of plant procyanidins, J. Chem. Soc. Perkin I., 11:1387.

Thrasher, D.M., Icaza, E.A., Ladd, W., Bagley, C.P., and Tipton, K.W., 1975, Bird resistant milo for pigs, Louisiana Agric., 19:10.

Tipton, K.W., Floyd, E.H., Marshall, J.G., and McDevitt, J.B., 1970, Resistance of certain grain sorghum hybrids to bird damage in Louisiana, Agron. J., 62:211.

Van Buren, J., 1970, Fruit phenolics, in: "The Biochemistry of
    Fruits and Their Products," Vol. 1, A.C. Hulme, ed., Academic
    Press, New York.

Vohra, P., Kratzer, F.H., and Joslyn, M.A., 1966, The growth
    depressing and toxic effects of tannins to chicks, Poultry
    Sci., 45:135.

Wah, C.S., Sharma, K., and Jackson, M.G., 1977, Studies on various
    chemical treatments of Sal seed meal or to remove inactive
    tannins, Ind. J. Anim. Sci., 47:8.

Walker, J.R.L., 1975, "The Biology of Plant Phenolics," Edward
    Arnold (Publ.) Limited, London.

Wallace, H.D., Combs, G.E., and Houser, R.H., 1974, Florida grown
    bird resistant grain sorghum for growing-finishing swine,
    Fla. Agric. Exp. Sta. Res. Rept. No. AL-1974-11.

Watson, T.G., 1975, Inhibition of microbial fermentation by sorghum
    grain and malt, J. Appl. Bacteriol., 38:133.

Weinges, K., 1968, Mechanism of the formation and the constitution
    of flavonoid tannin, J. Polymer Sci., Part C, 16:3625.

Weinges, K., Bahr, W., Ebert, W., Goritz, K., and Marx, H.D., 1969,
    Konstitution, entstehung und bedeutung der flavonoid-gerbstoffe,
    Fortschr Chem. Org. Natrustoffe, 27:158

White, T., 1957, Tannins - their occurrence and significance,
    J. Sci. Food Agric., 8:377.

Williams, A.H., 1966, Dihydrochalcones, in: "Comparative Phyto-
    chemistry," T. Swain, ed., Academic Press, New York.

Williams, R.T., 1959, "Detoxification Mechanisms," John Wiley
    and Sons, New York.

Yapar, Z., and Clandinin, D.R., 1972, Effect of tannins in
    rapeseed meal on its nutritional value for chicks, Poultry
    Sci., 51:22.                                                        ·

Zombade, S.S., Lodhi, G.N., and Ichhporani, J.S., 1979, The
    nutritional value of salseed (Shorea robusta) meal for growing
    chicks, Brit. Poultry Sci., 20:433.

## 23

GENETIC AND CARCINOGENIC EFFECTS OF PLANT FLAVONOIDS:
AN OVERVIEW

James T. MacGregor

United States Department of Agriculture
Western Regional Research Center
Berkeley, California  94710

I. Introduction

Flavonoids, a diverse group of naturally occurring plant constituents which share the common structural feature of two phenyl
rings linked by a three-carbon chain (Geissman and Hinreiner,
1952), are practically ubiquitous among vascular plants (Geissman,
1962), including food plants (Bate-Smith, 1954; Kühnau, 1976).
The three-carbon chain may be formed into a third six-membered
ring through an oxygen on one of these phenyl rings, and the
naturally occurring derivatives are frequently polyphenolic. The
structure and numbering of some common naturally-occurring flavonoids are illustrated in Figure 1.

Because of the widespread human exposure to dietary flavonoids, it was of considerable interest when it was first reported
in 1977 that quercetin (3,3',4',5,7-pentahydroxyflavone) and
kaempferol (3,4',5,7-tetrahydroxyflavone), two flavonols which
occur in glycosidic form in many food products, were mutagenic in
Salmonella typhimurium (Bjeldanes and Chang, 1977; Hardigree and
Epler, 1977; Sugimura et al., 1977). Since it had recently been
established that mutagenic activity in Salmonella typhimurium was

497

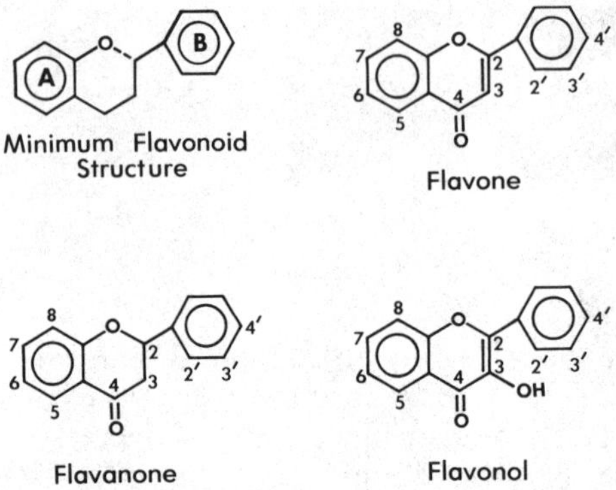

Figure 1.   Structure and numbering of some common flavonoids.

strongly correlated with carcinogenicity in mammals (McCann et al., 1975; McCann and Ames, 1975), these findings caused concern that these and related flavonoids in foods might pose a genetic and/or carcinogenic risk to humans. These concerns led to subsequent studies of the genetic and carcinogenic activity of flavonoids in various species of organisms. It is the purpose of this review to provide an overview of the information available on the genotoxicity of flavonoids and to consider the implications of the available data for human health risk.

## II. Structural and Metabolic Requirements for Flavone and Flavonol Mutagenicity in Salmonella Typhimurium

The initial work which followed the discovery of the mutagenicity of quercetin and kaempferol was directed primarily at elucidation of the structural and metabolic requirements for mutagenic activity in Salmonella, and investigation of the responses of other in vitro test systems to agents active in Salmonella. Between 1977 and 1979, more than 100 related compounds were tested for mutagenic activity in Salmonella (Bjeldanes and Chang, 1977; Brown and Dietrich, 1979; Brown, 1980; Hardigree and Epler, 1978; MacGregor and Jurd, 1978; Sugimura et al., 1977). Although it has recently been shown that there are at least two distinct classes of mutagenic flavonoids with different strain specificity and structural requirements for activity, these initial studies indicated that only flavonols (3-hydroxyflavones) were mutagenic. All active flavonols exhibited a characteristic strain specificity with maximum activity in strain TA98 and a lesser, but highly significant, response in strain TA100. The newly developed strain TA97 responds about the same as strain TA98 (unpublished observation). In strains lacking the pKM101 plasmid which codes for gene products responsible for error-prone repair of damaged DNA (Perry and Walker, 1982), activity was considerably less, but was frequently significant in frameshift tester strains TA1537 or TA1538. No activity was observed in the missense mutant strain TA1535. Table 1 illustrates the responses observed with a number of typical mutagenic flavonols.

A number of the key structural requirements for mutagenic activity of flavonols are illustrated by comparison of the structures of representative inactive compounds given in Figure 2 with the closely related active compounds in Table 1. The presence of a free hydroxyl group at position -3, the 2,3- double bond, and an intact flavone ring structure appear to be essential for activity. Flavonol itself is, however, nonmutagenic, indicating that the ring hydroxyls are necessary to permit conversion to the active mutagenic form. The compounds active without metabolic activation by mammalian liver supernatants (S9) all had free hydroxyl groups at positions 3'and 4' on the B-ring.

Table 1. Nutagenicity of Representative Flavonols in <u>Salmonella Typhimurium</u>[a]

| Name/Structure | Metabolic Activation | Strain | | | | |
| --- | --- | --- | --- | --- | --- | --- |
| | | TA98 | TA100 | TA1535 | TA1538 | TA1537 |
| QUERCETIN | S9[b] | +++ | ++ | 0 | + | + |
| | None | ++ | ++ | 0 | + | + |
| MYRICETIN | S9 | ++ | ++ | 0 | 0 | 0 |
| | None | + | + | 0 | 0 | 0 |
| RHAMNETIN | S9 | ++ | ++ | 0 | 0 | + |
| | None | + | ++ | 0 | 0 | + |
| KAEMPFEROL | S9 | ++ | ++ | 0 | 0 | + |
| | None | 0 | 0 | 0 | 0 | 0 |
| GALANGIN | S9 | ++ | ++ | 0 | 0 | + |
| | None | 0 | 0 | 0 | 0 | 0 |
| MORIN | S9 | + | + | 0 | 0 | + |
| | None | 0 | 0 | 0 | 0 | 0 |

[a]Data is from MacGregor and Jurd (1978) and Brown (1980). +++ = > 10 rev/nmole; ++ = 1-10 rev/nmole; + = significantly mutagenic but < 1 rev/nmole; 0 = not significantly mutagenic.

[b]9000xg supernatant from rat liver with cofactors required for NADPH-dependant microsomal oxidation.

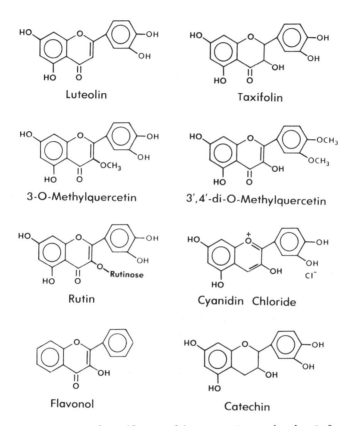

Figure 2.   Representative flavonoids nonmutagenic in <u>Salmonella</u>
<u>typhimurium</u> strains TA98 and TA100.   Based on data from references
8,9,43,55.

Compounds with no, one, or two meta hydroxyl groups have been
found to be mutagenic only in the presence of a mammalian-liver,
NADPH-dependent, metabolic activation system. These data suggest
that free ortho, or possibly para (no 2',5' hydroxy-substituted
flavonols which meet all other structural requirements for
activity have yet been tested), hydroxyl groups are required for
activity. Compounds without ortho or para hydroxyl groups may
have them introduced by the liver microsomal NADPH-dependent
oxidative metabolizing system. As first shown by Brown and
Dietrich (1979), directly mutagenic compounds with 3',4'-
hydroxyl groups exhibit enhanced activity in the presence of S9
preparations as a result of activation by nonmicrosomal enzymes
in the soluble (S100) fraction of mammalian liver. These soluble
factors are not induced by Aroclor or phenobarbital (Brown and
Dietrich, 1979), nor are they dependent on exogenously added
NADPH (unpublished observation). Although there is at present
no direct evidence which establishes the mechanism of flavonol
activation and interaction with DNA, the structural and metabolic requirements for activity, and the known chemical properties of flavonols, suggest an activation pathway involving
oxidation of the B-ring to quininoid intermediates. One scheme,
proposed by MacGregor and Jurd (1978), which appears to account
for the observed structure-activity relationships is illustrated
in Figure 3A. These structure-activity relationships and the
proposed quininoid intermediates have been discussed elsewhere
(MacGregor and Jurd, 1978; Brown, 1980).

The first suggestion of a distinct second class of flavonoids
with a different mechanism of mutagenicity was the report of
Nagao et al. (1981) that wogonin (5,7-dihydroxy-8-methoxyflavone)
was mutagenic in Salmonella typhimurium strain TA100, with little
activity in strain TA98. Not only was this strain specificity
different than that of the flavonols previously reported to be
mutagenic, but this compound does not contain the 3-hydroxyl
group which was previously thought to be necessary for activity.
We have recently examined a number of related compounds and have
found nine additional 8-hydroxy- or methoxy-substituted flavones
which exhibit a similar strain specificity in Salmonella (Elliger
et al., in press). Table 2 summarizes the activities of the ten
flavones exhibiting this type of activity. In each case, the
response of strain TA98 was either not significant, or was very
much weaker than that observed in strain TA100. The activities
of the more active compounds exceed those of quercetin, which
was previously the most active of the flavonols with reported
mutagenicity. Quercetin has a potency of 12.4 revertants/nmole
in strain TA98 and 6.1 rev/nmole in strain TA100 in the presence
of S9 mix, while norwogonin (5,7,8-trihydroxyflavone) and
sexangularetin (3,4',5,7-tetrahydroxy-8-methoxyflavone) have
potencies of 17.0 and 29.1 rev/nmole in strain TA100. Recent
unpublished observations with tester strain TA97 have shown

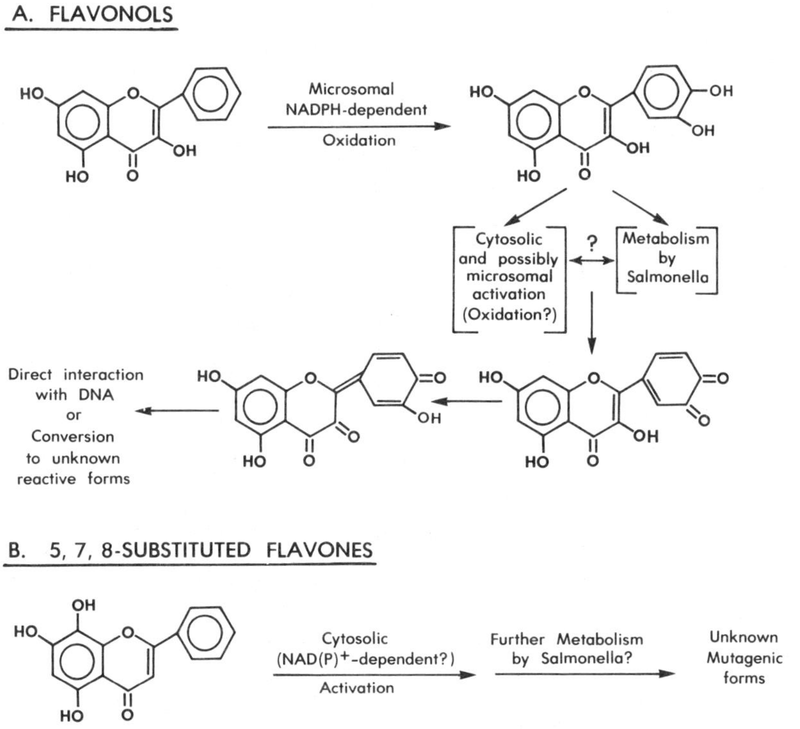

A. FLAVONOLS

B. 5, 7, 8-SUBSTITUTED FLAVONES

Figure 3. Possible routes of activation of flavones to genotoxic intermediates. See text for discussion.

that norwogonin also effectively reverts this strain.

None of these agents was significantly active in the absence
of S9 mix. The S9 activation is, however, not mediated by the
microsomal cytochrome P450 system, since activation is not diminished when microsomes are removed by centrifugation at 100,000xg
for 1 hr (Elliger et al., in press). It is likely that one or
more pyridine nucleotides may play a role in this S100-mediated
activation, since maximum activation is not observed if NADP is
omitted from the usual cofactor mix (Elliger et al., in press).

The contrasting activation requirements for a typical proto-
type of each of the three major known types of mutagenic flavonoids are illustrated in Figure 4: direct-acting flavonols with
3',4'-hydroxyl substitution are represented by quercetin; flavonols without hydroxyls at both the 3'- and 4'-positions are represented by kaempferol; and TA100-selective, S100-activated, 8-
hydroxy or -methoxy flavones by norwogonin. Quercetin is mutagenic in both strains TA98 and TA100 without an exogenous
metabolizing system, but addition of S9 markedly increases its
activity. The activation capacity of S9 is mainly due to
noninducible activity in the cytosol and is not dependent on
exogenous NADP or NADPH. Kaempferol, which has only one
hydroxyl group in the B-ring, requires an NADPH generating
system and the microsomal fraction for activity, and is much
more efficiently activated by Aroclor-1254-induced S9 than by
uninduced S9. Both quercetin and kaempferol show significantly
greater activity in strain TA98 than in strain TA100. Norwogonin is not mutagenic without metabolic activation, is activated
by factors contained in the cytosolic fraction, and has marked
activity in strain TA100 but little or no activity in strain
TA98. Addition of exogenous NADP or NADPH increases activity,
suggesting the possibility that a cytosolic redox reaction may
be responsible for the activation of norwogonin and its analogs.

The structural requirements for 8-hydroxyflavone activation
are distinct from those for flavonols such as quercetin or kaempferol. Whereas B-ring hydroxylation, 2,3-unsaturation, and 3-hy-
droxylation appear critical for flavonol mutagenicity, the B-ring
and 2,3-positions do not appear to be involved in the activation
of 8-substituted flavones. The 5,7,8-substitution pattern on the
A-ring is particularly effective at conferring high activity in
the flavone series. The seven most active compounds known at
present all have -OH or $-OCH_3$ substituents at these positions.
Chromone and acetophenone analogs with trihydroxy-substitution
analagous to the 5,7,8-A-ring positions of flavone also exhibit
S100-mediated mutagenicity which is selective for strain TA100
over TA98, although the strain specificities and activation

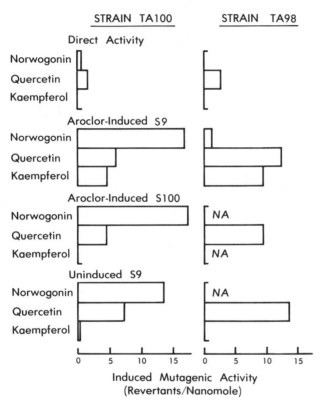

Figure 4.  Activation of norwogonin, quercetin and kaempferol in the <u>Salmonella his</u> reversion assay by rat liver fractions.

Data are from references 9,19,33 and 43.  Data for Aroclor-induced S100 and uninduced S9 are normalized by multiplying the observed ratio of activities at a single concentration of test agent by the known potency of the test agent in standard multi-dose plate assays with Aroclor-induced S9.  NA indicates that data were not available for that condition.

Table 2.  Mutagenic Activity of 8-Hydroxy- and 8-Methoxy-substituted Flavones in Salmonella Typhimurium Strains TA100, TA98 and TA1537.

| Substituent | Common Name | Revertants / Nanomole[a] | | | | | |
| --- | --- | --- | --- | --- | --- | --- | --- |
| | | TA100 | | TA98 | | TA1537 | |
| | | NoS9 | S9 | NoS9 | S9 | NoS9 | S9 |
| 5,7,8-triOH | Norwogonin | b | 17.0 | b | 1.2 | b | 1.7 |
| 5,7-diOH,8-OCH$_3$ | Wogonin | b | 13.8 | b | 0.39[d] | b | 1.6 |
| 3,4',5,7-tetraOH,8-OCH$_3$ | Sexangularetin | b | 29.2 | b | 6.6 | c | c |
| 3,5,7,8-tetraOH | 8-Hydroxygalangin | b | 5.8 | b | b | b | 0.97[e] |
| 5,8-diOH,7-OCH$_3$ | Isowogonin | b | 3.4 | b | b | b | 1.1 |
| 3,4',5,7-tetraOH,3',8-diOCH$_3$ | Limocitrin | b | 3.3 | b | b | c | c |
| 5,7,8-triOH,3-OCH$_3$ | --- | b | 2.4 | b | b | b | b |
| 8-OH | --- | b | 2.0 | b | b | b | 0.26 |
| 5,8-diOH | Primetin | b | 1.4 | b | b | b | 0.35 |
| 8-OH,5,7-diOCH$_3$ | --- | b | 0.6 | b | b | c | c |
| 3,3',4',5,7,8-hexaOH | Gossypetin | b | b | b | b | c | c |
| 3,5,8-triOH,3,4',7-triOCH$_3$ | --- | b | b | b | b | c | c |

[a] Data are from Elliger et al. (in press).  Values are initial slopes of dose-response curves fit by a least-squares linear regression.  Reported slopes differ from zero at $p < 0.005$.  Data fit the linear model at $p > 0.10$.

[b] Revertant frequency was less than twice the spontaneous background at the highest dose tested (1660 nmole/plate for 5,7-dimethoxy-8-hydroxyflavone; 166 nmole/plate for other compounds).

[c] Value not determined under this condition.

[d] Slope significant at $0.005 < p < 0.01$.

[e] Nonlinear dose-response curve.

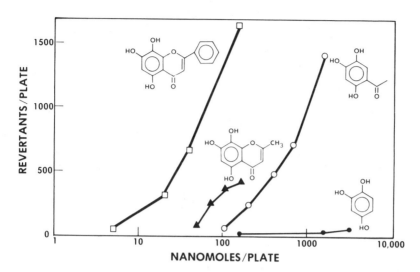

Figure 5.  Mutagenic activity of norwogowin (□), 5,7,8-trihy-droxy-2-methylchromone (▲), 2',4',5'-trihydroxyacetophenone (o), and 1,2,4-trihydroxybenzene (●) in _Salmonella_ _typhimurium_ strain TA100 with S9 activation.  From reference 19.

requirements of these acetophenones  and chromones do not paral-lel those of the flavones in all respects.  These data suggest that the B-ring and the 2,3- positions of the heterocyclic ring are not involved in the activation of these 5,7,8-substituted flavones (Figure 5).  The aryl ketone moeity, however, does appear to be critical, since 2',4',5'-trihydroxyacetophenone but not 1,2,4-trihydroxybenzene exhibits this type of activity (Fig-ure 5).

In summary, there are at least two distinct classes of flavo-noids mutagenic in _Salmonella_ _typhimurium_.  The first class is most active in strains TA98 and TA97, with lesser, but still marked, activity in strain TA100.  To date, only flavonols have exhibited this type of activity.  Activation of these flavonols may involve oxidation of hydroxyl groups in ring-B to quininoid intermediates.  The presence of a free hydroxyl at position-3 and a double bond at the 2,3-position are critical for activity.

Flavonols with free 3',4'-hydroxyl groups are frequently directly active, but liver supernatants enhance this activity, mainly via non-microsomal factors. Flavonols lacking 3',4'-hydroxyl groups are activated by a microsomal, NADP(H)-dependent step, presumably through the 3',4'-dihydroxy derivative. The second class is most active in strains TA100 and TA97, with little or no activity in strain TA98. Flavones with hydroxy- or methoxy- substitution at positions 5,7 and 8 of the A-ring are the most active known compounds of this type. Similarly-substituted chromones and acetophenones appear to have a similar type of activity. Metabolic activation is required for mutagenicity, and is mediated by cytosolic factors which may be at least partly dependent on pyridine nucleotide cofactors. The B-ring and 2,3-positions do not appear to be involved in the activation of flavones which exhibit this type of mutagenicity.

III.  Genetic Effects in Nonmammalian Species Other Than
      Salmonella Typhimurium

     Studies of mutagenic effects of flavonoids in nonmammalian species other than Salmonella typhimurium are limited. The information available, however, suggests that the mutagenic response of microorganisms exposed to flavonoids varies markedly among species and strains.

     Hardigree and Epler (1978) compared the response of the yeast Saccharomyces cerevisciae and of E. coli exposed to quercetin with that of Salmonella typhimurium. They found a three- to five-fold increase in the frequency of gene conversion in Saccharomyces and an increased reversion frequency at the nad locus in E. coli at levels of 1 mg/ml or above. No increases were observed in forward mutation to canavanine resistance or reverse mutation of the lys, his or hom loci in Saccharomyces, nor in forward mutation at the gal locus or reverse mutation at a missense arg locus in E. coli. In contrast, manyfold increases in reversion at the his D3052 locus in Salmonella typhimurium strain TA98 are obtained at quercetin concentrations of 10-20 µg/ml. In our own laboratory, we have been unable to obtain significant mutagenesis with quercetin using a variety of exposure conditions in the Bacillus subtilis multigene forward mutation assay for sporulation mutants (Sacks and MacGregor, 1982). Quercetin is the only mutagen tested thus far which is strongly mutagenic in Salmonella but negative in the B. subtilis sporulation assay (Sacks and MacGregor, in press). Thus, the mutagenic response to quercetin exposure of Salmonella typhimurium strains containing the pKM101 plasmid appears to be orders of magnitude greater than that of other microorganisms examined to date. The results taken as a whole indicate that quercetin may not be strongly genotoxic in most microorganisms.

In _Drosophila_ _melanogaster_, quercetin and kaempferol have
been reported to increase the frequency of sex-linked recessive
lethal mutations when fed at a dietary level of 0.04-0.17M in 5%
glucose solution (Watson, 1982). Since _Drosophila_ metabolizes
many promutagenic chemicals to the same mutagenic forms generated
in mammals (Vogel, 1975), and since germ cell development in
_Drosophila_ is similar to that in man, this report raises the
question of the potential of flavonoids to induce germ cell mutagenesis in man. Although the sparse pharmacokinetic data available suggest that germ cell exposure to intact flavonoids is
likely to be small, there are currently no published investigations of germ cell effects of flavonoids either in man or in
laboratory mammalian species.

## IV.  Genetic Effects in Mammalian Cells In Vitro and In Vivo

Reported studies of the genetic effects of flavonoids in
mammalian cells are limited to a few representative 3,5,7-
hydroxy- or methoxy-substituted flavones. Quercetin is the pro-
totype most frequently studied, and is the only compound for
which sufficient data exists to make meaningful comparisons
among different endpoints in various cell types. Quercetin has
been reported to increase the frequency of genetic damage in
several lines of cultured cells. The earliest studies were
carried out by Japanese workers in the late 1970's, and included
reports of mutation at the hgprt locus in V79 hamster fibroblasts
(Maruta et al., 1979), cell transformation in hamster embryo
cells (Umezawa et al., 1977), and increased frequencies of
chromosomal aberrations and sister chromatid exchange in human
and Chinese hamster cells (Yoshida et al., 1980). Other reports
of quercetin-induced genetic damage in mammalian cells in vitro
include induction of mutation at the thymidine kinase locus in
L5178Y mouse lymphoma cells (Amacher et al., 1977; Meltz and
MacGregor, 1981), chromosomal aberration in Chinese hamster ovary
cells (Carver et al., 1983), single-strand DNA breakage in L5178Y
cells (Meltz and MacGregor, 1981), and possible weak transformation of BALB/c 3T3 cells (Meltz and MacGregor, 1981).

To determine the relative importance of various types of fla-
vonoid-induced genetic damage in cultured mammalian cells, Carver
et al. (1983) measured the frequencies of chromosomal aberration,
gene mutation at four different loci, and sister chromatid exchange in single populations  of Chinese hamster ovary cells exposed to quercetin, kaempferol, and galangin (3,5,7-trihydroxy-
flavone). After fifteen hours of exposure, the frequency of
chromosomal aberration was markedly increased by all three compounds. Little or no increase in the frequencies of mutation at
the hgprt, aprt, or Na/K-ATPase loci, or in the frequency of

sister chromatid exchange, was observed in the same cell
populations. The frequency of mutation at the thymidine kinase
locus, which in other cell lines is thought to reflect chromosomal lesions as well as point mutations (Hozier et al., 1981),
was also increased significantly. The pronounced clastogenic
effects were not observed with shorter exposure periods, suggesting the possibility of secondary effects due to interference with
cell replication, rather than a direct covalent modification of
DNA by reactive intermediates. The spectrum of damage observed
in these hamster fibroblast studies is similar to that observed
with ionizing radiation, which causes marked increases in chromosomal aberrations with much weaker effects on specific locus
mutation and sister chromatid exchange (Perry and Evans, 1975;
Carrano and Thompson, 1982). This was an unexpected finding,
since interest in flavonol genotoxicity originated with the
discovery of specific locus mutation induction in Salmonella.
Although a marked increase in mutation at the hgprt locus had
earlier been reported in the closely related hamster V79 cell
line after exposure to quercetin or kaempferol (Maruta et al.,
1979), those results were obtained on an undefined population of
cells which survived two days of exposure to very high flavonol
concentrations. Those results therefore may not be representative of the initial treated population. More recently, van der
Hoeven et al. (in press) also failed to demonstrate increases in
either hgprt mutants or in the sister chromatid exchange frequency in V79 cells exposed to quercetin.

A cautious interpretation of the in vitro mammalian cell
studies reported to date is probably warranted until the characteristics of the induced genotoxicity are better understood. The
reported genetic effects are either weak, such as the effects on
cell transformation and the majority of reports of gene mutation
and sister chromatid exchange, or are observed at concentrations
known to have complex biochemical and cellular effects, which
could lead to secondary genetic effects due to disturbances in
cell replication. Effects on cell replication, DNA synthesis,
cellular cyclic AMP levels, membrane ATPases, protein kinases,
and other cellular functions have been reported at concentrations in the range of 1-20 ug/ml (see Table 4, below). Since the
reported genotoxic effects were obtained with doses of similar or
greater magnitude, often with relatively long exposure periods,
the possibility of secondary effects due to biochemical disturbances other than direct reaction with nucleic acids must be considered.

Only a few investigations of flavonoid genotoxicity in mammals in vivo have been reported. Quercetin did not increase the
frequency of sister chromatid exchange in the peripheral lymphocytes of rabbits given doses up to 250 mg/kg, i.p. (MacGregor
et al., 1983). Feeding 4% quercetin in the diet for 7 days

neither increased the frequency of nuclear anamolies in the colonic epithelium of mice, nor altered the frequency of dimethyl-hydrazine-induced colonic damage (Wargovich and Newmark, 1983).

Sahu et al. (1981) reported relatively large increases in the frequency of micronuclei in the bone marrow erythrocytes of mice following i.p. treatments with quercetin, kaempferol, or neohesperidin dihydrochalcone. These results, if confirmed, would provide evidence that flavonols are capable of causing either chromosomal breakage or anaphase chromosome lag in vivo, and would lend support to the positive carcinogenicity reports. However, we (MacGregor et al., 1983) have failed to observe increased frequencies of micronucleated erythrocytes in mice exposed to quercetin, rhamnetin, neohesperidin dihydrochalcone, or hesperetin dihydrochalcone under a variety of exposure and sampling conditions, including those used by Sahu et al. (1981). Aeshbacher et al. (1982) also failed to observe increased frequencies of micronuclei in bone marrow erythrocytes from male mice given oral doses of from 1.0 to 1000 mg/kg quercetin. Among the flavonoids tested in our laboratory, only galangin consistently increased the frequency of micronuclei in erythrocytes. The largest increases obtained, however, were only 2-3 times control values and were observed only at highly toxic doses. Galangin was much more toxic than the other flavonoids tested, and galangin-treated mice which had high micronucleus frequencies were often already moribund at the time of sampling. Since typical clastogens produce much greater frequencies of micronuclei at less toxic doses, it is uncertain whether this relatively weak effect of galangin is due to direct genotoxicity or is secondary to the marked systemic toxicity.

Cea et al.( 1983) recently reported induction of micronuclei in mouse bone marrow erythrocytes after i.p. treatment with 0.5 to 2.0 mg/kg 5,3',4'-trihydroxy-3,6,7,8-tetramethoxyflavone. These are striking results, since most flavonoids studied to date exhibit little in vivo toxicity even at considerably higer doses. It will be important to confirm these results and to test closely related flavonoids for similar genotoxic effects.


V.  Carcinogenicity Studies

The reports of flavonoid mutagenicity have prompted a number of carcinogenicity studies of quercetin and related flavonoids. The published reports of long-term dietary administration are summarized in Table 3. Quercetin is also currently under test in the Carcinogenesis Bioassay Program of the National Toxicology Program (U.S. Department of Health and Human Services, 1983), but results from this study are not yet available. Although the majority of the published studies report no carcinogenic effects

Table 3.    Carcinogenicity Studies of Quercetin and Related Flavonols

| Study | Flavonol Tested | Dietary Level | Species | Number of Each Sex | Effect on Tumor Frequency[a] |
|---|---|---|---|---|---|
| Pamukcu et al., 1980 | Quercetin | 0.1% | Rat (Norwegian-derived) | 7-18 | Increased Intestinal Tumors (20/25)<br>Increased Bladder Tumors (5/25) |
| Erturk et al., 1983 | Quercetin | 1,2% | Rat (Fisher 344) | 14-21[b] | Increased Hepatomas (9/14)<br>Increased Bile Duct Tumors (11/14)<br>Increased Liver Preneoplastic Foci (12/14) |
| Hirono et al., 1981 | Quercetin<br>Rutin | 1,5,10%<br>5,10% | Rat (ACI)<br>Rat (ACI) | 8-20<br>10-20 | No increase<br>No increase |
| Takanashi et al., 1983 | Quercetin<br>Kaempferol | 0.1%<br>0.04% | Rat (Fisher 344)<br>Rat (ACI) | 15<br>6[c] | No increase<br>No increase |
| Ambrose et al., 1952 | Quercetin<br>Quercitrin | 0.25-1%<br>0.25-1% | Rat[d]<br>Rat[d] | 5<br>5 | No increase<br>No increase |
| Saito et al., 1980 | Quercetin | 2% | Mouse (ddy) | 35 | No increase |
| Hosaka and Hirono, 1981 | Quercetin | 5% | Mouse (strain A (A/JJms)) | 23-24 | No increase in Lung Adenoma[e] |
| Morino et al., 1982 | Quercetin<br>Rutin<br>Quercetin + Croton Oil | 1,4,10%<br>10%<br>1% | Golden Hamster | 7-16<br>20<br>7 | No increase<br>No increase<br>No increase |
| Hirose et al., 1983 | Quercetin<br>Quercetin + BHBN[f] | 5%<br>5%<br>0.01% | Rat(F344) | 24<br>25 | No increase<br>No increase |

[a] Incidence in highest dose group is given in parentheses.
[b] Only females were tested.
[c] 22-30 in control groups.
[d] Strain not specified
[e] Only lung tissue was examined in this study.
[f] N-Butyl-N-(4-hydroxybutyl)nitrosamine.

of quercetin or its glycosides at dietary levels up to 10%, two
reports indicate relatively high frequencies of tumors due to
quercetin exposure. Pamukcu et al. (1980) reported an 80%
incidence of intestinal, mainly ileal, tumors and a 20%
incidence of bladder tumors in rats fed 0.1% quercetin for 58
weeks. Ertuck et al. (1983) reported hepatoma incidences of 24
and 64% and bile-duct tumor incidences of 62 and 79% in female
Fisher 344 rats fed 1% and 2% quercetin, respectively, in the
diet. The factors responsible for the discrepancy between the
positive and negative reports of quercetin carcinogenicity are
unknown at present, but it is essential to achieve an
understanding of these factors if any meaningful estimates of
human risk are to be derived from the data.

VI.  Flavonoid-Related Genetic Risk in Humans:
     Current Knowledge and Future Research Needs

     The widespread occurrence of flavonoids among food plants
(Bate-Smith, 1954; Kühnau, 1976; Mazaki et al., 1982) makes it
important to determine if the genotoxic activities of flavonoids
reported in various test systems can occur in humans. It has
been estimated that the average human dietary intake of flavonoids is approximately one gram per day, of which about 170 mg
comprises flavones, flavonols and flavanones (Kühnau, 1976).
There are at least 2000 different naturally-occurring flavonoids
(Wollenweber and Dietz, 1981), among which about 30 have so far
been shown to cause genetic damage in the Ames Salmonella assay
(Brown, 1980; Nagao et al., 1981; Elliger et al., in press).
Unfortunately, follow-up studies of the few most active compounds in other test organisms have not yielded uniform results
either in vitro or in vivo. Very few flavonoids have been tested
for genotoxicity in mammals in vivo, and widely varying results
have been reported in the few instances of independent studies
of the same agent in a single species (i.e., micronucleus induction in the mouse and tumor induction in the rat by quercetin).

     The chemistry and biology of flavonoids is complex, and both
beneficial and detrimental effects of flavonoids in mammals have
been predicted. Some of the reported cellular and biochemical
effects of flavones and flavonols are summarized in Table 4.
Many of the reported cellular effects of flavonoids could indirectly influence the biochemical processes related to DNA replication and/or tumor progression.

     The large number of naturally-occurring flavonoids, the com-
plexity of their cellular and biochemical effects, and the variability in response reported in different test organisms make
assessment of their long-term in vivo effects in man a formidable
problem. The available evidence, however, suggests some research

Table 4.  Some Reported Biological Properties of Flavones and Flavonols

| Effect | Flavonoid | Effective Concentration (In Vitro) or Dose (In Vivo) | Reference |
|---|---|---|---|
| Induction of chromosomal aberration, mutation, and cellular transformation in mammalian cells | Quercetin, kaempferol, galangin | $10^{-5} - 10^{-4}$M | 2*,12,47, 51,78,89 |
| Inhibition of glycolysis, cellular replication, and mitochondrial and plasma membrane ATPases | Quercetin, kaempferol, luteolin, and related compounds | $3 \times 10^{-5}$M | 25,41 73,74 |
| Inhibition of cellular DNA synthesis | Quercetin | $10^{-5} - 10^{-4}$ | 26,59 |
| Inhibition of viral replication in HeLa cells | 4',5-dihydroxy-3,3',7-trimethoxyflavone | $10^{-6}$M | 38 |
| Inhibition of various cellular enzymes: aldose reductase, histidine decarboxylase, prostaglandin synthetase, lipoxygenase, ascorbic acid oxidase, catechol-0-methyltransferase, phosphodiesterase | Various flavones | $10^{-7} - 10^{-3}$M | 16,21,22, 28,39,45,81 |
| Elevation of cellular cyclic-AMP levels, inhibition of phosphodiesterase | Quercetin | $10^{-4}$M | 26,28 |
| Inhibition of cyclic-AMP independent protein kinase | Quercetin | $3 \times 10^{-6} - 10^{-4}$M | 27 |
| Increased efficiency of ATP-dependent sodium pump | Quercetin | $2-6 \times 10^{-5}$M | 72 |
| Inhibition of lactate transport | Quercetin, morin fisetin, and related compounds | $3-4 \times 10^{-5}$M | 6 |

Table 4. Some Reported Biological Properties of Flavones and Flavonols (Continued)

| Effect | Flavonoid | Effective Concentration (In Vitro) or Dose (In Vivo) | Reference |
|---|---|---|---|
| Inhibition of epinephrine metabolism | Quercetin | 150 mg/kg | 4 |
| Stimulation of cytochrome P450-mediated metabolism | 7,8-Benzoflavone, flavone, tangeretin, nobiletin | $10^{-5}$M | 10,13 |
| Inhibition of cytochrome P450-mediated metabolism | Quercetin, kaempferol, morin, chrysin | $10^{-5}$M | 10,13 |
| Induction of cytochrome P450 in vivo, inhibition of polycyclic aromatic hydrocarbon carcinogenicity | β-naphthoflavone, tangeretin, nobiletin | 0.2-0.3% in diet or 150-170 mg/kg, p.o. | 85,86,87 |
| Antioxidant activity | Quercetin and related compounds | $1-2 \times 10^{-4}$M | 40,62,71 |
| Metal chelation | Quercetin, rutin, morin, and related compounds | Not specified | 15,18,40 |
| Reduction of capillary permeability and fragility (Vitamin P activity) | Hesperidin, rutin, epicatechin, and related compounds | 3-4000 mg/kg | 46,68,90 |
| Inhibition of antigen-induced histamine release | Quercetin, related flavones, and other flavonoids | $1-5 \times 10^{-5}$M | 52,53 |
| Inhibition of cytotoxic T-lymphocyte induction | Quercetin | $5-10 \times 10^{-6}$M | 70 |
| Altered phospholipid metabolism and inhibition of lysosomal enzyme release in polymorphonuclear leukocytes | Quercetin | $2-10 \times 10^{-5}$M | 42 |
| Inhibition of TPA-induced tumor promotion and ornithine decarboxylase activity | Quercetin | 10-30 μmole/mouse applied to skin | 39 |

directions which may help to explain the variable results obtained in the genotoxicity and carcinogenicity tests reported to
date, and which may lead to meaningful estimates of potential
human risk.

Because it is not practical to conduct comprehensive animal
toxicity studies on a significant fraction of the known flavonoids, it is important that the mechanism(s) of the observed
genotoxic effects be elucidated.  Although differences in genotoxic response among different cell types has made generalizations about genotoxic potential difficult, these differences may
prove useful in defining the mechanisms of flavonoid genotoxicity.  Variations in response among different species of bacteria,
for example, may permit correlation of these differences in response with differences in biochemical or physiological properties among these species.  This may provide valuable clues to
the importance of such factors as metabolism, lesion processing,
cellular uptake, and alterations in the biochemistry of DNA
replication.  Systematic investigations of model flavonoids
designed to determine the importance of these factors would
therefore be valuable.  A better definition of the molecular
nature of the mutations observed in bacteria is also needed.
Most of the bacterial mutagenicity studies of flavonoids carried
out to date have been directed mainly at screening for mutagenic
activity, and very little is known about the molecular basis of
the mutations observed.

It is important to determine if the observed in vitro genotoxic effects are due to direct interaction of flavonoid-derived
intermediates with DNA, or whether the genotoxicity is secondary
to the interaction of the flavonoid with biochemical processes
which affect the fidelity of DNA replication.  The structural
and metabolic requirements for flavonol mutagenicity are consistent with a requirement for generation of DNA-reactive quininoid
intermediates (see above), but no direct evidence for direct DNA
modification by flavonoids has yet been obtained.  In mammalian
cells, such structure-activity studies are not yet available.

Very little information on the in vivo genotoxic potential
of flavonoids in mammals is available.  Convincing evidence of
in vivo genotoxicity in mammals at dose levels comparable to
those likely to be ingested in the human diet is lacking, although in vitro genotoxic effects have been observed at concentration levels which could conceivably be achieved in the human
gut (Carver et al., 1983).  Available data suggest that flavones
and flavonols are extensively metabolized in the gut (Griffiths,
1975; Griffiths, 1982; Kühnau, 1976; Scheline, 1978), are poorly
absorbed as intact flavones or flavonols (Gugler et al., 1975;

Ueno et al., 1983), and are extensively conjugated with glucu-
ronide or sulfate in vivo (Ueno et al., 1983). There is however, practically no direct data on the levels of specific
dietary flavones and flavonols, and their metabolites, in the
gut and other tissues of the human after dietary ingestion of
flavones as their naturally-occurring glycosides. Better
definition of the intestinal metabolism and absorption of flavonoids in man and other mammals is of particular importance
(Horowitz, 1981) because the intestinal flora plays a major role
in flavonoid metabolism (Griffiths, 1975, 1982; Kühnau, 1976),
and because Pamukcu et al. (1980) have reported that the intestine was the major site of tumor induction following quercetin
exposure. If metabolism by gut flora were a major determinant
of quercetin carcinogenicity, this might account for the divergent findings reported by various laboratories (see section IV,
above). The intestinal flora is extremely complex, with a composition which varies among species and depends on complex interactions involving environment and diet (Savage, 1977). This is
a very important issue to resolve, since the standard laboratory
animal species may provide poor experimental models for the intestinal flora of man. The composition of the bacterial flora
of man and rodents differ considerably, and marked differences
in flavonoid metabolism between man and laboratory animals have
been noted (DeEds, 1968, Griffiths, 1982). In order to address
human risk, it will be necessary to better understand the metabolic pathways associated with both activation and inactivation
of genotoxic forms, and to determine the relative importance of
these pathways in the human.

Epidemiological studies of the relationship between flavonoid
intake and human genetic endpoints or cancer incidences have
apparently not been conducted. This is unfortunate, since
readily identifiable populations with high intakes of certain
specific flavonoids exist. Flavonoid food supplements are sold
through health food stores, apparently as a result of the older
literature claiming a beneficial "vitamin P" activity. Proto-
types of all the "genotoxic" flavonoids described above are represented among the products commonly available. The representative products illustrated in Figure 6 were obtained at a nearby
health food store. Included are rutin tablets (recommended dose
1 gm/day), bioflavonoid complex from orange, lemon and grapefruit
(presumably from peels) (recommended dose 1 gm/day), bioflavonoid
complex plus rutin (recommended dose 500 mg citrus bioflavonoids
plus 500 mg rutin/day), skullcap tablets (recommended dose 425 mg
or more/day), and skullcap tea. Rutin is the 3-rhamnoglucosyl
derivative of quercetin; wogonin, norwogonin, and related compounds occur in skullcap (Scutellaria) species (Popova et al.,
1973, 1979); citrus is high in various flavones and flavanones,

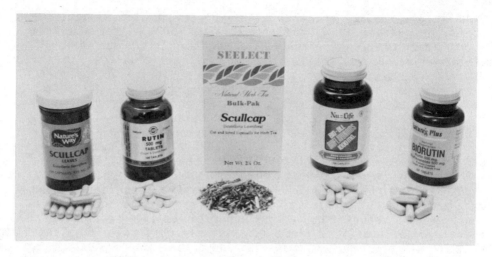

Figure 6.   Bioflavonoid products available through health food outlets.   From left to right:   Skullcap leaves, 425mg tablets; rutin, 500mg tablets; skullcap tea; citrus bioflavonoid mixture, 1000mg tablets; rutin (500mg) plus bioflavonoid (500mg) tablets.

with a somewhat lesser flavonol content (Horowitz, 1981).   Studies of the populations consuming these products for prolonged periods would be extremely useful in determining whether these flavonoids have harmful, beneficial, or neutral effects in humans.

REFERENCES

1. Aeschbacher, H-U., Meier, H. and Ruch, E. (1982).   Nonmutagenicity in vivo of the food flavonol quercetin.   Nutrition and Cancer 4, 90-98.

2. Amacher, D.E., Paillet, S. and Ray, V.A. (1979).   Point mutations at the thymidine kinase locus in L5178Y mouse lymphoma cells, I.   Application to genetic toxicology testing. Mutat. Res. 64, 391-406.

3. Ambrose, A.M., Robbins, D.J., and DeEds, F. (1952). Comparative toxicities of quercetin and quercitrin. _J. Am. Pharm. Assoc._ _41_, 119–122.

4. Axelrod, J. and Tomchick, R. (1959). Activation and inhibition of adrenaline metabolism. _Nature (Lond.)_ _184_, 2027.

5. Bate-Smith, E.C. (1954). Flavonoid compounds in foods. _Adv. Food. Res._ _5_, 261–300.

6. Belt, J.A., Thomas, J.A., Buchsbaum, R.N. and Racker, E. (1979). Inhibition of lactate transport and glycolysis in Erlich ascites tumor cells by bioflavonoids. _Biochemistry_ _18_, 3506–3511.

7. Bjeldanes, L.F. and Chang, G.W. (1977). Mutagenic activity of quercetin and related compounds. _Science_ _197_, 577–578.

8. Brown, J.P. (1980). A review of the genetic effects of naturally occurring flavonoids, anthraquinones and related compounds. _Mutat. Res._ _75_, 243–277.

9. Brown, J.P. and Dietrich, P.S. (1979). Mutagenicity of plant flavonols in the Salmonella/mammalian microsome test-Activation of flavonol glycosides by mixed glycosidases from rat cecal bacteria and other sources. _Mutat Res._ _66_, 223–240.

10. Buening, M.K., Chang, R.L., Huang, M-T., Fortner, J.G., Wood, A.W. and Conney, A.H. (1981). Activation and inhibition of benzo(a)pyrene and aflatoxin $B_1$ metabolism in human liver microsomes by naturally occurring flavonoids. _Cancer Res._ _41_, 67–72.

11. Carrano, A.V. and Thompson, L.H. (1982). Sister chromatid exchange and single gene mutation. _Cytogenet. Cell Genet._ _33_, 57–61.

12. Carver, J.H., Carrano, A.V. and MacGregor, J.T. (1983). Genetic effects of the flavonols galangin, kaempferol, and quercetin on Chinese hamster ovary cells _in vitro_. _Mutat. Res._ _113_, 45–60.

13. Conney, A.H. (1982). Induction of microsomal enzymes by foreign chemicals and carcinogenesis by polycyclic aromatic hydrocarbons. _Cancer Res._ _42_, 4875–4917.

14. Cea, G.F.A., Etcheberry, K.F.C. and Dulout, F.N. (1983). Induction of micronuclei in mouse bone marrow by the flavonoid 5,3',4'-trihydroxy-3,6,7,8-tetramethoxyflavone (THTMF). Mutat. Res. 119, 339-342.

15. Clark, W.G. and Geissman, T.A. (1949). Potentiation of effects of epinephrine by flavonoid ("vitamin P-like") compounds. Relation of structure to activity. J. Pharmacol. Exp.Ther. 95, 363-381.

16. Clemetson, C.A.B. and Anderson, L. (1966). Plant polyphenolics as antioxidants for ascorbic acid. Ann. N.Y. Acad. Sci. 136, 339-378.

17. DeEds, F. (1968). Flavonoid metabolism, in Comprehensive Biochemistry, Vol. 20, ed. M. Florkin and E.H. Stotz, Elsevier, Amsterdam, pp. 127-171.

18. Detty, W.E., Heston, B.O. and Wender, S.H. (1955). Amperometric titrations of some flavonoid compounds with cupric sulfate. J. Am. Chem. Soc. 77, 162-165.

19. Elliger, C.A., Henika, P.R. and MacGregor, J.T. (In press). Mutagenicity of flavones, chromones and acetophenones in Salmonella typhimurium: New structure-activity relationships. Mutat. Res. 00, 000-000.

20. Erturk, E., Nunoya, T., Hatcher, J.F., Pamukcu, A.M. and Bryan, G.T. (1983). Comparison of bracken fern and quercetin carcinogenicity in rats. Proc. Seventy-Fourth Ann. Mtg. Am. Assoc. Cancer Res. 24, 53 (March, 1983).

21. Fiebrich, F. and Koch, H. (1979). Silymarin, an inhibitor of lipoxygenase. Experientia 35, 1548-1550.

22. Fiebrich, F. and Koch, H. (1979). Silymarin, an inhibitor of prostaglandin synthetase. Experientia 35, 1550-1552.

23. Geissman, T.A. and Hinreiner, E. (1952). Theories of the biogenesis of flavonoid compounds. Bot. Rev. 18, 77-244.

24. Geissman, T.A. (1962). The Chemistry of Flavonoid Compounds, MacMillan, New York.

25. Graziani, Y. (1977). Bioflavonoid regulation of ATPase and hexokinase activity in Erlich ascites cell mitochondria. Biochim. Biophys. Acta. 460, 364-373.

26. Graziani, Y. and Chayoth, R. (1979). Regulation of cyclic AMP level and synthesis of DNA, RNA and protein by quercetin in Erlich ascites tumor cells. Biochem. Pharmacol. 28, 397-403.

27. Graziani, Y., Chayoth, R., Karny, N., Feldman, B. and Levy, J. (1982). Regulation of protein kinases activity by quercetin in Erlich ascites tumor cells. Biochim. Biophys. Acta. 714, 415-421.

28. Graziani, Y., Winikoff, J. and Chayoth, R. (1977). Regulation of cyclic AMP level and lactic acid production in Erlich ascites tumor cells. Biochim. Biophys. Acta 497, 499-506.

29. Griffiths, L.A. (1975). The role of the intestinal microflora in flavonoid metabolism, in Flavonoid Chem. Biochem., Proc. 4th Hung. Bioflavonoid Symp., Elsevier, Amsterdam, pp. 201-213.

30. Griffiths, L.A. (1982). Mammalian metabolism of flavonoids, in The Flavonoids: Advances in Research, ed. J.B. Harborne and T.J. Mabry, Chapman and Hall, Lond., pp. 681-718.

31. Gugler, R., Leschik, M. and Dengler, H.J. (1975). Disposition of quercetin in man after single oral and intravenous doses. Eur. J. Clin. Pharmacol. 9, 229-234.

32. Hardigree, A.A. and Epler, J.L. (1977). Mutagenicity of plant flavonols in microbial systems. Abstr. 8th Ann. Mtg. Environmental Mutagen Society, Colorado Springs, Colorado, p. 48.

33. Hardigree, A.A. and Epler, J.L. (1978). Comparative mutagenesis of plant flavonoids in microbial systems. Mutat. Res. 58, 231-239.

34. Hirono, I., Ueno, I., Hosaka, S., Takanashi, H., Matsushima, T., Sugimura, T. and Natori, S. (1981). Carcinogenicity examination of quercetin and rutin in ACI rats. Cancer Lett. 13, 15-21.

34a. Hirose, M., Fukushima, S., Sakata, T., Inui, M. and Ito, N. (1983). Effect of quercetin on two-stage carcinogenesis of the rat urinary bladder. Cancer Lett. 21, 23-27.

35. Horowitz, R.M. (1981). Flavonoids, mutagens, and citrus, in Quality of Selected Fruits and Vegetables of North America, ed. R. Teranishi and H. Barrera-Benitez, American Chemical Society, Washington, D.C., pp. 43-59.

36. Hosaka, S. and Hirono, I. (1981). Carcinogenicity test of quercetin by pulmonary adenoma bioassay in strain A mice. Gann 72, 327–328.

37. Hozier, J., Sawyer, J., Moore, M., Howard, B. and Clive, D. (1981). Cytogenetic analysis of the L5178Y/TK$^{+/-}$ → TK$^{-/-}$ mouse lymphoma mutagenesis assay system. Mutat. Res. 84, 169–181.

38. Ishitsuka, H., Ohsawa, C., Ohiwa, T., Umeda, I. and Suhara, Y. (1982). Antipicornavirus flavone Ro 09-0179. Antimicrob. Ag. Chemother. 22, 611–616.

39. Kato, R., Nakadate, T., Yamamoto, S. and Sugimura, T. (1983). Inhibition of 12-0-tetradecanoylphorbol-13-acetate-induced tumor promotion and ornithine decarboxylase activity by quercetin: possible involvement of lipoxygenase inhibition. Carcinogenesis 4, 1301–1305.

40. Kühnau, J. (1976). The flavonoids. A class of semi-essential food components: Their role in human nutrition. Wld. Rev. Nutr. Diet. 24, 117–191.

41. Lang, D.R. and Racker, E. (1974). Effects of quercetin and F$_1$ inhibitor on mitochondrial ATPase and energy-linked reactions in submitochondrial particles. Biochim. Biophys. Acta 333, 180–186.

42. Lee, T-P., Matteliano, M.L. and Middleton, E., Jr. (1982). Effect of quercetin on human polymorphonuclear leukocyte lysosomal enzyme release and phospholipid metabolism. Life Sci. 31, 2765–2774.

43. MacGregor, J.T. and Jurd, L. (1978). Mutagenicity of plant flavonoids: Structural requirements for mutagenic activity in Salmonella typhimurium. Mutat. Res. 54, 297–309.

44. MacGregor, J.T., Wehr, C.M., Manners, G.D., Jurd, L., Minkler, J.L. and Carrano, A.V. (1983). In vivo exposure to plant flavonols: Influence on frequencies of micronuclei in mouse erythrocytes and sister-chromatid exchange in rabbit lymphocytes. Mutat. Res. 124, 255–270.

45. Martin, G.J., Graff, M., Brendel, R. and Beiler, J.M. (1949). Effect of vitamin P compounds on the action of histidine decarboxylase. Arch. Biochem. 21, 177–180.

46. Martin, G.J., Szent-Györgyi, A., et al. (1955). Bioflavonoids and the capillary. Ann. N.Y. Acad. Sci. 61, 637–736.

47. Maruta, A., Enaka, K. and Umeda, M. (1979). Mutagenicity of quercetin and kaempferol on cultured mammalian cells. Gann 70, 273–276.

48. Mazaki, M., Ishii, T. and Uyeta, M. (1982). Mutagenicity of hydrolysates of citrus fruit juices. Mutat. Res. 101, 283–291.

49. McCann, J. and Ames, B.N. (1975). Detection of carcinogens as mutagens in the Salmonella/microsome test: Assay of 300 chemicals: Discussion. Proc. Nat. Acad. Sci. USA 73, 950–954.

50. McCann, J., Choi, E., Yamasaki, E. and Ames, B.N. (1975). Detection of carcinogens as mutagens in the Salmonella/microsome test: Assay of 300 chemicals. Proc. Nat. Acad. Sci. USA 72, 5135–5139.

51. Meltz, M.L. and MacGregor, J.T. (1981). Activity of the plant flavonol quercetin in the mouse lymphoma L5178Y TK$^{+/-}$ mutation, DNA single-strand break and BALB/c 3T3 chemical transformation assays. Mutat. Res. 88, 317–324.

52. Middleton, E., Jr. and Drzewiecki, G. (1982). Effects of flavonoids and transitional metal cations on antigen-induced histamine release from human basophils. Biochem. Pharmacol. 31, 1449–1453.

53. Middleton, E., Jr., Drzewiecki, G. and Krishnarao, D. (1981). Quercetin: An inhibitor of antigen-induced human basophil histamine release. J. Immunol. 127, 546–550.

54. Morino, K., Matsukura, N., Kawachi, T., Ohgaki, H., Sugimura, T. and Hirono, I. (1982). Carcinogenicity test of quercetin and rutin in golden hamsters by oral administration. Carcinogenesis 3, 93–97.

55. Nagao, M., Morita, N., Yahagi, T., Shimizu, M., Kuroyanagi, M., Fukuoka, M., Yoshihira, K., Natori, S., Fujino, T. and Sugimura, T. (1981). Mutagenicities of 61 flavonoids and 11 related compounds. Environ. Mutagenesis 3, 401–419.

56. Pamukcu, A.M., Yalciner, S., Hatcher, J.F. and Bryan, G.T. (1980). Quercetin, a rat intestinal and bladder carcinogen present in bracken fern (Pteridium aquilinum). Cancer Res. 40, 3468–3472.

57. Perry, K.L. and Walker, G.C. (1982). Identification of plasmid (pKM101)-coded proteins involved in mutagenesis and UV resistance. Nature (Lond.) 300, 278–281.

58. Perry, P. and Evans, H.J. (1975). Cytological detection of mutagen-carcinogen exposure by sister chromatid exchange. Nature (Lond.) 258, 121-125.

59. Podhajcer, O.L., Friedlander, M. and Graziani,Y. (1980). Effect of liposome-encapsulated quercetin on DNA synthesis, lactate production, and cyclic adenosine 3':5'-monophosphate level in Erlich ascites tumor cells. Cancer Res. 40, 1344-1350.

60. Popova, T.P., Litvinenko, V.I. and Kovalev, I.P. (1973). Flavones of Scutellaria baicalensis roots. Khim. Prir. Soedin 6, 729-733.

61. Popova, T.P., Litvinenko, V.I. and Pakaln, D.A. (1979). Study of phenol compounds of populations of Scutellaria sevanensis, Farm Zh. (Kiev) 6, 49-53.

62. Richardson, G.A., El-Rafey, M.S. and Long, M.L. (1947). Flavones and flavone derivatives as antioxidants. J. Dairy Sci. 30, 397-413.

63. Sacks, L.E. and MacGregor, J.T. (1982). The B. subtilis multigene sporulation test for mutagens: Detection of mutagens inactive in the Salmonella his reversion test. Mutat. Res. 95, 191-202.

64. Sacks, L.E. and MacGregor, J.T. (In press). The Bacillus subtilis multigene sporulation test for detection of environmental mutagens, in Chemical Mutagens, Vol. 9, ed. F.J. deSerres, Plenum Press, p. 000-000.

65. Sahu, R.K., Basu, R. and Sharma, A. (1981). Genetic toxicological testing of some plant flavonoids by the micronucleus test. Mutat. Res. 89, 69-74.

66. Saito, D., Shirai, A., Matsushima, T., Sugimura, T., and Hirono, I. (1980). Test of carcinogenicity of quercetin, a widely distributed mutagen in food. Teratogen. Carcinogen. Mutagen. 1, 213-221.

67. Savage, P.C. (1977). Microbial ecology of the gastrointestinal tract. Ann. Rev. Microbiol. 31, 107-133.

68. Scarborough, H. and Bacharach, A.L. (1949). Vitamin P. Vitamins and Hormones 7, 1-55.

69. Scheline, R.R. (1978). Mammalian Metabolism of Plant Xenobiotics, Academic Press, N.Y., pp. 295-329.

70. Schwartz, A., Sutton, S.L. and Middleton, E., Jr. (1982). Quercetin inhibition of the induction and function of cytotoxic T lymphocytes. Immunopharmacol. 4, 125-138.

71. Simpson, T.H. and Uri, N. (1956). Hydroxyflavones as inhibitors of the aerobic oxidation of unsaturated fatty acids. Chemistry and Industry, 956-957.

72. Spector, M., O'Neal, S. and Racker, E. (1980). Reconstitution of the $Na^+K^+$ pump of Erlich ascites tumor and enhancement of efficiency by quercetin. J. Biol. Chem. 255, 5504-5507.

73. Suolinna, E-M., Buchsbaum, R.N. and Racker, E. (1975). The effect of flavonoids on aerobic glycolysis and growth of tumor cells. Cancer Res. 35, 1865-1872.

74. Suolinna, E-M., Lang, D. and Racker, E. (1974). Quercetin, an artificial regulator of the high aerobic glycolysis of tumor cells. J. Natl. Cancer Inst. 53, 1515-1519.

75. Sugimura, T., Nagao, M., Matsushima, T., Yahagi, T., Seino, Y., Shirai, A., Sawamura, M., Natori, S., Yoshihira, K., Fukuoka, M. and Kuroyanagi, M. (1977). Mutagenicity of flavone derivatives. Proc. Jpn. Acad., Ser. B, 53, 194-197.

76. Takanashi, H., Aiso, S. and Hirono, I. (1983). Carcinogenicity test of quercetin and kaempferol in rats by oral administration. J. Food Safety 5, 55-60.

77. Ueno, I., Nakano, N. and Hirono, I. (1983). Metabolic fate of [$^{14}$C] quercetin in the ACI rat. Japan. J. Exp. Med. 53, 41-50.

78. Umezawa, K., Matsushima, T., Sugimura, T., Hirakawa, T., Tanaka, M., Katoh, Y., and Takayama, S. (1977). In vitro transformation of hamster embryo cells by quercetin. Toxicol. Lett. 1, 175-178.

79. United States Department of Health and Human Services (1983). Fiscal Year 1983 Annual Plan. National Toxicology Program. Research Triangle Park, N.C.

80. van der Hoeven, J.C.M., Bruggeman, I.M. and Debets, F.M.H. (In press). Genotoxicity of quercetin in cultured mammalian cells. Mutat. Res. 000, 000-000.

81. Varma, S.D. and Kinoshita, J.H. (1976). Inhibition of lens aldose reductase by flavonoids - Their possible role in the prevention of diabetic cataracts. Biochem. Pharmacol. 25, 2505-2513.

82. Vogel, E. (1975). Some aspects of the detection of potential mutagenic agents in Drosophila. Mutat. Res. 29, 241-250.

83. Wargovich, M.J. and Newmark, H.L. (1983). Inability of several mutagen-blocking agents to inhibit 1,2-dimethylhydrazine-induced DNA-damaging activity in colonic epithelium. Mutat. Res. 121, 77-80.

84. Watson, W.A.F. (1982). The mutagenic activity of quercetin and kaempferol in Drosophila melanogaster. Mutat. Res. 103, 145-147.

85. Wattenberg, L.W. and Leong, J.L. (1968). Inhibition of the carcinogenic action of 7,12-dimethylbenz(a)anthracene by beta-naphthoflavone. Proc. Soc. Exp. Biol. Med. 128, 940-943.

86. Wattenberg, L.W. and Leong, J.L. (1970). Inhibition of the carcinogenic action of benzo(a)pyrene by flavones. Cancer Res. 30, 1922-1925.

87. Wattenberg, L.W., Page, M.A. and Leong, S.L. (1968). Induction of increased benzpyrene hydroxylase activity by flavone and related compounds. Cancer Res. 28, 934-937.

88. Wollenweber, E. and Dietz, V.H. (1981). Occurrence and distribution of free flavonoid aglycones in plants. Phytochem. 20, 869-932.

89. Yoshida, M.A., Sasaki, M., Sugimura, K., and Kawachi, T. (1980). Cytogenetic effects of quercetin on cultured mammalian cells. Proc. Japan Acad. 56(B), 443-447.

90. Zaprometov, M.N. (1959). Vitamin P and its uses, in Vitamin P, Its Properties and Uses, Akademiya Nauk SSSR, Vitamin Sources and their Utilization, Collection 4, Trans. for Nat. Sci. Found., Washington, D.C. by Israel Program for Scientific Translations, Jerusalem, 1963, pp. 3-21.

# 24

FACTORS WHICH FACILITATE INACTIVATION OF QUERCETIN MUTAGENICITY

Mendel Friedman and G. A. Smith

Western Regional Research Center, Agricultural Research Service, U.S. Department of Agriculture, Berkeley, CA 94710 USA

ABSTRACT

Oxygen, oxidizing enzymes such as polyphenol oxidase (tryrosinase) and alkaline pH, irreversibly inactivate the mutagenicity of quercetin in the Ames test. The loss of mutagenic activity correlates with decreases in the ultraviolet absorption maximum of quercetin near 370 nm. The extent of inactivation increases with time but apparently not significantly with temperature of exposure, and decreases with quercetin concentration. Metal salts such as ferrous and copper sulfates also facilitate inactivation, but these effects may be reversible. Understanding the factors which minimize the mutagenic potential of food ingredients should lead to safer foods.

INTRODUCTION

Polyphenolic compounds such as quercetin are widely distributed in plant foods (Mabry and Mabry, 1975; Herrmann, 1976). Quercetin and several related compounds are mutagenic in the Ames test (Bjeldanes and Chang, 1977; MacGregor and Jurd, 1978; Brown, 1980; Nagao et al., 1980; Tamura et al., 1980; Watson, 1982; MacGregor, 1984).

Although many bacterial mutagens are also carcinogenic (McCann et al., 1975), the evidence for carcinogenicity of quercetin is equivocal. Hirono et al. (1981) reported no significant difference in incidence of tumors between control rats and rats fed a diet containing 1 or 5% quercetin or 5% rutin (a nonmutagenic quercetin glycoside) for 540 days, or 10% quercetin or 10% rutin for 850 days. This is in contrast to an earlier report by Pamakcu et al.

(1980) that quercetin induces bladder and intestinal cancer in rats. A human oral bacterium, Streptococcus milleri, can hydrolyze rutin in vitro to its genotoxic aglycone quercetin (Parisis and Pritchard, 1983).

Although more work is needed to clarify this problem, a need also exists to define conditions which reduce the mutagenic potential of quercetin because it is regularly consumed by animals and man. In this paper, we evaluate the influence of pH, time, temperature, quercetin concentration, oxygen, tyrosinase, and several related oxireductases and metal salts on quercetin mutagenicity in Salmonella typhimurium strain TA 98. Spectroscopic measurements were used to assess whether mutagenicity correlates with changes in the ultraviolet absorption spectra of quercetin and inactive compounds formed during the treatments.

EXPERIMENTAL

Mutagenicity assay

Quercetin solutions were prepared daily by dissolving 30 mg quercetin in 100 ml of absolute ethyl alcohol. The alcohol was presaturated with nitrogen to prevent oxidation. Fifty ml of the quercetin solution and 50 ml nitrogen-saturated appropriate buffer were mixed to obtain the desired quercetin solutions. The final concentration of quercetin was 15 mg/100 ml 50% ethanol in Tris buffer (0.5M).

Mutagenicity assays with Salmonella typhimurium strain TA98 have been described by Ames et al. (1975). Procedures for growth, storage, and verification of the presence of R-factor plasmid and rfa character were carried out according to Ames et al., (1975). Rat liver homogenate (S-9) fraction metabolizing system was used throughout the experiment. The liver S-9 was prepared from Aroclor 1254-induced male Sprague-Dawley-derived rats (Simonsen Laboratory, Gilroy, CA) and was used at a concentration of 0.1 ml S-9 per ml of medium. Samples of 0.1 ml of quercetin reaction mixture were incorporated into top agar, and colony assays were made in duplicate plates with tester strain TA98. A reference standard of aflatoxin $B_1$ in DMSO was added to positive control plates. (Cf. also, Friedman et al., 1982).

To demonstrate that the reaction mixture did not affect microsomal activity, quercetin was prepared in buffers of pH 8, 9, and 10 buffers at room temperature in the presence of oxygen. The pH of each solution was adjusted to 7.5 with 1 N HCl before the Ames test was performed. Parallel samples were then spiked with a second dose of quercetin and the mutagenicity assays were carried out.

To assess the reversibility of quercetin inactivation, a series of samples were treated at pH 10 for 24 hours at 37°C in the presence of oxygen. One sample was subjected to the Ames test to assure inactivation. Other samples were adjusted with 1N HCl to pH 6.0, 7.5, or 9.0, incubated for an additional 8 hours at 37°C in an atmosphere of nitrogen or oxygen, cooled to room temperature and then re-tested for mutagenicity.

Tyrosinase was dissolved in the reaction mixture before addition of quercetin. The reaction was carried out by adding 0.15 mg of enzyme to 15 mg of quercetin in 100 ml 50% ethanol-buffer.

Since it was necessary to use ethanol to keep quercetin in solution, we tested effects of ethanol on tyrosinase activity with its usual substrate, tyrosine (Worthington, 1982). The following results were obtained (in units/mg): pH 6.5 Tris buffer, 3,000; 50% pH 6.5 buffer-50% ethanol, 700; pH 7.0 buffer, 2,500; 50% pH 7.0 buffer-50% ethanol, 1,000; pH 7.5 buffer, 2,650; 50% pH 7.5 buffer-50% ethanol, 1,100. Although ethanol suppresses tyrosinase activity, it does not completely inactivate the enzyme under conditions of the present study. Additional studies revealed that the extent of suppression is directly related to the amount of ethanol in the mixed solvent.

Tests of effects of pH were carried out at room temperature. The influence of anaerobic and aerobic conditions were evaluated by purging the reaction mixture with oxygen or nitrogen (oxygen-free). Spectra were recorded after 15 min, 24 hours and 96 hours. Ames tests were done after 24 hours.

Reactions testing effects of temperature were carried out for 3 hours at pH 7.5 or 9.0 in the presence of oxygen or nitrogen at the following temperatures: 0, 25, 37, 50, 60, and 75°C.

Reactions testing effects of time on the inactivation of quercetin in the presence of oxygen at pH 7.5 and 9.0 were carried out at the following time intervals: 0, 1, 2, 4, 7, 12 and 24 hours.

Metal salt treatments were conducted in the solution of 50% ethanol Tris buffer (0.5M pH 7.5). A 2:1 molar ratio of mineral salt to quercetin was used in all experiments. The weight ratio of tyrosinase to quercetin was 1:100. The mineral salt and quercetin were dissolved in ethanol-buffer separately before mixing. Tyrosinase was also dissolved separately in a small amount of ethanol-buffer and added to the mineral-quercetin solution. The final volume was then adjusted in a volumetric flask to 10 ml, and the reaction mixture was left standing at room temperature for one hour. The Ames test was then carried out with 0.1 ml portions of the reaction mixture as previously described.

Reversibility of ferrous sulfate- and copper-sulfate-catalyzed inactivation was tested by repeating the Ames test after adjusting the reaction mixture to pH 2 with 1 N HCl to release the metal salt from its quercetin complex.

## RESULTS AND DISCUSSION

Quercetin is known to be susceptible to oxidative degradation by oxidizing agents including oxygen and oxidizing enzymes and high pH (Matsuura et al., 1970; Nishinaga et al., 1979; Rajandra and Brown, 1981; Brown et al., 1982; Blouin et al., 1982). A number of different products are produced depending on oxidative conditions. (Figure 1).

Several factors expected to modify quercetin mutagenicity were, therefore, studied to establish optimal conditions for inactivating or suppressing the mutagenic potential of quercetin.

### pH and oxygen

Table 1 presents data on the mutagenicity of quercetin in the Ames Salmonella typhimurium mammalian microsome test for quercetin in a series of buffers, saturated with either oxygen or nitrogen. The Table also includes values for negative controls (buffers) and a positive control (aflatoxin $B_1$).

Exposure of quercetin to buffers in an oxygen atmosphere had no observable effect on quercetin mutagenicity in the pH range 2 to 5. The transition point for the pH-oxygen inactivation appears to be near pH 6.5, since the number of revertants at this pH was about 440 compared to about 1000 at the lower pH values. Complete inactivation, where the number of revertants equals that observed with buffer controls, took place at pH values above 8.

To assess whether compounds formed during the inactivation process adversely affected the activity of the Salmonella typhimurium TA98 strain employed in this study, quercetin was first inactivated at 37°C for 24 hours by exposure to pH 8, 9, 10 buffers in the presence of oxygen. The pH of each solution was then adjusted to 7.5 with 1 N HCl before the Ames test was run. Parallel inactivated samples were spiked with a second dose of quercetin immediately before the Ames test was conducted.

Table 2 shows that the spiked samples were fully active. The microsomes and bacteria were not affected by the reaction medium or by inactivated quercetin products.

Since the Ames test data in Table 3 show nonregeneration of activity under conditions described in the experimental section, pH 10 inactivation was judged irreversible.

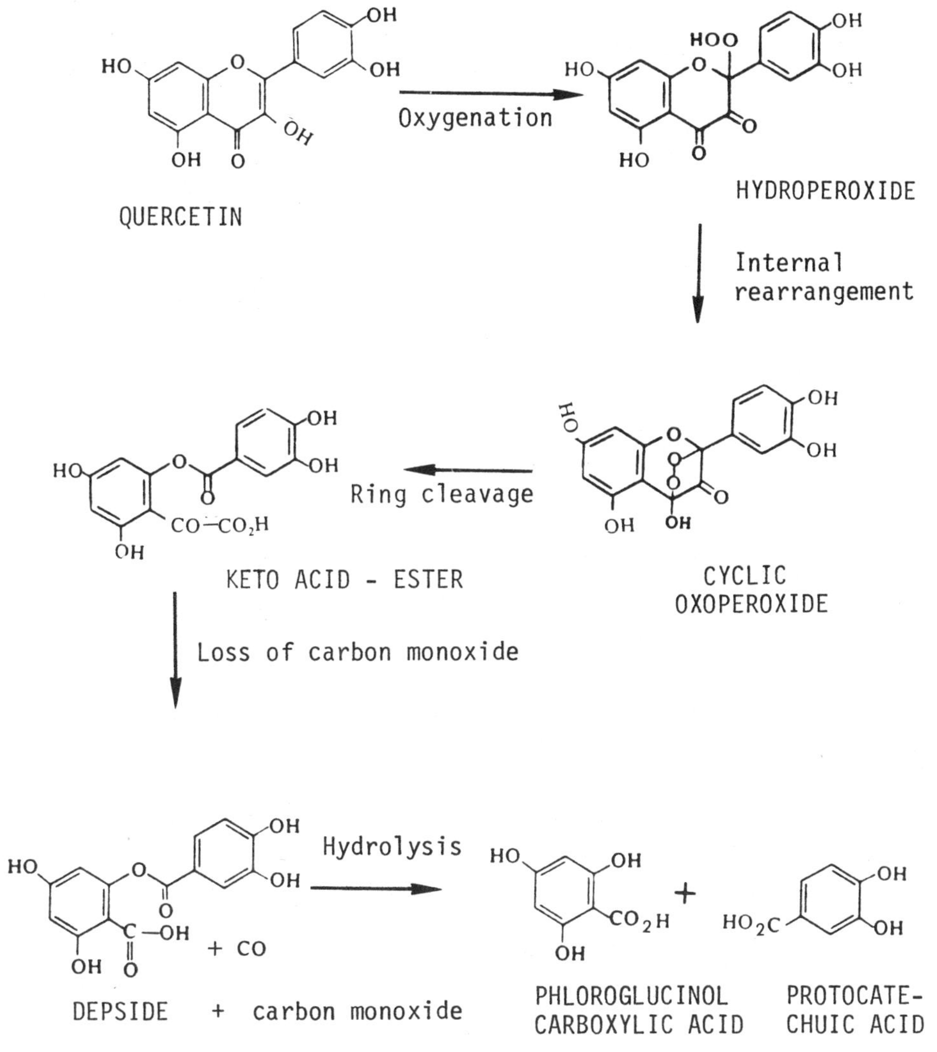

Figure 1. Effect of quercetin oxygenation. Adapted from Matsuura et al., 1970; Igrashi et al., 1978; Nishinaga et al., 1979; Rajandra and Brown, 1981; Brown et al., 1982.

Table 1.   Effect of pH and oxygen on the mutagenic activity of quercetin[a].

| pH of buffer | Atmosphere | Revertants/plate (TA 98) |
|:---:|:---:|:---:|
| 2.0 | buffer control | 54;   67[b] |
| 2.0 | $O_2$ | 982; 1009 |
| 2.0 | $N_2$ | 889;   963 |
| 5.0 | buffer control | 65;   74 |
| 5.0 | $O_2$ | 965; 1055 |
| 5.0 | $N_2$ | 986; 1012 |
| 6.5 | buffer control | 51;   58 |
| 6.5 | $O_2$ | 432;   452 |
| 6.5 | $N_2$ | 568;   620 |
| 7.5 | buffer control | 63;   66 |
| 7.5 | $O_2$ | 599;   626 |
| 7.5 | $N_2$ | 1056; 1065 |
| 8.0 | buffer control | 50;   63 |
| 8.0 | $O_2$ | 54;   57 |
| 8.0 | $N_2$ | 1018; 1073 |
| 9.0 | buffer control | 68;   68 |
| 9.0 | $O_2$ | 55;   53 |
| 9.0 | $N_2$ | 811;   871 |
| 10.0 | buffer control | 57;   65 |
| 10.0 | $O_2$ | 50;   60 |
| 10.0 | $N_2$ | 552;   563 |
| Aflatoxin (0.15 $\mu$g) | | 1151; 1458 |

[a]Conditions:  15 mg quwercetin in 100 ml 50% ethanol-buffer; 24 hr; 25°C.
Buffers:  pH 2, HCl; pH 5-9, 0.5M Tris; pH 10, 0.05M sodium borate.

[b]Duplicate values.

Table 2.  Effect of pH on the activity of <u>Salmonella</u> typhimurium in the presence of quercetin (15 mg in 100 ml 50% ethanol-buffer). All pH values are adjusted to 7.5 for the mutagenicity assays.

| Treatment conditions | Sterility | Revertant/plate (TA 98) | |
|---|---|---|---|
| A. Quercetin; <u>pH 8</u>; 37°C; 24 hrs; $O_2$ | no growth | 63; | 67 |
| B. Solution A spiked with quercetin | no growth | 773; | 878 |
| C. Same as in A except at <u>pH 9</u> | no growth | 59; | 68 |
| D. Solution C spiked with quercetin and reassayed | no growth | 803; | 844 |
| E. Same as in A except at <u>pH 10</u> | no growth | 62; | 66 |
| F. Solution E spiked with quercetin and reassayed | no growth | 801; | 802 |
| G. Aflatoxin (0.1 µg) | no growth | 1062; | 1125 |

Table 3.  Nonreversibility of quercetin inactivation at 37°C (15 mg in 100 ml 50% ethanol-buffer; 24 hours). See text.

| Buffer pH | Treatment | Revertants/plate (TA 98) |
|---|---|---|
| 10.0 | $O_2$ | 58; 61 |
| 9.0 | $O_2$ | 51; 60 |
| 7.5 | $O_2$ | 65; 65 |
| 6.0 | $O_2$ | 54; 62 |
| 9.0 | $N_2$ | 58; 66 |
| 7.5 | $N_2$ | 61; 66 |
| 6.0 | $N_2$ | 61; 76 |

Table 4.  Effect of temperature on quercetin mutagenicity (exposure for 3 hours at pH 7.5 and 9.0 in the presence of oxygen).

| Temperature (°C) | Revertants/plate (TA 98) | |
|---|---|---|
| | pH 7.5 | pH 9.0 |
| | ---- | |
| 0 | | 287;  319 |
| 25 | 841;  876 | 74;  79 |
| 37 | 707;  801 | 61;  65 |
| 50 | 718;  776 | 57;  66 |
| 60 | ---- | 72;  82 |
| 75 | 644;  824 | 57;  61 |
| Buffer control | 56;  61 | 82;  83 |
| Aflatoxin (0.15 μg) | 981;  1171 | 1592;  1613 |

## Temperature effects

Two series of experiments were carried out to evaluate the influence of temperature on quercetin inactivation. In the first, quercetin solutions were kept for three hours at pH 7.5 in the temperature range 0 to 75°c in the presence of oxygen. Table 4 shows that temperature had only a minor influence on the mutagenic activity of quercetin at pH 7.5.

In contrast exposure of quercetin to identical conditions but at pH 9.0 led to complete inactivation in the entire temperature range with one exception. Inactivation at 0°C proceeded to about 70% of the original value at pH 9.0.

These results strikingly demonstrate the minor influence of temperature compared to the major effect of pH on quercetin mutagenicity.

## Quercetin concentration

Table 5 shows that exposure of 15 mg of quercetin in 100 ml of pH 7.5 ethanol-buffer for 24 hours at 37°C produced an inactive product. However, activity was retained when the concentration of quercetin was increased to 30, 60, 75 mg in 100 ml of ethanol-buffer. In contrast, at pH 9.0, inactivation took place at all four concentrations. These results suggest that, in addition to the paramount importance of hydroxide ions in the destruction of quercetin, quercetin oxidation appears to be governed by concentration, at least at low concentrations.

A possible explanation for the apparent non-dependence of the inactivation at pH 7.5 on quercetin concentration, is that destruction of part of the quercetin at the high concentrations still leaves enough quercetin to produce the maximum number of revertants. Additional studies are needed to define the dependence of the inactivation on concentration at different pH values.

## Time of exposure

A kinetic study was performed at pH 7.5 and 9.0 to ascertain the dependence of quercetin inactivation on time at 37°C. Table 6 shows that at pH 7.5, inactivation starts after about 4 hours and is complete after about 12 hours. In contrast, inactivation at pH 9.0 starts within minutes and is complete after about two hours. Time of exposure also appears to be an important variable governing quercetin inactivation.

Table 5.  Effect of concentration on the mutagenicity of quercetin
          (exposure for 24 hours at 37°C at pH 7.5 and 9.0 in the
          presence of oxygen).

| Quercetin (mg/100 ml ethanol-buffer) | pH | Revertants/plate (TA 98) |
|---|---|---|
| 15 | 7.5 | 82;  82 |
| 30 | 7.5 | 1465; 1517 |
| 60 | 7.5 | 2100; 2200 |
| 75 | 7.5 | 2200; 2200 |
| 15 | 9.0 | 58;  72 |
| 30 | 9.0 | 51;  67 |
| 60 | 9.0 | 66;  74 |
| 75 | 9.0 | 82;  73 |
| Buffer control | 7.5 | 62;  74 |
| Buffer control | 9.0 | 73;  76 |
| Aflatoxin (0.1 µg) | | 1230 |

Table 6.

Effect of time on the mutagenicity of quercetin at pH 7.5 and 9.0.[a]

| Time of exposure (hours) | pH | Revertants/plate (TA 98) |
|---|---|---|
| 0 | 7.5 | 906; 962 |
| 1 | 7.5 | 998; 1020 |
| 2 | 7.5 | 857; 1000 |
| 4 | 7.5 | 509; 512 |
| 7 | 7.5 | 550; 604 |
| 12 | 7.5 | 64; 70 |
| 24 | 7.5 | 74; 76 |
| 0 | 9.0 | 609; 780 |
| 1 | 9.0 | 131; 142 |
| 2 | 9.0 | 60; 81 |
| 4 | 9.0 | 63; 63 |
| 7 | 9.0 | 69; 82 |
| 12 | 9.0 | 67; 79 |
| 24 | 9.0 | 69; 82 |
| Buffer only | 7.5 | 58; 81 |
| Buffer only | 9.0 | 80; 84 |
| Aflatoxin $B_1$ (0.075 µg) | | 1500; 2050 |

[a]Conditions: 15 mg quercetin in 100 ml 50% ethanol-Tris buffer (0.5M), at 37°C in the presence of exygen.

Effect of polyphenol oxidase (tyrosinase)

Polyphenol oxidase catalyzes the enzymatic browning of fruits and vegetables; e.g. the conversion of phenolic compounds in the presence of oxygen to brown melanins (Eskin et al., 1971; Leathan et al., 1980; Schwimmer, 1981; Vamos-Vigyazo, 1981; Barbeau and Kinsella, 1983). It was shown that tyrosine is first hydroxylated to 3, 4-dihydroxyphenylalanine, which is then oxidized to ortho-quinone phenylalanine. The latter molecule is highly reactive and interacts with amino groups of lysine and other protein functional groups to produce brown pigments (Hurrell et al., 1982; Price et al. 1979). Since quercetin is a phenolic flavonoid that may be suscep-tible to undergo similar transformations, it was of interest to study whether treating quercetin with tyrosinase would abolish mutagenic activity. Since the ultraviolet spectrum of quercetin has two absorption maxima near 260 and 370 nm (Markham and Mabry, 1975), we expected that exposure of quercetin to tyrosinase would alter these absorption maxima, if the enzyme destroyed the integrity of the quercetin molecule.

Figure 2. suggest that these expectations were indeed observed. Both absorption maxima decreased with increasing pH on exposure of an oxygen-saturated solution of quercetin to tyrosinase. The ex-tent of the decrease in the absorption maximum near 370 nm corre-lated with decreased mutagenicity. Spectroscopy may therefore be useful for monitoring the mutagenic inactivation of quercetin and possibly other flavonoids.

Limited studies showed that the following enzymes had no or only minor effects on quercetin mutagenicity under conditions similar to those used for tyrosinase (1: 100 weight ratio enzyme to quercetin; 50% pH 7.5 buffer ethanol; 37°C; 2 hours): catalase, lipoxygenase, peroxidase, and superoxide dismutase. Additional studies are needed to define the ability of these oxidizing enzymes to modify flavonoids over a broader pH range.

Metal salts

Because metal salts have a strong affinity for and seem to alter the genotoxicity of phenolic compounds (Eskin et al., 1971; Kuhnau, 1976; Stich et al., 1981; Narasinga Rao and Prabhavathi, 1982), experiments were carried out to assess the influence of a series of mineral salts in the presence of nitrogen, oxygen, and tyrosinase on the mutagenic activity of quercetin. Salts selected for this study include mainly those which are present in foods as trace elements, and that, therefore, might exert similar modulating effects on quercetin mutagenicity in foods.

Table 7 shows that (a) copper, ferrous and ferric sulfates have the ability to inactivate quercetin mutagenicity at pH 7, even in a

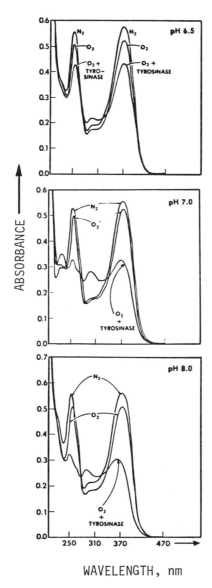

WAVELENGTH, nm

Figure 2. Effect of nitrogen, oxygen, and oxygen plus tyrosinase
on quercetin spectra at three pH values. Upper curve, pH 6.5;
middle curve, pH 7.0; lower curve, pH 8.0. Note relative decreases
in the absorption spectra near 370 nm. Conditions of treatment:
0.15 mg tyrosinase and 15 mg quercetin in 100 ml 50% ethanol-Tris
buffer (0.5M); 37°C; 2 hrs.

Table 7.

Reversibility of copper sulfate and ferrous sulface catalyzed inactivation of quercetin mutagenicity[a].

| Treatment | Revertants/plate (TA 98) | |
|---|---|---|
| | pH 7.0 | pH 2.0 (acidified pH 7) |
| $CuSO_4$ + buffer | 38;  40 | |
| $FeSO_4$ + buffer | 42;  52 | |
| Quercetin + $O_2$ | 1015; 1368 | |
| Quercetin + $N_2$ | 1038; 1195 | |
| Quercetin + tyrosinase + $O_2$ | 674;  741 | |
| Quercetin + tyrosinase + $N_2$ | 973; 1028 | |
| Quercetin + tyrosinase + $O_2$ + $CuSO_4$ | 161;  196 | 461;  478 |
| Quercetin + tyrosinase + $N_2$ + $CuSO_4$ | 231;  235 | 464;  538 |
| Quercetin + $O_2$ + $CuSO_4$ (no tyrosinase) | 134;  147 | 337;  389 |
| Quercetin + $N_2$ + $CuSO_4$ (no tyrosinase) | 156;  169 | 454;  466 |
| Quercetin + tyrosinase + $O_2$ + $FeSO_4$ | 85;  89 | 638;  803 |
| Quercetin + tyrosinase + $N_2$ + $FeSO_4$ | 488;  531 | 709;  769 |
| Aflatoxin $B_1$ (0.05 $\mu g$) | 1332; 2000 | |

[a]Conditions:  0.15 mg tyrosinase and 15 mg quercetin in 100 ml 50% ethanol-
pH 7 Tris buffer (0.5M); 25°C; 1 hour.  Weight ratio of metal salt to
quercetin = 2:1.

Table 8. Effect of metal salts on quercetin mutagenicity.[a]

| Treatment | Revertants/plate (TA98) |
|---|---|
| Buffer + $CuSO_4$ | 64 |
| Buffer + $FeSO_4$ | 30 |
| Buffer + $Fe_2(SO_4)_3$ | 50 |
| Quercetin + $O_2$ | 1048; 1056 |
| Quercetin + $N_2$ | 1168; 1330 |
| Quercetin + tyrosinase + $O_2$ | 958; 1006 |
| Quercetin + tyrosinase + $N_2$ | 775; 914 |
| Quercetin + tyrosinase + $FeSO_4$ + $O_2$ | 92; 113 |
| Quercetin + tyrosinase + $FeSO_4$ + $N_2$ | 681; 644 |
| Quercetin + $FeSO_4$ + $O_2$ | 457; 445 |
| Quercetin + $FeSO_4$ + $N_2$ | 629; 770 |
| Quercetin + tyrosinase + $Fe_2(SO_4)_3$ + $O_2$ | 99; 112 |
| Quercetin + tyrosinase + $Fe_2(SO_4)_3$ + $N_2$ | 120; 125 |
| Quercetin + $Fe_2(SO_4)_3$ + $N_2$ | 110; 114 |
| Buffer + $ZnSO_4$ | 106; 116 |
| Quercetin + tyrosinase + $ZnSO_4$ + $O_2$ | 674; 819 |
| Quercetin + tyrosinase + $ZnSO_4$ + $N_2$ | 922; 1062 |
| Aflatoxin $B_1$ (0.05 µg) | 1212; 1545 |

[a]Conditions: 0.15 mg tyrosinase and 15 mg quercetin in 100 mg 50% ethanol-pH 7.0 Tris buffer (0.5M); 25°C; 1 hour. Weight ratios of metal salt to quercetin = 2:1.

Table 9.  Effect of cobaltous chloride and manganous chloride on mutagenic activity of quercetin.

| Treatment | Revertants/plate (TA98) |
|---|---|
| Buffer + $CoCl_2$ | 46;  46 |
| Buffer + $MnCl_2$ | 58;  60 |
| Quercetin + $O_2$ | 805;  823 |
| Quercetin + $N_2$ | 1038; 1142 |
| Quercetin + tyrosinase + $O_2$ | 656;  705 |
| Quercetin + tyrosinase + $N_2$ | 843;  992 |
| Quercetin + tyrosinase + $CoCl_2$ + $O_2$ | 783;  796 |
| Quercetin + tyrosinase + $CoCl_2$ + $N_2$ | 900;  972 |
| Quercetin + $CoCl_2$ + $O_2$ | 744;  839 |
| Quercetin + $CoCl_2$ + $N_2$ | 770;  780 |
| Quercetin + tyrosinase + $MnCl_2$ + $O_2$ | 531;  638 |
| Quercetin + tyrosinase + $MnCl_2$ + $N_2$ | 638;  640 |
| Quercetin + $MnCl_2$ + $O_2$ | 672;  674 |
| Quercetin + $MnCl_2$ + $N_2$ | 648;  711 |
| Aflatoxin $B_1$ (0.05 μg/plate) | 1198; 1282 |

REFERENCES

Aeshbacher, H. U. (1982).  The significance of mutagens in foods. In "Mutagens in Our Environment", M. Sorsa and H. Vainio, Eds., Alan R. Liss, Inc., New York, pp. 349-362.
Ames, B. N., McCann, J. & Yamasaki, E. (1975).  Methods for detecting carcinogens and mutagens with the Salmonella/mammalian-microsome test. Mutation Res. 31, 347-364.

Arai, S. (1980). Deterioration of food proteins by binding unwanted compounds such as flavors, lipids and pigments. In "Chemical Deterioration of Proteins", J. R. Whitaker and M. Fujimaki, Eds., American Chemical Society Symposium Series, Washington, D.C. 123, 195-209.

Barbeau, W. F. and Kinsella, J. F. (1983). Factors affecting the binding of chlorogenic acid to fraction 1 leaf protein. J. Agric. Food Chem., 31, 67-91.

Blouin, F. A., Narins, Z. M. and Cherry, J.P. (1982). Discoloration of proteins by binding with phenolic compounds. In "Food Protein Deterioration: Mechanisms and Functionality", J. P. Cherry, Ed., American Chemical Society Symposium Series, Washington, D.C. 206, 67-91.

Brown, J. P. (1980). A review of genetic effects of naturally occurring flavonoids, anthraquinones and related compounds. Mutation Res., 75, 244-277.

Brown, S. B., Rajananda, V., Holroyd, J. A. and Evans, E. G. V. (1982). A study of the mechanism of quercetin oxygenation by $^{18}O$ labeling. Biochem. J., 205, 239-244.

Eskin, N. A. M., Henderson, H. M. & Townsend, R. J. (1971). "Biochemistry of Foods", Academic Press, New York, p. 71.

Friedman, M., Wehr, C.M., Schade, J.E. and MacGregor, J.T. (1982). Inactivation of aflatoxin $B_1$ mutagenicity by thiols. Food Chem. Toxicol. 20, 887-892.

Herrmann, K. (1976). Flavonols and flavones in food plants: A review. J. Food Technol. 11, 433-448.

Hirono, I., Ueno, I., Hosaka, S., Takanashi, H. Matsushima, T., Sugimura, T. & Natori, S. (1981). Carcinogenicity examination of quercetin and rutin in aci rats. Cancer Letters, 13, 15-21.

Hurrell, R. F., Finot, P. A. & Cua, J. L. (1982). Protein-polyphenol reactions. British J. Nutr., 47, 191-211

Igrashi, K., Sassa, T., Ikeda, M. and Yasui, T. (1978). Aerial oxidation products of quercetin in acidic solution. Agric. Biol. Chem., 42, 1617-1619.

Kuhnau, J. (1976). The Flavonoids. A class of semi-essential food components: Their role in human nutrition. World Rev. Nutri. Dietetics, 24, 117-191

Leatham, G. F., King, V. and Stahmann, M. A. (1980). In vitro protein polymerization by quinones or free radicals generated by plant or fungal oxidative enzymes. Phytopathology, 70, 1134-1140.

MacGregor, J. T. (1984). Genetic and carcinogenic effects of plant flavonoids. This volume.

MacGregor, J. T. & Jurd, L. (1978). Mutagenicity of plant flavonoids: Structural requirements for mutagenic activity in Salmonella typhimurium. Mutation Res. 54, 297-

Mabry, R. T. J. and Mabry, H. (Eds.) (1975). "The Flavonoids", Academic Press, New York.

Markham, K. R. & Mabry, T. J. (1975). Ultraviolet-visible and proton magnetic resonance spectroscopy of flavonoids. In "The Flavonoids", T. J. Mabry and H. Mabry, Eds., Academic Press, New York, pp. 44-76.

Masuura, T., Matsushima, H. and Nakashima, R. (1970). Photosensitized oxygenation of 3-hydroxyflavones as an nonenzymatic model
for quercetinase. Tetrahedron., 25, 435-443.

Nagao, M., Morita, N., Yahagi, T., Shimizu, M., Kuroyanagi, M.,
Fukuoka, M., Yoshihira, K., Natori, S., Fujino, T. and Sigumura,
T. (1981). Mutagenicities of 61 flavonoids and 11 related
compounds. Environ. Mutagenesis, 3, 401-419.

Narasinga, Rao, B.S. and Prabhavathi, T. (1982). Tannin content
of foods commonly consumed in India and its influence on ionisable
iron. J. Sci. Food Agric, 33, 89-96.

Nishinaga, A., Tojo, T., Tomita, H. and Matsuura, T. Y. (1979).
Base-catalyzed oxygenolysis of 3-hydroxyflavones. J. Chem. Soc.
Perkin I, 2511-1516.

Nishinaga, A., Tojo, T. and Matsuura, T. Y. (1974). A model cata-
lytic oxygenation for the reaction of quercetinase. J. Chem. Soc.
Chem. Commun., 896-897.

Pamukcu, A.M., Wang, C.Y., Hatcher, J. and Bryan, G.T. (1980).
Carcinogenicity of tannin-free extracts of bracken fern in rats.
J. Nat. Cancer Inst, 65, 131.

Parisis, D. M. and Pritchard, E. T. (1983). Activation of rutin by
human oral bacterial isolates to the carcinogen-mutagen quercetin.
Archivs. Oral Biol., 28, 583-590.

Pierpoint, W. S. (1969). O-Quinones formed in plant extracts.
Their reaction with bovine serum albumin. Biochem. J., 112,
619-629.

Rajandra, V. and Brown, S.B. (1981). Mechanism of quercetin oxygenation. A possible mode for haem degradation. Tetrahedron
Letters, 43, 4331-4334.

Saito, D., Shirai, A., Matsushima, T., Sugimura, T. and Hirona, I.
(1980). Test of carcinogenicity of quercetin, a widely distributed mutagen in food. Teratogenesis, Carcinogenesis, Mutagenesis, 1, 213-221.

Schwimmer, S. (1981). "Source Book of Food Enzymology", p. 284,
AVI, Westport, Connecticut. pp. 284, 650.

Seino, Y., Nagao, M., Yahagi, T., Sugimura, T., Yasuda, T. and
Nishimura, S. (1979). Identification of a mutagenic substance
in a spice, sumac, as quercetin. Mutation Res., 58, 225-229.

Stich, H.F. Rosin, M. P., Wu, C. H. and Paovalo, C. (1981). The
action of transition metals on the genotoxicity of simple phenols,
phenolic acids and cinnamic acids. Cancer Letters, 14, 251-260.

Tamura, G., Gold, C., Ferro-Luzzi, A. and Ames, B. N. (1980. Feca-
lase: a model for activation of dietary glycosides to mutagens by
intestinal flora. Proc. Natl. Acad. Sci. USA, 77, 4961-4965.

Vamos-Vigyazo, L. (1981). Polyphenol oxidase and peroxidase in
fruits and vegetables. Crit. Rev. Food Sci., 15, 49-127.

Watson, W. A. (1981). The mutagenicity of quercetin and kaemferol
in Drosophila Melangaster. Mutation Research, 103, 145-147.

Worthington[TM]. (1982). Worthington Enzymes and Biochemical. sp.
173.

25

MUTAGENS IN COOKED FOOD: CHEMICAL ASPECTS

L.F. Bjeldanes*, J.S. Felton and  F.T. Hatch

Department of Nutritional Sciences, University of California, Berkeley, CA  94720 (L.F.B.), and Biomedical and
Environmental Research Program, Lawrence Livermore
Laboratory, University of California, Livermore, CA
94550 (J.S.F. and F.T.H.).

This paper is a review of published and unpublished work which
was presented at the Pacific Conference on Chemistry and Spectroscopy in San Francisco, CA on October 27, 1982.  Part of this work
was also presented at the National Meeting of the American Chemical
Society, September 12-17, 1982 in Kansas City, Missouri.

With the development of simple and rapid methods for detecting
agents which modify genetic material of cells (i.e., Ames bacterial
assay) has come the realization that mutagens are widespread in our
environment (1).  The attention of several research groups has
turned toward examining mutagenic activity of foods.  As a result
of this extensive research effort the occurrence of mutagenic substances in certain heated foods and food components is now well
documented.  Proteins and several amino acids yield highly mutagenic
substances when pyrolyzed at temperatures above 300°C (2-4).
Similar temperatures were reported to be necessary to generate
appreciable mutagenicity in pyrolyzed food samples (5).  Under py-
rolytic conditions, carbohydrate and lipid components and foods rich
in them tend to exhibit less mutagenic activity than samples rich
in proteins or amino acids.  Evidence compiled by several research
groups indicates that mutagenic activity is also induced in certain
foods cooked under conditions less severe than required for pyrolysis (6-9).  In these studies mutagenic activity is detected in
fried and broiled protein-rich food (beef), and at a higher level
than in cooked carbohydrate-rich food (10).  Broiled fish samples
also contain mutagens (11).  In addition, extended boiling of beef
stock results in mutagen formation (6,8).

Reviewed herein are the results of our initial work to determine the identities of mutagenic substances present in fried ground
beef and fried egg.  Results of surveys of mutagenic activity of
cooked protein-rich foods have appeared (12, 13).

## Isolation of Mutagens From Fried Ground Beef

Prior to proceeding with a large scale extraction and fractionation of mutagens from fried beef, an effort was made to maximize
mutagen yield from the initial extraction procedure (14).  Total
mutagenic activity in ground beef fried under standard conditions
(6 min/side, 200°C) was estimated to be in the range of 300-400 re-
vertants/gE based on corrected measurements of activity in an
initial aqueous acid extract.  By contrast, methods employing $(NH_4)_2$
$SO_4$ to precipitate proteins from an acid extract or an initial extraction with acetone provided mutagenicity yields of only about 25
revertants/gE and 60 revertants/gE, respectively.  Use of a mixed
solvent ($CH_3OH:CH_2Cl:H_2O$, 45:45:10, v:v:v) improved the mutagenicity
yield to approximately 100 revertants/gE.  Our best recovery of
mutagenicity was obtained with the use of an XAD-2 resin to absorb
active substances from an initial aqueous acid extract.  By this
method, recoveries in the range of 200-250 revertants/gE were obtained.  An outline of this fractionation scheme is presented in
Figure 1.

The procedure which included the XAD-2 resin to remove active
substances from an initial aqueous acid extract was scaled up to
process a 40 kg batch of ground beef.

The active eluent from the XAD-2 column was stripped of volatile solvents and the remaining aqueous mixture was made basic with
a few drops of concentrated NaOH.  This mixture was extracted with
petroleum ether and the combined organic phases were dried with a
single partition against saturated NaCl (aq.) and reduced in volume.

Samples were then chromatographed on a preparative C-8 bonded
phase column at low pressure under the conditions specified in
Figure 2.  Results of the Salmonella assay of collected fractions
are also given in Figure 2.  The results shown in the figure indicate two major bands of mutagenic activity which are resolved under
these preparative conditions.  These bands are designated CE (for
earlier, more polar) and CL (for later, less polar).

All material eluting in the E-band region was pooled to give
a combined early (CE) fraction.  The principal mutagenic material
of this fraction was obtained as a single subfraction following two
chromatographic sequences using a C-8 semi-preparative column with
methanol/water containing 100 ppm triethylamine as the eluting solvent.  The composition of this subfraction (α3) was then investigated in comparison with the known mutagens IQ(3-methyl-1,3,6,9-tetra-

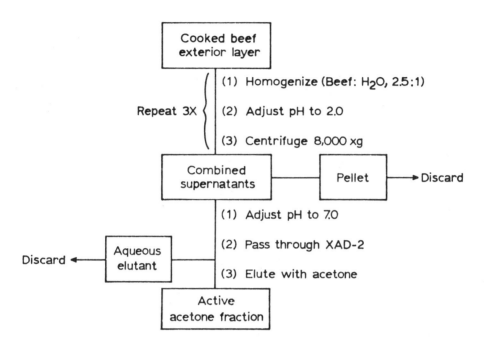

Figure 1.   Acid-Xad Extraction Scheme

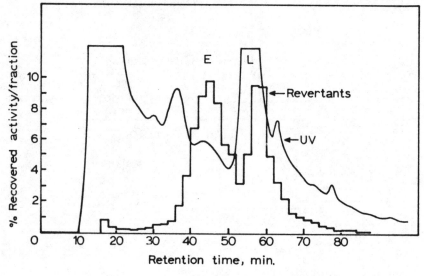

Figure 2.

azacyclopenta[a]naphthalen-2-amine) and MeIQx(3,8-dimethyl-1,3,6,9-tetraazacyclopenta[a]naphthalen-2-amine.

Co-injection of MeIQx with α3 gave a chromatogram showing an enhancement in the latest UV peak of α3 and thus suggesting the presence of this compound. Results of further fractionation of the CE-derived fraction on a normal phase HPLC system indicated the presence of two distinct areas of mutagenic activity. The major active substance coeluted with MeIQx. Ultraviolet ($\lambda_{max}$274) and mass spectral (m/e; 213,197,185,144) properties of this substance are consistent with the MeIQx assignment. The minor active component did not co-elute with any of the standard mutagens isolated from food or pyrolysis experiments and is unidentified.

The active fractions corresponding to the later band of activity (CL) from the original low pressure chromatography were combined. This material was rechromatographed on the C-8 semipreparative column with methanol/water as the eluting solvent. The three major bands of mutagenic activity were combined into fractions designated CL-I, CL-II and CL-III.

Two consecutive chromatographic runs of CL-II resulted in the isolation of a single mutagenic substance. Spectral characterization of the active material (CL-II-B) is in progress.

It should be noted that IQ, MeIQ (the 4-methyl derivative of IQ) and MeIQx elute earlier than CL-II-B (around 30 min) and Trp-P-1 (9H-1, 4-dimethyl-3, 9 diazafluoren-2-amine) and the product of soybean pyrolysate (9H-1, 9-diazafluoren-2-amine) both elute much later (greater than 90 min). Only Glu-P-1 (8-methyl-1, 4b, 9-triazafluoren-2-amine) elutes near the active compound (around 47 min).

Band CL-III was also submitted to rechromatography by reversed phase HPLC. The mutagenic profile of the corresponding fractions indicates that this fraction contains many substances at least several of which appear to be mutagenic.

At this stage of our studies it is possible to estimate the contribution of the major mutagenic fractions to the total activity of the fried meat. The total activity is nearly equally divided between the CE and CL fractions. Of the mutagenic activity in the CE fraction perhaps 80-90% is due to MeIQx with 10-20% due to a less polar, unidentified substance. Approximately 70% of the activity of the CL fraction is due to a single unidentified substance (CL-II-B). The remaining 30% of the mutagenic activity of this later eluting fraction appears to arise from several unidentified components.

Extraction of Mutagens From Fried Egg Patties

Results of initial experiments indicated that reproducible

Table 1.   Mutagenicity of Initial Extracts from Fried Eggs.

| Solvent | gE range | Rev/gE* |
|---|---|---|
| (1) Acetone | 0.05 - 0.50 | 148 |
| (2) Methanol | 0.07 - 0.70 | 108 |
| (3) $CHCl_3$:MeOH (1:1) | 0.05 - 0.45 | 14 |
| (4) $CHCL_3$:MeOH:$H_2O$ (10:10:1) | 0.03 - 0.29 | 92 |
| (5) $Ch_2Cl_2$:MeOH (1:1) | 0.06 - 0.63 | 228 |
| (6) $Ch_2Cl_2$:MeOH:$H_2O$ (10:10:1) | 0.04 - 0.37 | 578 |
| (7) $Ch_3CN$:$H_2O$ (10:1) | 0.00 - 0.50 | 538 |
| (8) Lyophilization ($-H_2O$) (MeOH extraction) | 0.13 - 1.30 | 245 |
| (9) Soxhlet MeOH 64.7°C for 24 hrs. | 0.40 - 0.44 | 771 |
| (10) XAD-2 | 0.00 - 10.0 | 5 - 10** |

*Dioxane used as solvent.  <u>Salmonella typhimurium</u> TA1538 revertant frequencies obtained from four doses in linear portion of dose response curves.

**XAD-2 isolation techniques repeatedly showed very low levels of mutagenic activity.  Although the indicated dose range is higher than for other systems studied, the dose-response curve over this range was linear.

levels of mutagenic activity were obtained by frying homogenized eggs
in the form of patties. These patties were prepared by heating egg
material for one hour at 95°C in petri plates. This process by
itself produces no detectible mutagenic activity in the eggs. Detectible levels of activity were induced in egg patties fried at pan
temperatures above 225°C. Degree of browning appeared to correspond
to levels of mutagenic activity of the egg patties. The highest
levels of activity (>10,000 revertants/100gE) were detected in samples which began to show some charring (samples fried >300°C for
>4 min/side). Subsequent larger scale mutagen isolation work was
conducted with a batch of egg patties (5 kg) fried for 4 min/side
with a pan temperature of 325°C.

A series of experiments was conducted to determine the best
conditions for extraction of mutagens from fried egg crusts. Mutagenic activities for initial extracts or fractions are presented in
Table I. The highest levels of activity were obtained with the mixed
solvents methylene chloride/methanol/water and acetonitrile/water and
the soxhlet extraction procedure using methanol. The procedure using
acetonitrile/water was used in subsequent fractionation experiments
because the methylene chloride/methanol/water procedure produced a
inhomogenous mixture on further work-up and possible artifact production during the prolonged heating of the soxhlet extraction procedure was to be avoided. The procedure employing an XAD-2 resin for
isolation of mutagens from an aqueous solution, was unsuitable for
isolation of mutagens from fried eggs because of low mutagen yield.
This result was in marked contrast to the results of mutagen isolation studies from ground beef.

Mutagen frequency (revertant/gE) was critically dependent on
the range of doses of egg extracts used in the biological assay.
Thus, revertant frequencies determined for the acetonitrile/water
extract increased from 20 to 105 to 538 revertants g/E as the dose
range was decreased from 0-25gE to 0-2.5gE to 0-0.5gE, respectively.
Results of subsequent experiments support the idea that this dependence of mutagenic activity on dose range may arise from the
presence of inhibitory substances present in the crude initial
extract.

Further studies were conducted to obtain a fraction sufficiently concentrated in mutagenic activity and free of interfering substances for subsequent chromatographic separation of active components. As indicated in Scheme I, a procedure was developed, which
incorporated an acid/base partition and a final diethyl ether
extraction of mutagens. By this procedure, diethyl ether extracts
are produced with good recoveries of activity from the initial
extracts. The resulting diethyl ether extracts are homogenous,
clear solutions suitable for HPLC analysis. It is of interest to
note that the dose range over which mutagenic activity is determined is not as limited for the diethyl ether extract as it is for

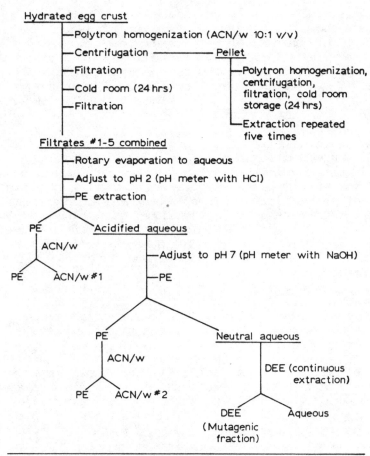

ACN -Acetonitrile, W-Water, PE-Petroleum ether, DEE-Diethyl ether

Figure 3.

the initial acetonitrile/water extract. Thus, good linearity of
the dose-response curve is observed for the diethyl ether extract
over a dose range of 0-2.5gE. It appears that the acid-base
partition/diethyl ether extraction procedure has separated the
mutagenic substances from other substances which mask mutagenic
activity. Possibly relevant in this regard is the report of inhibitory effects of certain fatty acids on mutagenic activity (15).

REFERENCES

1. M. Nagao, T. Sugimura, and T. Matsushima, Environmental mutagens
   and carcinogens, Ann. Rev. Genet. 12:117 (1978).
2. T. Matsumoto, D. Yoshida, S. Mizusaki, and H. Okamoto, Mutagenic
   activity of amino acid pyrolysate in Salmonella typhimurium,
   Mutation Res. 48:279 (1977).
3. T. Matsumoto, D. Yoshida, S. Mizusaki, and H. Okamoto, Mutagenicities of pyrolysates of peptides and proteins, Mutation
   Res. 5:281 (1978).
4. M. Nagao, M. Honda, Y. Seino, T. Yahagi, T. Kawachi, and
   T. Sugimura, Mutagenicities of protein pyrolysates, Cancer
   Lett. 2:335 (1977).
5. M. Uyeta, T. Kanada, M. Mazaki, and S. Taue, Studies on mutagenicity of food (I). Mutagenicity of food pyrolysates,
   J. Food Hyg. Soc. Japan (Shokuhin Eiselgaka Zasshi) 19:226
   (1978).
6. B. Commoner, A. J. Vithayathil, P. Dolara, S. Nair, P. Madyastha,
   and G. C. Cuca, Formation of mutagens in beef and beef extract
   during cooking, Science 201:913 (1978).
7. M. W. Pariza, S. H. Ashoor, F. S. Chu, and D. B. Lund, Effects
   of temperature and time on mutagen formation in pan-fried
   hamburger, Cancer Lett. 7:63 (1979).
8. N. E. Spingarn and J. H. Weisburger, Formation of mutagens in
   cooked foods. I. Beef, Cancer Lett. 7:259 (1979).
9. J. Felton, S. Healy, D. Stuermer, C. Berry, H. Timourian,
   F. T. Hatch, M. Morris, and L. F. Bjeldanes, Mutagens from
   the cooking of food. I. Improved extraction and characterization of mutagenic fractions from cooked ground beef,
   Mutation Res. 88:33 (1981).
10. N. E. Spingarn, L. A. Slocum, and J. H. Weisburger, Formation of
    mutagens in cooked foods. II. Foods with high starch content,
    Cancer Lett. 9:7 (1980).
11. H. Kasai, S. Nishimura, M. Nagao, Y. Takahashi, and T. Sugimura,
    Fractionation of a mutagenic principle from broiled fish by
    high-pressure liquid chromatography, Cancer Lett. 7:343
    (1979).
12. L. F. Bjeldanes, M. M. Morris, J. S. Felton, S. Healy,
    D. Stuermer, P. Berry, H. Timourian, and F. T. Hatch,
    Mutagens from the cooking of food II. Survey by Ames/
    Salmonella test of mutagen formation in the major protein-

rich foods of the American diet, <u>Food Chem. Toxicol.</u> 20:357 (1982).

13. L. F. Bjeldanes, M. M. Morris, J. S. Felton, S. Healy, D. Stuermer, P. Berry, H. Timourian, and F. T. Hatch, Mutagens from the cooking of food III. Secondary sources of cooked dietary protein, <u>Food Chem. Toxicol.</u> 20:365 (1982).

14. L. F. Bjeldanes, K. R. Grose, P. H. Davis, D. H. Stuermer, S. K. Healy, and J. S. Felton, An XAD-2 resin method for efficient extraction of mutagens from fried ground beef, Mut. Res. Lett. 105:43 (1982).

15. H. Hayatsu, S. Arimoto, K. Togawa, and M. Makita, Inhibitory effect of the ether-extract of human feces on activities of mutagens: Inhibition of oleic and linoleic acids, <u>Mut. Res.</u> 81:287 (1981).

## 26

MUTAGENS IN COOKED FOODS-METABOLISM AND GENETIC TOXICITY

J.S. Felton, L.F. Bjeldanes and F.T. Hatch

Biomedical Sciences Division
Lawrence Livermore National Laboratory and
Department of Nutritional Sciences
University of California, Berkeley
Livermore, CA 94550

ABSTRACT

Recently developed in our laboratories is an efficient extraction procedure incorporating XAD resin adsorption which yields from 200°C grilled ground beef an extract containing 230 Salmonella TA1538 revertants per g fresh weight of original ground beef. These mutagenic components are specific for frameshift-sensitive Salmonella strains and have an absolute requirement for metabolic activation. S9 activation by cytochrome P-448 inducers, Aroclor 1254 (PCB), 3-methylcholanthrene (3-MC) and B-naphthoflavone(BNF), resulted in the largest mutagenic response. Phenobarbital induction gave 20% of the PCB response and Pregnenolone-16a-carbonitrile and corn oil were inactive. Human liver microsomes and BNF-induced rodent intestinal S9 were also active metabolizing fractions. Normal-phase HPLC separation of methanol-extractable metabolites generated from reaction of 2-amino-3-methylimidazo [4,5-f]quinoline (IQ), a mutagenic component of broiled food, rat liver microsomes and cofactors resulted in one direct-acting mutagenic peak and a second more polar peak still requiring metabolic activation. Two potent thermally-produced bacterial mutagens, Trp-P-2 and IQ, were examined in mammalian cells. In excision repair-deficient CHO cells, Trp-P-2 exposure caused cytotoxicity, mutagenicity (thioguanine and azaadenine resistances), sister chromatid exchange, and chromosomal aberrations at concentrations more than 30-fold lower than those for IQ. In normal repair-proficient CHO cells Trp-P-2 was one-half as active and IQ was inactive. Relative to Trp-P-2, IQ is much more potent in the Salmonella bacterial system than in mammalian CHO cells.


GENETIC TOXICOLOGY

Microbial Mutagenicity of Cooked Beef Extract

Dose-response curves for bacterial mutagenicity were assessed in five Salmonella strains. The frameshift sensitive strains TA98, TA1538, and TA1537 all showed positive linear response with the extract from the cooked beef. The base substitution sensitive strains TA100 and TA1535 were negligible in their response (Felton, et al., 1981).

Two food mutagen standards, 3-amino-1-methyl-5H-pyrido [4,3-b] indole (Trp-P-2) and 2-amino-3-methylimidazo [4,5-f] quinoline (IQ) were assessed at two different S9 concentrations in seven different Ames/Salmonella strains(see Table I). They were both strongly positive in TA1538, TA98, and TA97, all frameshift sensitive strains. Weak responses were seen with TA96 and TA104 for both compounds. TA96 reversion requires a frameshift correction like the strongly positive strains but is specific for bases A or T, not G or C (Levin et al., 1982). Thus, these two potent Salmonella mutagens appear to be not only frameshift specific but are specific for G/C changes. Weak or no significant response was seen for both compounds with TA100 and TA102. Both of these strains require base-substitution changes for reversion; and in the case of TA102, which gave no response for Trp-P-2 and a very weak one for IQ, reversion requires a base substitution for a T or A (Levin et al., 1982).

Effect of Metabolic Activation on Microbial Mutagenicity

Interestingly, the amount of S9 or microsomes needed for the metabolic activation affects the response of the two compounds in opposite fashions. IQ responds in a very linear way, with an increasing number of revertants with increasing S9 up to 4 mg of protein per plate. Trp-P-2, on the other hand, shows a maximum response at 1 mg per plate and shows approximately half the response at 2 mg protein per plate for all the strains in which a positive response was seen. This confirms the finding of Nagao et al. (1981) and is very similar to the response for another aromatic amine 2-acetylaminofluorene (Felton et al., 1976). What differences in the metabolic pathways account for this difference in response to S9 concentration are not clear. It is not a case of nonspecific receptors present in the S9 binding the active intermediates at the higher protein concentration; since either boiled S9 or addition of BSA do not reduce the mutagenic response. Both compounds appear to go

TABLE I

COMPARISON OF THE MUTAGENICITY OF

TRP-P-2 AND IQ WITH 8 SALMONELLA STRAINS

| COMPD. | S9 (mg/plate) | TA1538 | TA98 | TA96 | TA97 | TA100 | TA102 | TA104 | TA1978 |
|---|---|---|---|---|---|---|---|---|---|
| | | | | | Rev/µg (in thousands)* | | | | |
| IQ | 1 | 164 | 52 | -- | 96 | -- | NS* | 6.5 | -- |
| IQ | 2 | 272 | 94 | 1.9 | 104 | 3.5 | 0.6 | 11.6 | 3.6 |
| Trp-P-2 | 1 | 47 | 89 | 0.2 | 9.3 | -- | NS* | NS* | -- |
| Trp-p-2 | 2 | 24 | 52 | 0.1 | 4.5 | 1.6 | NS* | 0.7 | 0.9 |

*Calculated from the linear portion of the dose response curve by the method of Moore and Felton (1983).

through an N-OH metabolic intermediate (Yamazoe et al. 1981; Yamazoe et al., 1983), but the competing detoxifying reactions may be quite different.

## Requirement of Repair Deficiency

Dose-response curves were compared between Salmonella strains TA1538, a repair deficient strain having a deletion in the uvrB locus, and TA1978, a strain sufficient for uvrB repair. Although both strains produced mutations with Trp-P-2 and IQ, the repair deficient TA1538 was 30-50 fold more sensitive (Thompson et al., 1983). This result suggests that the excision-repair pathway operates on the types of damage produced by these compounds. The difference in response between the strains was not due to toxicity of the compounds because killing did not occur for either strain unless the dose was at least 10 times higher than that giving mutagenicity. With IQ the repair capacity modified killing and mutagenesis to a similar degree. With Trp-P-2 there was relatively little difference in the dose needed for killing between the two strains.

## CHO Cell Mutagenicity

Trp-P-2 and IQ have been evaluated in a repair-deficient Chinese hamster cell forward mutation system (Thompson et al., 1983). Trp-P-2 was very potent for producing mutations at both the hprt and aprt loci in both the repair competent AA8 cell line and the repair incompetent UV-5 line (see figure 1). In contrast to these results, IQ was quite weak. Equivalent mutation and toxicity effects were seen only at doses of IQ that were 70-100 fold higher than used for Trp-P-2. With IQ, toxicity differences were seen between the repair proficient and deficient strains. These results indicate that some of the damage leading to toxicity produced by IQ is not repairable by the excision repair pathway. Also, only the repair deficient strain showed a mutagenic response. Thus, in contrast to Salmonella, the CHO cells are much more responsive to Trp-P-2 than to IQ.

## Cytogenetic Effects

Both the cytogenetic end-points, SCEs and chromosomal aberrations, were assessed in the same cells in which mutagenesis was tested. As was seen for for mutation, Trp-P-2 was very potent at inducing SCEs and aberrations in both cell types at doses of approximately 1.0 ug per ml (see figure 1). IQ showed a very weak response for SCEs in the repair deficient strain only and no significant response for abberations in either cell line.

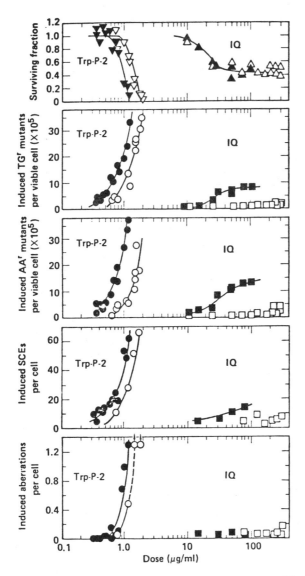

Fig. 1. The effects of Trp-P-2 and IQ on cell killing, mutations, SCEs, and chromosomal aberrations in normal and repair-deficient CHO cells. Data for the normal cells (AA8) is given by open symbols and for UV-5 cells by closed symbols. The applied dose of compound is indicated on the left side and te data for IQ on the right. Cultures were exposed as described (Thompson et al., 1983) and for each genetic endpoint the data are presented as induced values. Reprinted with permission of author and published from Thompson et al. (1983).

It appears that in contrast to bacterial cells, the CHO
cells are relatively inactive towards IQ for a number of
different genotoxic end-points, and that possibly the active
metabolite of IQ is so reactive that it never reaches the
nuclear target to cause the cytogenetic or mutational damage.
Studies by Thompson et al. (1983) support this point because
active metabolites of IQ are produced in the media which can
subsequently mutate Salmonella. In addition, binding of
tritium labeled IQ to CHO cell DNA is quite low compared to
Trp-P-2 but the two compounds show roughly the same mutagenic
efficiency (mutations per adduct bound) (Brookman et al,
manuscript in preparation; Hatch et al., abstracts of the 15th
Annual Meeting of the Environmental Mutagen Society, Montreal,
February 1984).

## Comparison of the Ames/Salmonella Reversion Assay to Two Forward Mutation Assays

The finding that Trp-P-2 and IQ behave differently in the
mammalian forward mutation assay and the Salmonella reversion
assay prompted us to compare the mutagenicity of both compounds
in two additional microbial assays, both testing for forward
mutations. One of these measures arabinose resistance in
Salmonella strains having the same basic cell wall and repair
mutations as the Ames strains (Whong et al ,1981). This assay
depends on the toxic intermediate, L-ribulose-5-P, building up
in the cell because of a mutation in the araD gene which codes
for the epimerase enzyme. Any mutation in loci controlling
enzymes earlier in the pathway will make the cell no longer
arabinose resistant. At least 3 and possibly 4 loci, if
mutated, could prevent killing in the presence of arabinose.
The second assay measures resistance to 8-azaguanine in TM677, a
strain having the standard Ames/Salmonella genotype including
the R-factor plasmid, pKM101 (Skopek et al, 1978).

In both assays IQ and Trp-P-2 were very potent. IQ, as was
seen in the Ames/Salmonella reversion assay, was more potent
than Trp-P-2 (see Table II). It is clear from these data that
neither the expression of the genetic damage nor the site that
is mutated is specific for one compound in bacteria and the
other compound in mammalian cells. Thus, the lack of a potent
response of IQ in CHO cells is probably not related to the
absence of a hypersensitive DNA sequence, but is more likely due
to differences in uptake, transport, or metabolism of the
mutagens.

It is interesting that the response in the arabinose forward
mutation assay is somewhat larger than the Ames/Salmonella
response but not as large as one would expect. A much bigger
difference might be expected because of the large target size

TABLE II

COMPARISON OF REVERSE AND FORWARD MICROBIAL MUTAGENESIS ASSAYS

| Test System* | Mutations/$10^9$ surviving cells/μg | |
|---|---|---|
| | IQ | Trp-P-2 |
| Ames (TA1538) | 140,000 | 65,000 |
| Arabinose resistance (SV50) | 34,000 | 39,000 |
| 8-AG resistance | 88,000 | 6,200 |

*2 mg of Aroclor-induced rat liver S9 per plate.

for DNA damage in the Arabinose system.  This brings up the
possibility that the few sequences responsible for the TA1538
reversion (Isono and Yourno, 1974) are extremely sensitive to IQ
and Trp-P-2 or, alternatively,  that there is a much larger
target such as a t-RNA suppressor gene.  In any case, the actual
size of the genetic targets for these compounds needs further
study.


METABOLISM

In vitro Animal Metabolism of the Basic Fraction from Cooked Beef

The basic fraction from fried hamburger (see Acid-XAD
extraction scheme, Bjeldanes et al., this proceedings) which is
made-up of 8-10 separate mutagens (Felton et al., 1984) has an
absolute dependence on metabolic activation (Felton et al.,
1981).  Not only is there a requirement for activation but
specific monooxygenases must be induced for metabolism to
ultimate mutagens.  Aroclor 1254, 3-methylcholanthrene,
B-naphthoflavone preinduced liver S9 fractions that are capable
of near maximal metabolism of the basic fraction to mutagenic
intermediates (see Table III).  This suggests that the

metabolism is due to cytochrome P-448 type induction. This is
confirmed by Yamozoe et al.,1983, who used purified forms of
cytochromes P-448 to produce the N-OH derivatives of Trp-P-2 and
IQ. In our laboratory mouse, hamster, and rat Aroclor-induced
hepatic S9 all metabolized the basic fraction to mutagenic
intermediates. The hamster S9 and microsomal fractions were
slightly more active than the rat and mouse preparations.

TABLE III

COMPARISON OF BASIC FRACTION* MUTAGENESIS USING DIFFERENT

INDUCED CYTOCHROME P-450 ACTIVATION SYSTEMS (MICE)

| Microsomal Fraction[†]<br>(Pretreatment) | TA1538 rev/plate**/gE*** |
|---|---|
| Corn oil | NS[††] |
| Pregnenolone-16α-Carbonitrile | NS |
| Phenobarbital | 10 |
| 3 methylcholanthrene | 24 |
| B-naphthoflavone | 36 |
| Aroclor 1254 | 51 |
| Aroclor 1254 (rat) | 42 |

*Basic fraction from ground beef fried at 200°C, 6 min per side.
**Revertants are calculated from the linear portion of the dose
response curve.
***gE = gram equivalents of fresh uncooked meat.
[†]Microsome concentration 0.5 mg protein/plate.
[††]NS= Not significantly different from zero slope.

## In vitro Human Metabolism

Most importantly, one would like to know whether human tissue is
capable of producing mutagenic metabolites from the cooked meat
fractions. We analyzed the ability of six human liver
microsomal preparations to metabolize the basic fraction from
the fried meat (Felton et al. 1981b). These preparations, which
were all autopsy specimens, were more active than the uninduced
mouse or rat liver microsomal fractions. Microsomes from one
individual were nearly as active as those from Aroclor-induced
rats and mice. The human metabolism appears to be due to
primarly cytochrome P-448 activation; since alpha-napthoflavone
showed a dose dependent inhibition of the production of the
mutagenic intermediates (Felton et al., 1981b). This finding not

only supports the fact that the human can metabolize the potentially mutagenic components produced from the cooking of the beef, but that through exposure to environmental and dietary inducers there is induction of specific cytochrome P-450(s) capable of this activation.

## Metabolism of IQ and Trp-P-2

Yamazoe et al., (1981) have described the characterization of the active metabolite of Trp-P-2 in much detail. It appears to be an N-hydroxy derivative which has the specific activity of about 4 X 10E4 Salmonella revertants per nmol N-hydroxy-Trp-P-2. More recently, the metabolism of IQ has been assessed by the same laboratory (Yamazoe et al., 1983). As was found for Trp-P-2, IQ is metabolized to the N-hydroxy derivative. A specific form of purified cytochrome P-450 designated P-448 IIa appears to be responsible for the metabolic activation of IQ. The activity of this N-hydroxy metabolite is calculated to be 5 X 10E5 revertants per nmole. This metabolite is the most mutagenic compound ever tested in the Ames/Salmonella assay, three times more potent than 1,8 dinitropyrene (Rosenkranz et al. 1980).

In our laboratory (Skiles and Felton, unpublished results), we have reacted 3H-IQ with Aroclor-induced rat liver microsomes and extracted the metabolites. These metabolites were then separated by normal phase HPLC and assessed for their mutagenicity. One metabolite representing less than 20% of the total metabolism was acid labile and direct acting (no requirement for S9 or microsomes) in the Ames/Salmonella test. Estimates for the specific activity of this compound are 1-3 X 10E6 revertants per ug, very similar to the figure calculated by Yamazoe et al. (1983) for N-hydroxy IQ. Identification of this metabolite plus at least two other polar metabolites is presently underway.


CONCLUSIONS

Extracts from cooked beef are clearly mutagenic in the frameshift sensitive Ames/Salmonella strains. Aroclor-induced metabolic activation is required for mutagenicity, but the human liver appears to be environmentally induced so that it is capable of metabolic activation. The synthetic mutagens IQ , a minor component of cooked beef basic extract (Felton et al. 1984) and Trp-P-2, a pyrolysate product from some broiled foods but not fried ground beef, are both very potent Salmonella mutagens (causing both forward and reverse mutations). In contrast, Trp-P-2 is a potent inducer of mutations, SCEs, and chromosome aberrations in CHO cells, but IQ is not. DNA binding studies support the hypothesis that the transport and activation of IQ do not allow ready access of active intermediates to the DNA. Metabolism appears to be specific for cytochromes P-448 and results in N-hydroxy derivatives. Carcinogenesis studies with Trp-P-2 and IQ (Matsukura et al., 1981; Sugimura, personal communication) are definitely positive in mice and rats; but more work with IQ and the major cooked food mutagens, such as MeIQx, are needed for estimates of cancer risk from eating well-done fried ground beef.

ACKNOWLEDGMENTS

Work performed under the auspices of the U.S. Department of Energy by the Lawrence Livermore National Laboratory under contract number W-7405-ENG-48 and sponsored by the National Institute of Environmental Health Sciences under Interagency Agreement No. 22201-ES-10063.

The authors would like to thank Mark Knize and Sue Healy for their technical assistance and Leilani Corell for her secretarial support.

## Reference List

Felton, J., Healy, S., Stuermer, D., Berry, C., Timourian, H., Hatch, F.T., Morris, M. & Bjeldanes, L.F. (1981a). Mutagens from the cooking of food. (I). Improved isolation and characterization of mutagenic fractions from cooked ground beef. Mutation Res. 88, 33-44.

Felton, J.S. Nebert, D.W., and Thorgiersson, S.S. (1976). Genetic differences in 2-acetylaminofluorene mutagenicity in vitro associated with mouse hepatic aryl hyrocarbon hydroxylase activity induced by polycyclic aromatic compounds. Mol. Pharmacol. 12, 225-233.

Felton, J.S., Healy, S.K., Knize, M., Stuermer, D.H., Berry, P.W., Timourian, HJ., Hatch, F.T., Morris, M., and Bjeldanes, L.F. (1981b) In vitro human and rodent metabolism of mutagenic fractions from cooked ground beef. Environmental Mutagenesis 3, 342.

Felton, J.S., Knize, M.G., Wood, C., Wuebbles, B.J., Healy, S.K., Stuermer, D.H., Bjeldanes, L.F., Kimble, B.J., and Hatch, F.T. (1984). Isolation and characterization of new mutagens from fried ground beef. Carcinogenesis 5, 95-102.

Isono, K. and Yourno, J. (1974). Chemical carcinogens as frameshift mutagens: Salmonella DNA sequence sensitive to mutagenesis by polycyclic carcinogens. Proc. Natl. Acad. Sci. USA 71, 1612-1617.

Levin, D.E., Hollstein, M., Christman, M.F., Schwiers, E.A., and Ames, B.N. (1982). A new Salmonella tester strain (TA 102) with A-T base pairs at the site of mutation detects oxidative mutagens. Proc. Natl. Acad. Sci. USA 79, 7445-7449.

Matsukura, N., Kawachi, T., Morino, K., Ohgaki, H., Sugimura, T., and Takayama, S. (1982). Carcinogenicity in mice of mutagenic compounds from a tryptophan pyrolyzate. Science, 213, 346.

Moore, D. and Felton, J.S. (1983). A microcomputer program for analyzing Ames test data. Mutation Res. 119, 95-102.

Nagao, M., Wakabayashi, K., Kasai, H., Nishimura, S., and Sugimura, T. (1981). Effect of methyl substitution on mutagenicities of 2-amino-3-methylimidazo[4,5-f]quinoline, isolated from broiled sardine. Carcinogenesis, 2 1147-1149.

Rosenkranz, H.S., McCoy, E.D., Sanders, D.R., Butler, M., Kiriazides, D.K., and Mermelstein, R. (1980). Nitropyrenes: Isolation, identification, and reduction of mutagenic impurities in carbon black and toners. Science 209, 1039-1043.

Skopek, T.R., Liber, H.L., Kaden, D.A., Thilly, W.G. (1978). Relative sensitivities of forward and reverse mutation assays in Salmonella typhimurium. Proc. Natl. Acad. Sci. USA 75, 4467-4469.

Thompson, L.H., Carrano, A.V., Salazar, E., Felton, J.S., and Hatch, F.T. (1983). Comparative genotoxic effects of the cooked-food-related mutagens Trp-P-2 and IQ in bacteria and cultured mammalian cells, Mutation Res. 117, 243-257.

Whong, W.Z., Stewart, J., and Ong, T.M. (1981). Use of the improved arabinose-resistant assay system of Salmonella typhimurium for mutagenesis testing. Environ. Mutagenesis 3, 95-99.

Yamazoe, Y., Ishii, K., Kamataki, T. and Kato, R. (1981). Structural elucidation of a mutagenic metabolite of 3-amino-1-methyl-5H-pyrido[4,3-b]indole. Drug Metabolism and Disposition 9, 292-296.

Yamazoe, Y., Shimada, M., Kamataki, T., and Kato, R. (1983).
    Microsomal activation of
    2-amino-3-methylimidzao[4,5-5]quinoline, a pyrolysate of
    sardine and beef extracts, to a mutagenic intermediate.
    Cancer Res. 43, 5768-5774.